For the updated syllabus

Complete Biology for Cambridge IGCSE®

Third edition

Ron Pickering

Oxford excellence for Cambridge IGCSE®

OXFORD

OXFORD
UNIVERSITY PRESS

Great Clarendon Street, Oxford OX2 6DP

Oxford University Press is a department of the University of Oxford. It furthers the University's objective of excellence in research, scholarship, and education by publishing worldwide in

Oxford New York

Auckland Cape Town Dar es Salaam Hong Kong Karachi Kuala Lumpur Madrid Melbourne Mexico City Nairobi New Delhi Shanghai Taipei Toronto

With offices in

Argentina Austria Brazil Chile Czech Republic France Greece Guatemala Hungary Italy Japan Poland Portugal Singapore South Korea Switzerland Thailand Turkey Ukraine Vietnam

First published in 2006
Second edition published 2011
Third edition published 2014

British Library Cataloguing in Publication Data

Data available

ISBN 978-0-19-839911-7

10 9 8 7 6 5 4 3

Printed in Great Britain by Bell and Bain Ltd., Glasgow

Paper used in the production of this book is a natural, recyclable product made from wood grown in sustainable forests. The manufacturing process conforms to the environmental regulations of the country of origin.

Acknowledgments

®IGCSE is the registered trademark of Cambridge International Examinations.

The publisher would like to thank Cambridge International Examinations for their kind permission to reproduce past paper questions.

Unless otherwise indicated, exam-style questions and sample answers were written by the author.

Cambridge International Examinations bears no responsibility for the example answers to questions taken from its past question papers which are contained in this publication.

We are grateful to our reviewers and project managers for getting Complete Biology for IGCSE® from the ideas stage to the finished product.

Thanks go to my wife Janet and our two sons Chris and Tom, for keeping me on the straight and narrow, and to our new daughter-in-law, Rie, for her veneration of the elderly! I should also like to acknowledge the part played by my Mum and Dad – reference to Section 3 will make it clear that I owe them a great deal!

Ron Pickering

The publisher would like to thank the following for permission to reproduce photographs:

Cover: Gary Yim/Shutterstock; **p3:** Ron Pickering; **p8:** Heather Angel/Natural Visions; **p12:** Ron Pickering; **p13t:** Ron Pickering; p13b: Ron Pickering; **p16l:** Dennis Kunkel Microscopy, Inc./Visuals Unlimited/Corbis; p16r: Claude Nurisdsany & Marie Perennou/Science Photo Library; **p25l:** Ed Reschke/Photolibrary/Getty Images; **p25r:** London Scientific Films/Oxford Scientific/Getty Images; **p30:** OUP Owned; **p19:** Nigel Cattlin/Alamy; **p61t:** Rex Features; **p61b:** Wesley Bocxe/Science Photo Library; **p67:** BSIP/Universal Images Group/Getty Images; **p74:** Claude Nuridsany & Marie Perennou/Science Photo Library; **p76:** Dr. Jeremy Burgess/ Science Photo Library; **p81t:** Slowfish/Shutterstock; **p81b:** Olegusk/Shutterstock; **p81r:** John Clegg/Science Photo Library; **p82t:** Daryl H/Shutterstock; **p82b:** Sebastien Burel/Shutterstock; **p83:** Ed Reschke/Photolibrary/Getty Images; **p85l:** Adam Gaul/Science Photo Library; **p85r:** Eye Of Science/Science Photo Library ; **p88:** Steve Gschmeissner/Science Photo Library; **p89l:** PR. H. Bezes/CNRI/Science Photo Library; **p89r:** Dr. P. Marazzi/Science Photo Library; **p96t:** Hans-Ulrich Osterwalder/Science Photo Library; **p96m:** London Scientific Films/Oxford Scientific/Getty Images **p96b:** Claffra/Shutterstock; **p101l:** Bill Gentile/Corbis; **p101r:** Ron Pickering; **p102:** CNRI/Science Photo Library; **p103:** Biology Media/ Science Photo Library; **p104:** Dr. Gopal Murti/Science Photo Library; **p116:** Photodisc/OUP; **p121:** Arthur Glauberman/Getty Images; **p122l:** Gemma Levine/ Premium Archive/Getty Images; **p122r:** Mark Edwards/Still Pictures/Robert Harding; **p126:** Fredrick Sierakowski/Rex Features; **p133:** Cordelia Molloy/Science Photo Library; **p143:** Image Broker/Rex Features; **p150:** Archive Photos/Stringer/ Moviepix/Getty Images; **p151:** Durand-Giansanti-Perrin/Sygma/Corbis; **p153:** Rex Features; **p154:** JLP/Jose Luis Pelaez/Corbis; **p159l:** Sielan/Shutterstock; **p159r:** Jared Wilson/Shutterstock; **p161:** Nigel Cattlin/Visuals Unlimited/Corbis; **p165:** Dr. Jeremy Burgess/Science Photo Library; **p166l:** Guyerwood/Bigstock; **p166m:** Alan Marsh/Design Pics/Corbis; **p166r:** Zilli Roberto/Dreamstime; **p168:** Marie C Fields/ Shutterstock; **p170t:** Vladoleg/Shutterstock; **p170b:** Olga Popova/Shutterstock; **p179:** Sarah Jones (Debut Art)/The Image Bank/Getty Images; **p180:** Adam Hart-Davis/Science Photo Library; **p181t:** Saturn Stills/Science Photo Library; **p181r:** Areeya_Ann/Shutterstock; **p181b:** Scott Camazine/Sue Trainor/Science Photo Library; **p181l:** Siu Biomed Comm/Custom Medical Stock Photo/Science Photo Library; **p188l:** Sf2301420max/Shutterstock; **p188m:** Absolute-India/Shutterstock; **p188r:** Solent News/Rex Features; **p191:** Reflektastudios/Bigstock; **p192:** Steve Gschmeissner/Science Photo Library; **p204:** Ron Pickering; **p206:** Cnri/Science Photo Library; **p223:** Cre8tive Images/Shutterstock; **p212r:** Ed Reschke/ Photolibrary/Getty Images; **p214l:** Noam Wind/Shutterstock; **p214r:** Ron Pickering; **p215t:** Science Photo Library; **p215l:** Science Photo Library; **p215r:** Science Photo Library; **p215bl:** Nils Jorgensen/Rex Features; **p215b:** Science Source/Science Photo Library; **p218:** Trappe/Caro/Alamy; **p222:** Marek Gucwa/Shutterstock; **p223:** Ron Pickering; **p229:** Nigel Cattlin/Science Photo Library; **p230:** Ron Pickering; **p236:** Zent/Dreamstime; **p246:** Tim Hazael/Science Photo Library; **p250t:** OUP owned; **p250m:** OUP owned; **p250b:** OUP owned; **p254:** Professor Stanley Cohen/Science Photo Library; **p257:** Piotr Tomicki/Shutterstock; **p259t:** Andrew Orlemmann/123rf; **p259b:** Ron Pickering; **p260t:** Vitaly Titov & Maria Sidelnikova/Shutterstock; **p260m:** E. Sweet/Shutterstock; **p260b:** Greg Baker/AP Images; **p271:** Ron Pickering; **p274l:** Ron Pickering; **p274b:** Davemhuntphotography /Shutterstock; **p289:** Ron Pickering; Cover: Gary Yim/Shutterstock.

Introduction

Biology is the study of life and living organisms. During the past few hundred years biology has changed from concentrating on the **structure** of living organisms (often by examining dead specimens!) to looking more at how they work or **function.** Over this time we have discovered much about health and disease, about the interactions of different organisms making up food chains, about the genes which control the activities of our bodies and how humans can control the lives of other organisms. These advances in biological knowledge raise new issues. We need to understand how our activities affect the environment, how humans can take responsibility for their own health and welfare, and how we must make appropriate rules for the use of our genetic information.

In *Complete Biology for Cambridge IGCSE* you will study a range of living organisms, the life processes they carry out, the effects that these life processes might have on our health, and the responsibilities which we have towards other organisms.

This book has been organised to help you to find information quickly and easily. It is written in two-page units or 'spreads'. Each unit is a topic which forms part of the IGCSE Biology syllabus, and the 'spreads' are collected into the four major sections of the syllabus. There is also a fifth section which will help to prepare for the practical assessment (Paper 5 or 6). Your teacher will be able to help you to relate the units to the sections of your syllabus.

Each person has their own way of working, but the following tips might help you to get the most from this book:

- Use the **contents** page — this will provide information on large topics, such as reflexes or water pollution.
- Use the **index** — this will allow you to use a single word such as 'neurone' or 'eutrophication' to direct you to pages where you can find out more about that word.
- Use the **questions** — this is the best way of checking whether you have learned and understood the material on each spread. Questions are to be found on most spreads and within or at the end of each section. Harder questions are identified by the icon ●. There are more questions, and revision 'tips', in the IGCSE Biology Revision Guide.

When you are using the book, keep a look out for these marks:

A line down the side of the text means that the material is only required for Extended Level. For simplicity, lines like this have *not* been put next to related diagrams or panels in the margin.

An asterisk indicates a spread or part of a spread that is providing extension material to set biology in a broader context. You would not normally be tested on this material in an IGCSE examination.

I hope that you enjoy using this book, and that it helps you to understand the world of biology. You, like every other living organism, are a part of this world – perhaps one day you will find yourself working to help others to understand more about it.

Ron Pickering
2014

Contents

3 Development of organisms and the continuity of life

4 Organisms and their environment

5 Practical biology

Support website: www.oxfordsecondary.com9780198399117

1.1 Biology is the study of life and living organisms

OBJECTIVES

- To understand that living things differ from non-living things
- To be able to list the characteristics of living things
- To understand that energy must be expended to maintain life

The dawn of life

Scientists believe that the Earth was formed from an enormous cloud of gases about 5 billion years ago. Atmospheric conditions were harsh (there was no molecular oxygen, for example), the environment was very unstable, and conditions were unsuitable for life as we know it.

Many scientists believe that the first and simplest living organisms appeared on Earth about 2.8 billion years ago. These organisms probably fed on molecules in a sort of 'soup' (called the **primordial soup**) which made up some of the shallow seas on the Earth at that time. A question that has always intrigued scientists, philosophers and religious leaders is:

> What distinguishes these first living organisms from the molecules in the primordial soup?

In other words, what is life?

Characteristics of living organisms

You know that a horse is alive, but a steel girder is not. However, it is not always so obvious whether something is alive or not – is a dried-out seed or a virus particle living or non-living? To try to answer questions like this, biologists use a list of characteristics that living organisms show.

Living organisms:

- **Respire**
- Show **irritability** (sensitivity to their environment) and **movement**
- **Nourish** themselves
- **Grow** and **develop**
- **Excrete**
- **Reproduce**

The opposite page gives more detail of the characteristics of life.

You may see other similar lists of these characteristics using slightly different words. You can remember this particular list using the word RINGER. It gives **Ringer's solution** its name. This is a solution of ions and molecules that physiologists use to keep living tissues in – it keeps the cells alive.

As well as the characteristics in the 'ringer' list, living things have a **complex organisation** that is not found in the non-living world. A snowflake or a crystal of quartz is an organised collection of identical molecules, but even the simplest living cell contains many different complex substances arranged in very specific structures.

Living things also show **variation** – the offspring are often different from one another and from their parents. This is important in adaptation to the environment and in the process of evolution.

How the characteristics of life depend on each other

Each of the characteristics of life is linked to the others – for example, organisms can only grow if they are nourished. As they take nourishment from their environment, they may also produce waste materials which they must then excrete. To respond to the environment they must organise their cells and tissues to carry out actions. Because of the random nature of reproduction, they are likely to show variation from generation to generation.

Depending on energy

The organisation in living things and their ability to carry out their life processes depends on a supply of **energy**. Many biologists today define life as a set of processes that result from the organisation of matter and which depend on the expenditure of energy.

In this book we shall see:

- how energy is liberated from food molecules and trapped in a usable form
- how molecules are organised into the structures of living organisms
- how living organisms use energy to drive their life processes.

Respiration: the chemical reactions that break down nutrient molecules in living cells to release energy for metabolism. The form of respiration that releases the most energy uses oxygen. Many organisms have **gaseous exchange** systems that supply their cells with oxygen from their environment.

Irritability (or **sensitivity)**: the ability to detect or sense changes in the internal or external environment and to make appropriate responses. The changes are called stimuli and the responses often involve **movement** (an action by an organism or part of an organism causing a change of position or place).

Nutrition: the taking in of materials for energy, growth and development. Plants require light, carbon dioxide, water and ions and make their foods using the process of photosynthesis. Animals require organic compounds and ions (and usually water) and obtain their foods 'ready-made' by eating them.

Growth and **development**: the processes by which an organism changes in size and in form. For example, as a young animal increases in size (as it grows), the relative sizes of its body parts change (it develops). Growth is a **permanent** increase in size and dry mass, and results from an increase in cell number or cell size or both.

Excretion: removal from organisms of toxic materials, the waste products of metabolism (chemical reactions in cells including respiration) and substances in excess of requirements.

Reproduction: the processes that make more of the same kind of organism – new individuals. An organism may simply split into two, or reproduction may be a more complex process involving fertilisation. Reproduction makes new organisms of the same species as the parents. This depends on a set of chemical plans (the genetic information) contained within each living organism.

Q

1. Approximately how many years passed between the formation of the Earth and the appearance of the first living organisms?
2. What sort of molecules do you think might have been present in the primordial soup?
3. **RINGER** is a word that helps people remember the characteristics of living organisms. Think of your own word to help you remember these characteristics.
4. Suggest *two* ways in which reproduction is essential to living organisms.

1.2 The variety of life

OBJECTIVES

- To know that organisms can be classified into groups by the features that they share
- To appreciate why classification is necessary
- To understand the use of a key
- To be able to name the five kingdoms, and describe their distinguishing characteristics
- To understand the hierarchy of classification
- To know why a binomial system of nomenclature is valuable

The need to classify living things

Variation and natural selection lead to evolution. Evolution, and the isolation of populations, leads to the development of new species (see page 216). Each species has different characteristics, and some of these characteristics can be inherited by successive generations of this species. Observing these inherited characteristics allows scientists to put all living organisms into categories. The science of placing organisms into categories on the basis of their observable characteristics is called **classification**. There are so many different types of living organism (i.e. an enormous variety of life) that the study of these organisms would be impossible without an ordered way of classifying them.

For example, classification is important in

- **conservation**: scientists need to be able to identify different organisms in habitats which are being managed, and they need to control which organisms are used in breeding programmes.
- **understanding evolutionary relationships**: organisms which have many of the same features are normally descended from common ancestors. The more features shared by different organisms the more recently they separated from one another during evolution.

Classification keys

Taxonomists (people who study classification) place organisms into groups by asking questions about their characteristics, such as 'Does the organism photosynthesise?' or 'Does the organism contain many cells?'. A series of questions like this is called a **classification key**. Examples of such keys are shown below and on the opposite page.

The characteristics of living organisms used to make classification keys have traditionally been based on **morphology** and **anatomy** (the shape and structure of organisms) because this was what the scientists could easily observe and measure.

This kind of key, with only two answers to each question (in this case, YES or NO), is called a dichotomous key ('dichotomous' means branching). It can be written as a branching or spider key, using the same questions:

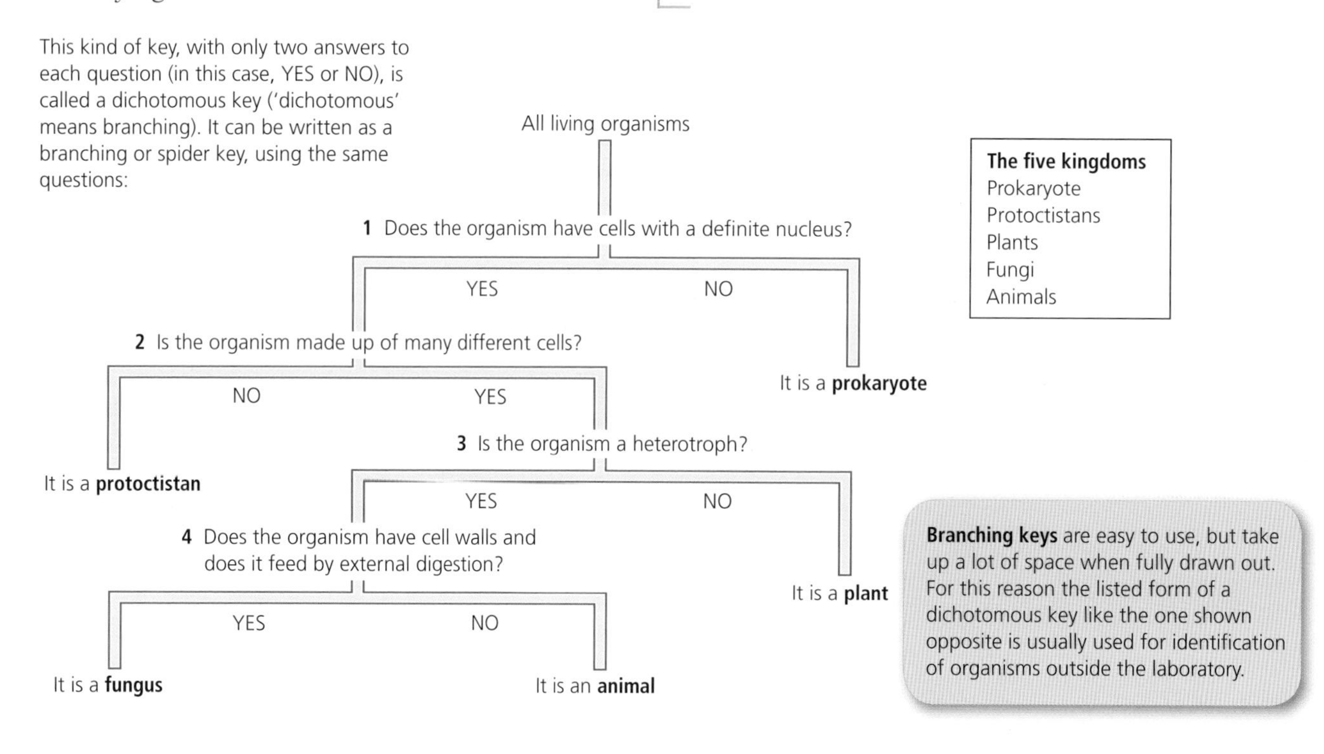

The five kingdoms
Prokaryote
Protoctistans
Plants
Fungi
Animals

Branching keys are easy to use, but take up a lot of space when fully drawn out. For this reason the listed form of a dichotomous key like the one shown opposite is usually used for identification of organisms outside the laboratory.

Question		
1 Does the organism have cells with a definite nucleus?	YES NO	Go to question 2 It is a prokaryote
2 Is the organism made up of many different cells?	YES NO	Go to question 3 It is a protoctistan
3 Is the organism a heterotroph?	YES NO	Go to question 4 It is a plant
4 Does the organism have cell walls and does it feed by external digestion?	YES NO	It is a fungus It is an animal

▲ A key may be used to place an organism in one of the five kingdoms

Five Kingdoms

Using the key opposite, it is possible to place any living organism into one of five very large groups. These groups, distinguished from one another by obvious characteristics of morphology and anatomy, are called the **five Kingdoms**. Each of these kingdoms contains an enormous number of different species, and keys can be used within a kingdom to place any individual species into further groups. The diagram below shows the names of these groups, and how the lion is classified within the Animal Kingdom.

Hierarchy of classification*

The sequence of kingdom, **phylum, class, order, family, genus** and **species** is called a **hierarchy of classification**.

Notice that each classification group is given a name. Lions belong to the class Mammalia and the order Carnivora, for example. The final two group names are written in *italics* – this is a worldwide convention amongst scientists. The lion is called simba in Swahili, león in Spanish and leu in Romanian but is known as *Panthera leo* to scientists in each of these countries. This convention of giving organisms a two-part name made up of their genus and species was introduced by the Swedish biologist Carolus Linnaeus. He gave every organism known to science a two-part name based entirely on the body structure of the organism. This binomial system of nomenclature is still in use today (binomial = 'two name').

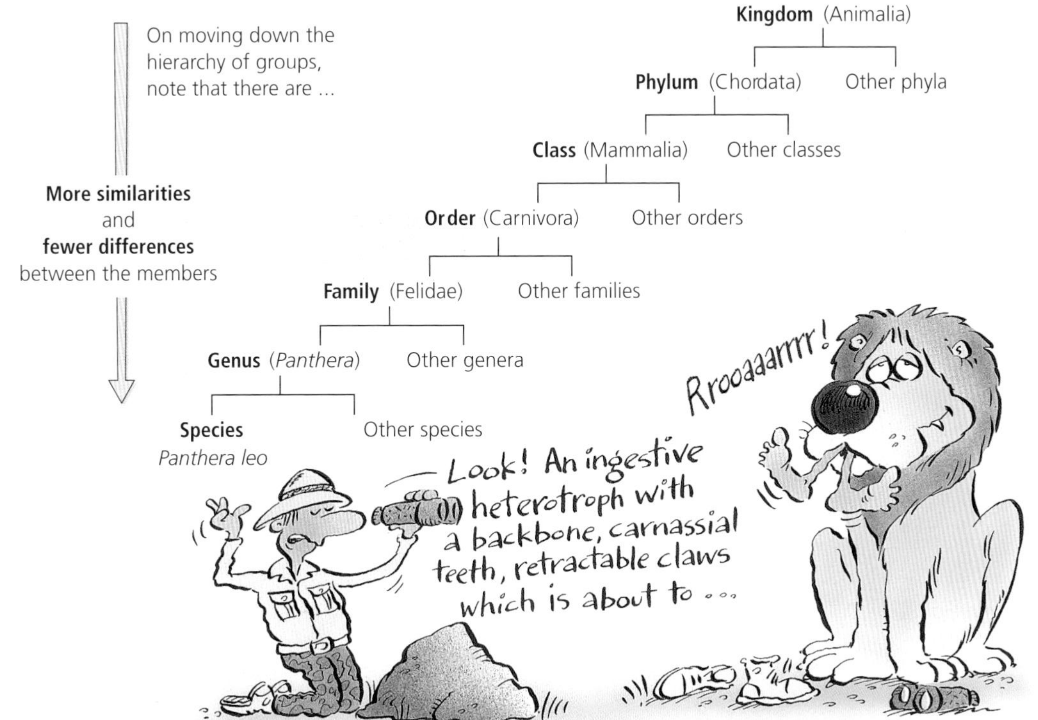

All animals are ingestive heterotophs.

All chordates have a notochord (becomes the backbone).

All mammals have fur and mammary glands.

All carnivores have well developed carnassial (flesh-cutting) teeth.

All Felidae have retractable claws.

All *Panthera* (big cats) can roar but cannot purr.

All lions belong to the same species. Members of the same species can mate and produce fertile offspring with other members of the same species.

▲ The hierarchical classification of the lion

New species today may be classified based on characteristics such as protein structure, chromosome number or gene (DNA or RNA) sequence, which Linnaeus knew nothing about. Each organism, even each individual, has its own DNA profile. Scientists can compare DNA profiles of different species.

How this helps in classification

- **Protein structure:** organisms which are closely-related (share a more recent ancestor) have very similar amino acid sequences in proteins such as haemoglobin.
- **DNA structure:** closely-related organisms have very similar base sequences in DNA (see page 196) because there has been less 'evolutionary time' for mutation to change these base sequences.

Organisms that are closely related have very similar DNA profiles – humans and chimpanzees, for example, share 98.6% of their genes!

Viruses

When the five-kingdom system of classification was devised, no one was able to find a place for the group of organisms called the **viruses**. This is because viruses do not show the typical features of living things – respiration, nutrition and reproduction, for example – unless they are inside the cells of another living organism. In other words, all viruses are parasites and therefore cause harm to their host. Some taxonomists have suggested that viruses belong in a sixth kingdom. There is great variation in the structure of viruses, but they all have certain common features. The structure of a typical virus is shown below.

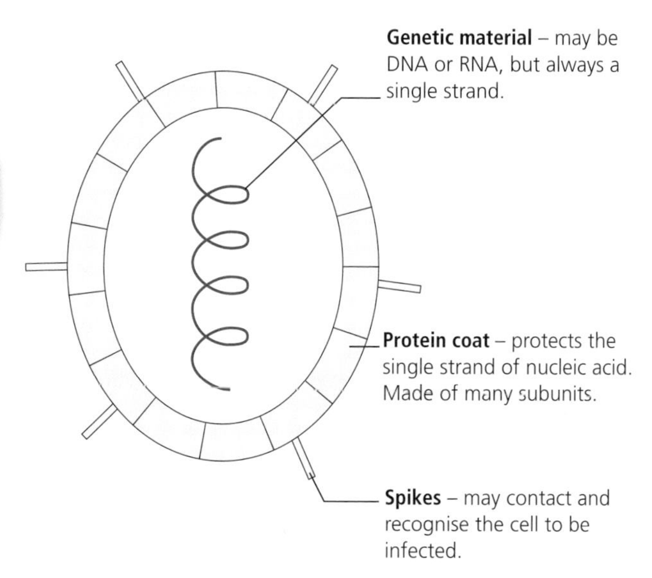
Genetic material – may be DNA or RNA, but always a single strand.
Protein coat – protects the single strand of nucleic acid. Made of many subunits.
Spikes – may contact and recognise the cell to be infected.

▲ A typical virus has genetic material and a protein coat, but cannot carry out its life processes. It has no cytoplasm.

Most viruses cause disease – they may infect humans, domestic animals or plants.

It is important not to confuse viruses with **bacteria**. The structure of bacteria and their importance to humans are described on page 244.

The pages that follow describe the characteristics that distinguish living organisms in some of the most important phyla and classes.

1 The scientific names for the weasel and mink are *Mustela nivalis* and *Mustela vison*, respectively. Both of these animals belong to the order Carnivora, as do the fox (*Vulpes vulpes*) and otter (*Lutra lutra*). The otter, mink and weasel all belong to the family Mustelidae.

a Which feature must they have in common to belong to the order Carnivora?

b Which two animals are most closely related?

c Which animal is the most different from the other three?

d Suggest one feature that places all of these organisms in the Animal Kingdom.

2 The scientific name for the human is *Homo sapiens*. Try to find out the meaning of this name.

3 Why is it difficult to classify viruses into one of the five kingdoms of living organisms?

4 The figure shows five different seaweeds.

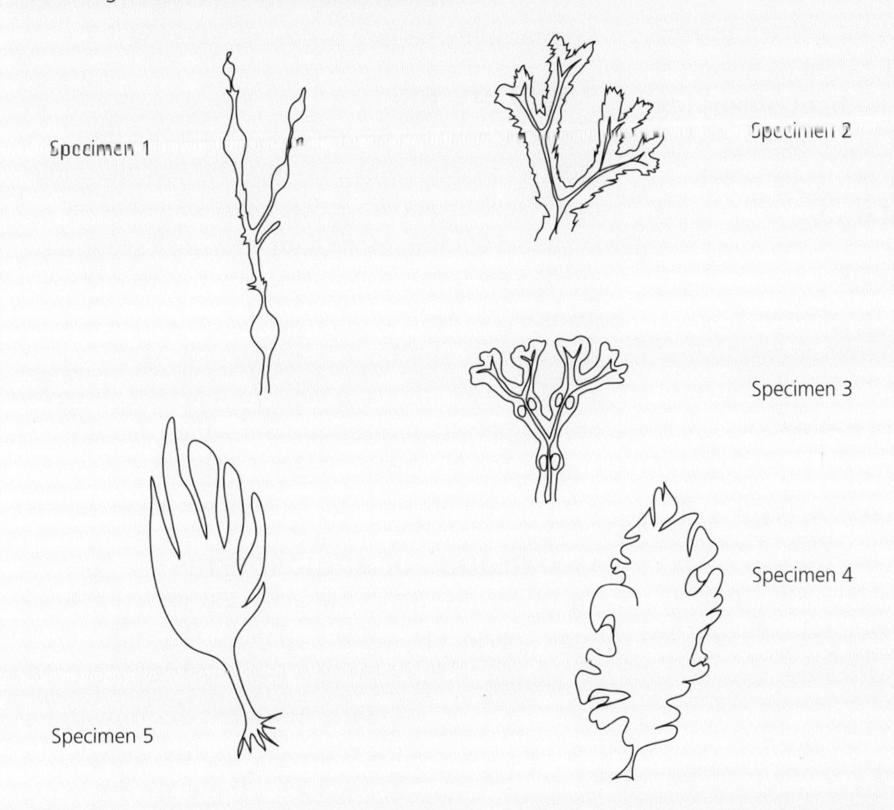

a Describe **ONE** feature of EACH seaweed which is NOT present in any of the others.

b Use your answers to part **a** to construct a dichotomous key which can be used to distinguish between the five seaweeds.

1.3 Plants

OBJECTIVES

- To recall that all plants are autotrophs, and are able to absorb light energy to drive photosynthesis
- To understand some of the steps in the adaptation of plants to life on dry land
- To recall the characteristics of the four main plant groups

Plants are autotrophs

As **autotrophs**, plants manufacture food molecules from simple, inorganic sources by the process of photosynthesis using light as a source of energy. Plants all **contain the light-absorbing pigment chlorophyll** (or similar molecules which perform the same function) inside cells which **have a definite cellulose cell wall**.

Adaptations to life on land

The first plants lived in water, but as living organisms evolved, plant forms developed that could live on land. The classification of plants into groups follows this sequence of evolution.

The Plant Kingdom may be divided into four main groups (phyla): **algae**, **mosses**, **ferns** and **seed plants**.

Algae and mosses can not grow far away from water, but ferns and flowering plants (angiosperms) are much better adapted to life on land.

Ferns

Ferns are much better adapted to life on land than either mosses or algae. They have roots, stems, complex leaves and vascular tissues. They are able to produce spores for wide dispersal. However, they do not have very thick cuticles and can only survive in shady, humid areas. The gametes of ferns, like those of mosses, must swim through a film of moisture to reach the site of fertilisation. An example of a fern is described below.

Angiosperms

The angiosperms or flowering plants are the most successful of plants – they have evolved into many species and have colonised almost every available habitat. More than 80% of all plants are angiosperms (plants with enclosed seeds). Many features of the lives of flowering plants are covered elsewhere in this book (see pages 38–49, 74–81, 156 and 160–169, for example). The diagram at the top of the opposite page summarises these features, and emphasises the adaptations of flowering plants to a successful life on land, including warmer habitats.

Two groups of angiosperms

There are two major subgroups within the angiosperms. In one group there is a single cotyledon in the seed (see page 167) – these are the **monocotyledons**. In the other group there are two cotyledons – these are the **dicotyledons** (eudicotyledons). There are other differences between monocotyledons and dicotyledons, as shown in the diagram on the next page.

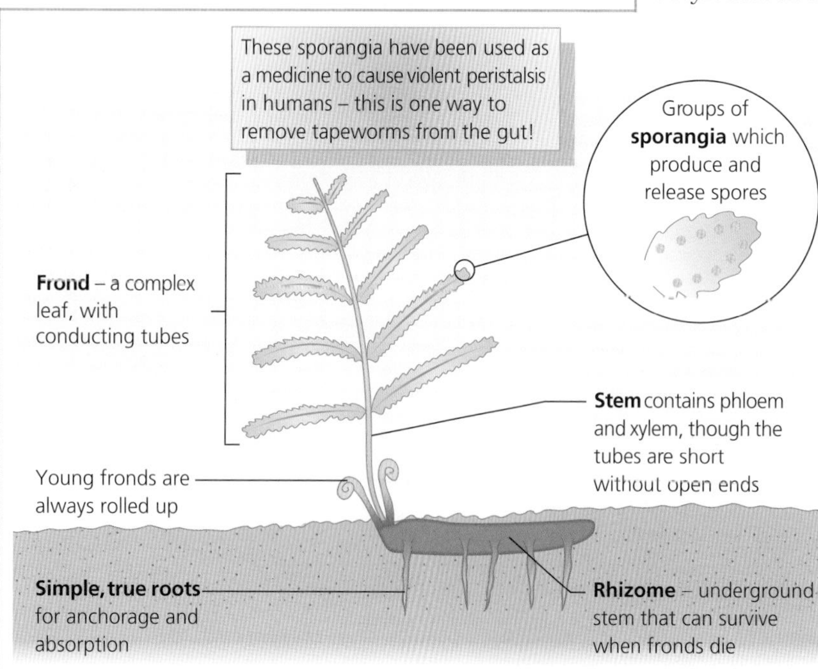

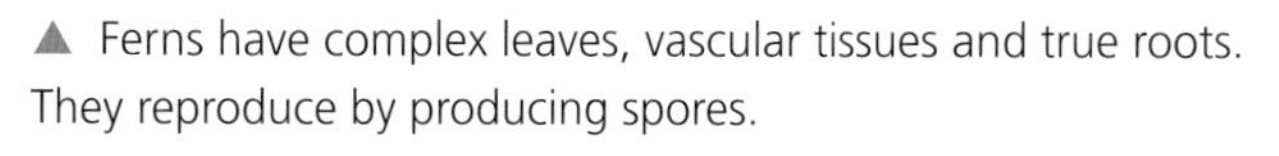
▲ Ferns have complex leaves, vascular tissues and true roots. They reproduce by producing spores.

▲ Each brown patch on the underside of the leaf is made up of many sporangia, which produce and release spores.

Angiosperm features adapt them for life on dry land

Flowers – the colour, pattern, shape, scent or nectar of the flower may attract insects, birds or mammals.

The **ovary** protects the ovules and developing embryo, particularly from drying out. ('Angiosperm' means 'enclosed seed').

Large **leaf surface** allows high rate of photosynthesis to supply energy for growth and fruit production. However, water losses by evaporation and diffusion through stomata are high.

Fruits are formed from ripened ovaries. Their specialised shapes, colours, smells and textures aid seed dispersal by wind, water and animals.

Stomata with guard cells regulate loss of water vapour and exchange of oxygen and carbon dioxide between plant and atmosphere.

Vascular system transports water, ions and organic solutes.

Extensive root systems anchor the shoot systems and absorb water and ions.

Monocotyledons and dicotyledons (eudicotyledons) – two groups of angiosperms (flowering plants)

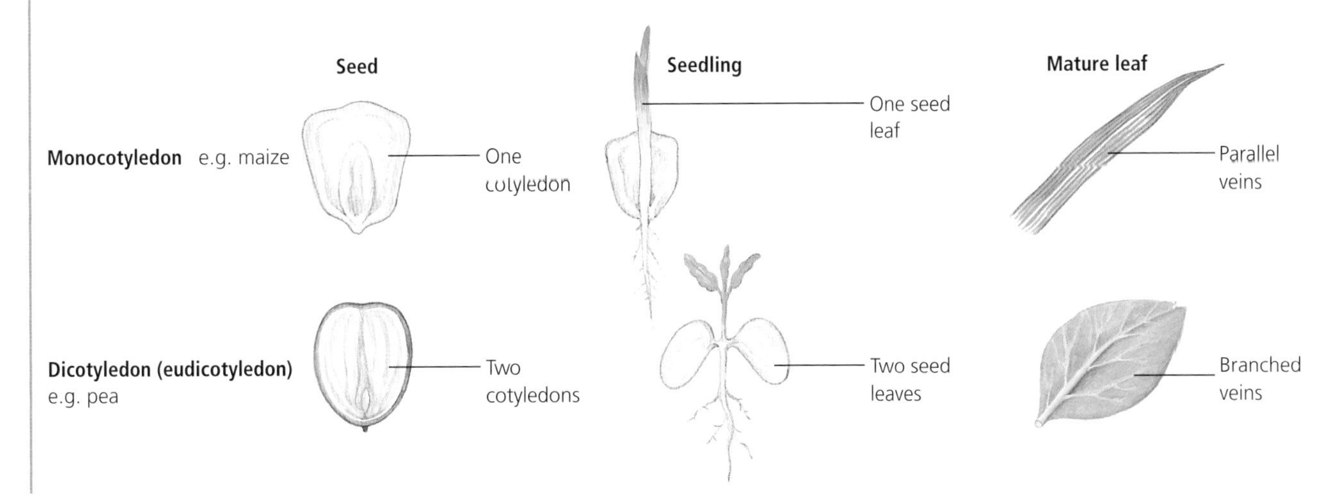

Q

1. In what ways are ferns well adapted to life on land?
2. Seed plants are well adapted to live and to reproduce in dry environments. What major adaptation allows reproduction on dry land?

1.4 Invertebrate animals

OBJECTIVES

- To know the difference between a vertebrate animal and an invertebrate animal
- To be able to describe the main characteristics of four invertebrate groups – annelids, nematodes, molluscs and arthropods
- To be able to distinguish between different classes of arthropods
- To understand the importance of metamorphosis in insects

Vertebrates and invertebrates

All animals share one characteristic – **they feed on organic molecules** (see page 28). Members of the Animal Kingdom can be divided into two large groups based on whether they have a backbone as part of a bony skeleton. Animals with a backbone are called **vertebrates** and those without a backbone are called **invertebrates**.

Four groups of invertebrates are described here:

Nematodes*

Hookworms are nematodes with bodies that are specialised for feeding and reproducing; often they are parasites inside the gut of another animal.

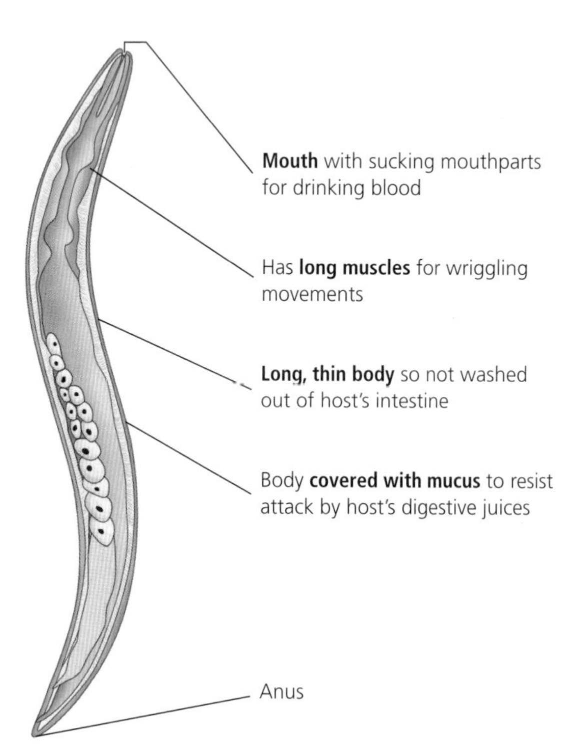

Annelids*

Annelids such as the earthworm have a long segmented body and chaetae.

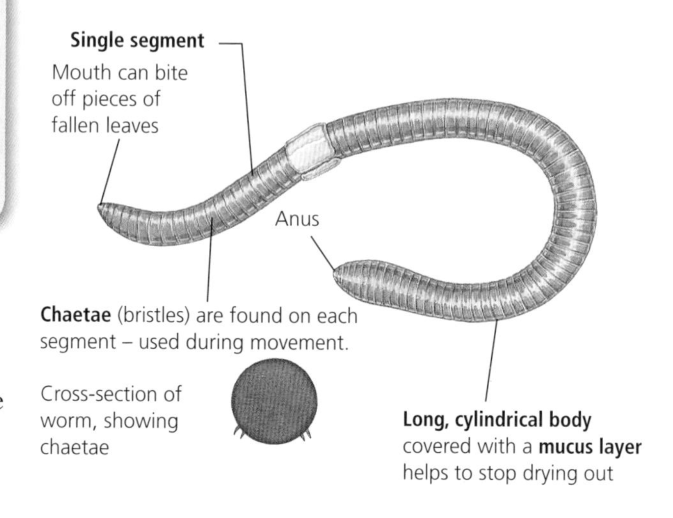

Molluscs*

Molluscs have a hard shell protecting a soft body with no limbs.

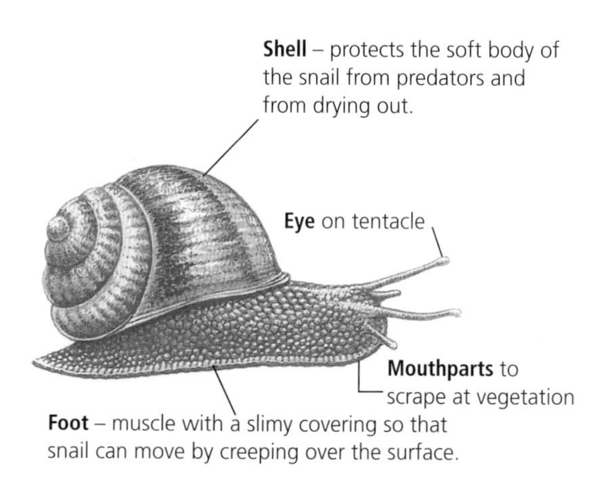

Shell – protects the soft body of the snail from predators and from drying out.
Eye on tentacle
Mouthparts to scrape at vegetation
Foot – muscle with a slimy covering so that snail can move by creeping over the surface.

Annelid, nematode or mollusc?*

	Annelid	Nematode	Mollusc
Body covering	Hard, slightly waterproof	Soft, not waterproof	Soft – shell helps to save water
Segments visible	Yes	No	No
Movement	Uses chaetae (bristles) to move from place to place	Wriggles but lives in one place	Creeps on foot from place to place
Feeding method	Herbivores	Mainly parasites	Mainly herbivores – some carnivores

Arthropods

The arthropods are the most numerous of all animals, both in terms of the number of different species and the number of individuals in any one species. The insects are arthropods that show an interesting adaptation in their life cycle called **metamorphosis** that allows them to use the resources of their habitat to the maximum.

Apart from insects, the arthropod phylum includes three other classes:

Crustacea

Crabs are slightly unusual because many of their segments are tucked under their body.

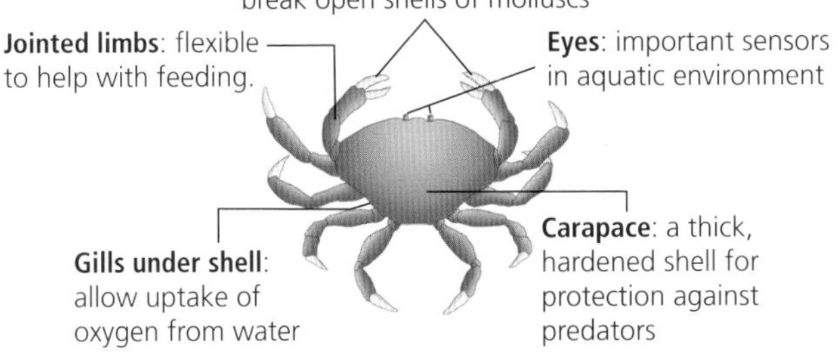

Hardened, serrated edge to claw: can hold onto slippery food, and break open shells of molluscs
Jointed limbs: flexible to help with feeding.
Eyes: important sensors in aquatic environment
Gills under shell: allow uptake of oxygen from water
Carapace: a thick, hardened shell for protection against predators

Myriapods

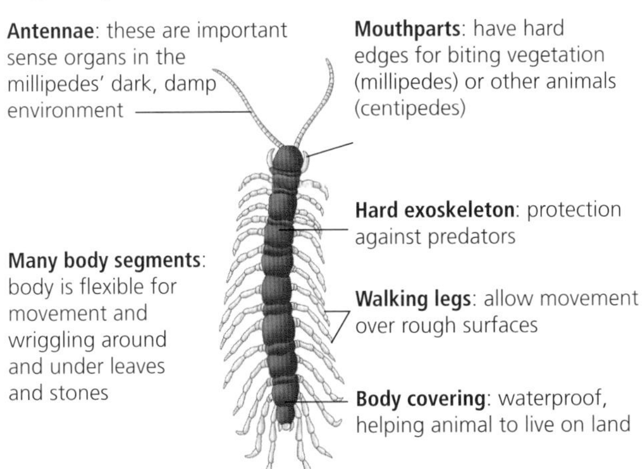

Antennae: these are important sense organs in the millipedes' dark, damp environment
Mouthparts: have hard edges for biting vegetation (millipedes) or other animals (centipedes)
Hard exoskeleton: protection against predators
Many body segments: body is flexible for movement and wriggling around and under leaves and stones
Walking legs: allow movement over rough surfaces
Body covering: waterproof, helping animal to live on land

Insects

e.g. housefly, mosquito

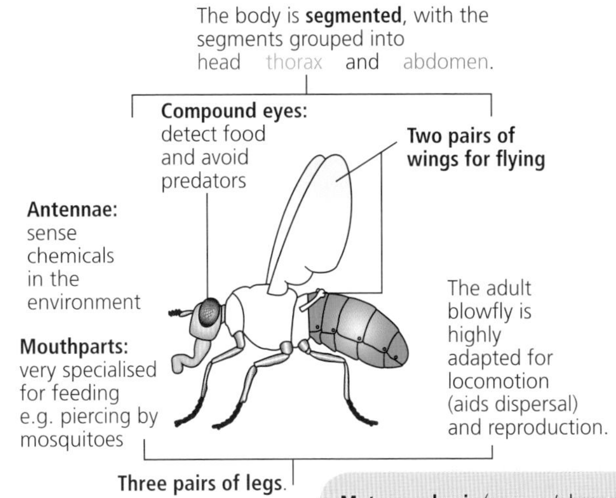

The body is **segmented**, with the segments grouped into head thorax and abdomen.
Compound eyes: detect food and avoid predators
Two pairs of wings for flying
Antennae: sense chemicals in the environment
Mouthparts: very specialised for feeding e.g. piercing by mosquitoes
The adult blowfly is highly adapted for locomotion (aids dispersal) and reproduction.
Three pairs of legs.

Metamorphosis (means 'change of body form') allows different stages which:

- do not compete for the same food sources
- can be highly specialised for different functions. The **larva** is adapted for **feeding** and **growth**, and the **adult** for **locomotion** and **reproduction**.

Arachnids

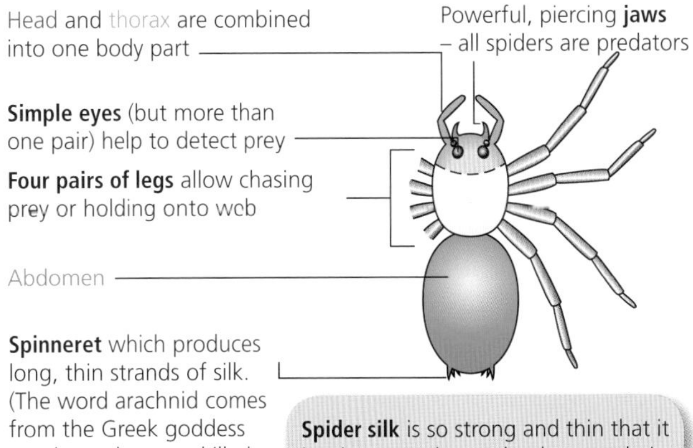

Head and thorax are combined into one body part
Powerful, piercing **jaws** – all spiders are predators
Simple eyes (but more than one pair) help to detect prey
Four pairs of legs allow chasing prey or holding onto web
Abdomen
Spinneret which produces long, thin strands of silk. (The word arachnid comes from the Greek goddess Arachne who was skilled at spinning.)

Spider silk is so strong and thin that it has been used to make the cross hairs in the telescopic sights of rifles.

Q

1 Copy and complete the following paragraph.
All animals have one common characteristic – ______.
The invertebrates are animals that do not have ______.
______, which are the most numerous of all animals.

2 The arthropods include four classes – insects, arachnids, crustaceans and myriapods.

a List three features that all of these classes possess.

b List three features that only insects possess.

c Compare insects and spiders under the headings 'Number of legs', 'Number of body sections', Number of wings' and 'Type of eyes'.

3 Insects are the most abundant of all animals on land. Many of them show an adaptation called complete metamorphosis. What does this term mean, and how does it help to explain why there are so many insect species?

1.5 Vertebrate animals: five classes

OBJECTIVES

- To know the characteristics of the vertebrates
- To understand how different classes of vertebrates show increasing adaptation to dry land
- To know the five classes of vertebrate, and to provide examples of each

If asked to name an animal, most people would probably name a mammal because these are the most familiar animals to us. Mammals are just one class of the phylum **Chordata**. The chordates are often called the **vertebrates**, although strictly speaking there are a few chordates that aren't vertebrates. Vertebrates have a hard, usually bony, internal skeleton with a backbone. The backbone is made up of separate bones called **vertebrae** which allow these animals to move with great ease.

There are five classes of vertebrates, which, like the members of the Plant Kingdom, show gradual adaptations to life on land. The classes are **fish**, **amphibians**, **reptiles**, **birds** and **mammals**.

Fish

Scales covered in mucus help streamlining for swimming

Lateral line contains sense organs to detect vibration

Operculum covering gills: gills have a large surface area for gas exchange

Fins for movement and stability

Amphibians

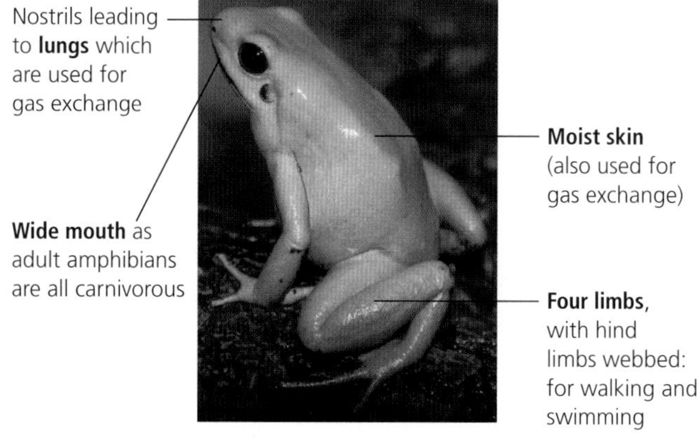

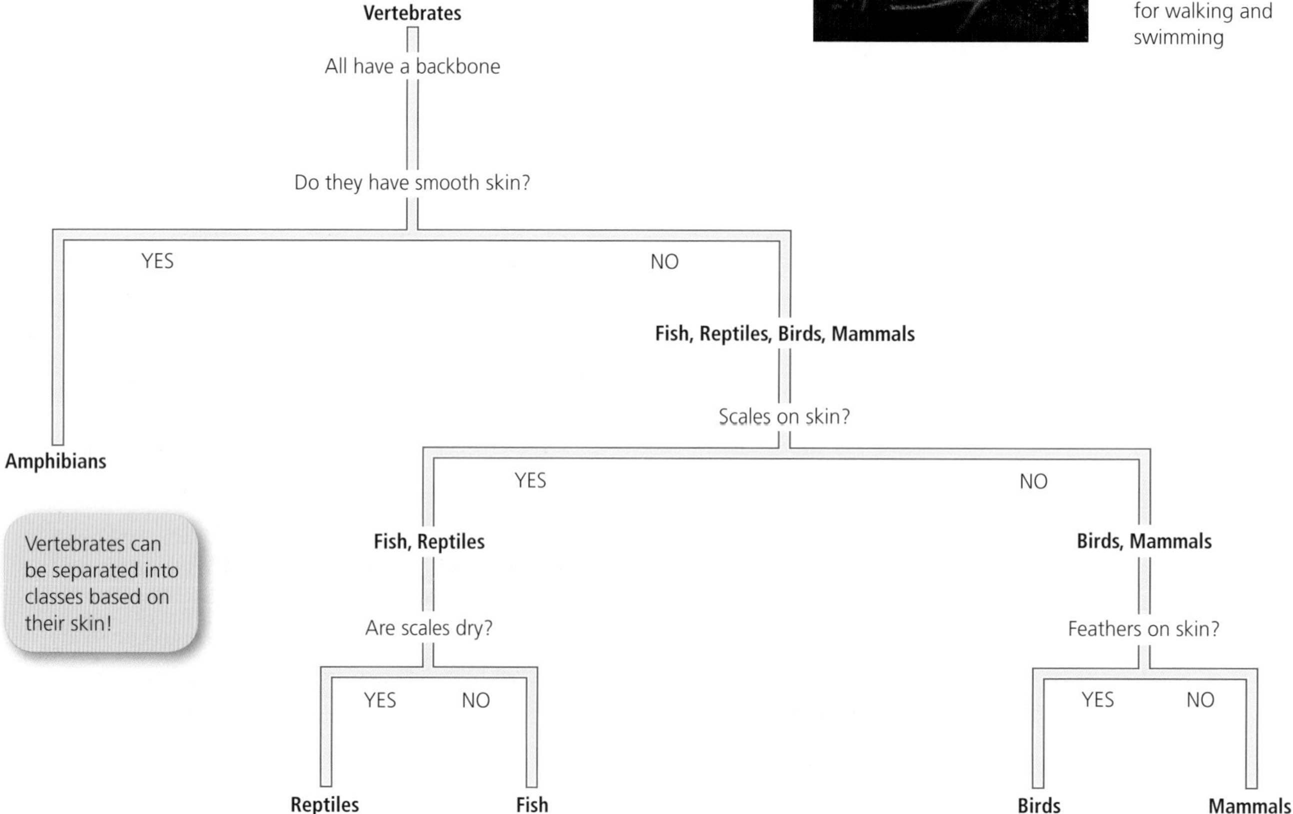

Vertebrates can be separated into classes based on their skin!

Reptiles

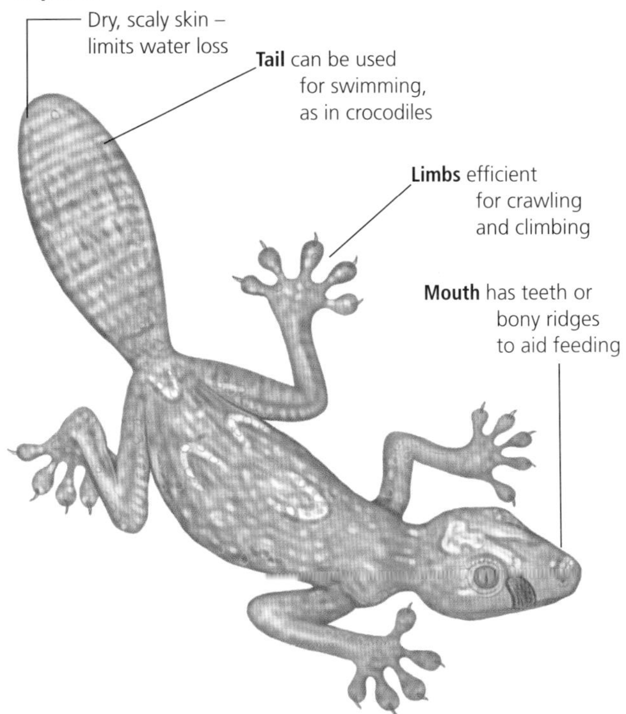
Dry, scaly skin – limits water loss
Tail can be used for swimming, as in crocodiles
Limbs efficient for crawling and climbing
Mouth has teeth or bony ridges to aid feeding

▲ The crocodile has the typical dry scaly skin of reptiles, and the eyes on the top of its head and its sharp, pointed teeth adapt it for catching prey in water.

Birds

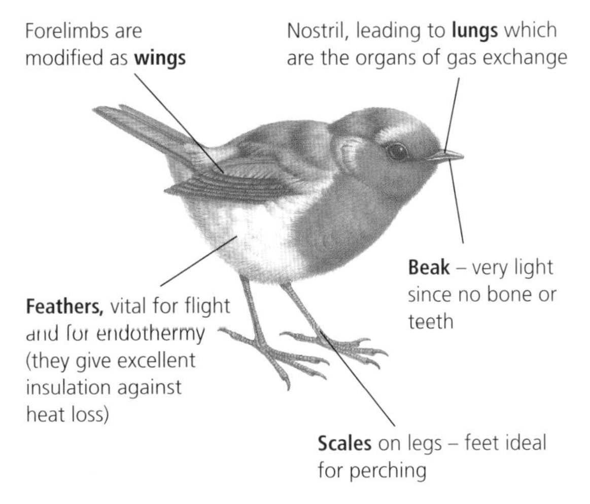
Forelimbs are modified as **wings**
Nostril, leading to **lungs** which are the organs of gas exchange
Beak – very light since no bone or teeth
Feathers, vital for flight and for endothermy (they give excellent insulation against heat loss)
Scales on legs – feet ideal for perching

▲ The heron has typical bird features of feathers and a beak. It is well adapted to capture fish and frogs as it has large eyes to spot its prey, a long pointed beak to grab its prey and large feet for walking over soft, muddy ground.

Mammals

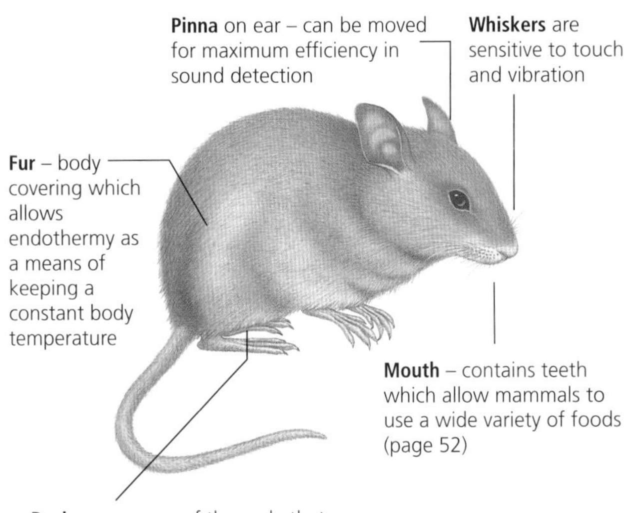
Pinna on ear – can be moved for maximum efficiency in sound detection
Whiskers are sensitive to touch and vibration
Fur – body covering which allows endothermy as a means of keeping a constant body temperature
Mouth – contains teeth which allow mammals to use a wide variety of foods (page 52)
Penis – an organ of the male that enables efficient internal fertilisation

Mammals are endothermic vertebrates that have the characteristics shown in the diagram below.

A wide range of adaptations has allowed mammals to colonise habitats as diverse as the polar wastes and the Arabian desert.

Humans are mammals

Humans show the typical mammalian characteristics of hair, mammary glands and a diaphragm, for example. Humans, though, are unique amongst all animals in that the adaptations they show allow them to modify their environment so that it is suitable for human occupation. As a result humans have been able to live and work in many habitats – no animal has a wider range. Human adaptation has allowed advanced development of the brain, and of all the complex activities that the brain can coordinate. The human brain is extremely sensitive to changes in temperature. Human adaptations include many that are concerned with the fine regulation of blood temperature (see page 130). Other features that make humans very special mammals include an upright posture, freeing the hands for complex movements including the use of tools.

Questions on characteristics and classification

1 This figure shows six arthropods, each of which could carry disease organisms.

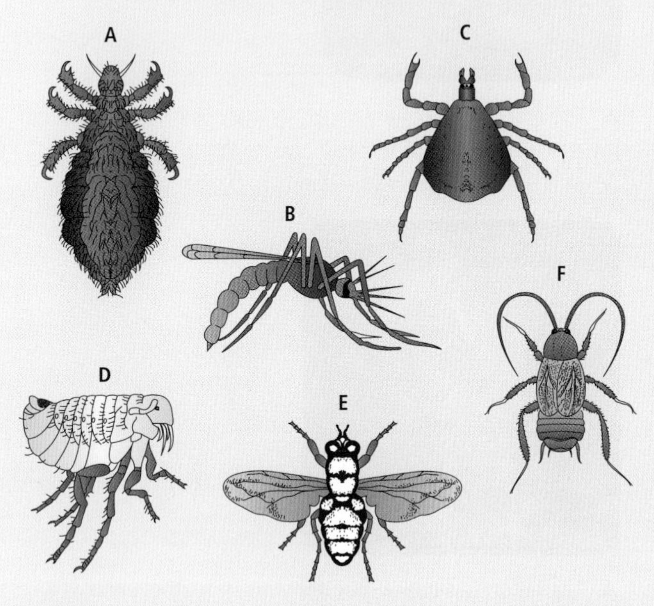

Key	
	arthropod
1 (a) Wings present (b) Wings absent	go to 2 go to 4
2 (a) Wings shorter than abdomen (b) Wings longer than abdomen	go to 3 Musca
3 (a) Abdomen long and narrow (b) Abdomen short and broad	Anopheles Periplaneta
4 (a) Has three pairs of legs (b) Has four pairs of legs	go to 5 Ornithodorus
5 (a) One pair of legs shorter than the other pairs (b) All pairs of legs of similar length	Pulex Pediculus

Use the key to identify each of the arthropods. Write the name of each arthropod in the correct box of the table. As you work through the key, tick (✓) the boxes in the table to show how you identified each arthropod.

Arthropod **A** has been completed for you as an example.

	1 (a)	1 (b)	2 (a)	2 (b)	3 (a)	3 (b)	4 (a)	4 (b)	5 (a)	5 (b)	Name of arthropod
A		✓					✓			✓	*Pediculus*
B											
C											
D											
E											
F											

Cambridge IGCSE Biology 0610
Paper 2 Q1 May 2009

2 a The binomial naming system used to identify all living things gives the Indian elephant a scientific name of *Elephas maximus*.

Which part of this name refers to the genus and which part refers to the species?

b The list gives the names of eight members of the cat family. The common or English name is followed by the binomial name.

Bobcat – *Lynx rufus*
Cheetah – *Acinonyx jubatus*
Jaguar – *Panthera onca*
European lynx – *Lynx lynx*
Leopard – *Panthera pardus*
Lion – *Panthera leo*
Iberian lynx – *Lynx pardinus*
Tiger – *Panthera tigris*

i State the common or English names of two members of the same genus.

ii Name the genus that has only one species.

Cambridge IGCSE Biology 0610
Paper 2 Q1 November 2008

3 a What are you? Follow the branch points at 1, 3, 4, 8, 9 and 11 on the next page to identify yourself as a mammal.

b Name **two** additional characteristics of mammals.

c In what way are humans **special** mammals?

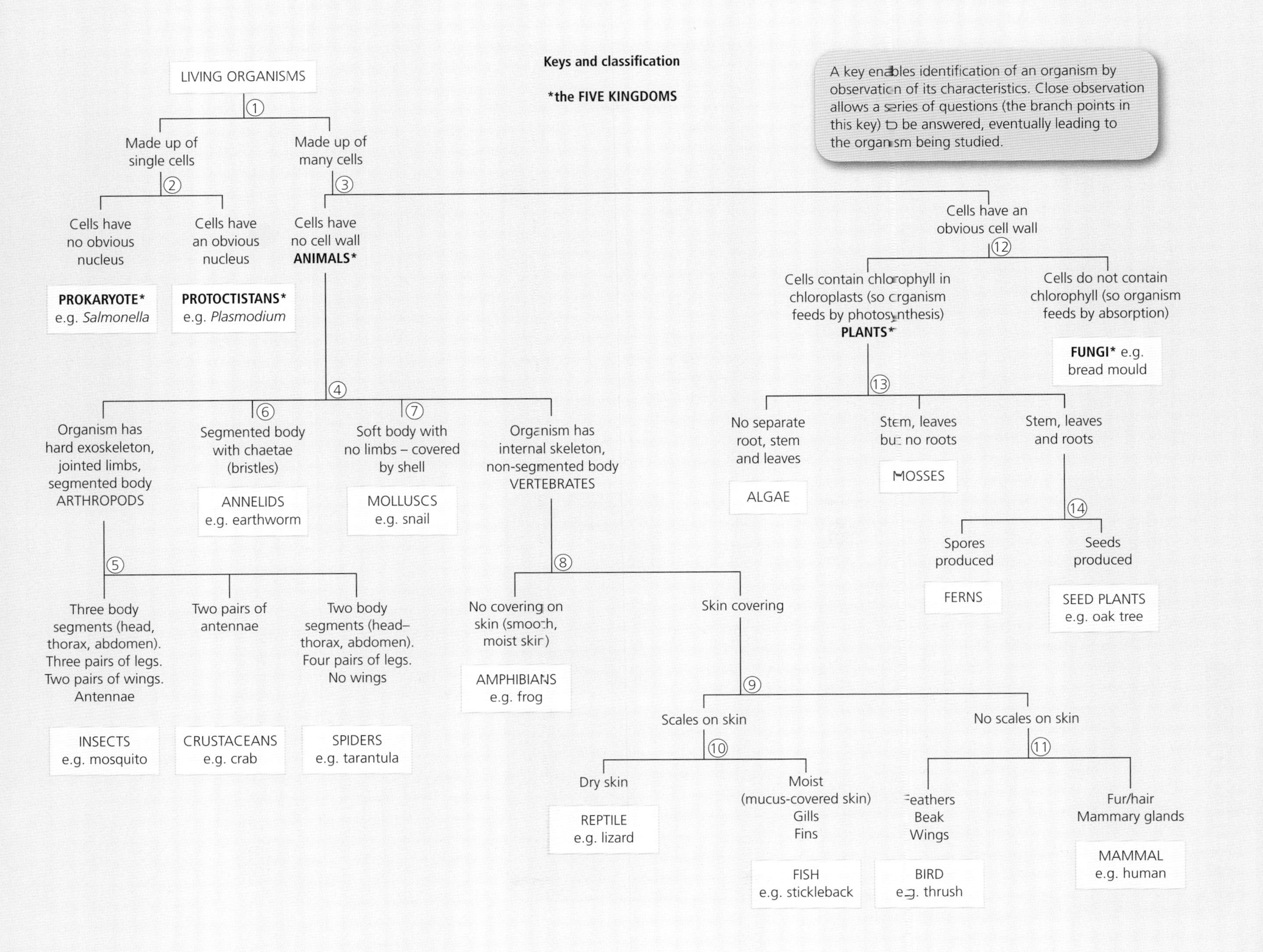
Keys and classification
*the FIVE KINGDOMS
A key enables identification of an organism by observation of its characteristics. Close observation allows a series of questions (the branch points in this key) to be answered, eventually leading to the organism being studied.
LIVING ORGANISMS
1
Made up of single cells
Made up of many cells
2
Cells have no obvious nucleus
Cells have an obvious nucleus
PROKARYOTE* e.g. Salmonella
PROTOCTISTANS* e.g. Plasmodium
3
Cells have no cell wall ANIMALS*
Cells have an obvious cell wall
4
Organism has hard exoskeleton, jointed limbs, segmented body ARTHROPODS
6
Segmented body with chaetae (bristles)
ANNELIDS e.g. earthworm
7
Soft body with no limbs – covered by shell
MOLLUSCS e.g. snail
Organism has internal skeleton, non-segmented body VERTEBRATES
5
Three body segments (head, thorax, abdomen). Three pairs of legs. Two pairs of wings. Antennae
Two pairs of antennae
Two body segments (head–thorax, abdomen). Four pairs of legs. No wings
INSECTS e.g. mosquito
CRUSTACEANS e.g. crab
SPIDERS e.g. tarantula
8
No covering on skin (smooth, moist skin)
Skin covering
AMPHIBIANS e.g. frog
9
Scales on skin
No scales on skin
10
Dry skin
Moist (mucus-covered skin) Gills Fins
REPTILE e.g. lizard
FISH e.g. stickleback
11
Feathers Beak Wings
Fur/hair Mammary glands
BIRD e.g. thrush
MAMMAL e.g. human
12
Cells contain chlorophyll in chloroplasts (so organism feeds by photosynthesis) PLANTS*
Cells do not contain chlorophyll (so organism feeds by absorption)
FUNGI* e.g. bread mould
13
No separate root, stem and leaves
Stem, leaves but no roots
Stem, leaves and roots
ALGAE
MOSSES
14
Spores produced
Seeds produced
FERNS
SEED PLANTS e.g. oak tree

1.6 Organisms are made up of cells

OBJECTIVES

- To know that the basic unit of living organisms is the cell
- To know that all cells have certain features in common, but that there are differences between plant and animal cells
- To understand that the study of cells requires the use of a microscope

All living organisms are made up of units called **cells**. Although cells may take on very specialised functions, they have certain common features. These are shown on the opposite page. Both animal and plant cells have a **cell surface membrane**, **cytoplasm** and a **nucleus**. These three features can be seen on the photograph of a liver cell below. In addition, plant cells have a **cellulose cell wall**, a **vacuole** and may have **chloroplasts**. These features can be seen on the photograph of the palisade cell below.

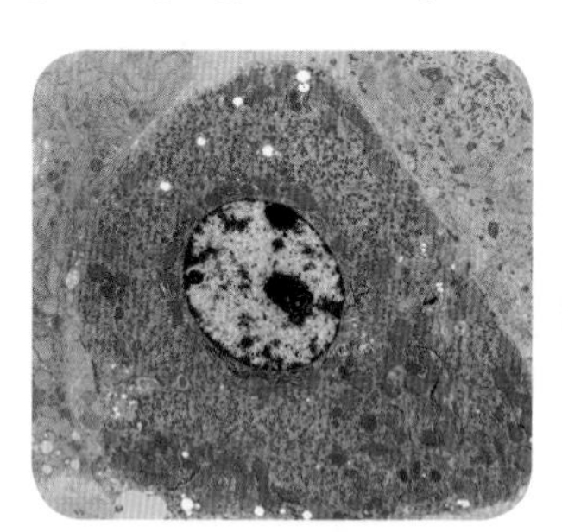

▲ A cell from the inside of the liver, viewed using a light microscope (magnified × 1500 times).

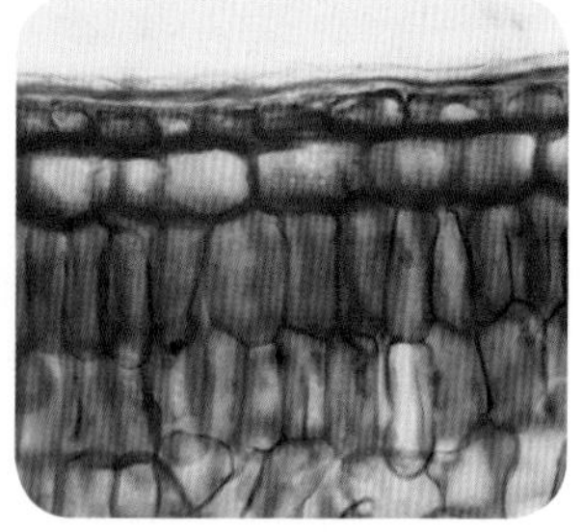

▲ A palisade cell from a leaf, viewed using a light microscope (magnified × 500 times).

The light microscope

Cells are too small to see with the naked eye so a **microscope** is used to study them. Visible light passes through a suitable specimen, and a series of lenses magnify the image that is formed. A light microscope can give a useful magnification of about 400 times, which means the image seen is actually 400 times larger than the specimen. The contrast between different structures in the image can be improved by using dyes or stains. The nucleus of an animal cell, for example, shows up particularly well when stained with a dye called **methylene blue**, and plant cells often show up better when stained with iodine solution.

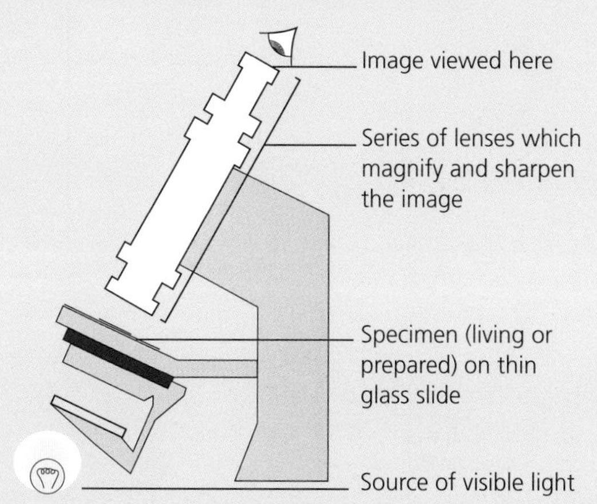

A typical animal cell is about one-fortieth of a millimetre in diameter. This is rather a clumsy term, so scientists use smaller units: one metre (m) contains 1000 millimetres (mm), and one millimetre contains 1000 micrometres (μm). So a typical animal cell is about 25 μm in diameter.

The **size** of a structure or an organism is measured in **units of length** (such as mm or m). When a diagram is made, or a photograph taken, it may not be easy to directly show the correct size – for example, when a structure is extremely small or very large. The correct (or true) size of an organism can be calculated using a combination of actual measurement and a known magnification.

Magnification $= \frac{\text{Measured length}}{\text{Actual length}}$

or **Actual (true) length** $= \frac{\textbf{Measured length}}{\textbf{Magnification}}$

We can also use a **scale line** to work out magnification

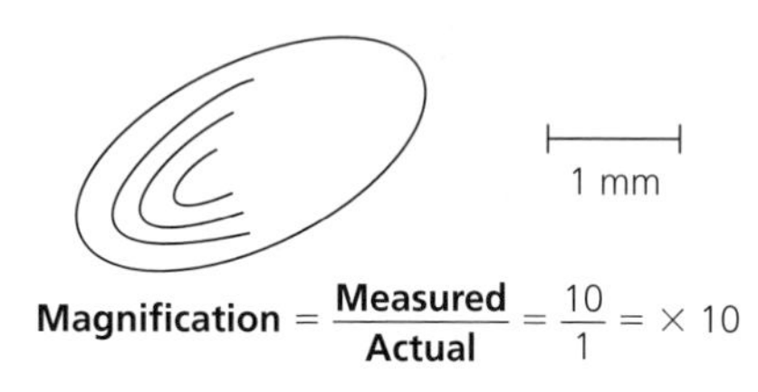

Magnification $= \frac{\textbf{Measured}}{\textbf{Actual}} = \frac{10}{1} = \times 10$

Calculating magnification and size

For example, look at this poppy seed.

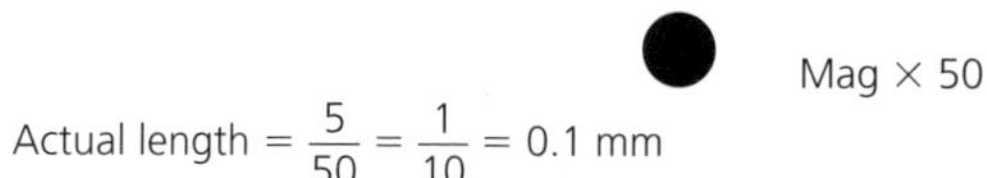

Actual length $= \frac{5}{50} = \frac{1}{10} = 0.1$ mm

Tips!

1. make sure that measured and actual lengths are given in the same units
2. to help remember the formula
   Magnification $= \frac{\text{Measured}}{\text{Actual}}$
3. Core students can use millimetres as a unit. Extended students may need to use millemetres and micrometres.

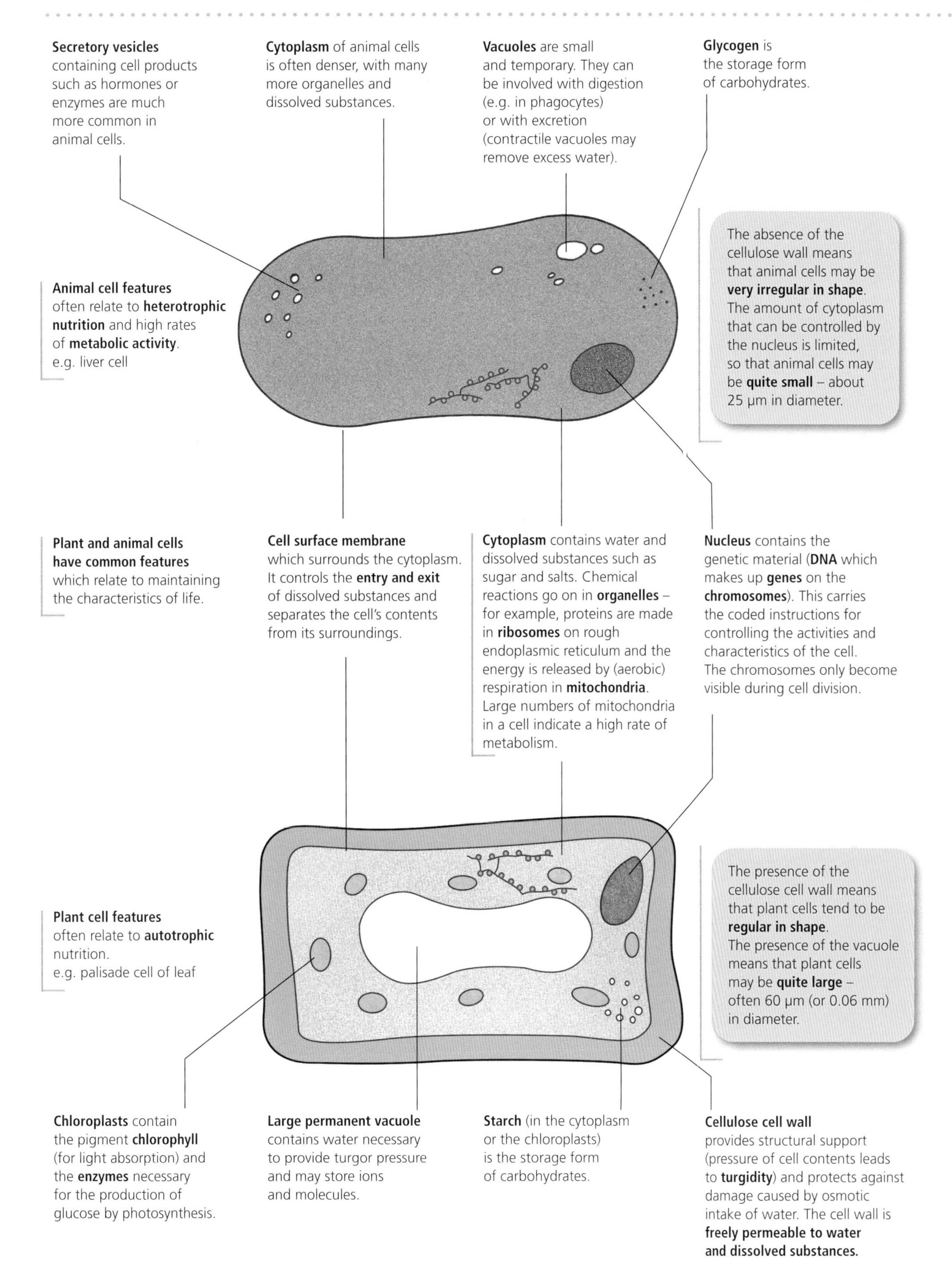

▲ The features of plant and animal cells allow these cells to carry out the basic processes of life. The differences between plant and animal cells are due to the differences in lifestyle between animals and plants, especially to their different methods of nutrition.

1.7 The organisation of living organisms

OBJECTIVES

- To understand that the body of a living organism is a highly organised structure
- To understand that cells, tissues, organs and systems represent increasing degrees of organisation in living organisms

Specialised cells

Large organisms are **multicellular** – they are made up of many cells. Different types of cell have particular structures designed to help them carry out different tasks and functions – they have become **specialised**. Some examples of specialised cells, and their functions, are shown in the table.

Cell type	Appearance	Functions and adaptations
Animal cells		
Red blood cell (page 82)		**Transports** oxygen from the lungs to the tissues where aerobic respiration occurs. The cytoplasm is filled with the pigment haemoglobin, which carries oxygen. The cells have no nucleus, leaving more space for haemoglobin, and they are very flexible (they can be forced through even the narrowest of blood vessels).
Muscle cell (page 108)		**Contracts** so that structures can be brought closer together. Muscle cells are long, and have many protein fibres in the cytoplasm. These fibres can shorten the cell when energy is available.
Ciliated cell (page 114)		Has a layer of tiny hairs (cilia) which can **move and push mucus** from one place to another. The mucus can transport trapped dust and microbes when it is pushed by the cilia.
Motor nerve cell (page 138)		**Conducts nerve impulses.** The cell has a long fibre called an axon along which impulses travel, a fatty sheath which gives electrical insulation and a many-branched ending which can connect with many other cells.
Other important specialised animal cells are the gametes, sperm and egg (page 172)		
Plant cells		
Root hair cell		**Absorbs minerals and water** from the soil water. The cell has a long extension (a root hair) which increases the surface area for the absorption of materials.
Xylem vessel		**Transports water and supports the plant.** The cell has no cytoplasm (so water can pass freely), no end wall (so that many cells can form a continuous tube) and walls strengthened with a waterproof substance called lignin.
Another important specialised plant cell is the palisade mesophyll cell (page 42)		

Specialised cells combine to form tissues ...

Cells with similar structures and functions are massed together in **tissues**. Some plant and animal tissues are shown in the tables below.

Animal tissue	Main functions
Epithelium	Lines tubes such as the gut and covers surfaces such as the skin
Connective tissue	Binds and strengthens other tissues, such as tendons
Blood	Transports substances around the body, and defends against disease
Skeletal tissue	Supports and protects softer tissues, and allows movement
Nervous tissue	Sets up nerve impulses and transmits them around the body
Muscle tissue	Contracts to support and move the body

Plant tissue	Main functions
Epidermis	Protects against water loss, and may be involved in absorption of water and ions
Mesophyll	Photosynthesis
Parenchyma	Fills spaces between other plant tissues and may be involved in storage, as in the potato tuber
Vascular tissue	Transports materials through the plant body
Strengthening tissue	Supports the plant

... tissues combine to form organs ...

Several tissues may be combined to form an **organ**, a complex structure with a particular function, such as the small intestine shown right.

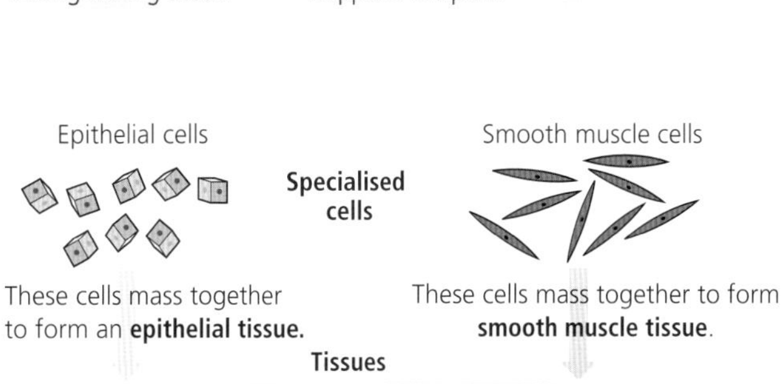

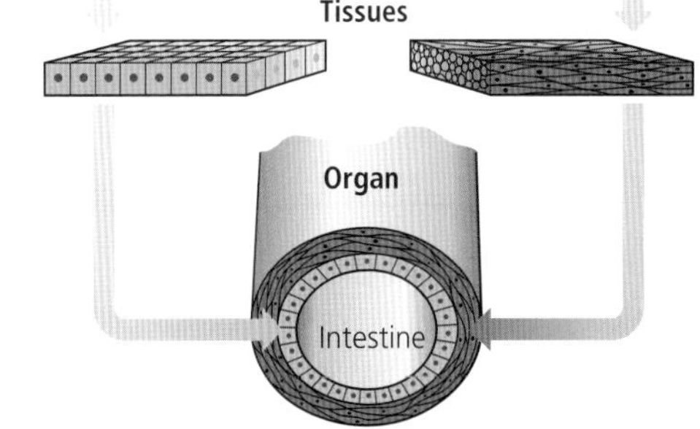

... organs combine to form organ systems

In complex organisms, several organs work together to perform a particular task. These organs form an **organ system**.

Each cell, tissue and organ in an organism has a specialised part to play (there is **division of labour**) but their activities must be coordinated.

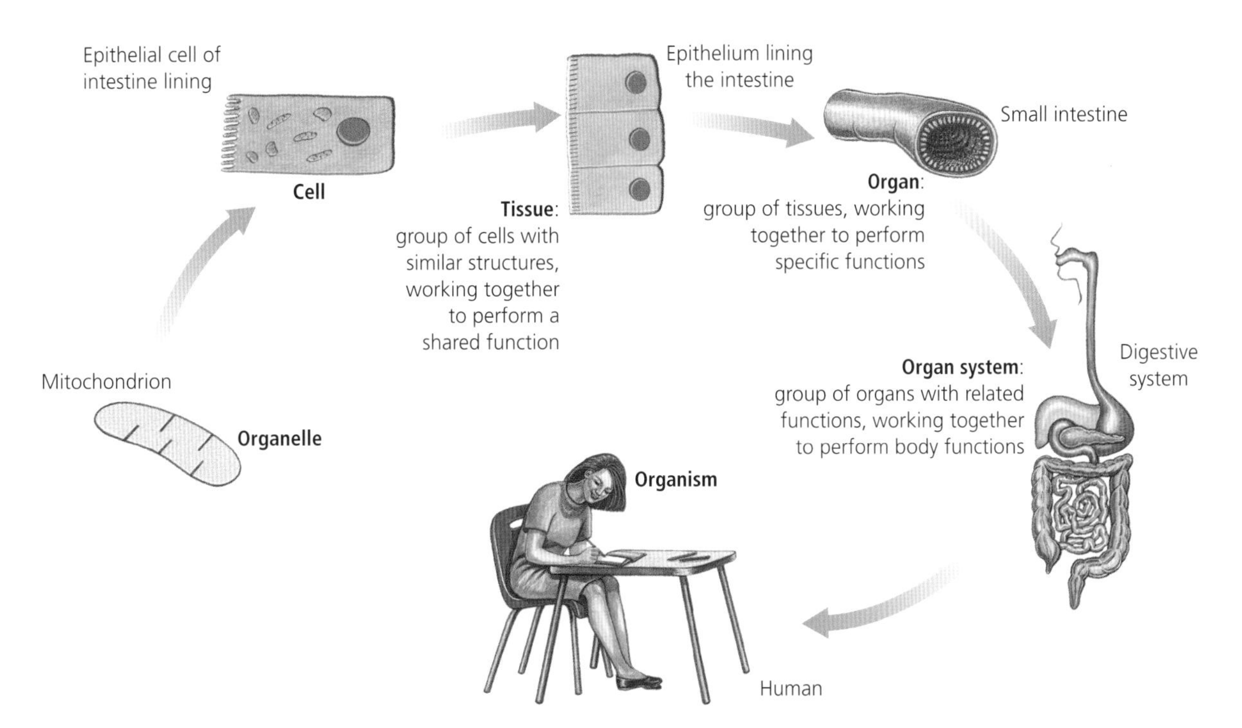

Questions on cells and organisation

1 Animals and plants are made up of cells, tissues and organs. This list contains some cells, tissues and organs:

phagocyte, sperm, epidermis, xylem, liver, blood, heart, leaf, ileum, ovary, neurone, brain, stem.

Make a table with these headings:

Cell	Tissue	Organ

Place each of the structures from the list into the correct column.

2 Arrange these biological terms in order of size (from the smallest to the largest):

organ, cell, organism, organelle, tissue, system.

3 Arrange these units of length in order, starting with the largest and ending with the smallest:

kilometre, micrometre, metre, millimetre.

An average plant cell is 50 micrometres long. How many plant cells could fit into one millimetre? Show your working.

4 The table below shows the percentage of different cells in a human body.

a Complete the table to show the percentage of 'other' cells in the body.

b Use the figures in the table to draw a pie chart to show the proportions of different cells in the body.

c The body contains about 100 000 000 000 000 cells. Work out how many red blood cells there would be – show your working.

Cell type	Percentage in body
Bone	23
Red blood cell	13
White blood cell	0.5
Muscle	24
Nerve	8
Skin	11
Egg	0.1

5 Copy and complete this table by placing a tick if the structure is present and a cross if it is not.

Structure	Liver cell	Palisade cell
Cell surface membrane		
Chloroplasts		
Cytoplasm		
Cellulose cell wall		
Nucleus		
Starch granule		
Glycogen granule		
Large, permanent vacuole		

6 A biology book said that 'the structure of a cell is closely related to its function'. Using both plant and animal examples, comment on the validity of this statement.

7 The following list of the characteristics of living organisms can be matched with a series of definitions of these characteristics. Match up the two lists, so that you write out the definition of each of the characteristics.

Characteristics: respiration, sensitivity, growth, reproduction, nutrition, excretion, development

Definitions:
- a change in size
- the release of energy from organic molecules
- the generation of offspring
- the ability to detect changes in the environment
- a change in shape or form
- the supply of food
- the removal of the waste products of metabolism

8 Identify these cell structures from their descriptions:

a structures which contain chlorophyll

b cavity that is only found in plant cells

c contains cellulose, and surrounds a plant cell

d jelly-like material which fills most of the cell

e coloured threads found inside the nucleus

f the sites of aerobic respiration

g controls the entry and exit of substances

h carries the genetic information, and controls the activities of the cell

Cell structures: cell membrane; cell wall; chloroplasts; chromosomes; cytoplasm; mitochondria; nucleus; vacuole

9 Match these cells with their adaptations and/or functions.

Cells: red blood cell; white blood cell; cell lining bronchioles of lungs; motor nerve cell; palisade cell; root hair cell; phloem sieve tube; sperm cell; egg cell

Adaptations:

- **a** has long extension to increase surface area for absorption of minerals and water
- **b** contains haemoglobin for oxygen transport
- **c** carries male parent's genetic information ready for fertilisation
- **d** transports sugars through the body of a plant
- **e** elongated, and transmits information from the central nervous system to a muscle
- **f** has cilia to trap and remove mucus and dust particles
- **g** contains powerful enzymes to digest bacteria
- **h** carries female parent's genetic information, and has food stores in cytoplasm
- **i** contains many chloroplasts

10 Say whether each of the following is an organelle, a cell, a tissue, an organ, a system or an organism:

- **a** stomach
- **b** mitochondrion
- **c** cat
- **d** onion bulb
- **e** a carrot plant
- **f** chloroplast
- **g** neurone
- **h** lung
- **i** brain
- **j** heart and circulation
- **k** xylem
- **l** the lining of the lung

11 On this diagram of a liver cell, identify

- **a** which structure carries out aerobic respiration
- **b** which structure controls the movement of salts into the cell
- **c** which structure would carry the genes
- **d** which feature increases the surface area of the cell
- **e** which structure is a food store

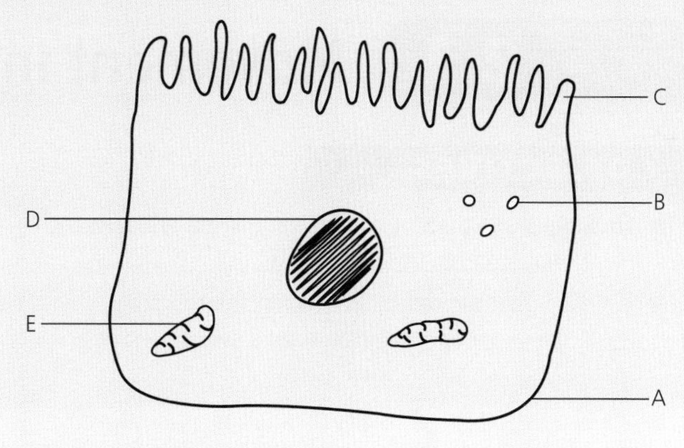

12 Use words from this list to complete the following paragraphs. The words may be used once, more than once or not at all.

palisade cell, epidermis, tissues, excretory system, specialized, cells, blood, kidney, chloroplasts, leaf, red blood cell, division of labour, xylem, phloem, nervous, systems, endocrine, organ.

- **a** Large numbers of ____ that have the same structure and function are grouped together to form ____, for example ____. Several separate tissues may be joined together to form an ____ which is a complex structure capable of performing a particular task with great efficiency. In the most highly developed organisms these complex structures may work together in ____, for example the ____ in humans is responsible for the removal of the waste products of metabolism.
- **b** The structure of cells may be highly adapted to perform one function, i.e. the cells may become ____. One excellent example is the ____ which is highly adapted to carry oxygen in mammalian blood. If the different cells, tissues and organs of a multicellular organism perform different functions they are said to show ____. One consequence of this is the need for close co-ordination between different organs – this function is performed by the ____ and ____ systems in mammals.
- **c** In plants an example of a cell highly specialized for photosynthesis is the ____ which contains many ____. These cells are located in the organ called the ____ which also contains other tissues such as ____ which limits water loss and ____ which transports water and mineral ions to the leaf.

2.1 Movement in and out of cells: diffusion

OBJECTIVES

- To understand that the contents of a living cell must be kept separate from its surroundings
- To know that the cell surface membrane can act as a barrier to some substances which might pass between a cell and its surroundings
- To understand the principles of diffusion, osmosis, active transport and phagocytosis

On page 16 we saw that the cell cytoplasm is surrounded by a **cell surface membrane**. This acts as a boundary between the cell contents and its surroundings – it has very little strength, but it plays a vital role in regulating the materials that pass in and out of the cell. Materials may pass in and out of cells by:

- **diffusion**
- **osmosis**
- **active transport**
- in special cases, **phagocytosis**.

Diffusion – 'mixing molecules'

Molecules and ions in a liquid or a gas move continuously. The movement is quite random, and the particles change direction as they bump into one another. The particles collide more often when they are close together (when they are **concentrated**) and so they tend to **diffuse**, or spread out, until they are spaced evenly throughout the gas or liquid. The random movement of the particles is due to their own **kinetic energy**. When diffusion happens in living cells, the cells themselves do not have to expend any energy for it to take place.

If there is a region of high concentration and a region of low concentration, we can say that there is a **concentration gradient** between the regions. We can therefore define **diffusion** as:

- **the net movement of molecules within a gas or liquid**
- **from a region of high concentration to a region of lower concentration (down a concentration gradient)**
- **as a result of their random movement**
- **until an equilibrium is reached.**

Partially permeable membranes

Not all particles can diffuse through cell surface membranes. Sometimes the particles are too big, or they have the wrong electrical charge on them, or the chemical composition of the membrane prevents them passing across. The diagram below shows a **partially permeable membrane** – it is **permeable** to glucose and water but **impermeable** to protein.

Diffusion and life processes

Diffusion is the main process by which substances move over short distances in living organisms. Some of the life processes that involve diffusion are shown in the diagram above right.

Living organisms have certain adaptations to speed up diffusion:

- **Diffusion distances are short** – the membranes in the lungs, for example, are very thin so that oxygen and carbon dioxide can diffuse between the blood and the lung air spaces.
- **Concentration gradients are maintained** – glucose molecules that cross from the gut into the blood, for example, are quickly removed by the circulating blood so that their concentration does not build up and equilibrium is not reached.
- **Diffusion surfaces are large** – the surface of the placenta, for example, is highly folded to increase the surface area for the diffusion of molecules between a pregnant female and the developing fetus in her uterus.

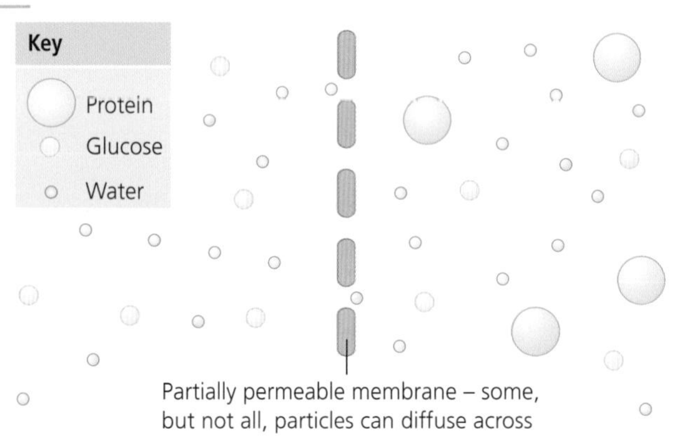

▲ The overall (net) movement of glucose and water molecules depends on their concentration gradient. Protein molecules cannot diffuse across this membrane, even though the concentration gradient suggests that they should move from right to left.

Life depends on the exchange of materials between different cells, and between cells and their surroundings. For example, a plant absorbs carbon dioxide from its surroundings by diffusion, and the carbon dioxide passes through the leaf to the photosynthesising cells by the same process.

From the lungs, oxygen enters the blood by diffusion. The continual movement of the blood keeps up a high concentration gradient between the air and the blood.

Glucose and amino acids pass from inside the gut into the blood, partly by the process of diffusion.

Oxygen produced by photosynthesis diffuses out of the plant into the air. It enters the boy's lungs as he breathes in. The lungs are adapted to speed up the diffusion of oxygen into the blood, since they have thin surfaces with a very large surface area.

Mineral ions from the soil solution are absorbed by plant roots. This process depends upon diffusion as well as on active transport.

▲ Many life processes depend on diffusion to move substances around. Diffusion has no 'energy cost' to a living organism.

How surface area affects rate of diffusion

In an investigation to measure the speed of diffusion, cubes of gelatin which had been stained purple with an indicator solution were placed in dilute acid as shown in the diagram. The time taken for each cube to turn completely orange is shown in the table.

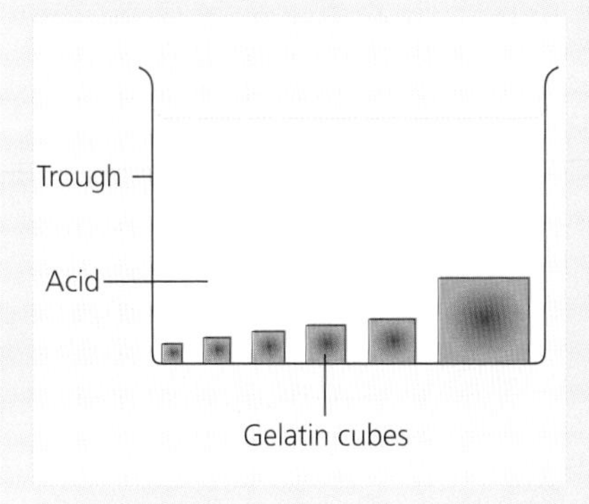

Length of side of cube / mm	Time taken to turn orange / s	Surface area of cube (total of 6 sides) / mm^2	Volume of cube / mm^3	Surface area to volume ratio (surface area ÷ volume)
1	20			
2	41			
3	76			
4	104			
5	188			
10	600			

a Copy and complete the table. Plot a graph of surface area to volume ratio against time taken to turn orange. Plot time taken on the vertical axis.

b What do the results suggest about the efficiency of diffusion in supplying materials to the centre of an organism's body?

c Suggest methods which organisms might use to improve the supply of materials by diffusion. Try to provide examples of these methods.

2.2 Movement in and out of cells: osmosis

Osmosis is a special case of diffusion

The biochemical processes in living cells always take place in a **solution**. A solution is made up of a **solvent** (the dissolving fluid) and a **solute** (the particles dissolved in the solvent). In living organisms, the solvent is water and the solution is called an aqueous solution.

Living cells are separated from their surroundings by the **partially permeable cell surface membrane**. The contents of the cell, the cytoplasm, is one aqueous solution and the surroundings of the cell, for example pond water, is another aqueous solution. If the two solutions do not have the same concentrations of various substances, molecules may move from one to the other by diffusion, if the membrane is permeable to these substances.

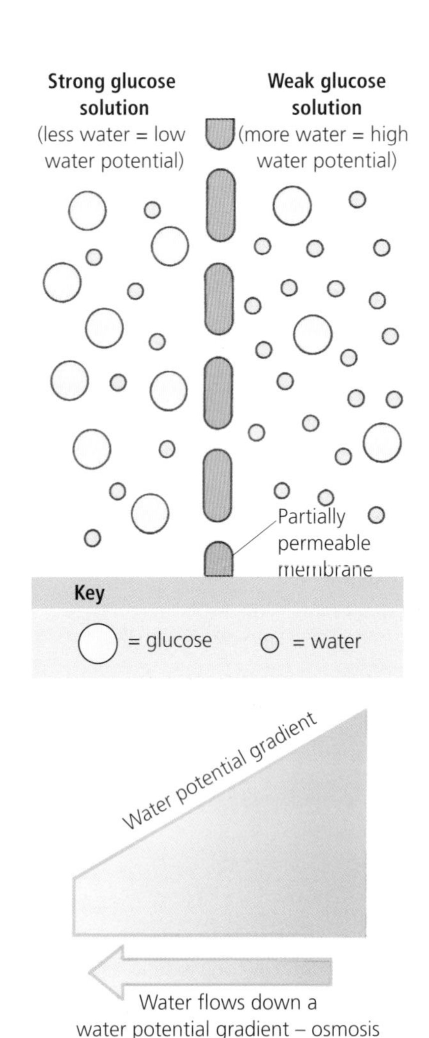

The diagram on the left shows two glucose solutions separated by a partially permeable membrane – this membrane will allow the diffusion of water molecules but not glucose (the solute) molecules. As a result water can move from the right, where there is a high concentration of water molecules, to the left, where there is a lower concentration of water molecules, by the process of diffusion. This diffusion of water is called **osmosis**, and will continue until a water equilibrium has been reached.

Because it is sometimes confusing to talk about water 'concentration', biologists use the term **water potential** instead. A solution with many water molecules has a **high water potential**, and a solution with few water molecules has a **low water potential**. In the diagram, a **water potential gradient** exists between the two solutions, and water molecules can flow down this gradient from right to left.

Osmosis can be defined as:

- **the diffusion of water molecules**
- **from a region of higher concentration of water molecules to a region of lower concentration of water molecules**
- **down a water potential gradient**
- **through a partially permeable membrane.**

Cells and osmosis

A cell is surrounded by a partially permeable membrane, and water may cross this membrane easily. If a cell is placed in a solution of lower water potential, water leaves the cell by osmosis. If the cell is placed in a solution of higher water potential, water enters by osmosis.

Plant cells and osmosis

If water enters a plant cell by osmosis the cytoplasm will swell, but only until it pushes against the cellulose cell wall, as shown below. A plant cell will not be permanently damaged by the entry of water. If water leaves a plant cell by osmosis the cytoplasm will shrink, but the cellulose cell wall will continue to give some support. Plant cells rarely suffer permanent damage by the loss of water.

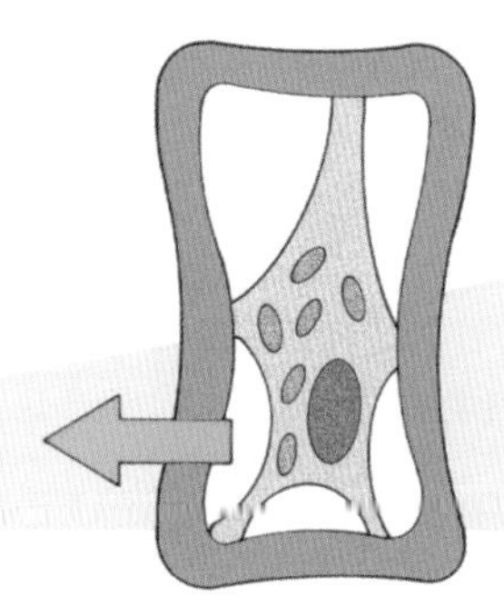

Cell in solution of lower water potential than cell contents – water leaves by osmosis. The cytoplasm pulls away from the cell wall and the cell becomes **flaccid** ('floppy').

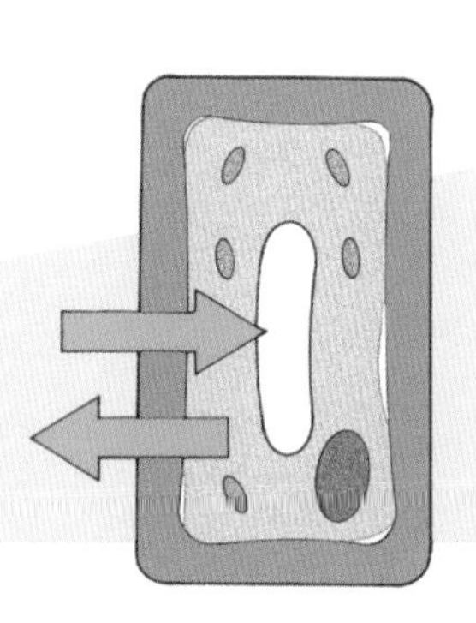

Cell in solution of equal water potential – no *net* movement of water; cytoplasm just presses against cell wall.

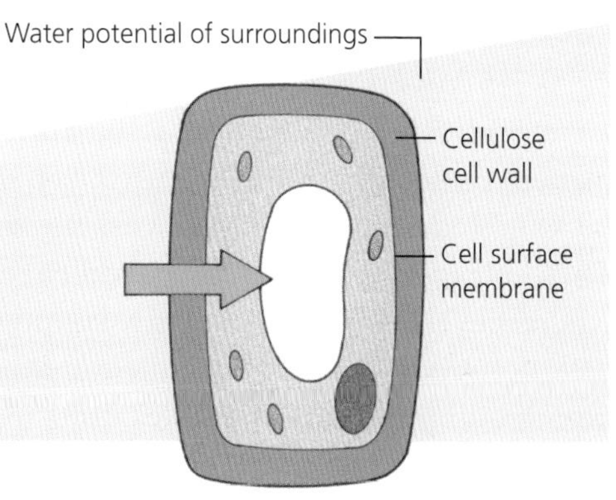

Cell in solution of higher water potential – water enters by osmosis. The cytoplasm pushes hard against the cell wall and the cell becomes **turgid** (firm).

Animal cells and osmosis

Animal cells have no cell wall, just a membrane. They are likely to suffer damage as a result of osmosis, as shown in the diagram below.

Osmosis is potentially damaging to animal cells, and animals have mechanisms to keep the blood plasma and the body fluids at the same water potential as the cytoplasm of cells. In mammals the kidney plays a vital part in this process of **osmoregulation** (see page 124).

Red blood cell

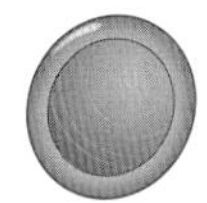

In a solution of lower water potential, the cell loses water, shrinks and becomes **crenated**

In a solution of the same water potential as the inside of the cell, the cell is in equilibrium

In a solution of higher water potential, the cell takes in water, swells and bursts (**haemolysis**)

The 'ghost' of a red blood cell; just the membrane is left behind

▲ Crenated red blood cells

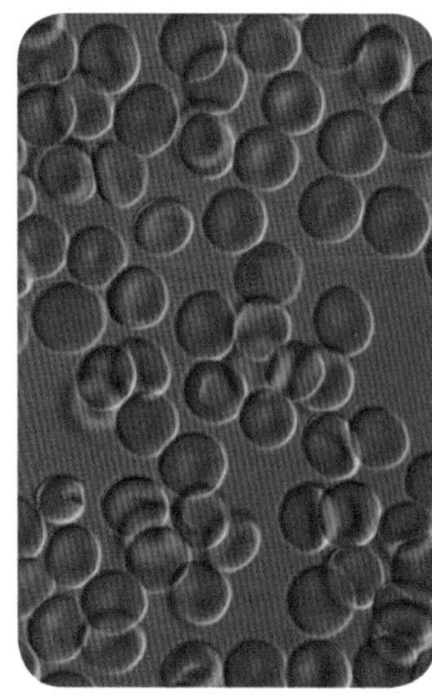

▲ Red blood cells in equilibrium

Active transport requires energy to move materials

Molecules and ions can move from one place to another by diffusion, but only until an equilibrium has been reached. If no concentration gradient exists between the two places, no diffusion can occur – this means that if an equilibrium has been reached, useful particles cannot be absorbed by diffusion. **Active transport** is a method by which particles can cross membranes even against a concentration gradient. In active transport, protein molecules in the cell surface membrane pick up and carry particles across the membrane. These protein molecules are called **carriers**, and when they work they use energy supplied by the cell. Active transport is explained in the diagram below, and important examples of its use are the uptake of ions by plant root hairs (page 74) and the uptake of glucose by epithelial cells of the villi. To summarise – active transport:

- **can move molecules against a concentration gradient** but
- **requires energy** and
- **involves protein carriers in membranes.**

Some cells use phagocytosis*

Some particles are too large to cross a membrane by diffusion or by active transport. A few very specialised cells have developed a method for taking up these particles – the particles are literally engulfed by the cell surface membrane flowing around them. This process of **phagocytosis** is used by white blood cells, and is described on page 102.

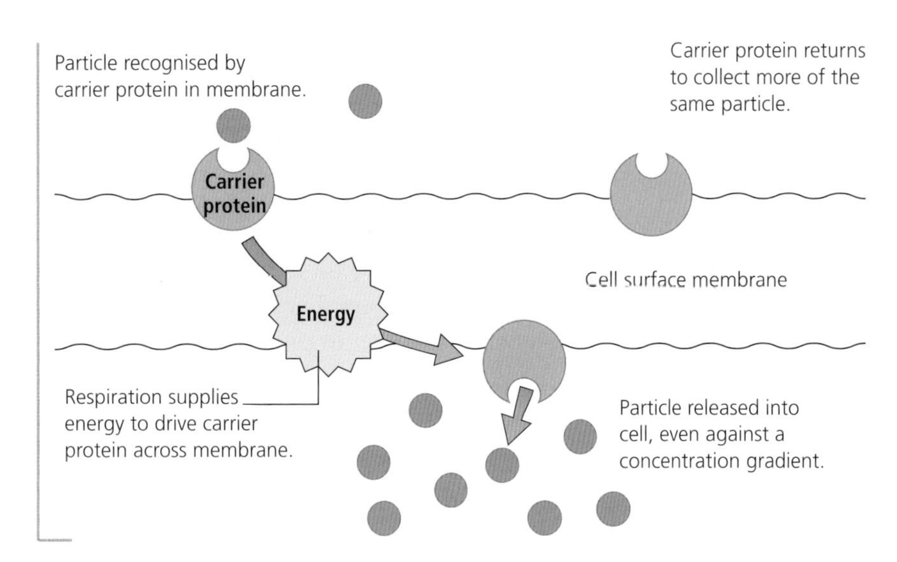

◀ Active transport uses energy to move substances against a concentration gradient. The protein carriers involved in active transport are rather like enzymes. They are able to recognise particular molecules and select them from a mixture. As a result active transport is specific, and the cell can 'choose' which molecules it will absorb from its surroundings.

Q

1. Using an example, explain what is meant by a partially permeable membrane.
2. What is a concentration gradient?
3. Make a list of the processes in living organisms that are dependent on diffusion.
4. How are living organisms adapted to increase the possibility of diffusion?
5. Copy and complete the following paragraphs.
   Animal cells contain ____, a semi-fluid solution of salts and other molecules, and are surrounded by a ____.
   When in distilled water, the animal cells ____ because the cell has a ____ water potential than the surrounding water. Plant cells do not have this problem because they are surrounded by a ____.
   In the gut, soluble food substances such as ____ cross the gut lining into the capillaries by the process of ____, which is the movement of molecules down a ____. When an equilibrium is reached between the gut contents and the blood, glucose may continue to be moved using the process called ____, which consumes ____ and can move molecules ____ a concentration gradient.
   The leaves of green plants obtain the gas ____, which they require for the process of photosynthesis, by the process of ____. They also lose the gas oxygen, produced during ____, by the same process.
6. Make a table that compares diffusion with active transport. Include one example of each process in your table. Under what circumstances would an organism use phagocytosis rather than diffusion or active transport?

Questions on diffusion and osmosis

1 Using **only** the information in the following passage and the figure below, answer questions **i** to **v**.

> **When a plant's cells are fully inflated by water the plant is said to be turgid. Each cell pushes against nearby cells and together they help to support the plant. If water is lost and not replaced each cell loses turgor. Cells in this state are said to be flaccid.**

i What does the term turgid mean?

ii How does turgor play a part in the support of plant tissues? The figure below shows the results of an experiment.

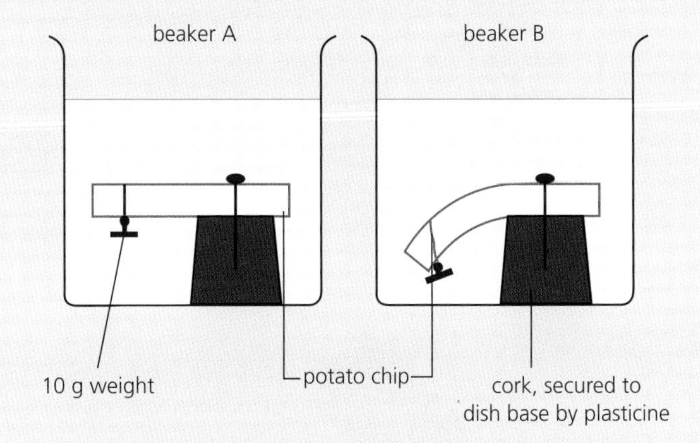

iii Which beaker is likely to contain pure water?

iv Which potato chip is flaccid?

v What would happen if the potato chip from beaker B was placed in beaker A for 24 hours? Explain your answer.

2 A student carried out an investigation based on the one in the diagram using a range of sugar solutions. He obtained the results below right.

a Copy and complete the table, and present the results in a suitable graph.

b Use the graph to calculate the sugar concentration inside the potato cells. Explain your answer.

c Why is it useful to present the results in terms of 'percentage change' in length?

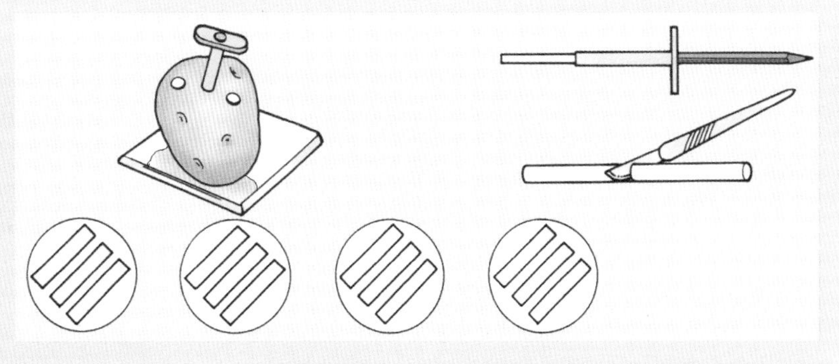

Concentration of sugar solution / mol/dm³	Length of cylinder at start / mm	Mean length / mm	Length of cylinder after 24 hours / mm	Mean length / mm	Percentage change
0.0	50, 50, 50		54, 52, 55		
0.1	50, 49, 50		53, 52, 53		
0.2	50, 50, 50		51, 52, 51		
0.3	50, 51, 49		49, 49, 49		
0.4	50, 50, 50		47, 47, 48		
0.5	50, 51, 51		45, 47, 46		

2.3 All living things are made up of organic molecules

OBJECTIVES

- To understand that the structures of living things depend on the molecules that make them up
- To list the types of molecule found in living organisms

Organic molecules

Biological molecules are often called **organic** molecules, since many of them were discovered in living organisms. Chemists have found that organic molecules all contain carbon and hydrogen atoms (often along with other elements). Carbon atoms bond strongly to other carbon atoms, so organic molecules can be large and show a wide variety of chain and ring structures, with many carbon atoms bonded together. Organisms need organic molecules to:

- provide **energy** to drive life processes
- provide **raw materials** for the growth and repair of tissues.

Nutrition supplies living organisms with the molecules that they need. There are four main groups of organic chemicals used by living things:

- **carbohydrates**
- **lipids**
- **proteins**
- **nucleic acids.**

The diagram on the next page shows the structures of these different groups of organic molecules: note which elements they contain.

Basic biochemistry*

Living organisms also contain inorganic molecules (such as water) and a number of ions. The study of the organic and inorganic molecules that make up living organisms is called **biochemistry**. The sum of all the chemical reactions in living organisms is sometimes called **metabolism**.

Large organic molecules are usually made up of lots of similar smaller molecules called **subunits**. The subunits can be split apart by a reaction called **hydrolysis**, which uses water. They can be joined together again, perhaps in new combinations, by a reaction called **condensation**, which produces water (see opposite page).

In this way living organisms can take molecules from their environment and rearrange them into shapes that suit their own particular requirements, as illustrated below.

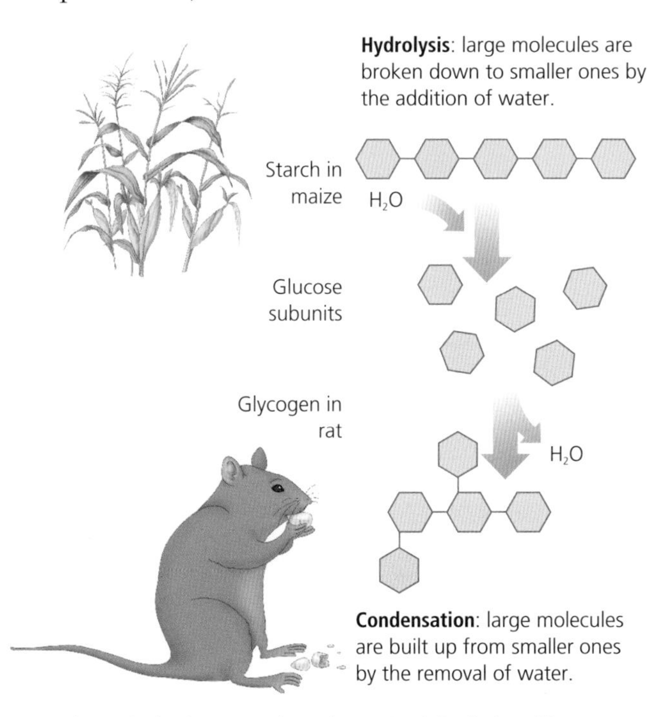

▲ The carbohydrate starch in the maize is hydrolysed in the rat's cells into subunits called glucose. These are then built up into the carbohydrate glycogen by condensation reactions.

Q

1. List the main groups of organic compounds found in living organisms. Suggest one important function for each group.
2. Some molecules such as glucose and amino acids are **soluble**, whereas others such as starch and fats are **insoluble**. Why is this physical property important in living organisms?

3. Some scientists would say that nucleic acids are the most important molecules in living cells; others might suggest that proteins are more important; and some might say that life could not continue without a supply of carbohydrates. Write a sentence in support of each of these points of view.

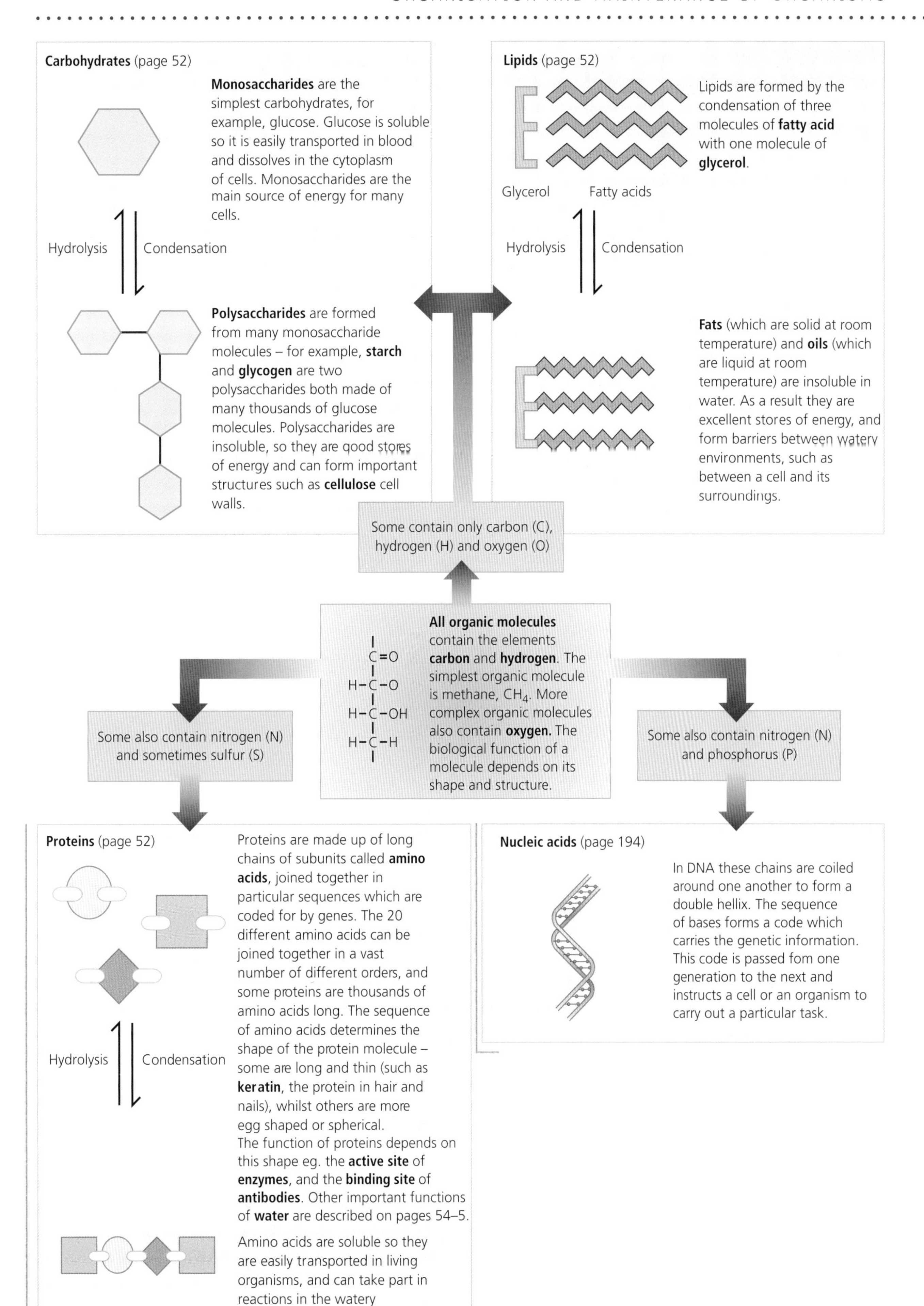
Carbohydrates (page 52)
Monosaccharides are the simplest carbohydrates, for example, glucose. Glucose is soluble so it is easily transported in blood and dissolves in the cytoplasm of cells. Monosaccharides are the main source of energy for many cells.
Hydrolysis
Condensation
Polysaccharides are formed from many monosaccharide molecules – for example, starch and glycogen are two polysaccharides both made of many thousands of glucose molecules. Polysaccharides are insoluble, so they are good stores of energy and can form important structures such as cellulose cell walls.
Lipids (page 52)
Lipids are formed by the condensation of three molecules of fatty acid with one molecule of glycerol.
Glycerol
Fatty acids
Hydrolysis
Condensation
Fats (which are solid at room temperature) and oils (which are liquid at room temperature) are insoluble in water. As a result they are excellent stores of energy, and form barriers between watery environments, such as between a cell and its surroundings.
Some contain only carbon (C), hydrogen (H) and oxygen (O)
C=O
H–C–O
H–C–OH
H–C–H
All organic molecules contain the elements carbon and hydrogen. The simplest organic molecule is methane, CH_4. More complex organic molecules also contain oxygen. The biological function of a molecule depends on its shape and structure.
Some also contain nitrogen (N) and sometimes sulfur (S)
Some also contain nitrogen (N) and phosphorus (P)
Proteins (page 52)
Proteins are made up of long chains of subunits called amino acids, joined together in particular sequences which are coded for by genes. The 20 different amino acids can be joined together in a vast number of different orders, and some proteins are thousands of amino acids long. The sequence of amino acids determines the shape of the protein molecule – some are long and thin (such as keratin, the protein in hair and nails), whilst others are more egg shaped or spherical.
The function of proteins depends on this shape eg. the active site of enzymes, and the binding site of antibodies. Other important functions of water are described on pages 54–5.
Hydrolysis
Condensation
Amino acids are soluble so they are easily transported in living organisms, and can take part in reactions in the watery cytoplasm of the cell.
Nucleic acids (page 194)
In DNA these chains are coiled around one another to form a double hellix. The sequence of bases forms a code which carries the genetic information. This code is passed fom one generation to the next and instructs a cell or an organism to carry out a particular task.

2.4 Testing for biochemicals

OBJECTIVE

- To describe simple chemical tests for the molecules of living organisms

Scientists often need to know whether or not a particular type of molecule is present in a solution. For example, a doctor might try to detect glucose in a urine sample (glucose in the urine suggests the patient has diabetes), or an environmental scientist might test for starch in the outflow from a food factory. There are a number of simple chemical tests that can be carried out on biological solutions. Some of these tests are described on the opposite page.

A special test for lipids

An important feature of fats and oils is that they are insoluble in water. This means that you cannot make an aqueous solution of a fat or oil on which to carry out a biochemical test. However, the fact that lipids are insoluble forms the basis of a **physical test**.

This is known as the **emulsion test**:

- 2 cm^3 of ethanol are added to the unknown solution, and the mixture is gently shaken.
- The mixture is poured into a test tube containing an equal volume of distilled water.
- If a lipid is present, a milky-white emulsion is formed.

▲ A milky emulsion shows that a lipid is present

1 **a** What is the difference between a fat and an oil?

b Both lipids and carbohydrates contain carbon, hydrogen and oxygen. How do they differ from one another?

c Draw a diagram to show a molecule of fat. Suggest why it is possible to have many kinds of fat.

2 Here are the results of a series of tests on biological solutions.

Suggest, giving your reasons, which of these solutions might be:

a the washings from a laundry

b milk

c crushed potato solution

d urine from somebody who has sugar diabetes

e sweetened tea.

Solution	Colour after testing with reagent:			
	Iodine solution	Benedict's reagent	Biuret reagent	Benedict's reagent after acidification and neutralisation
A	Blue-black	Clear blue	Clear blue	Clear blue
B	Straw yellow	Orange	Purple	Orange
C	Straw yellow	Clear blue	Clear blue	Orange
D	Blue-black	Clear blue	Faint purple	Clear blue
E	Straw yellow	Orange	Clear blue	Orange

3 Describe the two types of control which are used in food tests and explain why they are needed.

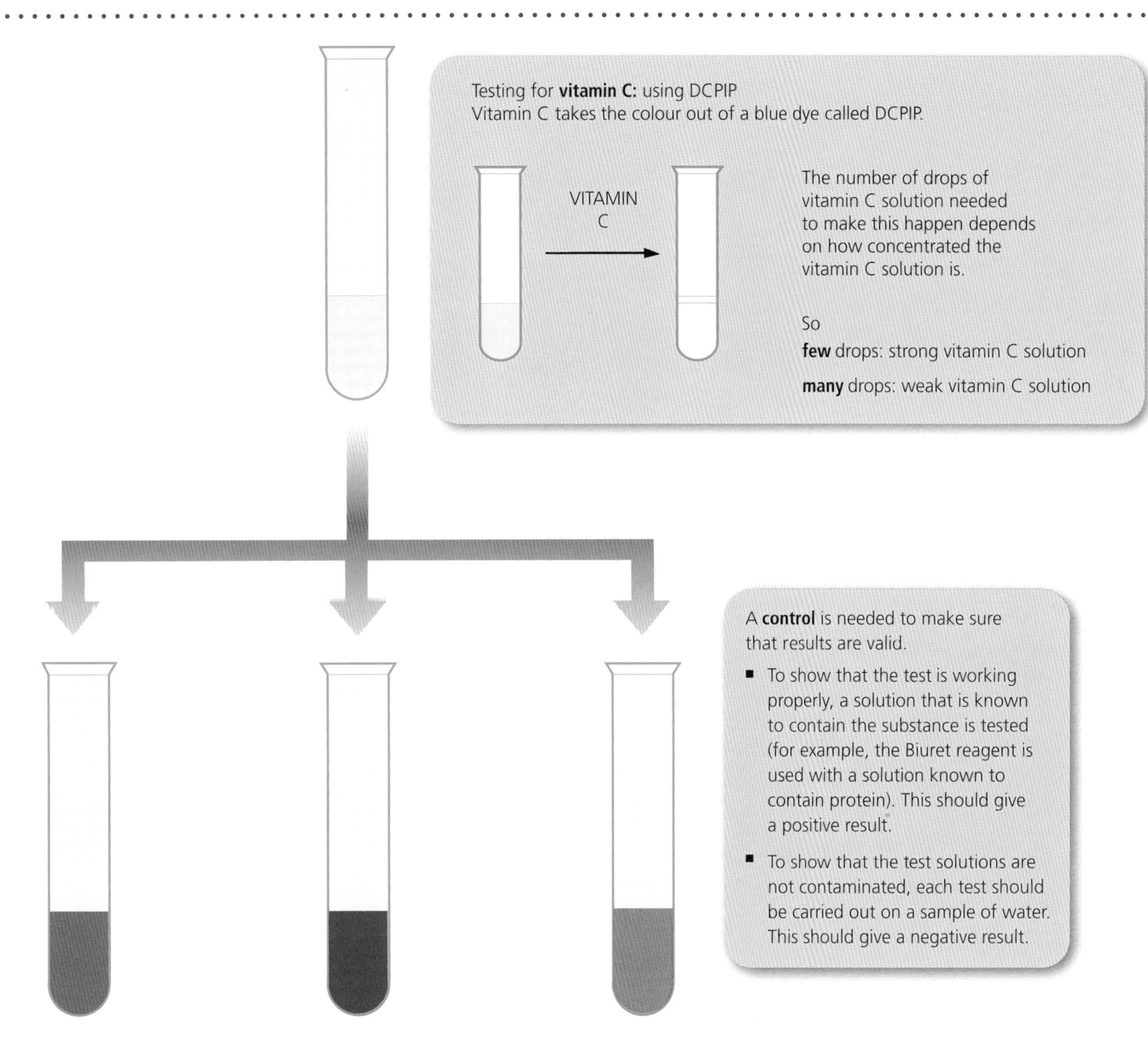

Testing for **vitamin C:** using DCPIP
Vitamin C takes the colour out of a blue dye called DCPIP.
VITAMIN C
The number of drops of vitamin C solution needed to make this happen depends on how concentrated the vitamin C solution is.
So
few drops: strong vitamin C solution
many drops: weak vitamin C solution
A **control** is needed to make sure that results are valid.
- To show that the test is working properly, a solution that is known to contain the substance is tested (for example, the Biuret reagent is used with a solution known to contain protein). This should give a positive result.
- To show that the test solutions are not contaminated, each test should be carried out on a sample of water. This should give a negative result.

To test for **protein,** a few drops of **Biuret reagent** are added to 2 cm^3 of the unknown solution, and the mixture is gently shaken. A **mauve/purple** colour is a positive result (protein is present).

To test for **starch**, a few drops of **iodine solution** are added to 2 cm^3 of the unknown solution, and the mixture is gently shaken. A **deep blue-black** colour is a positive result (starch is present).

To test for **glucose** (a **reducing sugar**), 2 cm^3 of **Benedict's reagent** are added to 2 cm^3 of the unknown solution, and the mixture is heated in a boiling water bath for 2–3 minutes. An **orange/brick-red** colour is a positive result (glucose is present).

When **making comparisons** between different solutions – for example, to compare the glucose content of different urine samples – it is important to carry out all tests under the same conditions. For example, a series of Benedict's tests should be performed:

- on equal volumes of unknown solutions
- using equal volumes of Benedict's solution
- with all mixtures heated to the same temperature
- for the same length of time.

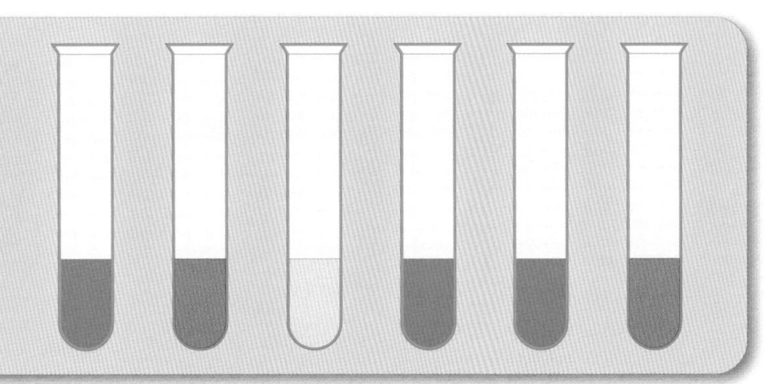

2.5 Enzymes control biochemical reactions in living organisms

OBJECTIVES

- To appreciate that biochemical reactions in living organisms must be controlled
- To understand how enzymes can act as biological catalysts
- To list and explain the factors that affect enzyme activity
- To list some examples of human exploitation of enzymes

Enzymes are biological catalysts

The sum of all the chemical reactions going on within a living organism is known as **metabolism**. **Anabolic** reactions build up large molecules from smaller ones, and usually require an input of energy. **Catabolic** reactions break down large molecules into smaller ones, and often release energy. The condensation of glucose molecules into the polysaccharide glycogen is an example of anabolism, and this happens in cells of the liver and skeletal muscle. The breakdown of glucose into carbon dioxide and water by respiration is an example of catabolism, and this also occurs in cells of the liver and skeletal muscle. What determines whether glucose molecules are built up into glycogen or broken down into carbon dioxide and water? The answer is **enzymes**. Enzymes are proteins that function as biological **catalysts** – (catalysts speed up reactions without themselves being changed by the reaction). The molecules that react in the enzyme-catalysed reaction are called **substrates**, and the molecules produced in the reaction are **products**. Different enzymes are involved in anabolic and catabolic reactions, and so the presence or absence of a particular enzyme controls what will happen to a particular molecule.

Enzymes and cells

Enzymes are synthesised in living cells. Most enzymes work inside the cell – examples of these **intracellular enzymes** are **catalase** (which breaks down harmful hydrogen peroxide in liver cells) and **phosphorylase** (which builds glucose into starch in plant storage cells). Other enzymes are made inside cells and then released from the cell to perform their function – examples of these **extracellular enzymes** include the digestive enzyme **lipase** (which breaks down fats to fatty acids and glycerol) and **amylase**, which converts starch to maltose during germination (see page 168). Enzymes are **specific** – most enzymes work on one kind of substrate only. For example, proteases break down proteins but have no effect on carbohydrates or lipids, and lipases break down lipids but do not affect proteins or carbohydrates.

The mechanism of enzyme action (the lock and key hypothesis)

An **enzyme** is a protein, folded into a complex three-dimensional shape. The **active site** is the part of the enzyme that allows it to act as a catalyst, as shown below.

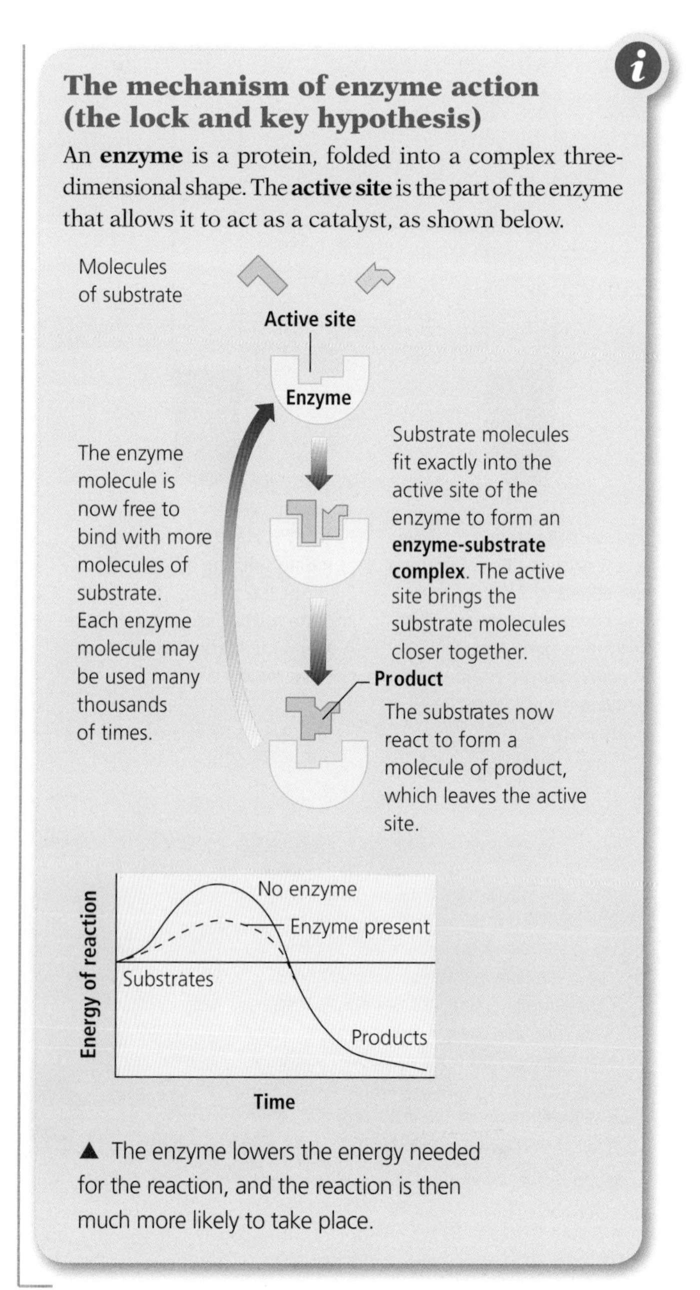

Molecules of substrate
Active site
Enzyme
The enzyme molecule is now free to bind with more molecules of substrate. Each enzyme molecule may be used many thousands of times.
Substrate molecules fit exactly into the active site of the enzyme to form an enzyme-substrate complex. The active site brings the substrate molecules closer together.
Product
The substrates now react to form a molecule of product, which leaves the active site.
No enzyme
Enzyme present
Substrates
Products
Energy of reaction
Time

▲ The enzyme lowers the energy needed for the reaction, and the reaction is then much more likely to take place.

Factors affecting enzyme activity

Temperature

Temperature affects the activity of enzymes. The enzyme activity increases with a rise in temperature, up to a point. This is because:

- a higher temperature speeds up the movement of substrate molecules, so that when they collide with the enzyme they have more energy and are more likely to bind to the active site
- the enzyme molecules themselves also gain in energy as the temperature rises so that they begin to vibrate. Eventually the enzyme molecules vibrate so much that they become **denatured** – they lose their three-dimensional shape and can no longer bind to their substrate. Because of this, high temperatures reduce enzyme activity.

  Each enzyme has an **optimum temperature**, which is a balance between these two effects, as shown in the graph below. Most human enzymes have an optimum temperature around 37°C, whilst for most plants the optimum is rather lower at around 25°C.

Denaturation is usually irreversible, and living cells make great efforts to keep the conditions suitable for their enzymes to work.

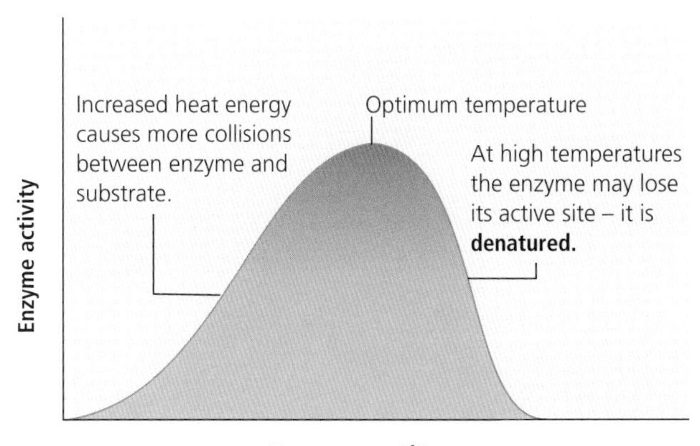

▲ An enzyme-catalysed reaction gets faster, reaches a maximum rate and then slows down again as you increase the temperature

pH

pH also affects enzyme activity.

Changing the acid or base conditions around an enzyme molecule affects its three-dimensional shape and can denature the enzyme.

Each enzyme has its own **optimum pH**, as shown in the graph below, which depends on the environment in which the enzyme is working – **pepsin** is an enzyme that works in the stomach, and has an optimum pH around pH 2.0 (very acidic), whereas **amylase** works in the mouth and small intestine and has an optimum pH around 7.5 (slightly basic).

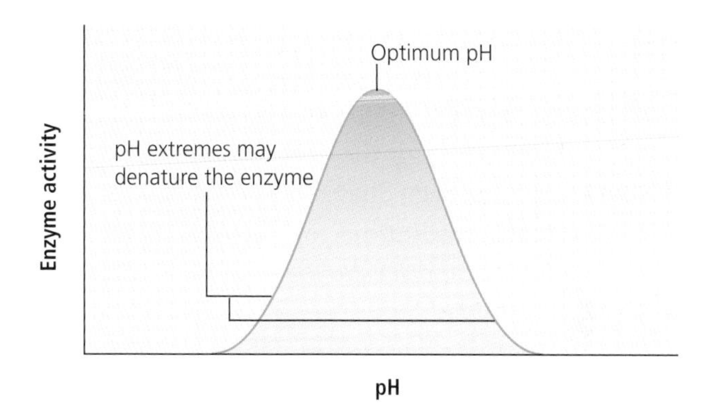

Activators and inhibitors

Some molecules change the likelihood of an enzyme being able to bind to its substrate. **Activators** make this binding more likely – for example, chloride ions are essential for the activity of salivary amylase. **Inhibitors** make it more difficult for the enzyme to bind to the substrate – for example, cyanide ions block the active sites of enzymes involved in respiration.

Humans exploit enzymes

Because enzymes are specific, and can be used over and over again, they are very useful in the fields of industry, food preparation and medicine – an enzyme from snake venom can break down blood clots, for example.

1 Copy and complete the following paragraph about enzymes.
Enzymes are ____ which speed up the biochemical ____ in living organisms. The enzymes themselves are not changed in these reactions, that is they are biological ____.
Enzymes are ____ – each one controls only one type of reaction. They are ____ by high temperatures and by extremes of pH.

2 Enzyme action is explained by the lock and key hypothesis (see box opposite). The example shows an enzyme that catalyses a condensation reaction – two small molecules are joined together to make a larger one. Redraw the diagram to show the action of an enzyme catalysing a hydrolysis reaction, and give an example of such an enzyme.

2.6 Enzyme experiments and the scientific method

OBJECTIVES

- To understand how enzymes can work as biological catalysts
- To understand and apply the scientific method

What is the scientific method?

When scientists are faced with a problem, they tackle it using the **scientific method**. This starts off with an **observation**; for example, a farmer might notice that all his cows have stopped producing milk. The next step is to produce a **hypothesis**, a possible explanation for the observation (perhaps the cows' diet has been changed). Following this hypothesis, **predictions** are made, such as: *adding more protein pellets to the cows' feed will increase their yield of milk.* Then **experiments** are designed and carried out to test whether or not the predictions are true. The **data** (results of the experiment) are analysed and **conclusions** drawn. These conclusions will allow the experimenter to accept or reject the original hypothesis.

An illustration of the scientific method

In an experiment, apparatus is used to measure the effect of changing one factor (variable) on the value of a second factor (variable). For example, the experiment illustrated below is designed to test the hypothesis: *temperature affects the activity of catalase.*

Catalase is an enzyme that catalyses the breakdown of hydrogen peroxide:

hydrogen peroxide ⟶ oxygen + water

Catalase is present in potato tissue. Discs of potato are put into a solution of hydrogen peroxide. The catalase breaks down the hydrogen peroxide into water and oxygen, and the oxygen passes into the manometer. The changing level of the manometer fluid shows how much oxygen is being produced, and gives a measure of the activity of the enzyme. The rate of oxygen production is measured at different temperatures.

An identical **control** experiment is also set up, in which the input variable (the temperature) is not changed. This confirms that the changes in temperature, and not an unknown variable, are causing the changes in enzyme activity. A control experiment ensures that the experiment is a **fair test**, and that the data collected is valid.

If the experimenter suspects that an error has been made, part of the experiment may be **repeated**. Taking a series of results and calculating the mean gives more accurate results than taking just one set, as any single inaccurate result then has less effect on the overall results.

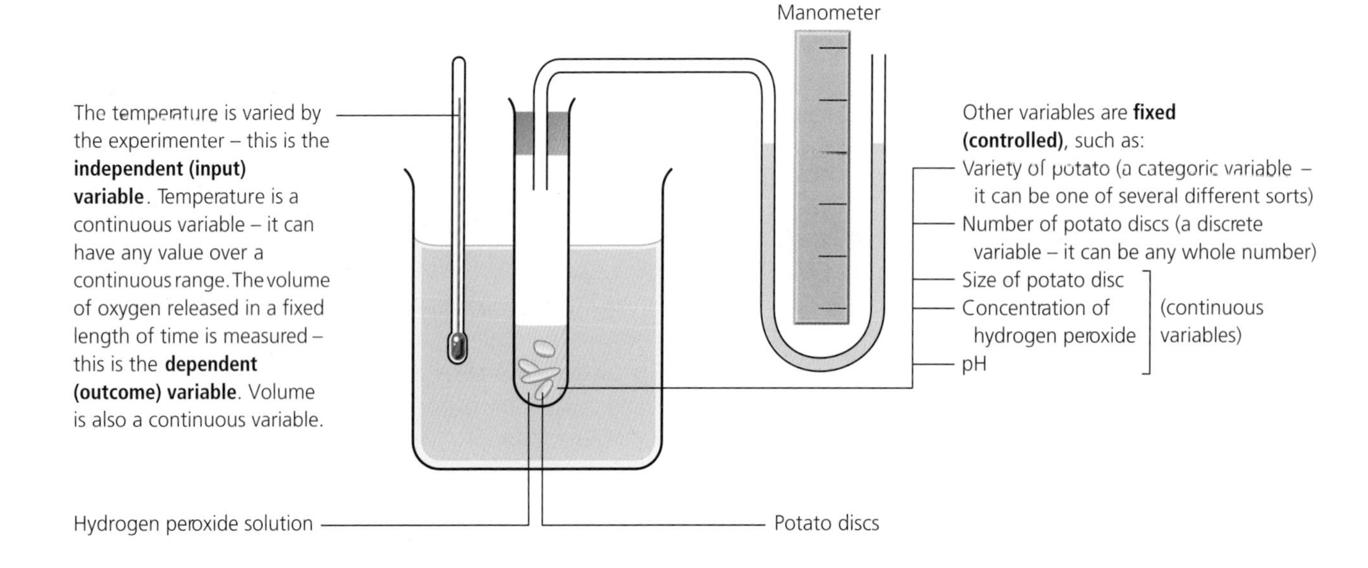

Dealing with data: recording results

Raw data are gathered in a table during the experiment. They may then be manipulated (converted into another form), and are often displayed as a graph to allow the experimenter to draw conclusions from the results.

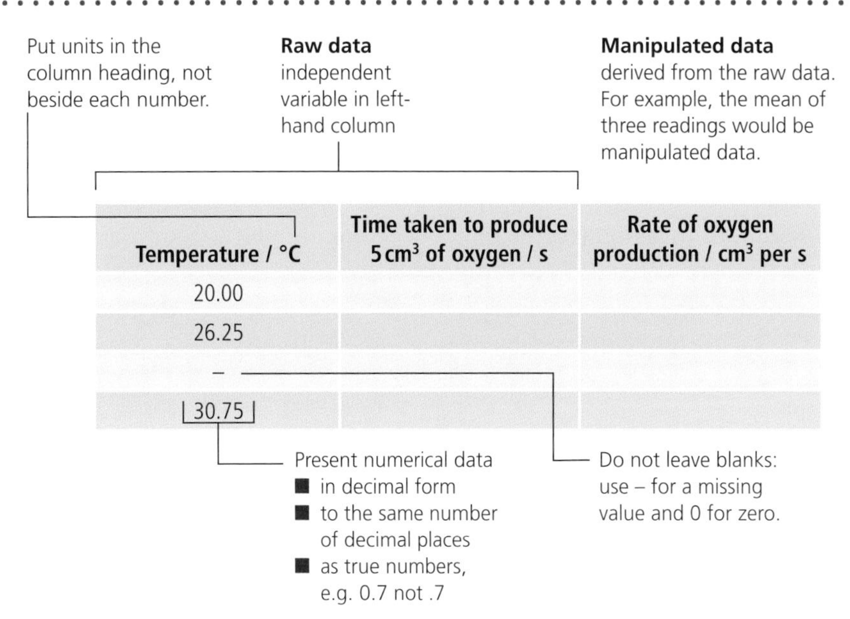

Temperature / °C	Time taken to produce 5 cm^3 of oxygen / s	Rate of oxygen production / cm^3 per s
20.00		
26.25		
–		
30.75		

▲ Results are recorded in a table during the experiment. Give an informative title, such as: *The effect of temperature on the activity of catalase.*

Drawing a graph

A graph is a visual representation of data which often helps to make a relationship more obvious.

Evaluating an experiment

Scientists often look back at the results of their experiments and the methods used to gather them. This **evaluation** process is an important part of the scientific method because the scientist needs to be sure that the techniques and apparatus used have given the most reliable results before he or she draws conclusions from them.

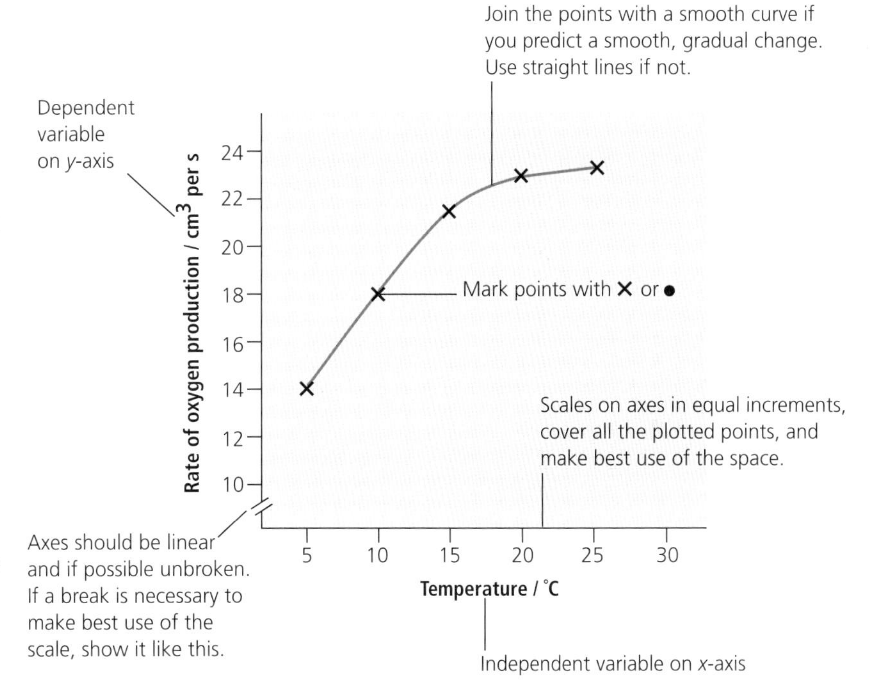

▲ Rule in the axes in black. Again, give an informative title.

Q

1 Using the apparatus shown opposite, a student investigated the effect of temperature on the activity of the enzyme catalase. Here are the results:

a Copy and complete the table by calculating the rate of oxygen release.

b Present the data in the form of a graph.

c Explain the shape of the graph.

Temperature / °C	Time taken to evolve 10 cm^3 of oxygen / s	Rate of oxygen release / cm^3 per s
15	40	
25	20	
35	5	
45	20	
55	40	
65	120	
75	No gas evolved	

Questions on enzymes and biological molecules

1 A student investigated the effect of changing pH on the rate of reaction of a digestive enzyme.

a Define the term *enzyme*.

The table shows the results of this investigation.

pH	1	2	3	4	5	6	7
rate of reaction / arbitrary units	10	15	9	6	3	1	0

b Plot the results as a line graph.

c Suggest where in the human digestive system this enzyme would have been most active.

d The investigation at pH 3 was repeated but the enzyme was boiled before its use. Suggest how and why the results would have been different.

Cambridge IGCSE Biology 0610
Paper 2 Q7 May 2007

2 Catalase is an enzyme found in plant and animal cells. It has the function of breaking down hydrogen peroxide, a toxic waste product of metabolic processes.

a State the term used to describe the removal of waste products of metabolism.

An investigation was carried out to study the effect of pH on catalase, using pieces of potato as a source of the enzyme.

Oxygen is formed when catalase breaks down hydrogen peroxide, as shown in the equation.

$$\text{hydrogen peroxide} \xrightarrow{\text{catalase}} \text{water} + \text{oxygen}$$

The rate of reaction can be found by measuring how long it takes for 10 cm^3 oxygen to be collected.

b i State the independent (input) variable in this investigation.

ii Suggest two factors that would need to be kept constant in this investigation.

The table below shows the results of the investigation, but it is incomplete.

pH	time to collect 10 cm^3 oxygen / min	rate of oxygen production / cm^3 min^{-1}
4	20.0	0.50
5	12.5	0.80
6	10.0	1.00
7	13.6	0.74
8	17.4	

c Calculate the rate of oxygen production at pH 8. Show your working.

d Plot the rate of oxygen production against pH.

e i Using data from the graph, describe the changes in the reaction rate between **pH 4** and **pH 8**.

ii Explain the change in the reaction rate between **pH 6** and **pH 8**.

Cambridge IGCSE Biology 0610
Paper 3 Variant 1 Q3 May 2008

3 A protein is used to hold other chemicals onto the clear plastic backing of photographic film, as shown below.

Trypsin is an enzyme which will digest the protein so that the coating on the photographic film is removed and the film becomes clear.

The table shows the results obtained by two students who investigated the effect of pH on the activity of this enzyme. They made up the solutions, set up the experiment and timed how long the enzyme took to digest the protein and clear the film.

pH	time for the protein to be digested / mins	
	student 1	student 2
2	12.0	14.0
4	8.0	9.0
6	2.0	3.0
8	0.5	1.0
10	8.0	9.0

a i Plot the results obtained by **student 2** in the form of a suitable graph.

ii Describe and explain the effect of pH on the activity of the enzyme.

b i Suggest reasons for the difference in the results for the two students.

ii If you were to carry out this investigation, describe what steps you would take to ensure that your results were as reliable and valid as possible.

Cambridge IGCSE Biology 0610
Paper 6 Q1 November 2007

4 a The bar chart opposite shows the percentage of each of the main elements in the human body. Convert this data into:

i a table **ii** a pie chart.

Which do you think is the better way of showing the data? Explain why.

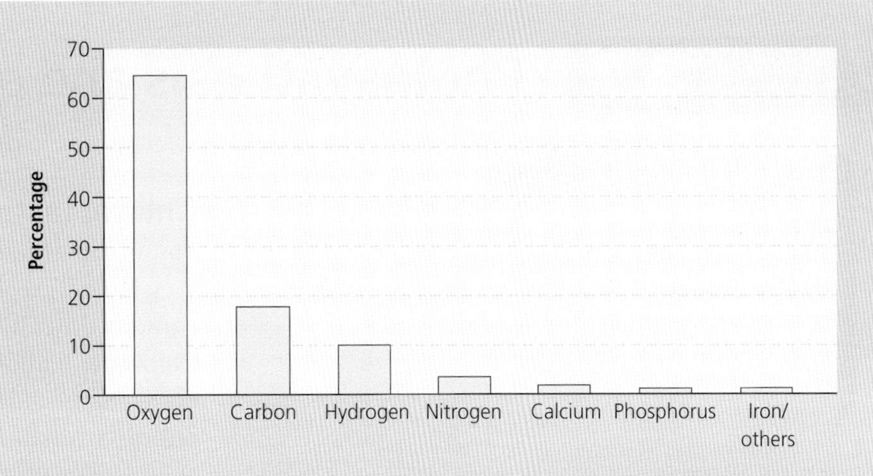

b These elements are mostly present in the body as part of compounds.

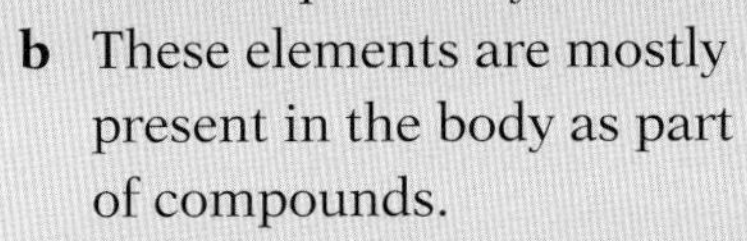

The proportions of the main groups of compounds in a human body are shown in the table.

Compound	Approximate percentage in the body
Fats	14%
Proteins	12%
Carbohydrates	1%
Water	70%

i Which is the most abundant compound?

ii Which compound contains most of the nitrogen in the body?

iii Which compound contains most of the oxygen?

iv Which compound contains most of the carbon?

v The total of these compounds does not add up to 100%. Suggest another organic compound that forms a proportion of the remainder.

vi Try to find out which structure(s) in the body contain most of the calcium.

5 a The properties of enzymes were investigated using the simple experiment shown opposite.

Which of the following conclusions are valid from these results?

i The enzyme reaction occurs more quickly at 30°C than at 20°C.

ii The enzyme in saliva is inactivated by boiling.

iii Saliva contains an enzyme which digests starch.

iv Boiled saliva contains an enzyme which digests starch.

b What is the purpose of tube D?

c How could you increase the validity of these results?

d What is the product of starch digestion present in tubes A and B after 20 minutes? Describe a simple biochemical test for this substance.

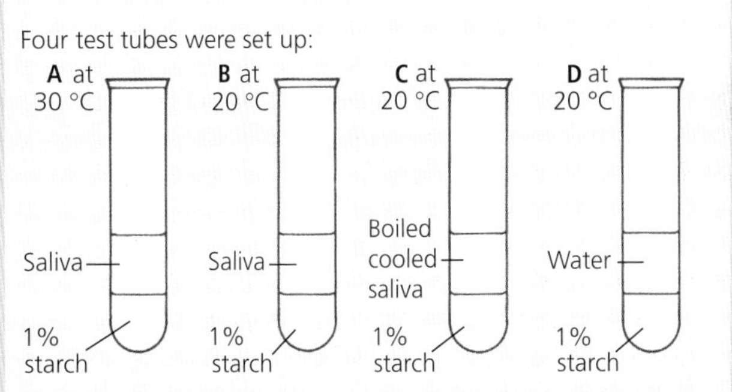

The tubes were shaken. They were immediately tested for starch by adding one drop of iodine solution to one drop of the mixture in a specimen tube. The test was repeated at intervals.

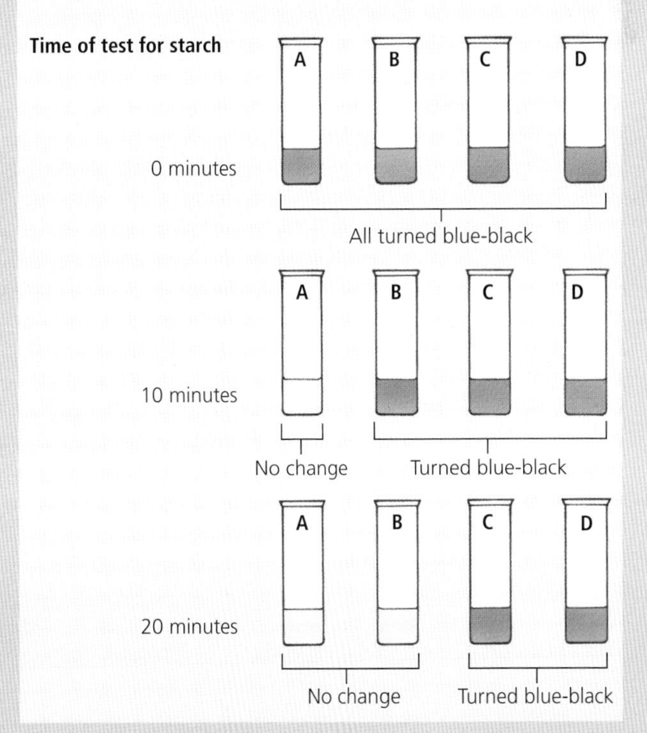

2.7 Photosynthesis and plant nutrition

OBJECTIVES

- To understand that plants, like all living organisms, must receive nourishment
- To know the basic definition for photosynthesis
- To know word and symbol equations for photosynthesis
- To remember that a plant body has organs that are well suited to carry out particular functions

Plants need food

Plants, like animals, require raw materials for building tissues and as a source of energy. They manufacture everything they need out of simple ions and compounds available in the environment. The building up of complex molecules from simpler substances (**synthesis**) requires energy and enzymes. The enzymes are in the plant's cells, and the energy comes from sunlight. The process is therefore called **photosynthesis**: this could be defined as the basic process by which plants manufacture carbohydrates from raw materials using energy from light. An outline of plant nutrition is shown in the diagram below.

Chloroplasts are energy transducers

Plants can absorb and use light energy because they have a green pigment, **chlorophyll**, contained in **chloroplasts** in some of their cells. Chlorophyll allows the energy in sunlight to drive chemical reactions. Chloroplasts act as **energy transducers**, converting light energy into chemical energy.

Defining photosynthesis

Photosynthesis is the process in which light energy, trapped by chlorophyll, is used to convert carbon dioxide and water into glucose and oxygen.

These equations summarise photosynthesis:

$$\text{carbon dioxide} + \text{water} \xrightarrow[\text{chlorophyll}]{\text{light energy}} \text{glucose} + \text{oxygen}$$

$$6CO_2 + 6H_2O \xrightarrow[\text{chlorophyll}]{\text{light energy}} C_6H_{12}O_6 + 6O_2$$

These equations are simplified, and show only glucose as a food product of photosynthesis. In fact plants can make all of their food compounds by photosynthesis and other chemical processes.

▼ Plants are **autotrophic** (self-feeding) – they take simple substances from their environment and use light energy to build them up into complex food compounds

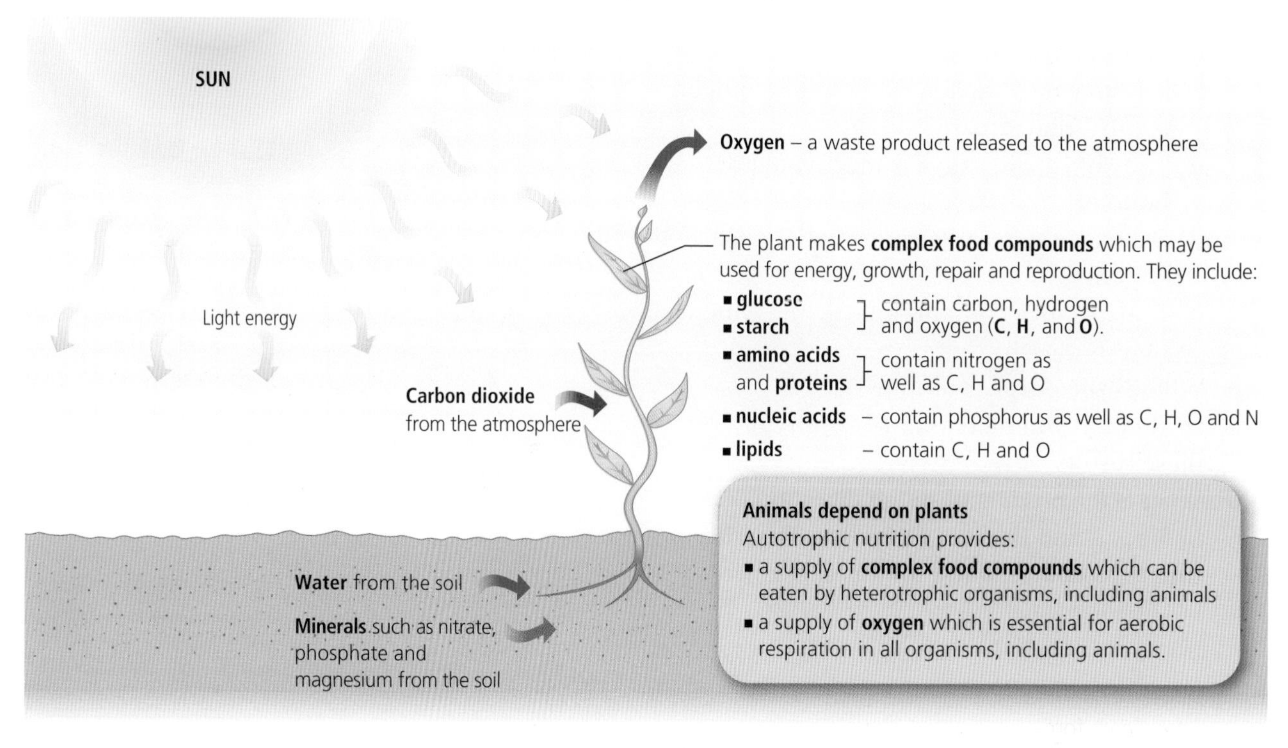

Some of the glucose produced is stored in the plant cells as **starch**. The experiment below shows how a starch test can be used to demonstrate the conditions needed for photosynthesis.

TESTING LEAVES FOR STARCH

To demonstrate that some factor is necessary for the production of starch, the plant must have no starch to begin with. The plant is placed in a dark cupboard or box for 48 hours. It uses any starch that is already in its leaves and is now **destarched**.

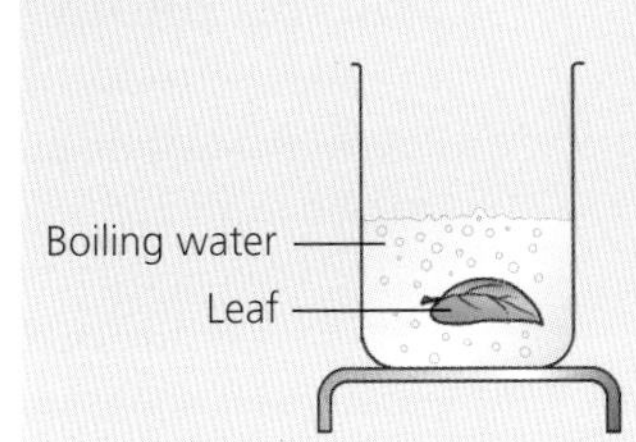

(a) Leaf is boiled in water for about 2 minutes. Purpose: to break down cell walls and to stop the action of enzymes within the leaf. Also allows easier penetration by ethanol.

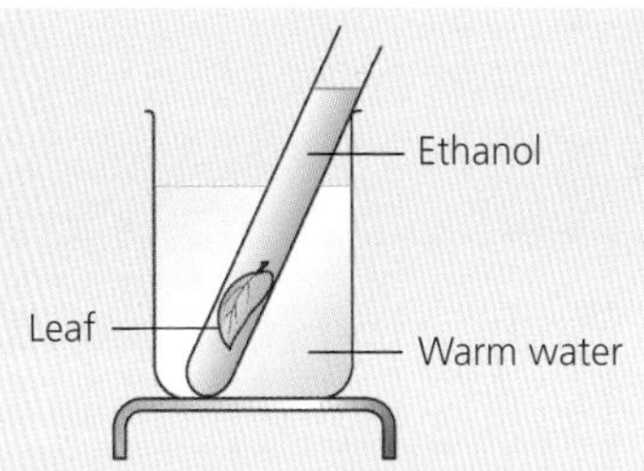

(b) Leaf is warmed in ethanol until leaf is colourless. Purpose: to extract the chlorophyll, which would mask observations later. (Chlorophyll dissolves in ethanol but not in water.)

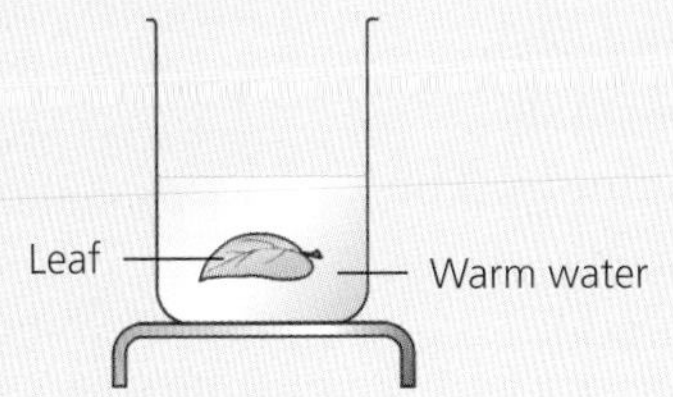

(c) Leaf is dipped into warm water (briefly). Purpose: to soften the now brittle leaf, and allow penetration by iodine solution.

(d) Leaf is placed on white tile and iodine solution added. Purpose: iodine shows the presence (blue-black) or absence (orange-brown) of starch; colours are shown against the white tile.

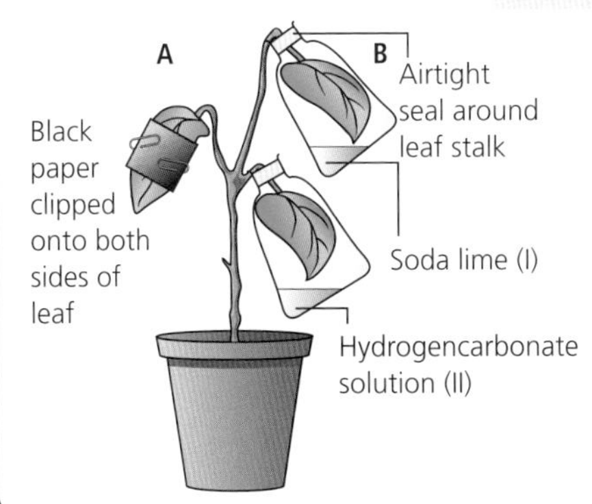

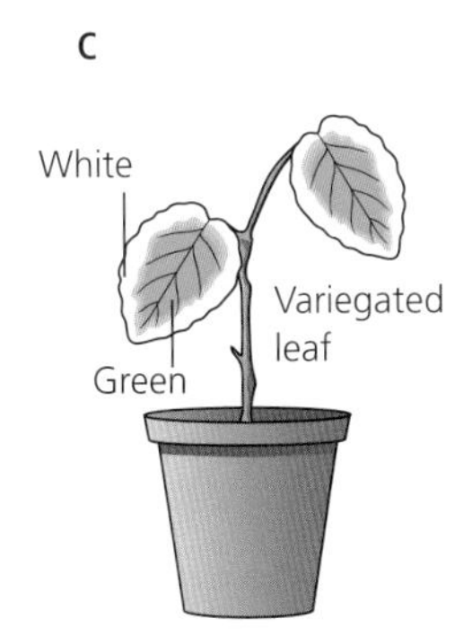

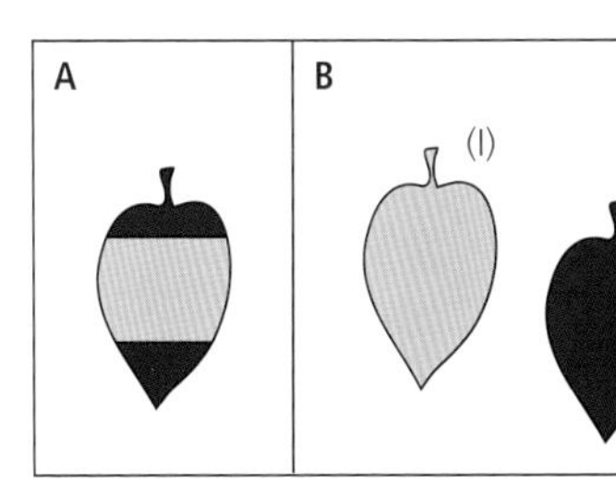

Q

These questions are about the diagram above.

1 Before testing for starch, the leaf is warmed in ethanol. The ethanol turns green. Why is this?

2 A group of students wanted to investigate the conditions needed for photosynthesis. They set up the three experiments shown as A, B and C. Each of the plants was given the same light conditions.

- **a** What does the result of experiment A suggest?
- **b** Soda lime removes carbon dioxide from the atmosphere, and hydrogencarbonate slowly releases it. What do the results of experiment B suggest about carbon dioxide and starch formation?
- **c** What is the purpose of experiment C?
- **d** From experiments A, B and C, list the factors necessary for starch formation by photosynthesis.

3 The students' teacher asked for some further tests to be completed before accepting their results as valid.

- **a** How could the students show that there was no starch in the leaves at the start of the experiment?
- **b** The teacher suggested that the black paper attached to the leaf in experiment A prevented the leaf from absorbing gases, and that was why no starch had been produced in the covered area.
  How could the students disprove this theory?

4 What additional experiment could be set up to show that the formation of starch by photosynthesis depends on the activity of enzymes? (There is a clue given in one part of the diagram above.)

2.8 The rate of photosynthesis

OBJECTIVES

- To understand a quantitative method of investigating photosynthesis
- To perform an exercise in experimental design

Measuring photosynthesis

The most straightforward way of showing whether or not photosynthesis has occurred is to test for the presence of starch. Starch is a product of photosynthesis and the blue-black colour of the iodine test is a simple method for detecting it. Unfortunately this is an 'all-or-nothing' test, that is, the colour is just as dark with a small amount of starch as it is with a large amount. The iodine test is said to be **qualitative** – it only shows whether or not starch is present. To find out *how quickly* photosynthesis is going on, we need to use a **quantitative** test.

The basic equation for photosynthesis is:

$$\text{carbon dioxide} + \text{water} \xrightarrow[\text{chlorophyll}]{\text{light energy}} \text{glucose (starch)} + \text{oxygen}$$

This shows that as well as starch, there is another product of photosynthesis – oxygen. It is quite easy to measure amounts (volumes) of oxygen, and so the production of oxygen can be used as a quantitative test for photosynthesis.

Tracing photosynthesis

The rate of photosynthesis can be estimated by measuring how much carbon dioxide a plant absorbs. This is done using a radioactively labelled form of carbon dioxide, $^{14}CO_2$. Labelled carbon dioxide is absorbed by the leaf cells and converted into labelled carbohydrates in exactly the same way as 'normal' carbon dioxide. A Geiger counter is used to detect how much radioactive material has been incorporated by the plant in a fixed length of time – the rate of photosynthesis. This method is useful for studying the rate of photosynthesis in a 'real-life' situation – in the middle of a crop field, for example – where it is not possible to collect and measure volumes of oxygen.

Using radioactive tracers can also provide information about other compounds that plants make following photosynthesis, and about how plants transport food substances from one place to another in the plant.

ANALYSING THE DESIGN OF AN EXPERIMENT

The simplest apparatus for measuring oxygen release is shown in the next column. This method depends on counting the number of oxygen bubbles given off in a fixed length of time.

Apparatus that can be used to measure the *volume* of oxygen released in a fixed length of time is shown on the opposite page. When looking at this, remind yourself of the principles of experimental technique.

SIMPLE ESTIMATION OF PHOTOSYNTHESIS

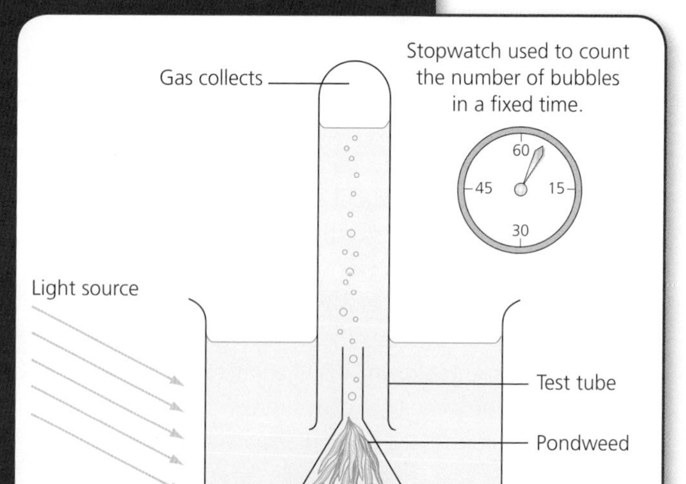

MEASURING THE RATE OF PHOTOSYNTHESIS

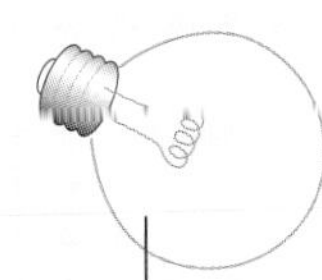

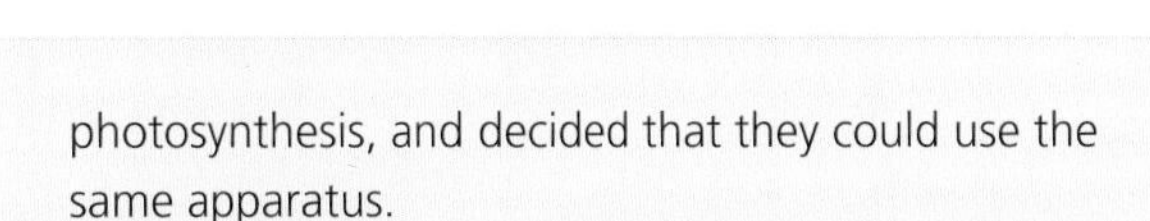

▶ A more accurate method for measuring the rate of photosynthesis. This apparatus can be used to investigate the effect of light intensity on the rate of photosynthesis.

Q

1 What important assumption about the bubbles is being made, when using the method shown on the page opposite, to measure the rate of photosynthesis?

2 Using the apparatus above, Samantha and Jane obtained the results shown in the table.

a Plot this information in the form of a graph.

Light intensity / arbitrary units	Volume of oxygen released / mm^3 per minute
1	7
2	14
3	21
4	28
5	34
6	39
7	42
8	44
9	45
10	45

b At what light intensity did the plant produce 25 mm^3 of oxygen per minute?

c What levels of light intensity had the greatest effect on the rate of photosynthesis? How could this information be useful to a grower of greenhouse tomatoes?

Samantha and Jane wanted to investigate whether the wavelength of the light would affect photosynthesis, and decided that they could use the same apparatus.

d What would be the input variable in this investigation? How could the two students change this variable?

e What would be the fixed variables in this investigation?

f The students decided to repeat their experiment, and then to pool their results with the results of other students in the same class. Why was this pooling of results important?

g The teacher said that if the students were going to pool their results then they must remove the plants from the apparatus and weigh them. Why should they do this?

h State the input, outcome and fixed variables in an investigation into the effect of temperature on the rate of photosynthesis.

3 This question is about the box 'Tracing photosynthesis' on the opposite page.

a What important assumption is being made when using the rate of uptake of $^{14}CO_2$ as a measure of the rate of photosynthesis?

b Why is it important that radioactively labelled carbon dioxide, $^{14}CO_2$, is treated by leaf cells in exactly the same way as 'normal' carbon dioxide?

c Which process might release $^{14}CO_2$ from the plant?

2.9 Leaf structure and photosynthesis

OBJECTIVES

- To understand how the structure of the leaf is adapted for photosynthesis
- To recall the structure of a palisade cell
- To understand how whole plants can be adapted to make the most of light energy

Features of the leaf

In order to photosynthesise efficiently a leaf needs:

- a method for **exchange of gases** between the leaf and its surroundings
- a way of **delivering water** to the leaf
- a system for the **removal of glucose** so that it can be transported to other parts of the plant
- an efficient means of **absorbing light energy**.

The diagram below shows how the structure of the leaf meets these requirements.

Waxy cuticle reduces water loss. It is thicker on the upper surface since this surface is usually more exposed to the warming rays of sunlight.

Upper epidermis – a complete covering which is usually one cell thick. It is transparent to allow the free passage of light, and has the major function of preventing the entry of disease-causing organisms such as bacteria and fungi.

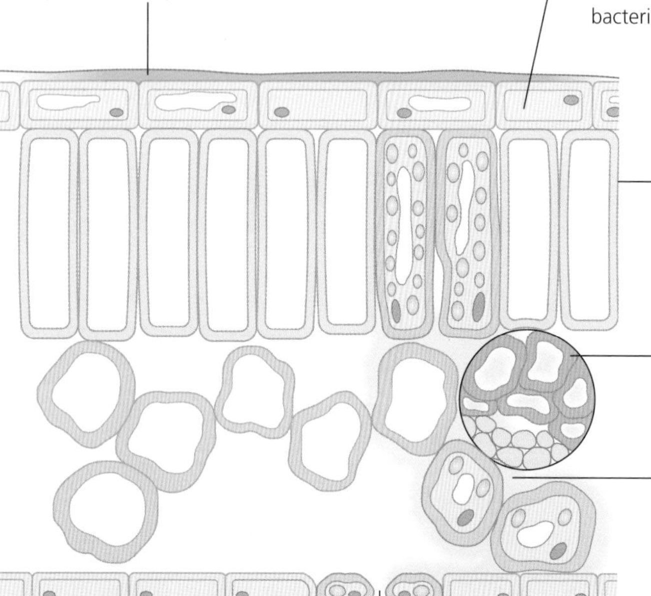

Palisade mesophyll – tall thin cells arranged in columns and separated by very narrow air spaces. Cells contain many chloroplasts, and the dense packing of these cells allows the **absorption of the maximum amount of light energy.**

Vein – the transport system in and out of the leaf. The **xylem** vessels **deliver water and mineral salts,** and the **phloem** sieve tubes **carry away the organic products of photosynthesis, such as glucose.**

Spongy mesophyll – these cells are rather loosely packed, and are covered with a thin layer of water. The air spaces between them aid the diffusion of gases through the leaf. The air spaces are saturated with water vapour so water diffuses out of the leaf.

Stomata – these minute pores **allow the entry of carbon dioxide and the exit of oxygen.** They are mainly present in the **lower epidermis.** This surface is less exposed to the Sun's radiation so that evaporation of water is kept to a minimum. The stomata can be closed when no carbon dioxide intake is needed (in the dark, for example).

When a plant is short of water, the guard cells become flaccid, closing the stoma.

When a plant has plenty of water, the guard cells become turgid. The cell wall on the inner surface is very thick, so it cannot stretch as much as the outer surface. So as the guard cells swell up, they curve away from each other, opening the stoma.

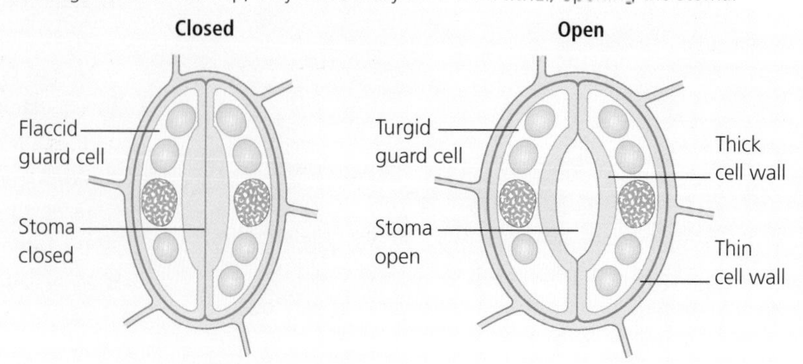

▲ Leaf structure is a compromise between maximising photosynthesis and minimising water loss

This diagram shows how leaf, palisade cell and chloroplast are adapted for photosynthesis

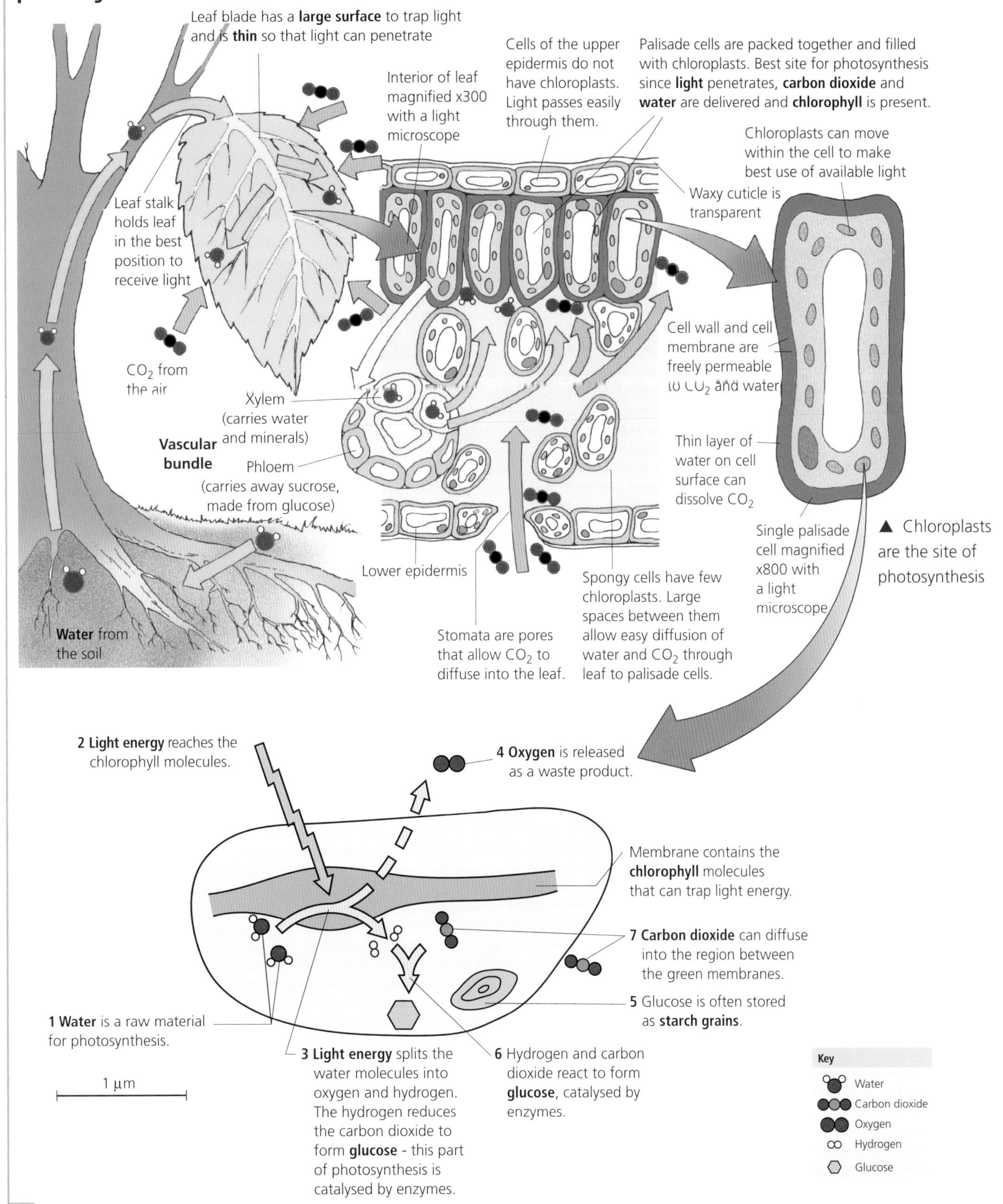

▲ Chloroplasts are the site of photosynthesis

Q

1 Work out the magnification of the single chloroplast. Show how you reached your answer.

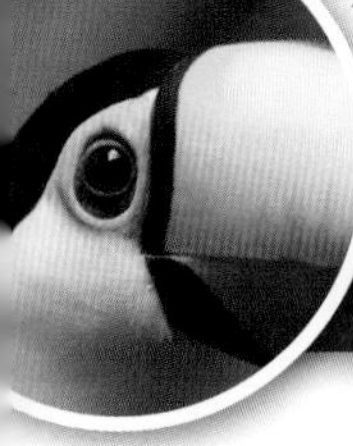

2.10 The control of photosynthesis

OBJECTIVES

- To understand that photosynthesis is affected by a number of different factors
- To appreciate that photosynthesis is affected most by the factor that is in shortest supply – the limiting factor
- To know that an understanding of limiting factors can be used in the efficient growth of greenhouse crops

Requirements for photosynthesis

The process of photosynthesis depends upon:

- the availability of light
- the presence of a pigment to absorb the light
- a supply of carbon dioxide and water
- a temperature suitable for enzyme activity.

The need for each of these requirements or **factors** is outlined in the diagram below.

If any of these factors is in short supply, the rate of photosynthesis will be less than its maximum possible rate. One factor can cause a 'bottleneck' in the overall process as shown at the top of the page opposite. (This factor is called the **limiting factor**.) The limiting factor is the one which controls the rate of the overall process because it is the factor nearest to its minimum value. The limiting factor varies at different times and under different conditions.

- In Britain, during the summer, light and temperature may be ideal for photosynthesis but the carbon dioxide concentration may be the limiting factor.

Availability of light – light provides the energy that drives photosynthesis (by splitting water molecules).
The light energy absorbed by a plant depends on:
- the **intensity** of the light source
- the **wavelength** of the light
- the **length of time** (duration) that the light is available.

Chlorophyl is essential for the absorption of light energy. The synthesis of chlorophyll requires **magnesium ions**, which must be supplied from the soil (see page 74).

Temperature affects the rates of enzyme activity. A 10°C rise in temperature can cause a doubling in the rate of enzyme activity (although higher temperatures cause denaturation). This is important both in **leaves** (enzymes involved directly in photosynthesis) and in **roots** (enzymes systems involved in active transport of mineral ions – see page 74).

Carbon dioxide concentration has a major influence on the rate of photosynthesis since it is the **substrate that is in shortest supply** (there is almost always enough water for photosynthesis).

Good greenhouse effect!
Burning fossil fuels:
- raises carbon dioxide concentration
- raises temperature.

Both of these factors increase the rate of photosynthesis (see page 40).

Poisons and photosynthesis
Photosynthesis may be inhibited by poisons, e.g.
- **paraquat** (a weed killer) prevents light energy being used to convert carbon dioxide to sugars
- **sulfur dioxide** (in acid rain) damages the palisade cells of the leaf.

Water availability – a shortage of water closes stomata (see page 80) which limits carbon dioxide uptake. There is always enough water as a substract for photosynthesis.

Well balanced plants! Plants are autotrophic ('self-feeding') and can carefully regulate their food production. Plants need nitrate to be able to convert glucose from photosynthesis into amino acids. If there is a shortage of nitrate, the plant will reduce its rate of photosynthesis since it will need fewer sugars to make amino acids.

▲ Factors affecting photosynthesis

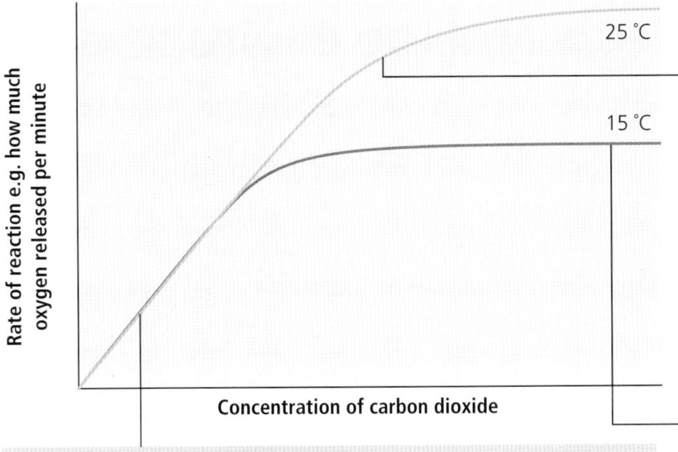

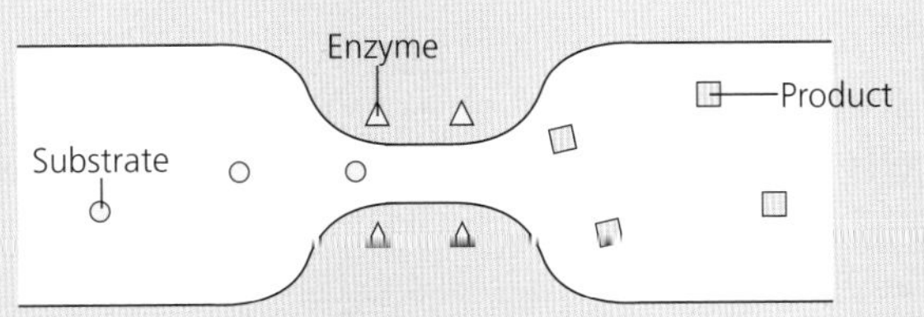

1 The **rate of the reaction**, i.e. how many molecules of product (□) are formed, is limited by the number of available substrate (carbon dioxide) molecules (○), since they can all pass through the enzyme 'channel'. The reaction will go faster if more substrate becomes available – in this situation **substrate concentration is the limiting factor**.

2 As the concentration of substrate is increased, a **second limiting factor** acts like a 'bottleneck'. No matter how many molecules of the substrate (○) are available, the number of molecules of product (□) formed is limited by how quickly the enzymes (△) can work. In this example the temperature might be the limiting factor, since enzymes work less quickly at lower temperatures.

3 If the bottleneck is widened, in this case by raising the temperature, more substrate molecules can be converted to product.

▲ Limiting factors control the rate of reactions in living organisms

- During any 24-hour period, light will be the limiting factor from dusk to dawn.
- During a British winter, plants may not photosynthesise on a bright, sunny day because temperature is the limiting factor.

The availability of water is rarely a limiting factor for photosynthesis, since there are so many other physiological processes in plants that depend on water that these processes will usually halt in a water shortage before photosynthesis does.

Controlling the limiting factors

In an open field, there is very little that farmers can do to speed up photosynthesis – they cannot change the degree of cloud cover or warm up the air, for example. However, in an enclosed environment such as a greenhouse, it is possible to control the factors affecting photosynthesis and so get the maximum yield from crops. This requires an understanding of the principle of limiting factors – it is no good simply increasing the light availability by having lights on in dull weather without making sure that carbon dioxide concentration and temperature are adequate, for example. A greenhouse grower will also try to use strains of plants selected for their high yield (see page 218), control any potential pests and will probably use automatic systems to control the factors that might limit the crop yield.

1 List the three most important factors that control the rate of photosynthesis.

2 The burning of fossil fuels can both help and hinder photosynthesis. Explain this statement.

3 Which ion deficiency is most likely to affect photosynthesis? Why?

4 What is a limiting factor? Which limiting factor is most likely to affect photosynthesis:

a on a cloudy, spring day

b on a bright, sunny day in winter

c in the middle of a crop field on a sunny, warm July day?

2.11 Photosynthesis and the environment

OBJECTIVES

- To understand that plants both photosynthesise and respire
- To understand that these processes affect the composition of the atmosphere
- To appreciate that photosynthesis plays an important part in the carbon cycle

Photosynthesis and respiration

From the equation for photosynthesis (see page 38) we know that this process removes carbon dioxide from the atmosphere and at the same time releases oxygen. This is the opposite of the exchange of gases in respiration:

glucose + oxygen ⟶ carbon dioxide + water

Green plants both photosynthesise and respire:

- if photosynthesis exceeds respiration (in the light) plants will, overall, remove carbon dioxide and add oxygen
- if photosynthesis is less than respiration (in the dark) plants will, overall, remove oxygen and add carbon dioxide.

Demonstrating gas exchange

The overall change in carbon dioxide levels in the atmosphere can be demonstrated using **hydrogencarbonate indicator**, as shown in the diagram below. Carbon dioxide produces a weak acid, **carbonic acid**, in water and this indicator is sensitive to the changes in pH caused by the acid.

In the dark, the rate of respiration greatly exceeds the rate of photosynthesis – the plant cells are living on the sugars they manufactured during previous periods of photosynthesis. As the light intensity increases after dawn, there comes a point where the rates of respiration and photosynthesis exactly balance one another and there is no net uptake or loss of carbon dioxide or oxygen. This is called the **compensation point**, and at this point the glucose consumed by respiration is exactly balanced by the glucose produced during photosynthesis. Beyond the compensation point the plant begins to gain glucose, as photosynthesis exceeds respiration.

EXPERIMENT TO DEMONSTRATE THE EFFECT OF LIGHT INTENSITY ON THE BALANCE BETWEEN PHOTOSYNTHESIS AND RESPIRATION

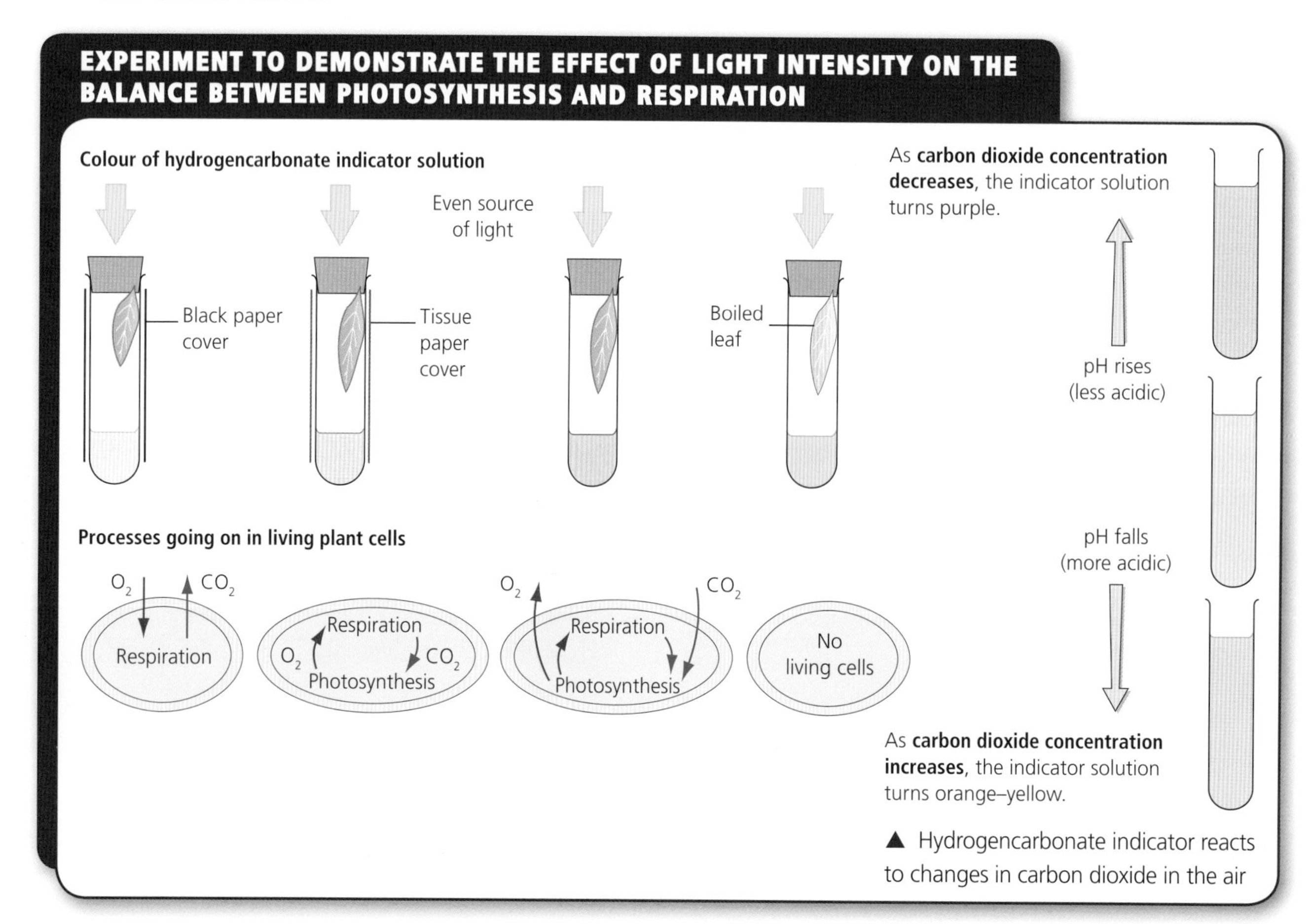

▲ Hydrogencarbonate indicator reacts to changes in carbon dioxide in the air

The products of photosynthesis

Plants manufacture all of their food requirements, starting from the glucose molecule. The diagram shows some of the products that the plant makes.

These products are available to the plant, and to any organism that eats the plant. An animal that eats plant material will use most of it to obtain energy by respiration. The carbon dioxide released as a waste product is then available for plants as a substrate for photosynthesis. An atom of carbon in carbon dioxide could be taken in by a plant, built up to a complex food molecule, and then eaten by an animal which breaks it down to carbon dioxide again. Thus atoms are **recycled** between simple and complex molecules, through the bodies of plants, animals and other living organisms. The **carbon cycle** is described simply in the diagram at the bottom of the page, and more fully on page 232.

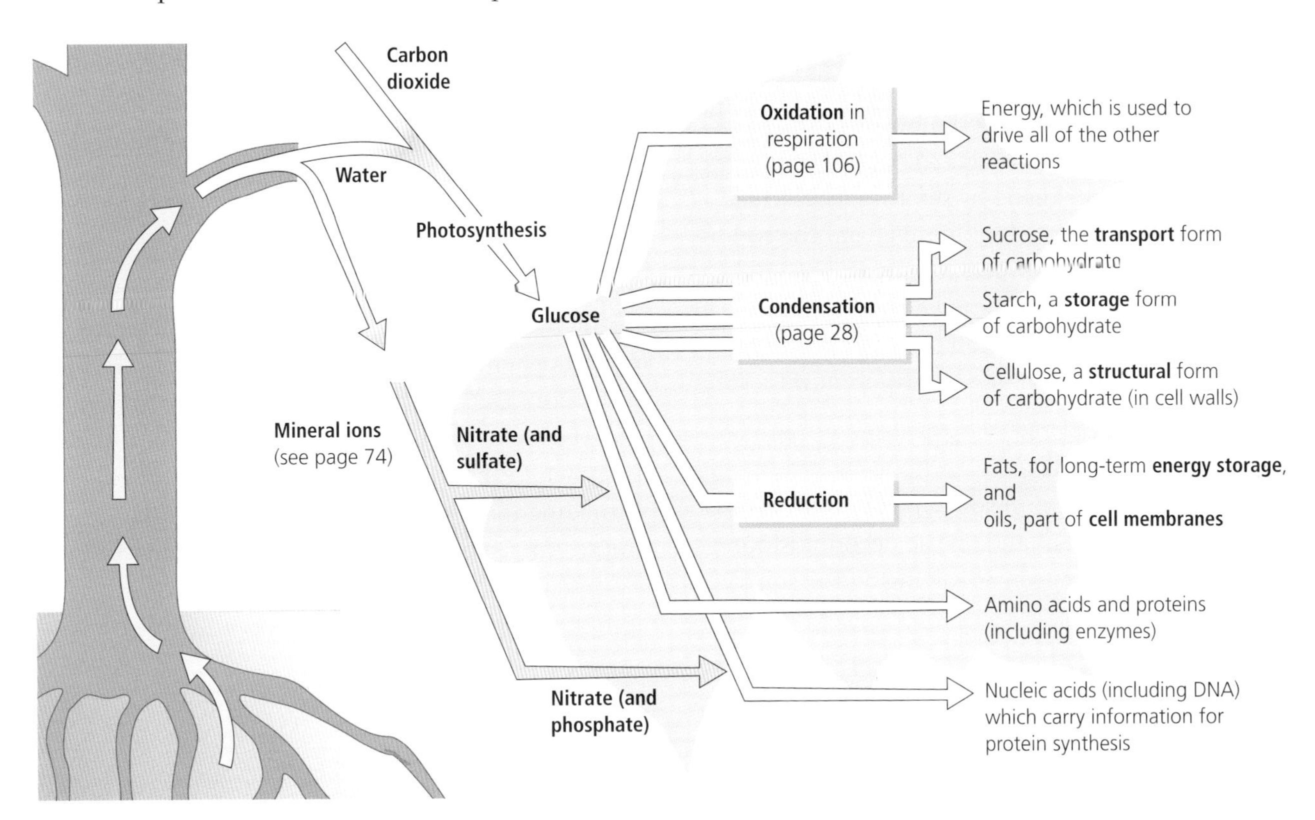

▲ Plants produce all their food molecules from the products of photosynthesis by metabolic reactions

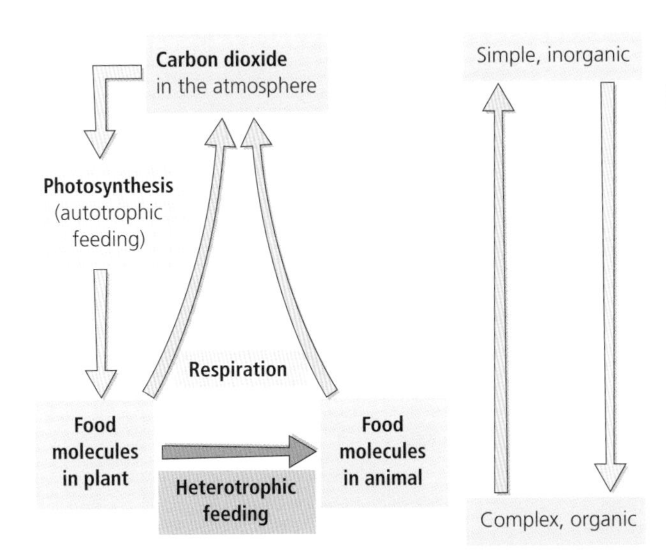

▲ Photosynthesis and the carbon cycle – carbon is recycled between simple inorganic and complex organic compounds

Q

1. Name an indicator that can be used to measure carbon dioxide concentrations in solution. What colour change would you predict for this indicator in the presence of an actively photosynthesising plant? Explain your answer.
2. What is meant by the term compensation point? Why is it important that a grower of greenhouse crops should understand this term?
3. 'Photosynthesis provides plants with sugars' – is this true, and is it the whole story?
4. Plants take up carbon dioxide, which they convert to carbohydrate. Explain how it is possible for the plant to take in the same molecule of carbon dioxide more than once.

2.12 Plants and minerals

OBJECTIVES

- To know that plants require minerals such as nitrate and magnesium
- To know the functions of minerals in plants
- To appreciate how information about plants' mineral requirements can be obtained

Plants need a number of minerals. The plant uses minerals to make food molecules such as amino acids, proteins and nucleic acids out of the carbohydrates made by photosynthesis (see page 39). Plants absorb minerals from the soil in the form of **ions**. The mechanism of ion uptake is described on page 74. Here we shall consider *why* the ions are required, and how scientists might work out the function of each different mineral nutrient.

Minerals in soil

The minerals present in soil depend on the type of rock beneath the soil and on the decomposition of animal and plant remains lying on the soil.

Minerals are taken out of the soil by plants, and are also washed out by rain. In natural, uncultivated soils there is a balance between the formation and the loss of mineral ions, as shown below.

In *cultivated* soils the ground is prepared and then the crop is harvested. There are few plant remains left to decompose and replace the minerals taken up into the plant body. Levels of minerals such as nitrate and phosphate fall, so farmers add these back in the form of **fertilisers**. These may be **natural fertilisers**, such as sewage sludge, animal manure or compost, or they may be **artificial fertilisers**. The most common artificial fertiliser is **NPK** fertiliser which contains three main nutrients – nitrogen (**N**), phosphorus (**P**) and potassium (**K**). The role of some of these minerals, and the effect on the plant of mineral deficiencies, is shown in the diagram on the opposite page. Providing these minerals is one of the most important aspects of preparing soil for agriculture.

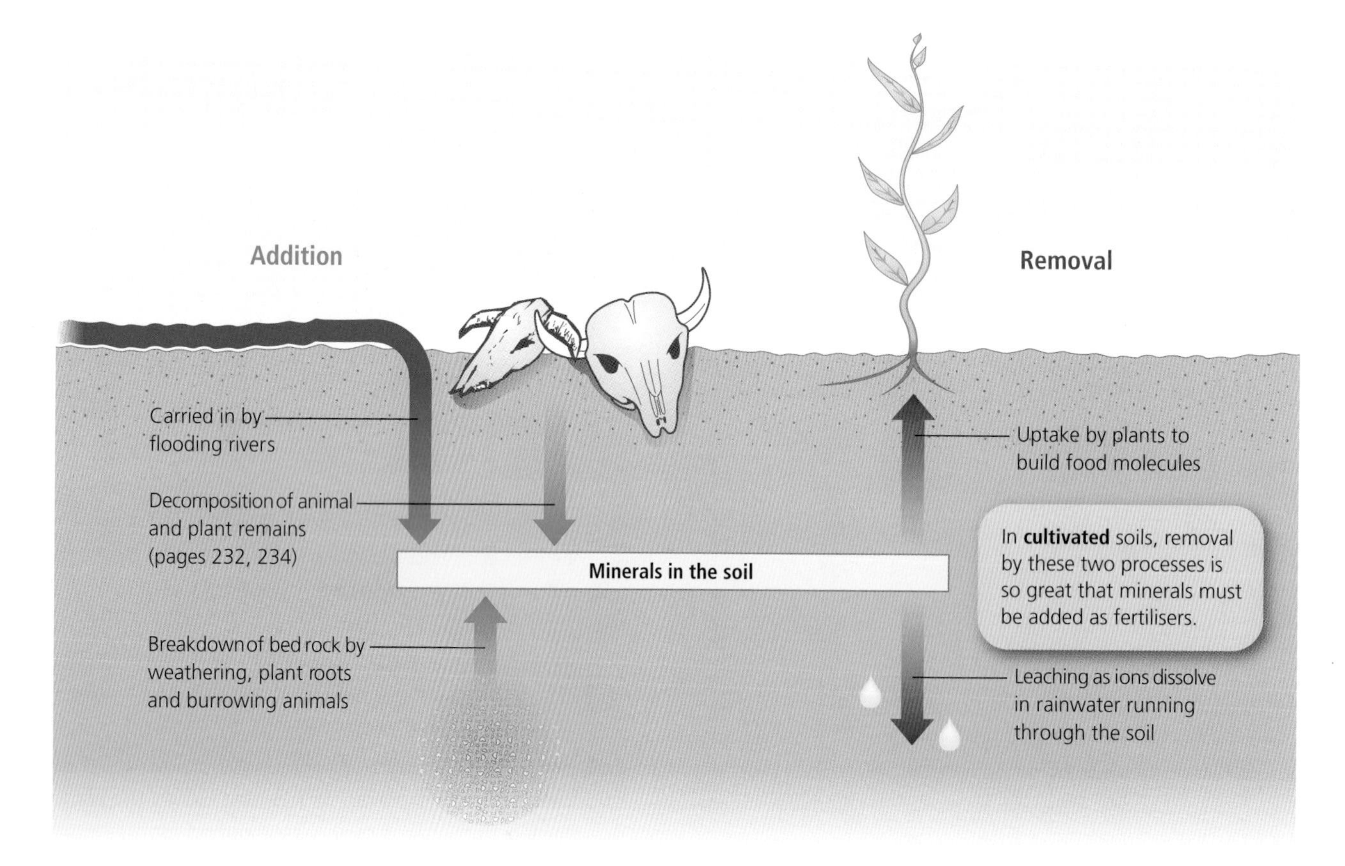

▲ Minerals are added to the soil and also removed by natural processes

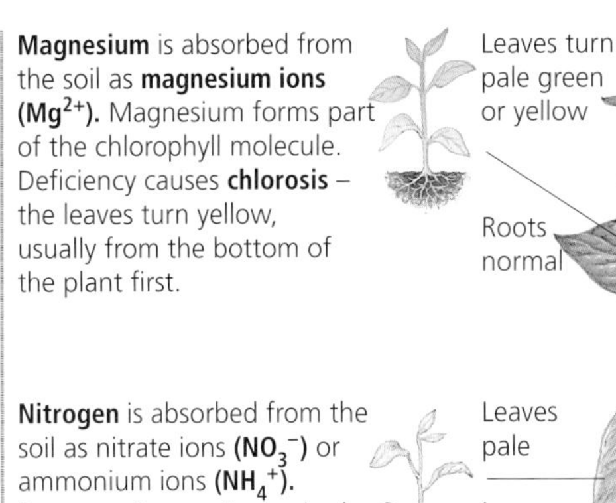

Magnesium is absorbed from the soil as **magnesium ions (Mg^{2+}).** Magnesium forms part of the chlorophyll molecule. Deficiency causes **chlorosis** – the leaves turn yellow, usually from the bottom of the plant first.

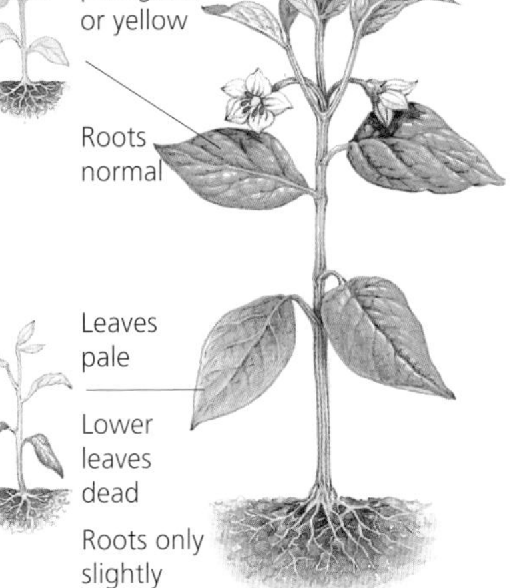
Leaves turn pale green or yellow
Roots normal
Leaves pale
Lower leaves dead
Roots only slightly affected

Nitrogen is absorbed from the soil as nitrate ions **(NO_3^-)** or ammonium ions **(NH_4^+).** Because nitrogen is required for so many food molecules, especially proteins (including enzymes), deficiency causes severe symptoms. The whole plat is stunted, with a weak stem and yellowing, dying leaves.

▲ Plant growth and mineral deficiencies

Nitrogen probably has the greatest effect on plant growth. Farmers use expensive NPK fertilisers to replace nitrogen removed from the soil when crops are harvested.

Problems with fertilisers

Excessive use of fertilisers can cause problems of eutrophication (see page 267). The fertiliser runs off into nearby streams, rivers and lakes and boosts the growth of algae. As algae die they are decomposed by bacteria, which use all of the oxygen dissolved in the water for aerobic respiration. As a result the water becomes oxygen-deficient, and larger animals such as fish and insects die.

PLANTS AND MINERALS – INVESTIGATION

The apparatus shown on the right can be used to investigate the effects of mineral deficiencies on plant growth. Cereal plants of equal age and size are grown in a series of culture solutions. One of the solutions contains all known mineral nutrients in the correct proportions; each of the others is missing a single mineral. All the vessels are placed in identical conditions of temperature and light intensity, and the plants are allowed to grow for equal lengths of time.

1. Why is the vessel surrounded by black paper?
2. The bubbled air supply provides oxygen to the roots. Why might this be important in mineral uptake?
3. What is the independent variable in this investigation, and what is the dependent variable?
4. Suggest some fixed variables, and state how you would attempt to control them.
5. Cereal plants such as grasses are useful in this sort of investigation, since many identical plants can be obtained from a single clump. Why is this important in providing valid data?
6. The plant nutrients used in this investigation are supplied as salts. How could you make up a culture solution that is only lacking a single element?

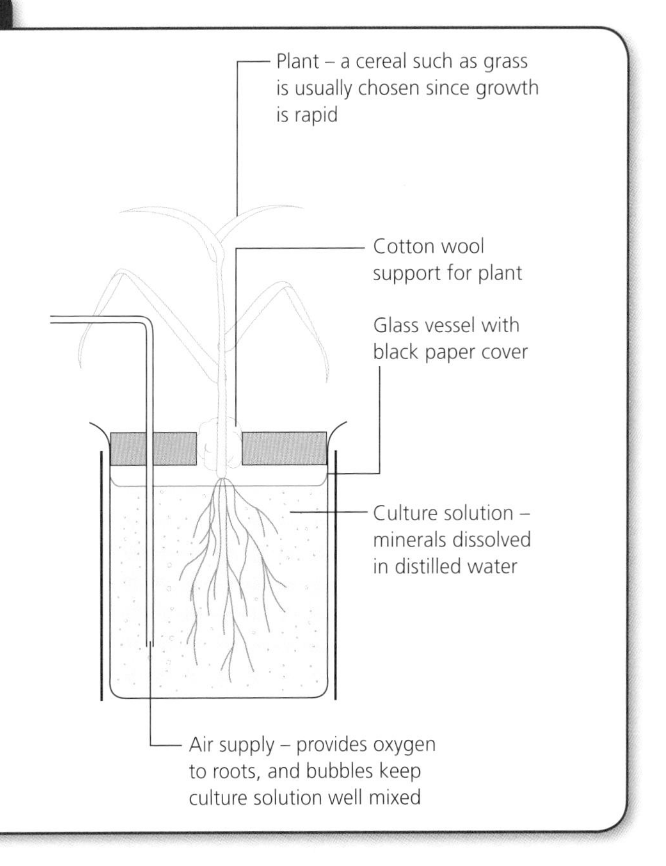
Plant – a cereal such as grass is usually chosen since growth is rapid
Cotton wool support for plant
Glass vessel with black paper cover
Culture solution – minerals dissolved in distilled water
Air supply – provides oxygen to roots, and bubbles keep culture solution well mixed

Questions on photosynthesis and plant nutrition

1 When a strip of filter paper is soaked in cobalt chloride solution, it is pink when damp and turns blue when dry.
The figure shows strips of dry cobalt chloride paper on the upper and lower surfaces of a leaf, pressed between dry glass slides.

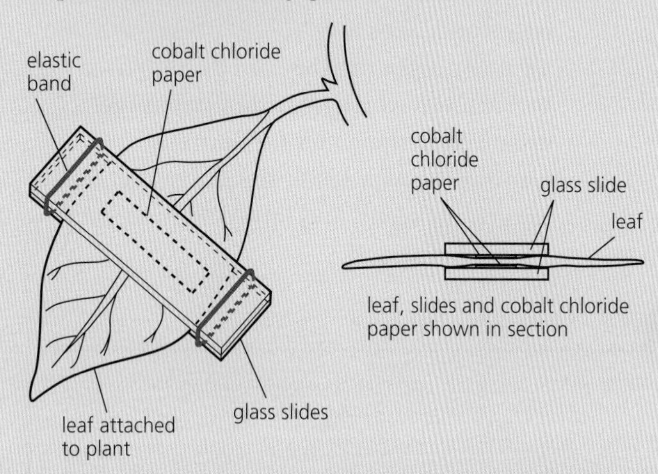

A student used this method to investigate the time taken for strips of cobalt chloride paper to turn from blue to pink. The experiment was performed five times. The results are shown in the table.

experiment number	time (in minutes) taken for paper to change from blue to pink	
	upper surface	lower surface
1	10	3
2	13	4
3	4	3
4	11	4
5	12	3

a **i** State the most likely conclusion that could be drawn from the results of experiments 1, 2, 4 and 5.

ii Suggest an explanation for the results of these experiments
for the upper surface of the leaf,
for the lower surface of the leaf.

b Suggest two possible reasons for the result obtained for the strip attached to the upper surface of the leaf in experiment 3.

c Suggest how the student could improve the reliability of this experiment.

d Suggest how the student could show that the concentration of water vapour coming out of a leaf is greater than its concentration in the atmosphere.

Cambridge O Level Biology 5090
Paper 2 Q3 May 2007

2 Copy and complete the following paragraphs.

a The cells of green plants absorb water by ____. Plant cells rely on water for ____, as a medium for biochemical reactions, as a ____ and as a raw material for ____.

b Water is obtained by plants from the soil solution. The water enters the plant through its ____. These structures are well adapted to the absorption of water. There are ____ growing on their epidermis which greatly increase the ____ over which water absorption can take place. In addition to water, these structures also absorb ____ such as ____ which is required for the synthesis of chlorophyll and ____ which is required for the manufacture of amino acids and proteins. These substances are absorbed by a process called ____ since it requires a supply of energy.

c Water is used as a transport medium for both ions and sugars. Mineral ions are transported within the main water flow through the plant in the ____ tissue. Sugars are transported in the living cells of the ____. These specialised tissues are grouped together into ____ bundles.

3 **a** **i** Name the two raw materials needed by plants for photosynthesis.

ii Name the gas produced by photosynthesis.

b The figure below shows a leaf, with white and green regions, that is attached to a plant. The plant had been kept in the dark for 48 hours and then a lightproof, black paper cover was placed over part of the leaf.

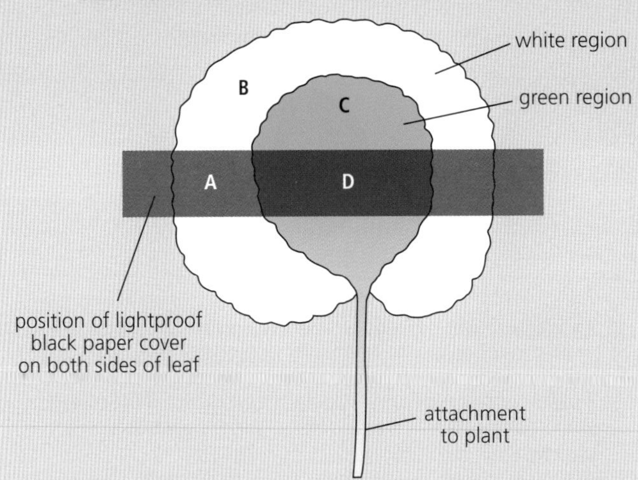

The plant is left under a light for 24 hours. After this time the leaf is removed from the plant and is tested for the presence of starch.

i Which chemical reagent is used to show the presence of starch?

ii Record the colour you would see, if you had carried out this test, in each of the areas **A**, **B**, **C**, and **D**.

iii Explain the results for areas **B** and **D**.

Cambridge IGCSE Biology 0610
Paper 2 Q4 May 2008

4 The figure below shows four test-tubes that were set up and left for six hours at a constant warm temperature.

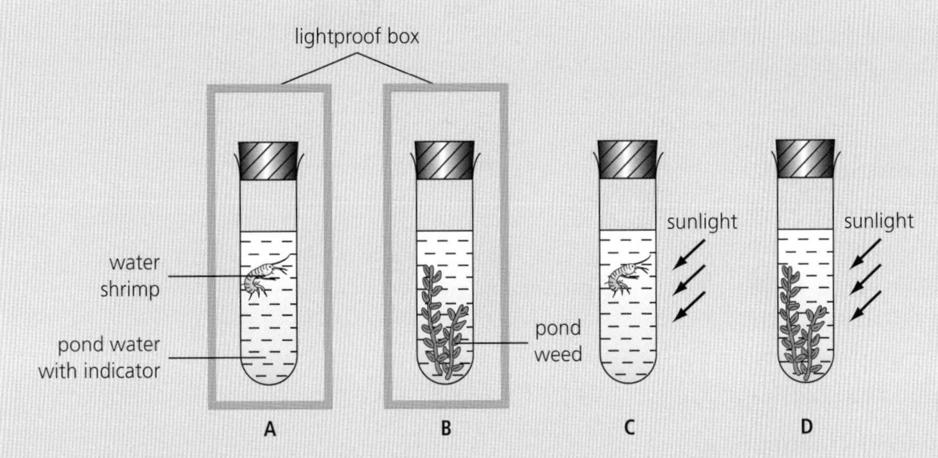

Hydrogencarbonate indicator (bicarbonate indicator) changes colour depending on the pH of gases dissolved in it, as shown below.

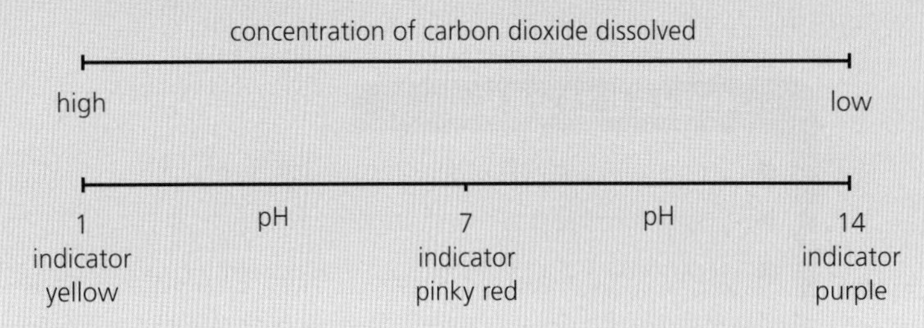

After six hours the colour of the indicator in all four tubes had changed.

a **i** Predict the colour of the indicator in each tube after six hours.

ii Suggest the reason for the change in colour of the indicator in each of tubes **A** and **D**.

b The figure below shows a fifth tube, **E**, set up at the same time and in the same conditions as tubes **C** and **D**.

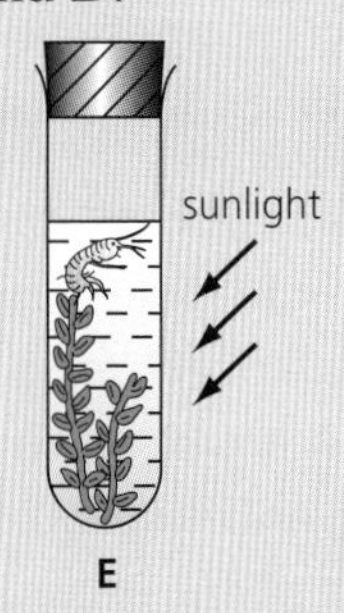

Suggest and explain the possible colour of the indicator in tube **E** after six hours.

Cambridge IGCSE Biology 0610
Paper 2 Q6 May 2009

2.13 Food and the ideal diet: carbohydrates, lipids and proteins

OBJECTIVES

- To understand why organisms require food
- To list the constituents of an ideal diet
- To know the functions of each component of an ideal diet

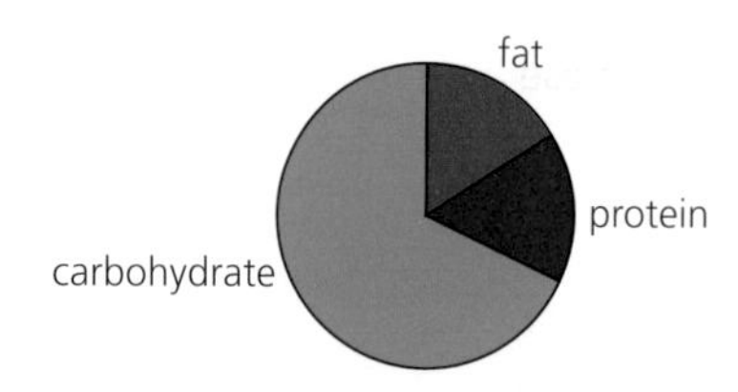

Food

All living organisms are made up of molecules, organised so that they can carry out the characteristics of life. Food supplies them with:

- molecules that are the **raw materials** for repair, growth and development of the body tissues
- molecules that can be oxidised in respiration, and act as a **source of energy**
- elements and compounds that enable the raw materials and energy to be used efficiently.

All living organisms need food, but some, green plants in particular, can make their own organic molecules. These organisms are said to be **autotrophic** (or 'self feeding'). Green plants use a form of autotrophic nutrition called **photosynthesis** to supply them with all the organic molecules they need (see page 38). Other organisms cannot make their own food, and must take in food molecules from their surroundings. These molecules have been made by another organism, so living things that feed in this way are said to be **heterotrophic** (or 'other feeding'). Humans, like all animals, are totally dependent on other organisms for their food.

An adequate diet provides sufficient **energy** for the performance of metabolic work, although the 'energy' could be in any form.

A balanced diet provides all dietary requirements **in the correct proportions**. Ideally this would be 1/7 **fat**, 1/7 **protein** and 5/7 **carbohydrate**.

In conditions of **under**nutrition the first concern is usually provision of an adequate diet, but to avoid symptoms of **mal**nutrition a **balanced diet** must be provided.

A balanced diet

The total of the molecules or **nutrients** that we need is called the **diet**. A **balanced diet** provides all the nutrients, in the correct amounts, needed to carry out the life processes. If the diet does not provide all the nutrients in the correct proportions, a person may suffer from **malnutrition** (see page 60, for example).

Food can be analysed to find out what chemicals it contains, using quite simple chemical techniques (see page 30). A balanced diet should contain the correct proportions of **carbohydrates**, **lipids**, **proteins**, **vitamins** and **minerals**, **water** and **dietary fibre**. We shall look at each in turn.

Q

1. Give three reasons why living organisms need food.
2.  Write one sentence to explain the difference between autotrophic and heterotrophic nutrition.
3. What is a balanced diet?
4. Which are the main energy foods for humans?
5. State four functions of proteins, using particular examples to illustrate your answer.

Carbohydrates, lipids and proteins: three food types

Good source	Functions in humans	Comments
Carbohydrates		
Rice, potatoes, wheat (e.g. pasta) and other cereals provide **starch**. Food sweetenings, such as those in desserts, sweets and soft drinks, and preservatives provide **refined sugars**, such as **sucrose (cane sugar)** and **glucose**.	A **source of energy**. Glucose is oxidised in **respiration** (see page 106) to release **energy** for active transport, cell division, muscle contraction and the manufacture of large biological molecules. Excess carbohydrate can be stored as **glycogen** (see page 132) and as **fat**.	Carbohydrates are digested in the mouth and small intestine and absorbed as **glucose.** Refined sugars are absorbed very rapidly, giving a sudden boost of 'energy source'. Starch is digested and absorbed more slowly, giving a steady supply of energy source: starches are called **slow release carbohydrates**.
Lipids		
Meat and animal foods (eggs, milk, cheese) are rich in **saturated fats** and **cholesterol**. Plant sources such as sunflower seeds and peanuts are rich in **unsaturated fats**.	Fats and oils are an important source of energy (see page 52). They are especially valuable as an **energy store** because they are insoluble in water. They also provide **insulation** – electrical insulation around nerve cells (see page 138) and thermal insulation beneath the skin (page 131) – and form part of cell membranes. **Steroid hormones**, including sex hormones, are made from cholesterol.	Fats and oils are digested in the small intestine and absorbed as fatty acids and glycerol. Some lipids contain **saturated fatty acids** and others contain **unsaturated fatty acids** (with at least one carbon–carbon double bond). The body can store unlimited amounts of fat, contributing to obesity (see page 60). The incorrect balance of saturated and unsaturated fatty acids, or an excess of cholesterol, can cause diseases of the circulation (see page 92).
Proteins		
Meat, fish, eggs from animals, and legumes (peas and beans) and pulses from plants. One of the best sources of protein is the soya bean. This contains very little fat (unlike most animal sources) and so is suitable for people with health problems caused by fat (see page 92). Soya beans can be flavoured and textured to make them taste and feel like meat – this **textured vegetable protein** is used as 'artificial' meat. **Mycoprotein** is also a low-fat substitute for meat.	Many functions, including ■ catalysts (enzymes) (page 32) ■ transport molecules, e.g. haemoglobin (page 82) ■ structural materials, as in muscles (page 108) ■ hormones, such as insulin (page 132) ■ in defence against disease, as antibodies (page 104)	Digested in the stomach and small intestine, and absorbed as amino acids. 20 different amino acids are needed to make up all of the different proteins in the human body. Some of these must be supplied in the diet as the body cannot make them – these are the **essential amino acids**. Proteins from animal sources usually contain all 20 amino acids, but plant proteins often lack one or two of the essential amino acids. Deficiency of protein causes poor growth – in extreme cases may cause **marasmus** or **kwashiorkor**.

Essential amino acids – cry baby

One factor that determines how comfortable babies feel is the supply of a 'pleasure chemical', called **serotonin**, in the brain (see neurotransmitters, page 137). An essential amino acid called **lysine** is needed to make serotonin. Lysine is in short supply in some artificial milks. Babies fed on these milk substitutes can't manufacture enough serotonin, they feel uncomfortable and they CCCRRRYYY! Natural mother's milk contains an adequate supply of lysine.

Vegetarians must ensure that their diet contains a wide range of protein sources since very few plants contain all the essential amino acids. An ideal vegetarian meal containing all the essential amino acids is baked beans on toast (which also contains a high proportion of dietary fibre).

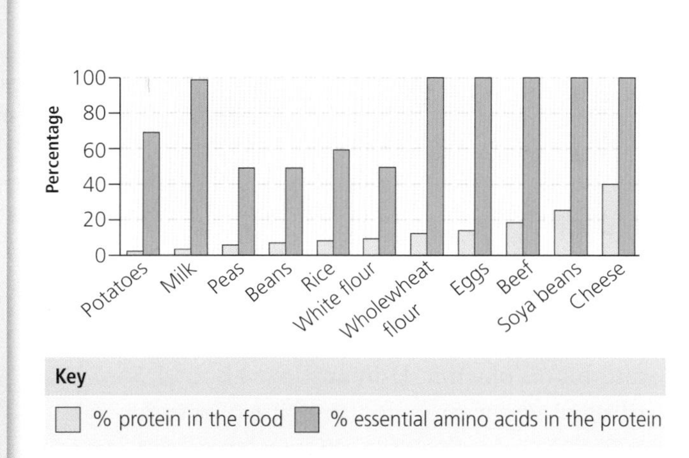

▲ The protein values of some common foods

2.14 Food and the ideal diet: vitamins, minerals, water and fibre

OBJECTIVES

- To list the constituents of an ideal diet
- To know the functions of each component of an ideal diet

Vitamins and minerals

Vitamins and minerals are essential for the body to be able to use the other nutrients efficiently. They are needed in only very small amounts. There are many different vitamins and minerals, and they are usually provided in the foods of a balanced diet.

Vitamins			
Water-soluble	**Food source**	**Symptoms of deficiency**	**Comments**
C (ascorbic acid)	Cherries, citrus fruits e.g. limes, lemons, oranges, and fresh green leafy vegetables	Scurvy – production of fibres in the body is affected.	Vitamin C also seems to protect cells from ageing.
Fat-soluble	**Food source**	**Symptoms of deficiency**	**Comments**
D (calciferol)	Liver, dairy products, eggs, fish liver oil	Rickets – bones are soft and may bend, because vitamin D is needed for the absorption of calcium.	Can be made by the body, just under the skin, but only if there is plenty of sunlight.

Minerals			
	Food source	**Symptoms of deficiency**	**Comments**
Iron	Red meat, liver, some leafy vegetables, e.g. spinach	Anaemia – iron is needed to produce **haemoglobin** for red blood cells. A shortage causes weakness as oxygen needed for respiration cannot be transported efficiently.	Iron is added to foods when metal utensils are used in cooking – the amount of iron in a piece of beef is doubled when the meat is minced in an iron mincer ready for making burgers!
Calcium	Milk, cheese and fish	Several problems ■ weak bones and teeth ■ poor clotting of blood ■ uncontrolled muscle contractions ('spasms').	Calcium shortage causes **rickets**, the same deficiency disease caused by insufficient vitamin D.

Water

Water forms about 70% of the human body. Two-thirds of this water is in the cytoplasm of cells, and the other third is in tissue fluid and blood plasma. Humans lose about 1.5 litres of water each day, in urine, faeces, exhaled air and sweat – this must be replaced

by water in the diet. It is obtained in three main ways:

- as a drink
- in food, especially salad foods such as tomatoes and lettuce
- from metabolic processes (think back to the equation for aerobic respiration – water is one of the products).

A loss of only 5% of the body's water can lead to unconsciousness, and a loss of 10% would be fatal. We shall look further at how the body conserves water on page 125.

Fibre

Dietary fibre is the indigestible part of food, largely cellulose from plant cell walls, which provides bulk for the faeces. Plenty of fibre in the diet stretches the muscles of the gut wall and helps push the food along by peristalsis (see page 66). A shortage of fibre can cause constipation, and may be a factor in the development of bowel cancer.

Summary of a balanced diet

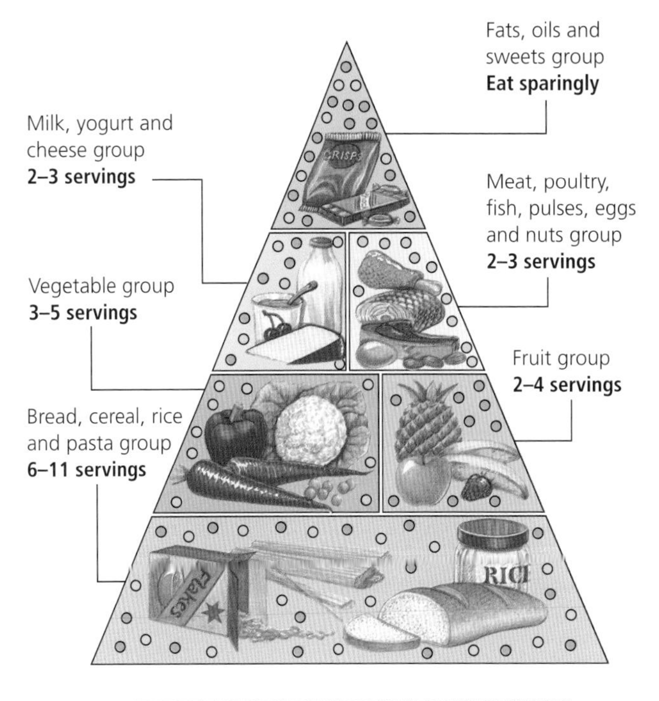

▲ Food guide pyramid: choosing to eat these amounts of different kinds of foods each day provides a balanced diet

1 Table 1 shows the daily energy needs of different people. Table 2 shows the energy content of four foods.

a What is the daily energy need for an active 8-year-old boy?

b How many grams of Food 3 would meet the energy needs of an office worker?

c How much more energy does a labourer need than a male office worker each day? Show your working.

d i From the foods given in Table 2, which ONE would be the best for the labourer to eat? Give a reason for your choice.

ii If the labourer ate only the food you have suggested in your answer to **i**, what is the least amount he should eat to meet his daily energy need? Show your working.

e The heat energy in foods can be measured experimentally by burning food under a known mass of water. The temperature rise of the water is recorded. The number of joules received by the water from 1 gram of food can be calculated using the following formula:

Heat gained by water from Y grams of food

$$= \frac{\text{mass of water} \times \text{temperature rise} \times 4.2}{Y} \text{ J per g}$$

Use the following data and the formula to work out the number of joules released from 1 gram of food which caused the rise in temperature of the water.

Mass of water = 20 g

Temperature rise = 18 °C

Mass of food = 2 g

Person	Occupation	Daily energy need / kJ
Active girl aged 8 years	Schoolgirl	8 000
Active boy aged 8 years	Schoolboy	8 400
Woman	Office worker	9 500
Man	Office worker	10 500
Active girl aged 15 years	School girl	11 800
Active boy aged 15 years	School boy	14 700
Man	Labourer	18 900

▲ Table 1

Food	1	2	3	4
Energy content / kJ per 100 g	3 800	130	1 050	400

▲ Table 2

Processed foods contain other substances*

Natural foods contain mixtures of carbohydrates, lipids, proteins, vitamins, minerals and water. Many foods consumed in the developed world also contain other substances. These 'additional' substances fall into two main categories:

- **fortifiers**: these are substances added to increase the nutrient value of the food ('fortification' means 'strengthening');
- **additives**: substances with no nutrient value but which are added to improve the appearance, flavour, texture and/or the storage properties of the food.

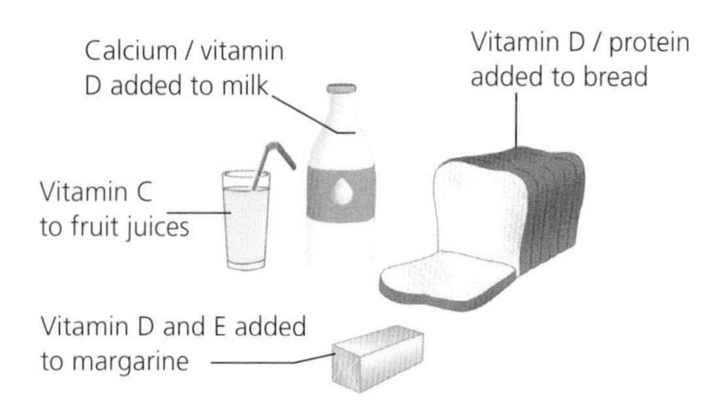

▲ Fortification adds extra nutrients to processed foods

All food packages in European Union countries must carry a full list of additives so that consumers know exactly what they are getting when they buy a processed food.

Food additives may be natural or synthetic.

Flavourings and flavour enhancers (usually E600 numbers)
- most common food additives
- some are natural (e.g. vanilla), some are artificial (e.g. propyl pentanoate – pineapple flavour)
- flavour enhancers make existing flavours stronger, by stimulating the taste buds (e.g. MSG – monosodium glutamate)

Colouring (E100 numbers)
- added to improve colour which may have been lost during cooking (e.g. peas become paler) or by dilution
- may be natural e.g. β carotene (E160) from tomatoes
- may be synthetic e.g. tartrazine (E102) from coal tar
- **cannot** be added to baby foods

Prevention of spoilage – Preservatives (E200 numbers)
- inhibit growth of fungi or bacteria e.g. sulfur dioxide (E220) controls the browning of potatoes

Anti-oxidants (E300 numbers)
- slow down deterioration caused by atmospheric oxidation e.g. vitamin C (ascorbic acid – E300) added to fruit drinks

Additives may be dangerous
For example
- tartrazine (E102) – linked to hyperactivity and poor concentration in children
- sulfur dioxide (E220) – may cause sensitivity in asthma sufferers
- monosodium glutamate (E621) – may cause vomiting, migraine and nausea

Improvement of texture
- thickeners alter the consistency of food e.g. cellulose E460
- stabilisers prevent components of emulsions separating out e.g. Xanthan E415 with yoghurt

All additives
- have an E (European) number
- must be pre-tested on animals

Ingredients		
Whole wheat, malted barley extract, sugar, salt, niacin, iron, thiamin (B1), riboflavin (B2), folic acid		
Allergens		
Contains gluten and wheat		
Nutrition		
Typical values	**per 37g serving with 150ml of semi-skimmed milk**	**per 100g**
Energy	831kJ	1427kJ
	197kcal	336kcal
Protein	9.4g	11.8g
Carbohydrate	32.7g	68.0g
of which sugars	9.0g	4.7g
Fat	3.1g	1.9g
of which saturates	1.7g	0.6g
Fibre	3.8g	10.1g
Sodium	0.18g	0.27g
Niacin	5.9mg	15.3mg
% RDA	33	85
Thiamin (B1)	0.5mg	1.2mg
% RDA	36	85
Riboflavin (B2)	0.8mg	1.4mg
% RDA	49	85
Folic Acid	73mg	170mg
% RDA	36	85
Iron	4.5mg	11.9mg
% RDA	32	85
per 37g serving with 150ml of semi-skimmed milk		
197 calories	3.1g fat	0.5g salt
*RDA = recommended daily allowance		
Guideline daily amounts for average adults		
	Women	Men
Calories	2000	2500
Fat	70g	95g
Salt	5g	7g

▲ Nutrition information from a packet of breakfast cereal

What about residues?*

Additives are used to improve food *quality*, as described above. **Residues**, on the other hand, are added to increase food *production*. Because of their widespread use, or because of poor food production techniques, these substances may remain in food presented to humans. Some chemicals which may remain as residues, and which cause concern for human health, are:

- **Pesticides**: may become concentrated in food chains, causing harm to the top consumers (including humans) – see page 266.
- **Fertilisers**: nitrates and phosphates may leach into water supplies (see page 268). This has caused problems (**blue baby syndrome** in which babies' blood does not become properly oxygenated) in babies fed on reconstituted powdered milk.
- **Antibiotics**: used to prevent infection of cattle and poultry. May lead to the selection of antibiotic-resistant strains of microbes (see page 248).
- **Hormones**: oestrogen is used to boost the growth rate of chickens. This may cause human males to develop feminine characteristics.

1 Table 1 shows the composition and energy content of four common foods.

Food	Energy content / kJ per g	Composition per 100g					
		Protein / g	Fat / g	Carbohydrate / g	Vitamin C / mg	Vitamin D / mg	Iron / mg
A	3700	0.5	80	0	0	40	0
B	150	1.2	0.6	7	200	0	0
C	400	2.0	0.2	25	10	0	8
D	1200	9.0	1.5	60	0	0	0

▲ Table 1

a Which food would be best to prevent rickets?
b Which food would be best for a young person training for cross-country running?
c Which food would be most needed by a menstruating woman?
d Which food would be the most useful to a body-builder?
e Which food would be most dangerous for a person with heart disease?

2 Breast milk contains all the nutrients a baby needs except for vitamin C and iron. However, the baby has sufficient iron stored in its liver for the first months of its life. The first milk a breast-fed baby receives is called colostrum. After a few days, normal breast milk is produced. Table 2 compares the composition of colostrum and normal breast milk.

	Nutrient / g per 100 cm^3		
	Fat	Protein	Sugar
Colostrum	2.5	8.0	3.5
Normal breast milk	4.0	2.0	8.0

▲ Table 2

a Use data from Table 2 to describe how the amounts of fat, protein and sugar are different in colostrum and normal breast milk.
b A baby feeding on normal breast milk drinks one litre of milk per day. Calculate how much protein the baby receives per day. Show your working.
c Suggest a suitable fruit juice a mother could give her baby to provide vitamin C.

2.15 Food is the fuel that drives the processes of life

OBJECTIVES

- To know that food has an energy value
- To know how to calculate the energy value of different foods
- To understand that different people have different demands for energy

Measuring the energy in food

Energy is released from food by respiration, which is an oxidation process, similar in some ways to combustion – if food is combusted (burned), it releases energy, mainly as heat. We can work out the energy value of any particular type of food by assuming that respiration releases the same amount of energy in total as combustion does. The energy value is found using a **food calorimeter** or **bomb calorimeter**, shown in the diagram opposite. A weighed sample of food is completely burned in an atmosphere of oxygen. The heat released by this combustion is transferred to a known volume of water, which rises in temperature as a result. The energy value of the food can be calculated as follows:

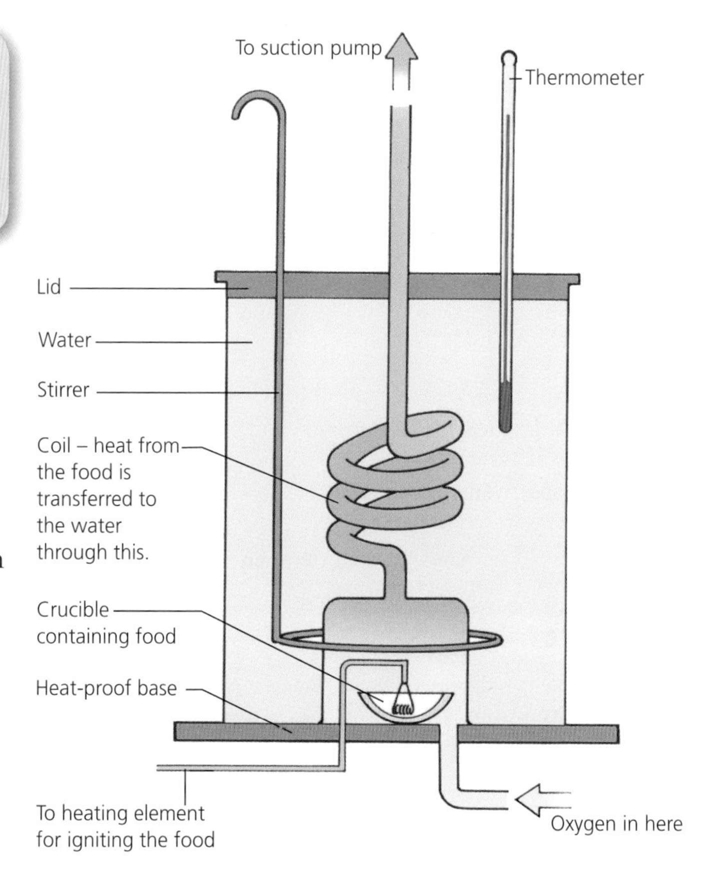

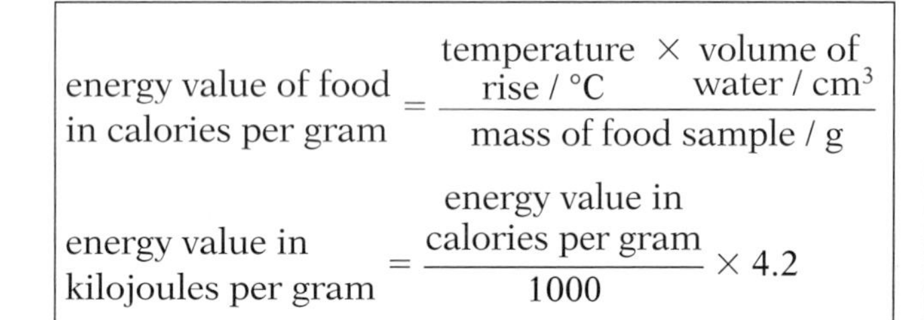

$$\text{energy value of food in calories per gram} = \frac{\text{temperature rise / °C} \times \text{volume of water / cm}^3}{\text{mass of food sample / g}}$$

$$\text{energy value in kilojoules per gram} = \frac{\text{energy value in calories per gram}}{1000} \times 4.2$$

Energy values of different foods

The three main energy-providing organic molecules found in food are fats, carbohydrates and proteins. Each has a different energy value:

Fats	39 kJ per g
Proteins	20 kJ per g
Carbohydrates	17 kJ per g

Carbohydrates provide most of our energy, not because there is most energy available per gram in carbohydrates, but because we eat more grams of carbohydrates than of proteins or fats.

Units of energy

The SI unit of energy is the joule (J). The amount of energy contained in food is measured in either joules (J) or calories (cal).

1000 J = 1 kJ (kilojoule) 1000 cal = 1 kcal (kilocalorie)

A calorie is an older unit of energy that is still often used. A calorie is the amount of energy needed to raise the temperature of 1 cm^3 of water by 1 °C.

1 calorie = 4.2 joules

Joules and calories are small amounts of energy so kilojoules and kilocalories are more frequently used to describe the energy values of foods. In popular nutrition guides, the calories referred to are actually kilocalories and are often spelt Cal, with a capital C. The energy content information given on the packaging of many foods has been calculated by burning a sample of the food in a food calorimeter.

Our energy requirements

The total of all the chemical reactions in the body is called **metabolism**. It is driven by the energy from respiration, so the amount of energy needed depends on how many metabolic reactions a person is carrying out. The amount of metabolism depends on three factors:

- The processes required simply to stay alive. These processes include breathing, excretion, thinking and keeping a constant body temperature. The energy consumed in one day by these processes is called the **basal metabolic rate** (BMR). The BMR is different for males and females, and gets less as a person gets older.
- The amount of **activity** or **exercise** – this requires muscular work and extra temperature regulation. This energy demand can be very little (in a sedentary office worker, for example) or very great (in a road worker, for example).
- The amount of food eaten – eating food means extra work for the digestive system and the liver. The energy demand simply caused by eating is called the **dynamic action of food**.

energy requirement per day = BMR + activity + dynamic action of food

The bar chart below illustrates the energy demands for different people. Three factors affect the body's energy requirements – **age**, **sex** and **occupation**.

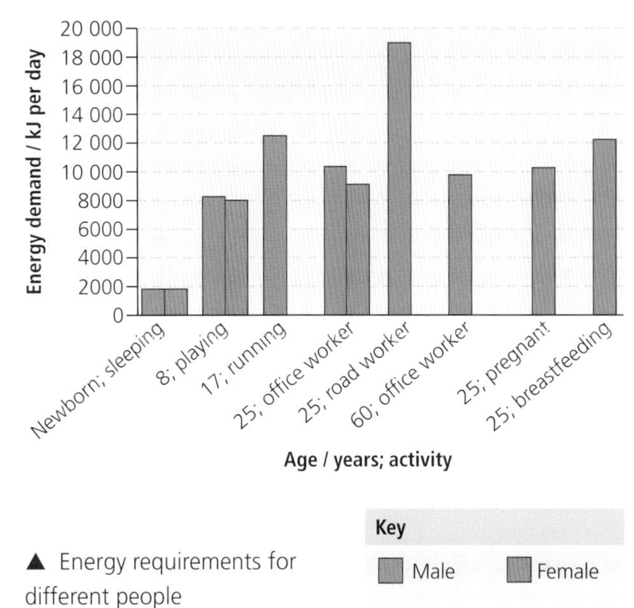

▲ Energy requirements for different people

ENERGY IN FOOD

A class of students used the apparatus shown here to investigate the energy content of a peanut.

They first recorded the temperature of the water (T_1) and weighed the nut.

The nut was held in a Bunsen burner flame until it caught fire, and then placed under the boiling tube.

The students recorded the maximum temperature the water reached (T_2), and then repeated the experiment twice more. A typical set of results from one pair of students is shown in the table at the bottom of the page.

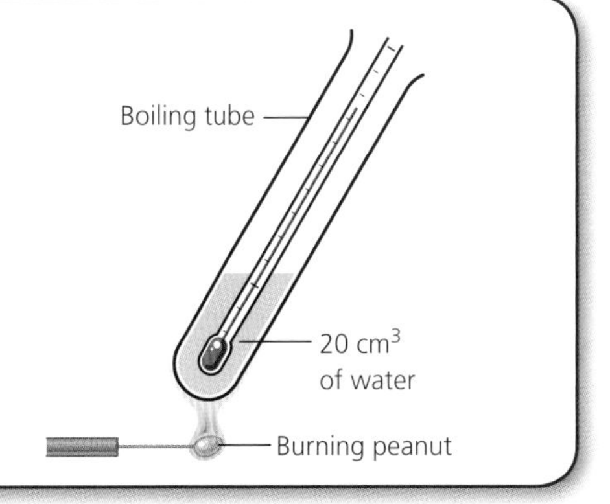

Q

1
- **a** Copy and complete the table of results.
- **b** Add a further column and include the energy value of the peanut in kJ per g.
- **c** Calculate the mean energy value of the peanut as found by these students. Why is the mean value more valid than any single value that they obtained?
- **d** Can you suggest any improvements to the apparatus so that the students' value might be closer to the 'professional' one (24.5 kJ per g)? Redraw the apparatus showing your suggestions.
- **e** What features of the method helped to improve the validity of the results? How could the students have treated the results differently to improve their validity further?

T_1 / °C	T_2 / °C	$T_2 - T_1$ / °C	Volume of water / cm³	Energy transferred to water / kcal	Mass of nut / g	Energy value / kcal per g
22	78				0.45	
22	74				0.52	
21	75				0.47	

2.16 Balancing energy intake and energy demand: problems causing malnutrition

OBJECTIVES

- To understand that different people have different demands for energy
- To understand that energy intake must be balanced by energy use

Malnutrition means literally 'bad feeding'. This 'bad feeding' could include:

- eating too much of all foods, so having a balanced diet but consuming more than is needed
- having too little food
- eating foods in the wrong proportions, for example, gaining too many kilojoules from fats and too few from carbohydrates.

Too much energy

The need for energy is related to **age**, **sex** and **activity**.

If the diet contains more energy than the body needs, the excess will be stored as **glycogen** or **fat**; if it contains less energy than the body needs, then the body's own tissues will be broken down to be respired. The diagram below shows the importance of an **energy balance**.

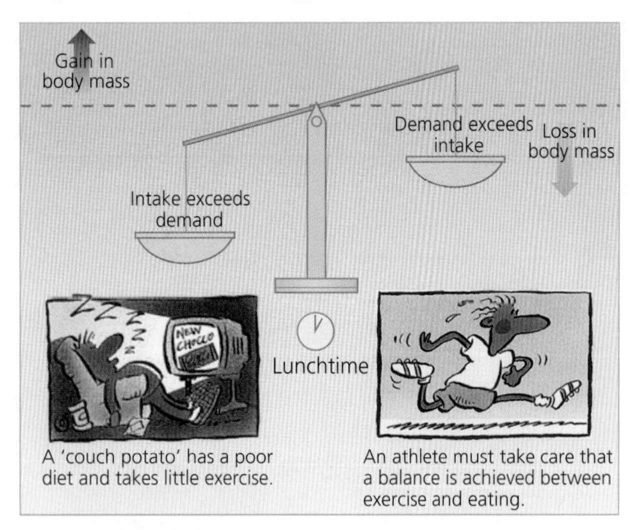

Storing excess food: overnutrition

The body can store a limited amount of glycogen, which allows the body to continue working even if the last meal was some time ago. However, the body can store an almost unlimited amount of fat to help the body survive periods without food. Our food intake should not be so great that we store an unhealthy amount of fat. A person whose fat storage is beyond a healthy limit is said to be **obese**.

The 'ideal' body mass differs from person to person, and depends on height and age. A person who is obese is at risk from a number of life-shortening diseases, including **diabetes**, **breathing difficulties**, **atherosclerosis** (narrowing of the arteries) and **arthritis**. Obesity is one of the most widespread results of malnutrition in the western world.

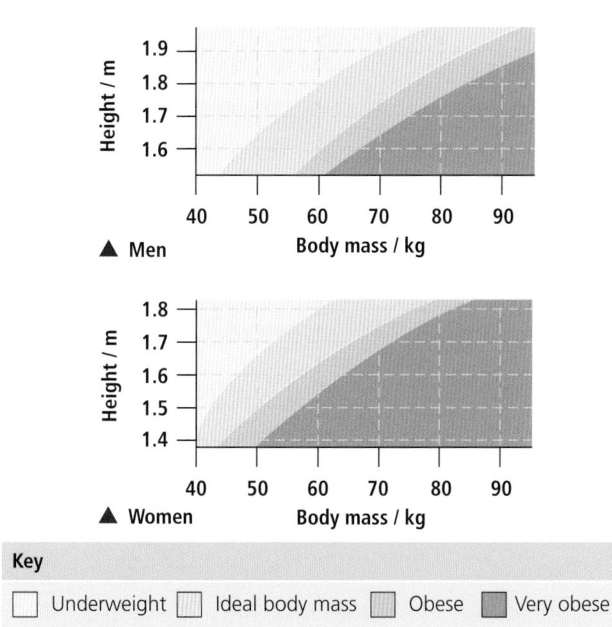

▲ The ideal body mass should fall within the range shown in these charts (designed for adults who have reached their full height)

Losing body mass

The energy balance diagram shows that to lose body mass, it is necessary to reduce energy intake below energy use. This can be done in two ways:

- by eating less 'high-energy' food, which will **reduce the energy intake**
- by taking more exercise, which will **increase the energy use**.

The best approach is to combine both methods, controlling the diet and also taking more exercise. Many people who rely on diet alone have great difficulty in controlling their body mass.

Malnutrition can also mean undernutrition: too little food

In the developing countries of the world, many people have diets which are neither adequate nor balanced. In other words, they do not receive enough energy to drive metabolism, nor do they take in the nutrients they need for growth and development. These people, often children, suffer from many deficiency diseases – shortage of iron means they are often anaemic, and a limited supply of vitamin C means that many of them have scurvy, for example. The most obvious signs of their malnourished state, however, are often caused by **protein deficiency**. There are two extremes of protein deficiency – **kwashiorkor** and **marasmus**.

Kwashiorkor

In kwashiokor, the child may not have received enough of its mother's milk (often because another child has been born) and may be forced onto a diet that is too high in carbohydrate (often maize). As a result the child may eat enough 'energy' food, but because the diet is poor in protein when its body should be developing quickly, the mental and physical development of the child may be impaired.

Marasmus

In marasmus, the child has symptoms of general starvation – there is not enough 'energy' food nor enough protein. All the body tissues waste away, and the child becomes very thin with a wrinkled skin.

Causes and treatments

Kwashiorkor and marasmus are both serious conditions. They are common in countries where drought conditions have led to poor harvests, or where people have left their homes because of civil war. Aid offered to these people must include both energy foods and the nutrients needed for growth. Powdered milk is often provided, because it is light and easy to transport. However it must be rehydrated with clean water otherwise water-borne diseases such as cholera may result.

Malnutrition in the developed world

As well as the problem of obesity, caused by taking in more energy than is needed for the body's metabolism and activity, there are other problems caused by an unbalanced diet for many people in developed countries. These problems, and the 'unbalanced diet' that causes them, are outlined in the figure.

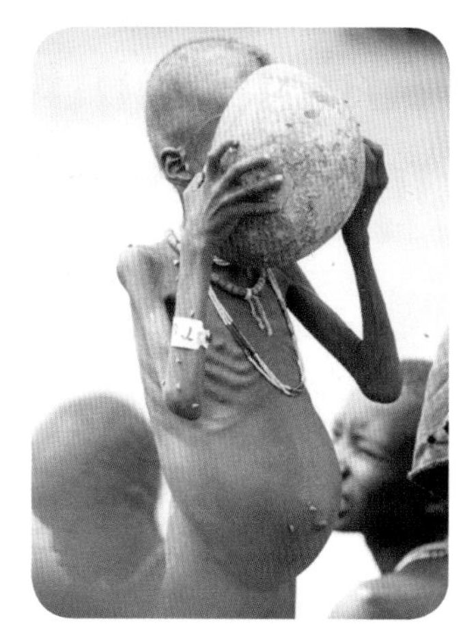

◀ This child has kwashiorkor. The absence of protein means that muscle development is very slow, and the limbs have a stick-like appearance. The swollen abdomen is caused largely by water from the blood plasma remaining behind in the body tissues. The liver is also swollen because it is working hard to make the proteins needed by the body from an inadequate dietary supply.

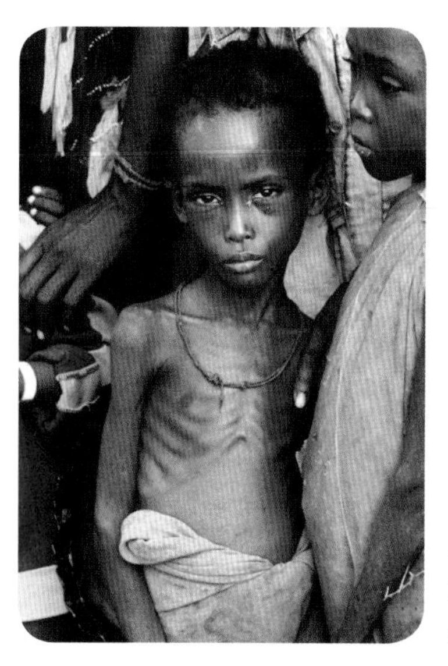

◀ Marasmus is caused by insufficient levels of all nutrients in the diet

▼ Dangers of an unbalanced diet

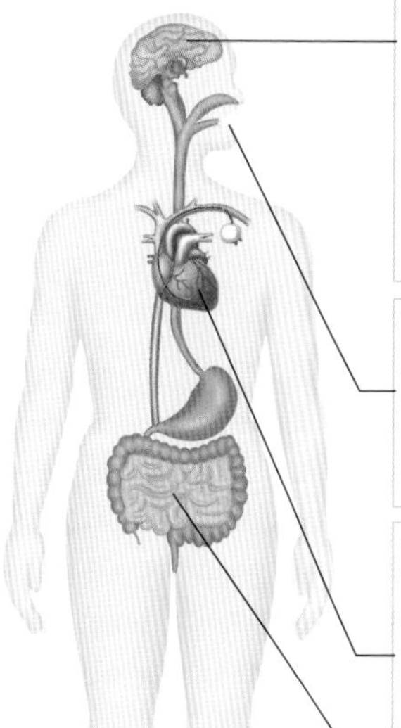

High blood pressure
Too much **salt** in the diet can cause water to be drawn into the blood and blood pressure to rise, eventually damaging delicate blood vessels (page 88) including those in the brain.

Tooth decay
A diet with a high content of **acidic**, **sugary foods** can cause damage to tooth enamel and dentine (see page 65).

Coronary heart disease
Too much **saturated fat** and **cholesterol** can cause blockage to blood vessels supplying oxygen to heart muscle (page 92).

Constipation and bowel cancer
Too little **fibre** means that faeces are not passed as regularly as they should be – the person is **constipated**. Bacteria can work on some of the trapped faeces and release chemicals that cause **colon (large bowel) cancer**.

Don't forget obesity!
Too many kilojoules – the most common problem of nutrition in the developed world (see page 55)!

2.17 Animal nutrition converts food molecules to a usable form

OBJECTIVES

- To know that the process of nutrition involves several stages
- To know the basic layout of the human alimentary canal

Definition: nutrition is the taking in of nutrients which are organic substances and mineral ions, then absorbing and assimilating them.

Nutrition involves a sequence of processes

Living things obtain food molecules from the environment. These molecules are not usually exactly the same as the biological molecules the organism needs to carry out its life processes. The processes of nutrition convert the food molecules into a form that can be used by the organism. In humans and many other animals, these processes take place in the **alimentary canal** (sometimes called the **gut**). The alimentary canal is a specialised tube running from the front of the animal (starting at its **mouth**) to the rear (ending at its **anus**). While the food is inside the tube it is not available to the body tissues. The food molecules must be changed into a form that can cross the gut wall and then be transported to the places where they will be used or stored. The processes of nutrition are described in the diagram below.

The alimentary canal is highly specialised in humans. The food molecules are converted to a usable form in a clear sequence, with each part of the gut being adapted to carry out particular functions. The layout of the alimentary canal is shown in the diagram opposite, and we shall see how its organs work in the next few pages.

1. List, in their correct sequence, the processes that make up nutrition.
2. What is the difference between the following?
   a. chemical and mechanical digestion
   b. absorption and assimilation
   c. egestion and excretion
3. Why are the epiglottis and the soft palate important in efficient feeding?
4. Name three glands that add juices to the alimentary canal.

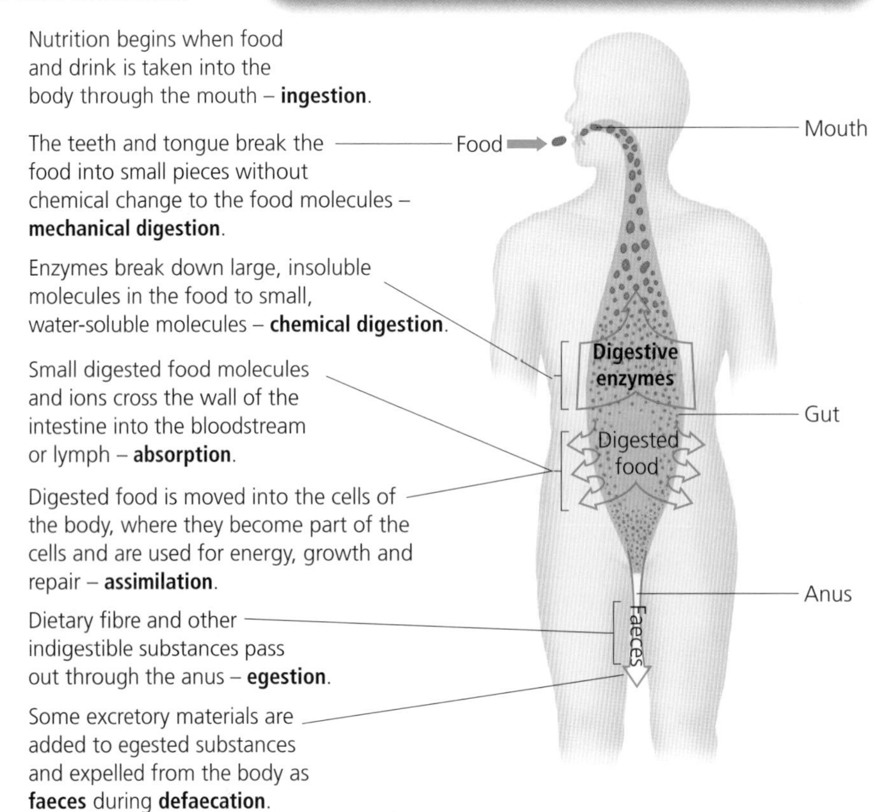

Nutrition begins when food and drink is taken into the body through the mouth – **ingestion**.
The teeth and tongue break the food into small pieces without chemical change to the food molecules – **mechanical digestion**.
Enzymes break down large, insoluble molecules in the food to small, water-soluble molecules – **chemical digestion**.
Small digested food molecules and ions cross the wall of the intestine into the bloodstream or lymph – **absorption**.
Digested food is moved into the cells of the body, where they become part of the cells and are used for energy, growth and repair – **assimilation**.
Dietary fibre and other indigestible substances pass out through the anus – **egestion**.
Some excretory materials are added to egested substances and expelled from the body as **faeces** during **defaecation**.
Food
Mouth
Digestive enzymes
Gut
Digested food
Anus
Faeces

▶ Nutrition is a sequence of processes involving ingestion, digestion, absorption and egestion

INGESTION

DIGESTION

ABSORPTION

EGESTION

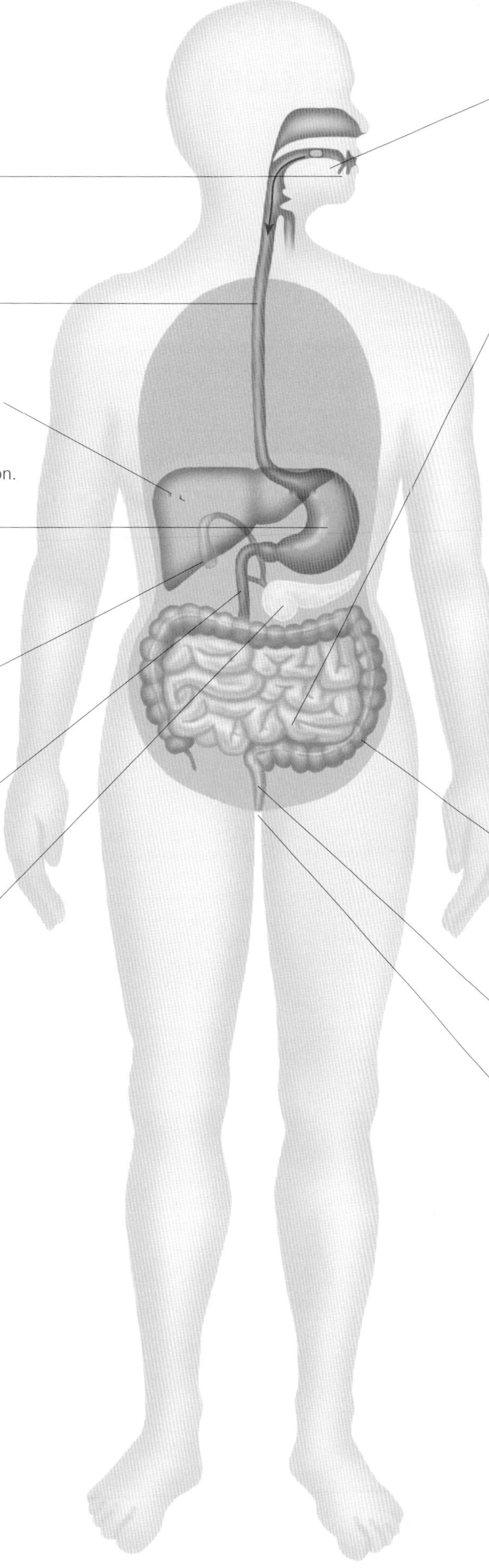

Salivary glands – three pairs, produce saliva and pour it into the mouth through **salivary ducts**.

Oesophagus – muscular tube which helps food move to stomach by peristalsis.

Liver – produces **bile**, which helps to neutralise acidic chyme and also emulsifies fats. Important in assimilation.

Stomach – muscular bag which stores food for a short time, and mixes food with acidic digestive juices to form the creamy liquid called **chyme**.

Gall bladder – stores bile before pouring it into the duodenum through the **bile duct.**

Duodenum – first part of the small intestine, where semi-liquid food is mixed with pancreatic juice and bile.

Pancreas – produces pancreatic juice (contains enzymes, mucus and hydrogencarbonate which neutralises acidic chyme) which is poured into the small intestine through the **pancreatic duct**.

Mouth – here food enters the alimentary canal. It is converted to a **bolus of food** – produced by teeth, tongue and saliva during **mastication** (see page 64) – ready for swallowing.

Ileum – longest part of the small intestine, where digested food is absorbed into the blood and lymphatic system.

Diarrhoea is the loss of watery faeces. This occurs if water is not reabsorbed by the walls of the small and large intestine. Cholera is a bacterial disease that affects water reabsorption. The bacterium releases a toxin that

- increases loss of water and chloride ions into the small intestine.
- prevents reuptake of water and chloride ions.

A person suffering from diarrhoea can be treated by **ORT (oral rehydration therapy)** – drinking a mixture of water and mineral salts.

Large intestine – (wider than the small intestine). The **colon** is part of the large intestine. It reabsorbs water from gut contents; also absorbs some vitamins and minerals.

Rectum – stores faeces before expelling them at a convenient time.

Anus – exit for faeces; defaecation is controlled by two anal sphincters. Constipation (the inability to expel the faeces) can result if there is not enough fiber in the diet. The faeces become too dry and hard to pass easily out of the rectum.

2.18 Ingestion provides food for the gut to work on

OBJECTIVE

- To understand the part played by teeth and the tongue in preparing food for the alimentary canal

Mastication produces a bolus

The first stage in processing food for use by the organism is taking the food into the gut. Once the food has been caught or collected, and perhaps cooked or processed in some other way, it is placed in the mouth. Here it is cut up by the teeth, and the pieces are mixed with saliva by the tongue. This cutting and mixing is called **mastication** (chewing), and produces a ball of food called a **bolus**. The bolus is swallowed and passed on to the next parts of the gut. The teeth play an important role in chewing, and it should be no surprise that:

- the structure of a tooth is closely related to its function
- there are different types of teeth adapted to deal with all types of food.

Human teeth and their function

The structure and function of a human tooth are shown in the diagram below. This type of tooth is called a molar, found towards the back of the jaw. The diagram above right shows the four different types of human teeth, and the part each of them plays in mastication.

The teeth in the skull

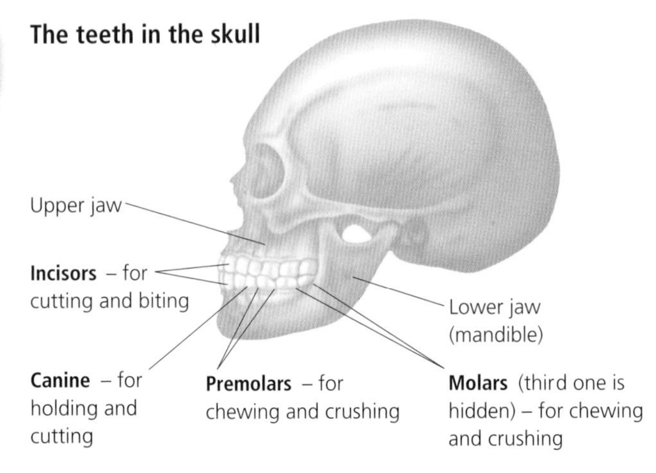

Side view of the four types of tooth

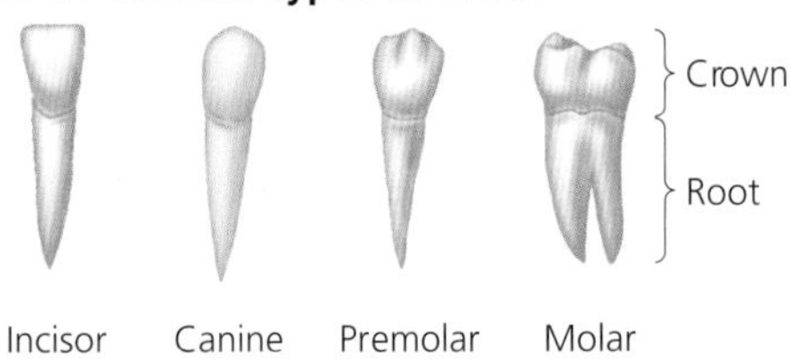

Surface view of the lower jaw

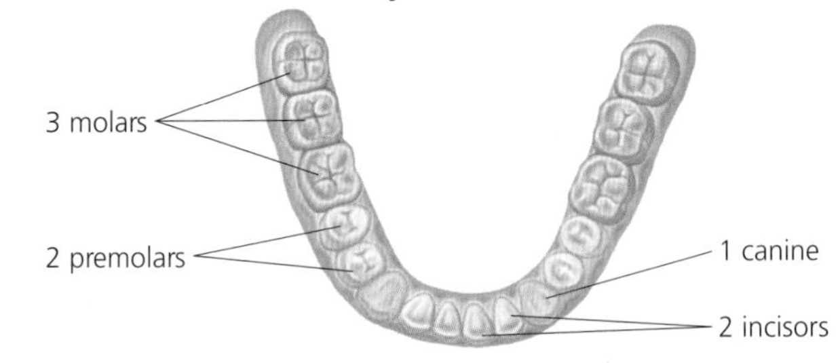

Enamel – the hardest tissue in the body. Produced by **tooth-forming cells** and made of calcium salts. Once formed, enamel cannot be renewed or extended.

Cement – similar in composition to dentine, but without any canals. It helps anchor the tooth to the jaw.

Pulp cavity contains:
- **tooth-producing cells**
- **blood vessels**
- **nerve endings** which can detect pain.

Crown

Neck

Root embedded in jawbone

Dentine – forms the major part of the tooth. Harder than bone and made of calcium salts deposited on a framework of **collagen fibres**. The dentine contains a series of fine canals which extend to the pulp cavity.

Gum – usually covers the junction between enamel and cement. The gums recede with age.

Periodontal membrane – bundles of collagen fibres, anchoring the cement covering of the tooth to the jawbone. The tooth is held firmly but not rigidly. The periodontal membrane has many nerve endings which detect pressures during chewing and biting.

Vitamin C deficiency impairs production of collagen fibres, including those in the periodontal membrane, so that the teeth become loose and may fall out – a classic symptom of scurvy (see page 54).

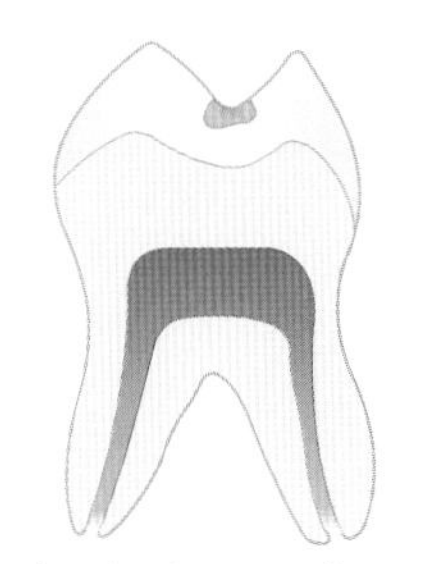
Decay begins in enamel – **no pain**.

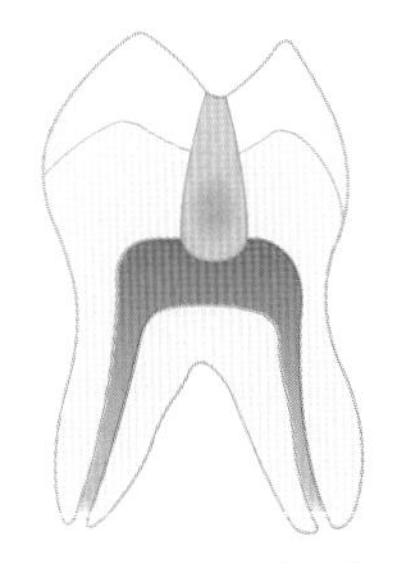
Decay penetrates dentine and reaches pulp – **severe toothache**.

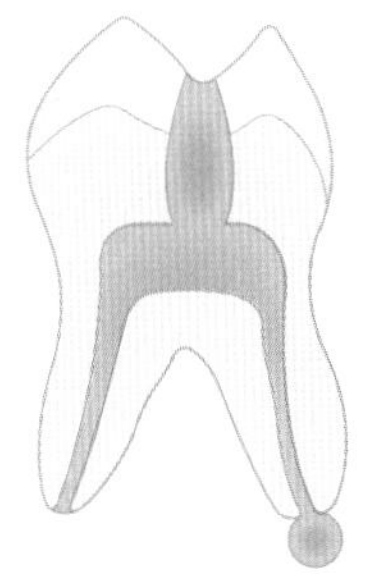
Bacteria now infect pulp and may form abscess at base of tooth – **excruciating pain**.

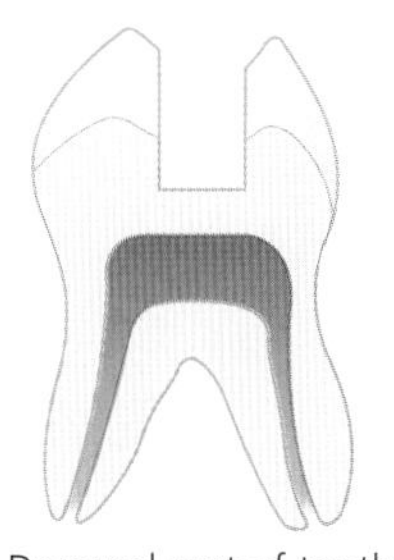
Decayed part of tooth is drilled out by a dentist.

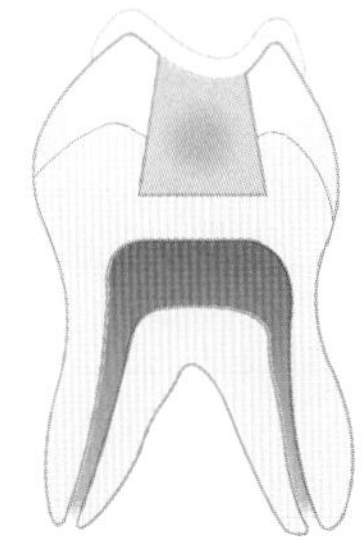
Hole is filled with amalgam or plastic, and may be coated with plastic.

Two sets of teeth

Humans, like other mammals, have two sets of teeth during their life. The first set is the milk dentition, which begins to grow through the gums one or two teeth at a time at around five months old. By about 18 months most children have a set of 20 teeth – there are fewer molars ('cheek teeth') than in an adult since the jaw is too small to hold any more. Between the ages of seven and twelve years these milk teeth fall out and are replaced by larger adult teeth. Eight new cheek teeth are added, with a further four (the wisdom teeth) appearing at the back of the jaw by about the age of 17 years. The individual now has the complete adult or permanent dentition of 32 teeth.

Western-style malnutrition may cause dental decay

A diet that provides too much energy can cause obesity (see page 60). Much of this energy-rich food will be in the form of refined sugars such as sucrose (cane sugar) which is used to sweeten many foods. These sugars may also be used by bacteria in the mouth to carry out their own life processes.

The bacteria produce a sticky matrix which traps food particles and forms a coating of **plaque** on the teeth. The bacteria in the plaque convert sugars in the food to acids. These acids remove calcium and phosphate from the enamel, allowing bacteria to reach the softer dentine beneath. This is the start of **dental decay** or **dental caries**. The dentine decays rapidly and the pulp cavity may become infected. The tooth will need dental treatment, as shown in the diagrams on the left.

Dental decay, and the gum disease that often goes along with it, can be prevented by:

- eating food with a low sugar content
- regular and effective brushing of teeth at least twice a day to prevent the build-up of plaque – plaque begins to reform after about 24 hours
- if brushing is not convenient, finishing a meal with a crisp vegetable or fruit, followed by rinsing with water.

Fluoride*

The mineral fluoride can be added to drinking water. Tests have shown that this mineral reduces the risk of dental decay, although some people say that they should not be *forced* to consume it.

- If tooth decay is reduced, costs of dental care will fall and general health will be improved.

  BUT
- Some people develop grey or brown spots on their teeth when they use fluoride.

Q

1. How do we know that a tooth is a living structure?
2. Imagine yourself on a British warship several hundred years ago. Explain, simply, why the sailors were losing their teeth.
3. **a** What is dental caries?

   b How does it begin?

   c Suggest **two** ways in which caries can be prevented or reduced.

2.19 Digestion prepares useful food molecules for absorption

OBJECTIVES

- To know that digestion converts large insoluble molecules into smaller, soluble molecules
- To understand that enzymes catalyse the breakdown of food
- To list examples of enzymes involved in digestion of carbohydrates, proteins and fats, and to know where they perform their tasks
- To understand that undigested food must be expelled from the gut

The diet contains three types of food molecule in large amounts – carbohydrates, proteins and fats. When ingested, these molecules may be too large to cross the gut wall and too insoluble to be transported in the watery blood plasma. **Digestion** is the breakdown of large, insoluble food molecules into small, water-soluble molecules. Digestion uses **mechanical** (chewing) and **chemical** (enzymatic) processes.

Digestive enzymes

In digestion, food molecules are broken down by **hydrolysis** reactions (breakdown with water), catalysed by a series of enzymes. There are different enzymes for the hydrolysis of each food type; each enzyme works in different regions of the gut. The basic process of hydrolysis is the same for all the food molecules (see page 28):

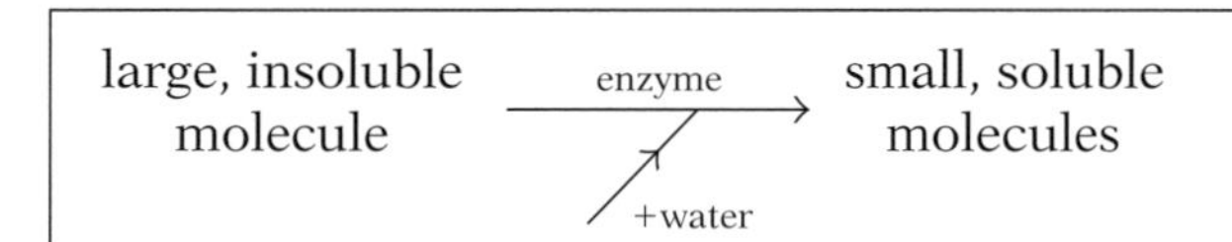

$$\text{large, insoluble molecule} \xrightarrow[\text{+water}]{\text{enzyme}} \text{small, soluble molecules}$$

Carbohydrate digestion begins in the mouth

The saliva contains an enzyme called **salivary amylase** (a carbohydrase). This enzyme catalyses the conversion of the insoluble polysaccharide starch to the soluble simpler sugar called maltose:

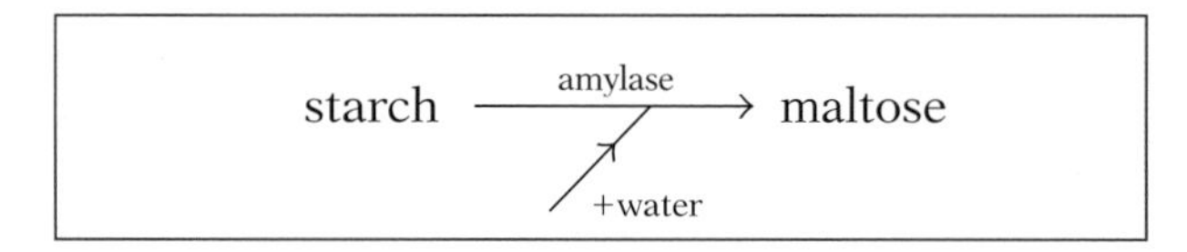

$$\text{starch} \xrightarrow[\text{+water}]{\text{amylase}} \text{maltose}$$

The saliva also contains mucus, which lubricates the food, and hydrogencarbonate, which provides the ideal conditions of pH (alkaline) for amylase to work. The starch is not usually all converted to maltose in the mouth since the food does not remain there for very long.

Passing to the stomach

Mastication produces a bolus of food, as we saw on page 64. The tongue pushes the bolus to the back of the mouth, and it is then swallowed (see opposite page) and enters the **oesophagus**. The oesophagus or gullet is a muscular tube leading from the mouth to the stomach. The bolus is forced down the oesophagus more quickly than can be explained by gravity alone. (It is even possible to swallow food when standing on your head!) Waves of muscular contraction push the bolus down towards the stomach, as shown in the diagram below. These waves of contraction are known as **peristalsis**, and they occur throughout the length of the gut. The reason why fibre is important in the human diet is that without it the gut contents are very liquid, and the muscles of the gut cannot squeeze the food along by peristalsis.

Cross-section of gut showing muscle layers

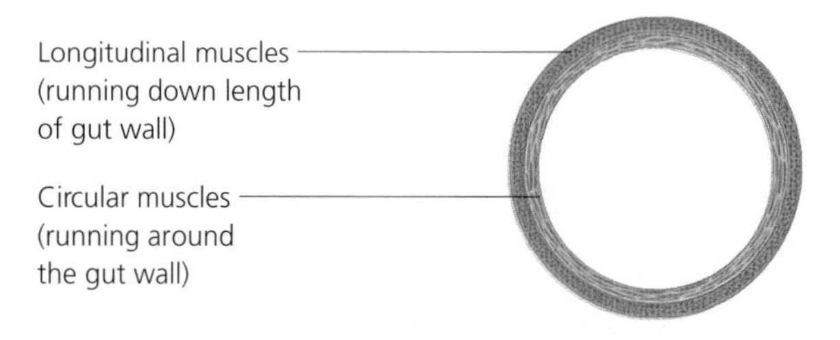

Longitudinal section of gut showing peristalsis

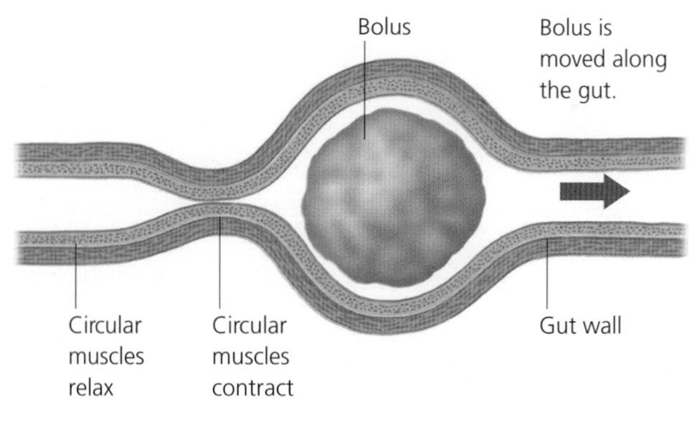

▲ Peristalsis is a wave of muscular contraction that moves food along the gut. Mucus lubricates the bolus, helping it to move along.

Protein digestion begins in the stomach

The stomach is a muscular bag with a lining that contains digestive glands. These glands produce three important secretions:

- **mucus**, which protects the walls of the stomach from attack by gastric (stomach) juices
- pepsin, a **protease** or protein-digesting enzyme
- **hydrochloric acid**, which provides the acidic conditions needed for the action of pepsin and denatures the enzymes in harmful microorganisms ingested with food.

Inside the stomach the food is churned up with the gastric juices. The long protein molecules are broken down by hydrolysis to smaller molecules called amino acids, as shown in the diagram below.

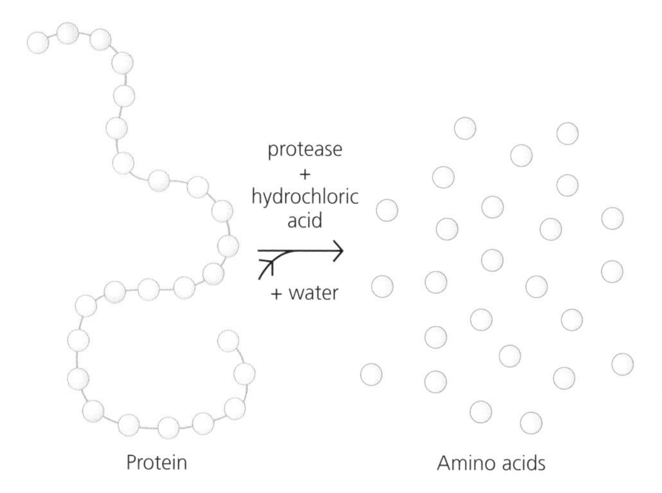

The pH in the stomach is too low (too acidic) for the action of amylase, so the digestion of carbohydrate comes to a halt whilst the food is in the stomach. The churning action of the stomach muscles mixes the food into a creamy liquid called **chyme**. Once the food is sufficiently liquid, it squeezes past a ring of muscle at the foot of the stomach, the **pyloric sphincter**, and enters the **duodenum**, the first part of the small intestine.

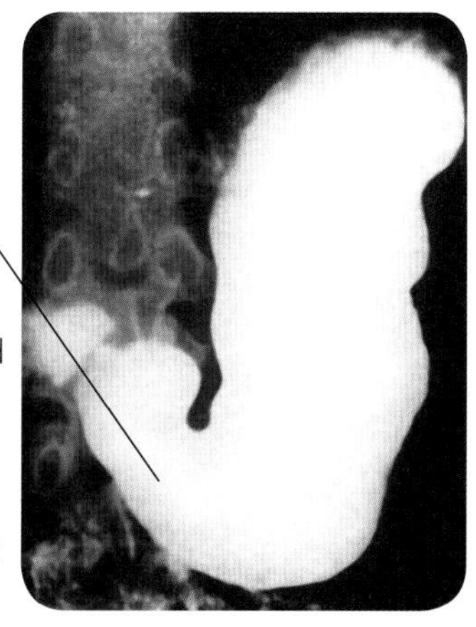

▶ This X-ray was taken after the patient swallowed a barium meal. The barium makes the stomach show up on the X-ray. Food is passing through the pyloric sphincter.

The swallowing reflex

It is important that the bolus of food travels down the oesophagus and not down the trachea (windpipe), or it might block the trachea and prevent breathing. A flap of muscle, the **epiglottis**, is forced across the top of the trachea whenever food is swallowed, ensuring food does not enter the trachea. We don't have to think about this swallowing reflex, but it can be overruled if we try to eat and talk at the same time!

At the bottom of the oesophagus, just where it joins the stomach, is a ring of muscle called the cardiac sphincter. When the bolus reaches this sphincter, the ring of muscle relaxes to allow the bolus through into the stomach. If the stomach contents pass upwards through this sphincter and make contact with the wall of the oesophagus, they cause a burning sensation known as 'heartburn'. This can be treated using an alkaline ('antacid') solution

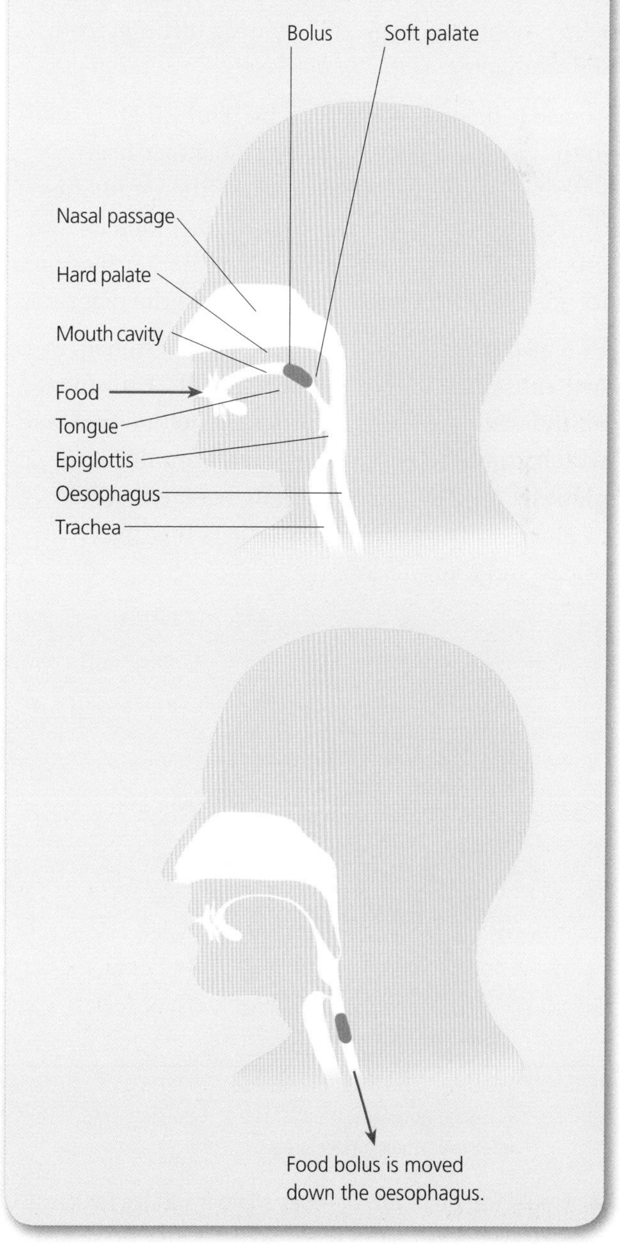

Digestion of food molecules is completed in the small intestine

The liquid chyme contains partly digested food molecules. The digestion of these food molecules is completed in the small intestine, using digestive juices which contain:

- enzymes from the pancreas
- enzymes from the intestine wall
- bile from the liver.

The enzymes from the pancreas carry out three tasks – **amylase** completes the conversion of starch to maltose, trypsin (another **protease**) completes the breakdown of proteins to amino acids, and **lipase** converts fats to fatty acids and glycerol. These enzymes work best at around pH 8 (slightly alkaline conditions), and the juice from the pancreas also contains hydrogencarbonate which neutralises the acid coming through from the stomach.

Lipase is helped in its action by **bile**, which is made in the liver and stored in the gall bladder from where it is released when needed. Bile **emulsifies** the fats – it converts them from large globules into much smaller droplets, giving a greater surface area for the lipase to work on, as shown in the diagram.

Bile also contains hydrogencarbonate to help neutralise the acid from the stomach. Enzymes on the wall of the small intestine include **maltase,** which completes the breakdown of maltose to glucose. All the starch has now been converted to glucose. The table opposite lists the digestive juices and their actions.

Water and digestion

The digestive juices are largely made up of **water** – this is one of the major requirements for water within the body. This water is the solvent for the biochemical reactions of digestion and is also used in the hydrolysis reactions that split up the large, insoluble food molecules. The juices also contain **mucus** which protects the wall of the gut from being digested by its own enzymes.

Egestion removes undigested food

Food may contain some molecules that cannot be digested by the enzymes of the human gut. Examples include substances in plant cells, such as the cellulose in cell walls and the lignin in wood. These make up '**dietary fibre**' (page 54). Water is first absorbed from the gut contents that remain after digestion, and this indigestible food is then expelled. This process is called **egestion**. Some excreted materials, such as salts in the bile, may be added to the indigestible foods to form the **faeces**. The faeces are stored temporarily in the rectum. When full, the rectum sets off a reflex action which causes its muscles to contract and squeeze the faeces out through the anus. Humans have a sphincter at the anus which can prevent this **defaecation** happening at an inconvenient time! The control of this sphincter has to be learned, and babies simply fill their nappies when the rectum is full.

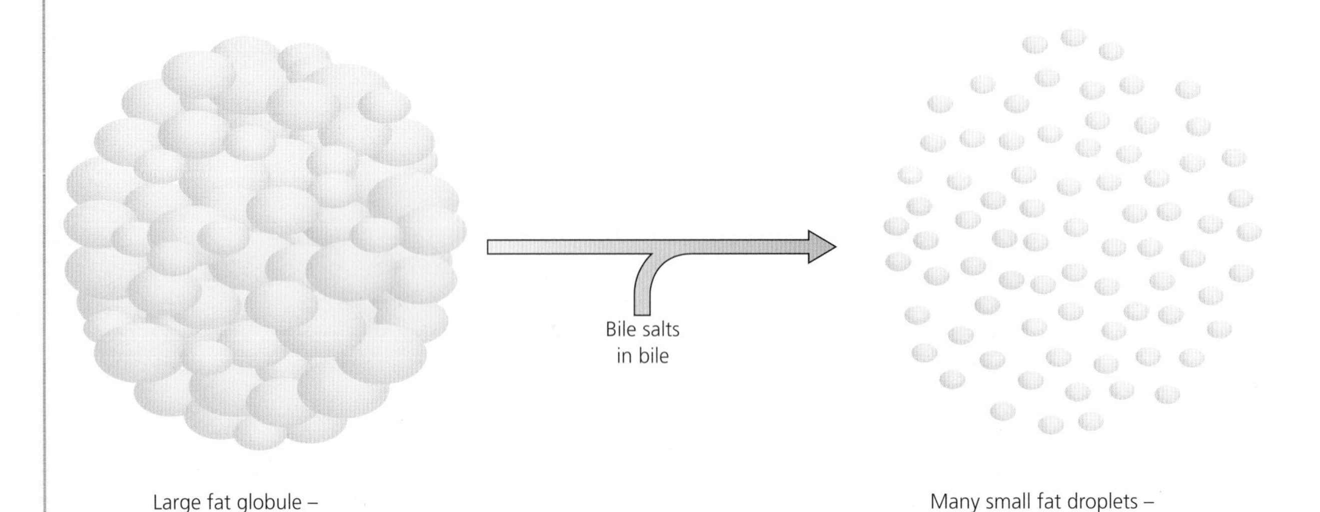

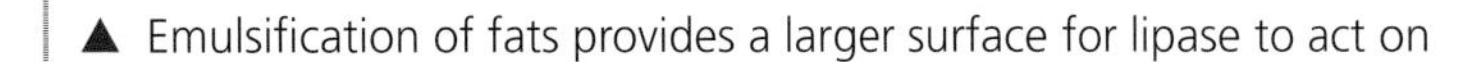
▲ Emulsification of fats provides a larger surface for lipase to act on

Region of gut	Digestive juice	Enzymes	Substrate	Product(s)	Other substances in juice	Function of other substances
Mouth	Saliva from salivary glands	Salivary amylase	Starch	Maltose	Hydrogencarbonate	Alkaline (pH 7.5) environment for amylase
Stomach	Gastric juice from glands in wall of stomach	Pepsin (protease)	Proteins	Amino acids	Hydrochloric acid	Acidic (pH 2) environment for pepsin; kills bacteria
Small intestine (duodenum)	Pancreatic juice from pancreas	Pancreatic amylase Trypsin (a protease) Lipase	Starch Protein Emulsified fats	Maltose Amino acids Fatty acids and glycerol	Hydrogencarbonate	Neutralises chyme: alkaline environment for enzymes
	Bile from liver (stored in gall bladder)	None			Bile salts Hydrogencarbonate	Emulsifies fats – converts globules to smaller droplets Neutralises chyme
Small intestine (ileum)	Intestinal juice from cells on villi	Maltase on the surface membrane of the epithelium lining the small intestine	Maltose	Glucose		

▲ The human digestive juices and their actions: note how starch and proteins are broken down in several stages catalysed by different enzymes

Q

1. In 1822 Alexis St Martin, a Canadian fur trapper, was wounded in his left side by a shotgun blast. Luckily the accident occurred close to an army fort where one of the surgeons, William Beaumont, was able to treat St Martin. The wound healed very slowly, and left a small hole in the side of the young man. Beaumont realised that this gave him a unique opportunity to study what was happening in his patient's stomach.

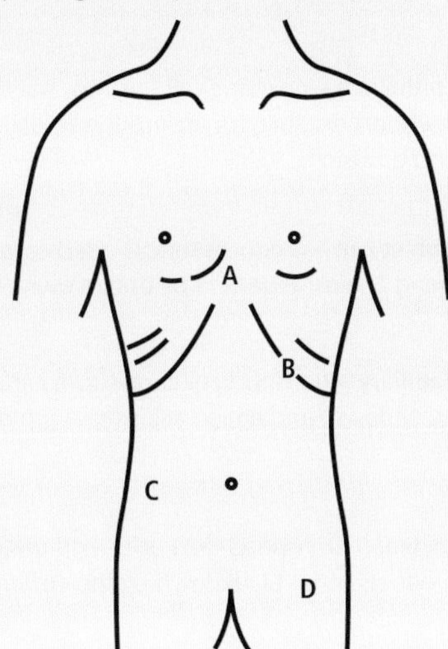

   a Which letter on the outline above represents the most likely position of the hole?

   b Beaumont described one of his investigations as follows: 'Juice was extracted from the stomach and placed in a small vial. A solid piece of boiled, recently salted beef weighing three drachms was added. The vial was corked and kept under controlled conditions. A similar piece of beef was suspended on a string into the man's stomach.'

   i Suggest one condition around the vial that Beaumont would have kept constant.

   ii What control experiment should Beaumont have performed?

   iii Why did Beaumont use 'boiled, salted beef'?

   c After two hours, Beaumont recorded the following results: 'Beef in vial – the cellular texture seemed to be entirely destroyed, leaving the muscular fibres loose and unconnected, floating about in fine, small shreds, very tender and soft. Beef in stomach – I drew out the string, but the meat was all completely digested and gone.'
   Use your knowledge of digestion to explain the difference between the changes in the vial and those in the stomach.

2. The liver produces a liquid which is added directly to the partly digested food in the small intestine.

   a Name the liquid.

   b Describe how it helps digestion.

3. Describe, in the correct sequence, how the protein and starch in a ham sandwich are broken down ready for absorption.

4. Digestive juices contain enzymes, water and some other substances. Name two of these other substances. State which digestive juice contains them, and state what function they perform.

2.20 Absorption and assimilation make food available

OBJECTIVES

- To understand that digested food in the gut is still 'outside' the body
- To know how the small intestine is adapted to the function of absorption of digested food
- To understand the part played by the liver in the assimilation and distribution of absorbed foods

Absorption of the products of digestion

Enzymes in the gut convert large, insoluble molecules to small, soluble molecules. These digested food molecules are transported across the lining of the gut into the blood or lymph, a process called **absorption**. They can then be distributed to the parts of the body where they will be used.

Most absorption happens from the **ileum**, the lower part of the small intestine. The ileum is very well adapted to perform this task:

- It is **very long**, about 6 m in an adult human, so food takes a long time to pass through it, and there is enough time for absorption to occur.
- The surface of the ileum is **highly folded**, which gives a much larger surface area for absorption than a simple tube would.

The lining of the ileum is folded into hundreds of thousands of tiny finger-like structures, the **villi**, which project out into the liquid digested food. The structure of a villus, and its adaptations to increase absorption, are shown in the diagram below.

Following absorption in the small intestine, the contents of the gut are little more than water and indigestible matter. Most of the remaining water is reabsorbed into the bloodstream from the **colon**, part of the large intestine. Some minerals and vitamins are also absorbed here.

The liver and assimilation

Digested food is absorbed into the bloodstream. Each type of absorbed food has a particular function in the body, so it is important that food molecules of the right type are available at the right time in the right place. The **liver** 'sorts out' digested food molecules, and all foods absorbed into the capillaries of the villi are sent first to the liver.

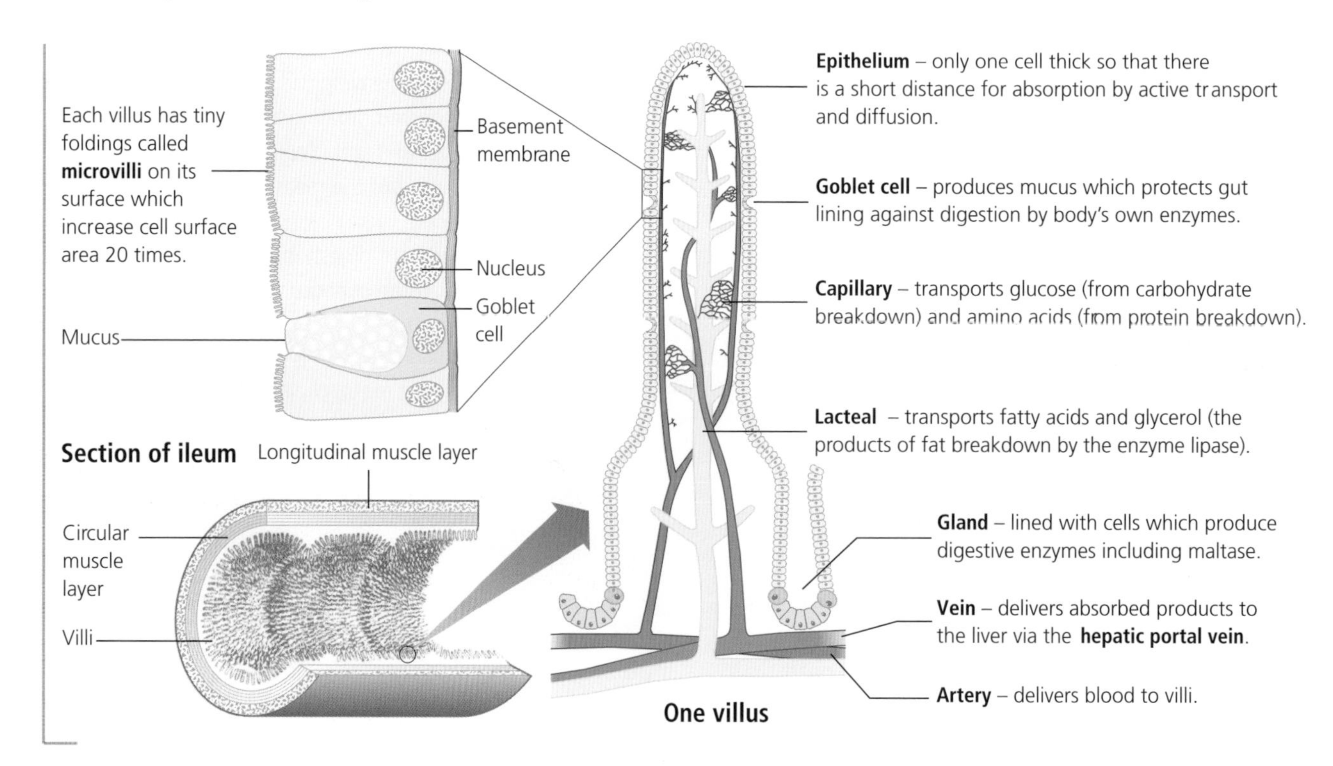

The liver has many functions – at least 500 different biochemical reactions go on inside its cells. These functions include:

- manufacture of bile, which is important for the digestion of fats
- storage of glucose as glycogen
- interconversion of glucose and glycogen, which keeps the glucose concentration constant for the working tissues of the body (see page 132)
- interconversion of amino acids – the liver can convert some amino acids into others that the body might require in a process called **transamination**
  The amino acids can be built up to proteins, including plasma proteins such as **fibrinogen**
- excretion of excess amino acids – the amino part of the amino acid is removed in a process called **deamination** (see page 124) and excreted in the urine as urea

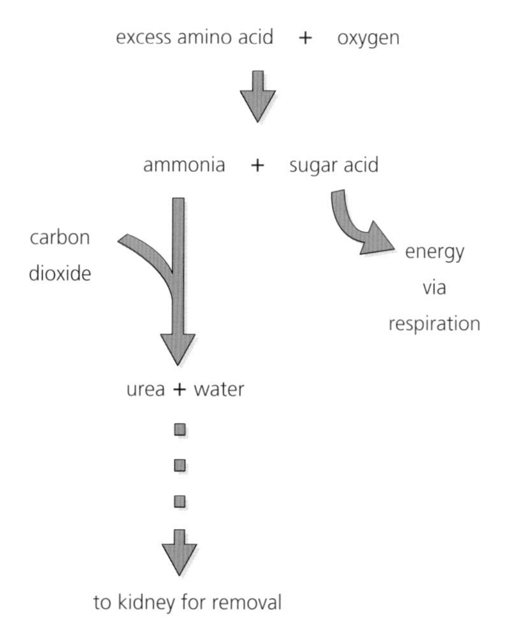

- removal of old red blood cells from the circulation and storage of the iron they contained
- breakdown of alcohol and other toxins, called **detoxification**.

A combination of transamination and deamination makes sure that there is always a 'pool' of amino acids available for use by the cells of the body.

As a result of these and other activities, the liver provides ideal concentrations of food molecules for the working of the body tissues. Each type of tissue uses food molecules for different purposes – for example, muscle cells manufacture muscle protein, bone cells take up calcium and phosphate to make bone, and all cells use glucose to release energy by respiration. The processes of moving food molecules into the cells where they are used is called **assimilation**.

Alcohol and the liver

The liver receives all the molecules that the gut absorbs from food. As well as useful molecules such as glucose and amino acids, this may include harmful molecules such as drugs or poisons. **Alcohol** is a drug (a molecule that affects the normal working of the body – see page 152) which passes quickly to the liver after being absorbed from the stomach and ileum. The cells of the liver convert this alcohol to another substance which does not pass through to the rest of the body's circulation. Unfortunately, in working to protect the other tissues, the liver is likely to harm itself. The substance produced from alcohol can be dangerous to liver cells in high concentrations and can cause a serious disease called cirrhosis of the liver. If the liver is damaged by excessive alcohol consumption, then the whole body is affected. For example, the blood glucose concentration cannot be controlled efficiently if the liver has been damaged.

Alcohol has other effects on the body – these are described on page 154.

Q

1. In what way is the blood supply to the liver unusual?
2. Name two substances stored in the liver, and two that are converted to different substances.
3. How is the structure of the villus adapted to its function?
4. What are goblet cells? Where else do you think they might be found apart from in the gut?
5. Some microorganisms infect the gut lining so that water cannot be absorbed. What effect(s) might this have on the infected person?

Questions on animal nutrition and health

1 Saliva may contain an enzyme, amylase, that acts on the substrate, starch.

a Describe how you would carry out a test to show the presence of

i the substrate,

ii the product of the action of amylase on starch.

A student carried out an investigation to find the effect of pH on the activity of amylase on the breakdown of starch.

The student carried out a second experiment but added 1 cm^3 of dilute sodium chloride (salt) solution to the amylase and starch.

The results are shown in the table.

pH	time taken for breakdown of starch / seconds	
	first experiment – no salt added	second experiment – salt added
3.5	600	450
4.0	65	50
4.5	90	60
5.0	150	100
6.0	300	200

b i Using the information in the table, construct a graph of the results using one set of axes.

ii Describe the effects of pH and of adding salt on the activity of amylase on the breakdown of starch.

c Describe how you would ensure that your results were as reliable and valid as possible if you were to carry out the above investigation.

Cambridge O Level Biology 5090 Paper 6 Q1 May 2009

2 The diagrams below show plans of the teeth in the upper and lower jaw. Plan A is from a 27-year-old woman called Hannah; plan B is from a 27-year-old woman called Caitlin.

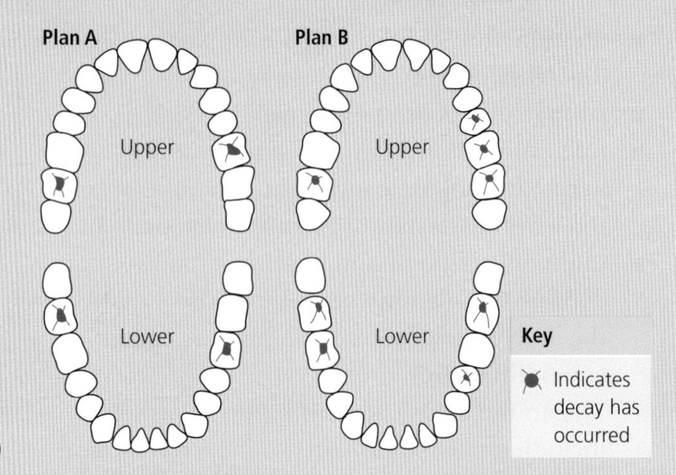

The two women live in different towns. In one town the natural drinking water contains fluoride salts, but there are no fluoride salts in the drinking water of the other.

a Which woman lives in the town with fluoride in the drinking water? Support your answer with numerical data.

b Suggest other reasons for the difference in the number of decayed teeth between the two women.

c Why is it important that both plans are from women, and each is aged 27?

d Explain why decay is more likely in the cheek teeth.

e How would these plans have been different if they had been taken when the women were 10 years old?

3 The table below shows the effect of pH on the time taken for the complete breakdown of a starch solution in the presence of an enzyme.

pH	5.0	5.5	6.0	6.5	7.0	7.5	8.0	9.0
time taken / min	20.00	15.00	8.00	4.00	1.25	1.25	3.00	8.00
rate of reaction								

a Copy and complete the table. Assume that the rate of reaction is the same as $^{1}/_{\text{time taken}}$.

b Plot the results in the form of an appropriate graph.

c What is the optimum pH for this enzyme?

d Name one region of the gut where this pH would be found. How is this pH kept constant?

e Suggest two chemical tests that could be carried out on samples of the solution to show that starch is being broken down.

f In this experiment pH is the input or independent variable, and rate of reaction is the outcome or dependent variable. Suggest three fixed variables which must be kept constant if these results are to be valid.

4 a The table below lists some of the food materials that need to be digested, the enzymes that carry out the digestion and the end products. Complete the table.

food material	digestive enzyme	end products of digestion
starch		simple sugars
		amino acids
fat	lipase	

b Amino acids and glucose are carried in the blood from the intestine to the liver. Describe the processes that occur in the liver when there is an excess of these materials arriving in the blood.

Cambridge IGCSE Biology 0610 Paper 2 Q3 May 2008

5 During the 16th and 17th centuries, sailors were as likely to die from the disease **scurvy** as from enemy action. A chance observation that sailors who were transporting citrus fruits did not develop scurvy led Admiral Nelson to insist that all British ships should carry plenty of limes on long voyages. (This is why British sailors were called 'Limeys'.)

Clearly limes and other citrus fruits contained a substance that prevented scurvy, but it was not until the middle of the 20th century that this substance was shown to be **ascorbic acid** (also called **vitamin C**). At that time an American surgeon called John Crandon carried out an experiment on himself:

1 He cut himself with a sterile scalpel – the cut healed in 3 or 4 days.
2 He had no vitamin C in his diet for 3 months – a similar cut took 3 months to heal.
3 He had no vitamin C for 6 months – a cut would not heal at all.

Do you think that Crandon's results were valid? How could this experiment have been improved? We now know that long-term deficiency in vitamin C makes it difficult for the body to manufacture the connective tissues that hold body structures together – thus skin damage is difficult to repair. The recommended daily amount (RDA) of vitamin C is the amount required to prevent scurvy, but many scientists believe that vitamin C has other functions in the body and that this RDA is probably too low. Higher doses of vitamin C help the immune system to fight off viruses, prevent cells ageing, and prevent anaemia by increasing the absorption of iron from food.

6 The table below shows the energy value of a number of products from the Mars company.

a Copy and complete the table to show the energy value of each product in kJ per 100 g (take care – some energy values are shown for 25 g; others are for whole bars or packets).

b Draw a bar chart showing the products in descending order of energy value (the one with the highest energy value on the left).

c For a 50 g portion which product would be:

i the least fattening

ii the most fattening?

d A Mars bar weighs 65 g. What is its energy value in kilojoules?

e While sitting watching television, a boy of 15 uses about 6 kJ per minute.

i How long would it take him to use all the energy obtained from the Mars bar?

ii What is he using this energy for?

f While playing football the same boy uses, on average, 30 kJ per minute. How long would it take him to use up the energy obtained from the Mars bar?

g A football match lasts for 90 minutes. If the boy ran for half of the game, would he have used more energy or less than the Mars bar provided? How could his body cope with any difference between energy demanded and energy supplied?

name of product	energy value / kJ per 25 g, unless otherwise stated	energy value / kJ per 100 g
Bounty, 60 g	490	
Maltesers	504	
Mars	454	
Milky Way	490	
Minstrels, 49 g	870	
Snickers	504	
Treets, 42 g	1040	
Twix	504	

2.21 Uptake of water and minerals by roots

OBJECTIVES

- To understand that water may enter and leave cells by osmosis
- To understand that dissolved substances may enter and leave cells by diffusion and active transport

Plants need water and minerals

Plants need to obtain certain raw materials from their environment. The roots of the plant are adapted to absorb both minerals and water from the soil. **Water** is essential to support the plant, as a reagent in many biochemical reactions and also as a transport medium (see page 236). The diagram below shows how water enters the plant through **root hair cells**.

Minerals have a number of individual functions and together have a great effect on the water potential of the plant tissues. Minerals from the soil are absorbed in the form of **ions**, for example, magnesium enters the root as Mg^{2+} ions and nitrogen enters as nitrate NO_3^- ions. If the soil solution contains higher concentrations of these ions than the root hair cell cytoplasm, the ions can enter by diffusion (see page 22). However, plants can continue to take up ions even if the concentration gradient is in the wrong direction, that is, if the concentration of the ions is higher inside the cell than in the soil solution.

Leaves have a large surface area for photosynthesis. When the stomata are open, water is lost by evaporation from spongy mesophyll cells (see page 80).

Roots have an enormous surface area and penetrate between the particles of soil.

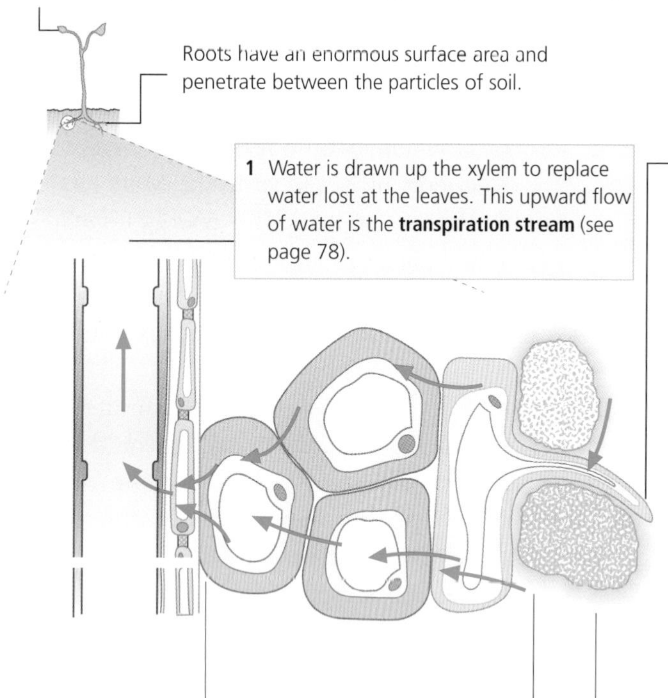

1 Water is drawn up the xylem to replace water lost at the leaves. This upward flow of water is the **transpiration stream** (see page 78).

Root hair cells have an enormous surface area. This helps them absorb water and minerals from the soil. The photograph shows root hair cells magnified × 300. The root hairs are very delicate and easily damaged. When plants are transplanted they recover much more quickly if the roots are kept in a ball of soil or compost so that the root hairs are not disturbed.

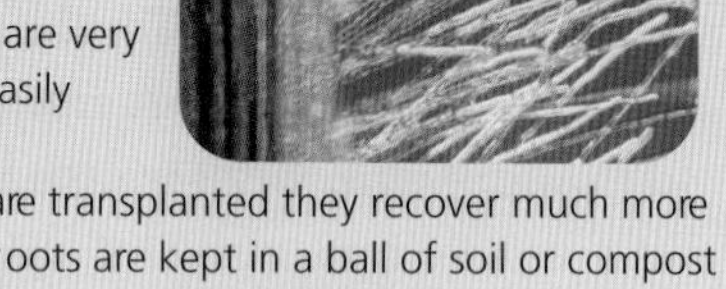

Note: Water movement through the plant occurs in the sequence 1–2–3–4: it **begins** with loss from the leaves, and is **completed** with water absorption from the soil solution.

2 Water (with any substances dissolved in it) is forced to cross the membrane and cytoplasm of cells. The cell walls contain a waxy material which makes them impermeable to water. This allows the membranes of the cells to **select** which substances can enter the xylem and be distributed through the body of the plant.

3 Water crosses the living cells of the **cortex** by (a) osmosis through the cells and (b) 'suction' through the freely permeable cellulose cell walls. Almost all of the water moves across the cortex by route (b).

4 Water enters root hair cells by **osmosis**, from the thin film of water surrounding the particles of soil. The soil water has a **higher water potential** than the cytoplasm of the root hair cell, so that water moves down a water potential gradient (see page 24).

▲ Uptake of water by root hair cells

Osmosis: a reminder

A cell's membrane controls the entry and exit of materials to and from the cell (see page 22). A typical plant cell such as that found in the mesophyll layer of the leaf has a high concentration of solutes. As a result water will enter a plant cell by osmosis from an environment with a high water potential, until the water inside the cell forces the cell membrane up against the cellulose cell wall.

When a plant cell contains plenty of water, the internal pressure of the cell contents against the cell wall supports the cell. The cell is said to be **turgid**, and turgidity helps support the plant. If the plant does not have a good supply of water, the cells lose their turgidity and slowly collapse. The cells are said to be **flaccid** and the plant is wilted.

Experiments on the uptake of ions also show that:

- the cells can select which ions enter from the soil solution
- any factor that affects respiration, for example lack of oxygen or low temperature, can reduce the uptake of ions. The diagram below shows some results that support these observations.

The explanation of these observations is that the root hair cells use **active transport** to carry out the selective uptake of ions against a concentration gradient, using energy from respiration (see page 26).

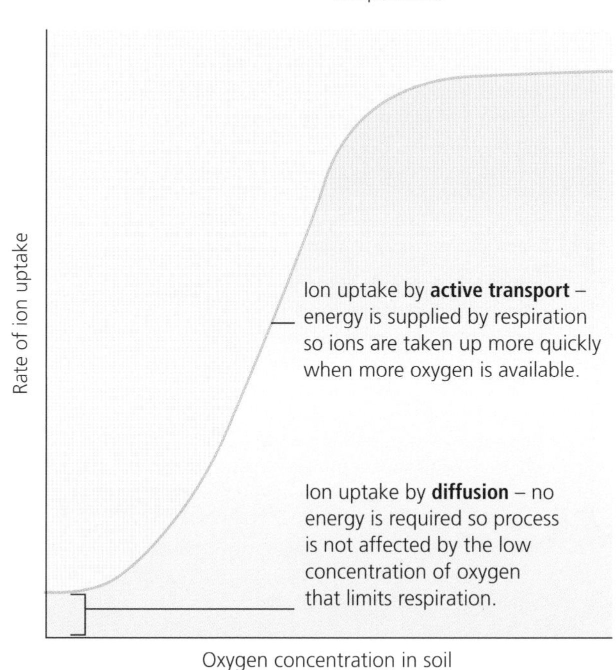

'Plateau' suggests that active transport of ions is limited by some other factor that limits respiration, such as temperature.
Rate of ion uptake
Ion uptake by **active transport** – energy is supplied by respiration so ions are taken up more quickly when more oxygen is available.
Ion uptake by **diffusion** – no energy is required so process is not affected by the low concentration of oxygen that limits respiration.
Oxygen concentration in soil

◀ Ion uptake depends on respiration

Active transport: application

To increase crop yields, farmers may drain fields that are liable to flooding. If the soil is not waterlogged, more oxygen in soil air spaces is available to the plants, so the rate of aerobic respiration in root cells is faster. This provides more energy for active transport, so that the growing plants will more quickly absorb mineral ions present in the soil. Farmers may also cover their fields with black polythene. This absorbs heat and helps to raise the soil temperature, so that seed germination and ion uptake by young roots will be faster.

1. Does a solution containing many molecules of dissolved sugar and amino acids have a high or a low water potential? Explain your answer.
2. Define osmosis in terms of water potential.
3. How does the strength of the cellulose cell wall help plants to support themselves?
4. A scientist investigated the uptake of magnesium ions by the roots of young cereal plants. He made the following observations:
   - **a** The rate of uptake was increased by raising the temperature, so long as it did not exceed 40 °C.
   - **b** Uptake stopped if the roots were treated with cyanide, an ion that prevents respiration.
   - **c** Ions were taken up even if they were present at a lower concentration in the solution around the roots than in the root cells themselves.
   - **d** If ion uptake continued for some time, the concentration of sugars in the root cells decreased.

   What conclusions can be drawn from each of these observations?

2.22 Transport systems in plants

OBJECTIVE

- To appreciate that water and dissolved substances are transported around the plant in specialised transport tissues

Xylem and phloem

Substances need to be transported for long distances throughout a plant's body – sugars, for example, are produced in the photosynthesising cells of the leaves and may need to be transported to storage cells in the roots. The water and ions absorbed by the roots may be required by cells at the growing tip (the **meristem**) of the shoot. These long-distance transport functions are carried out by two specialised plant tissues – the **xylem** and the **phloem**. These are tubes running through the plant, collected together in groups in the **vascular** (transport) **bundles**, as shown in the diagram opposite.

TRACING XYLEM

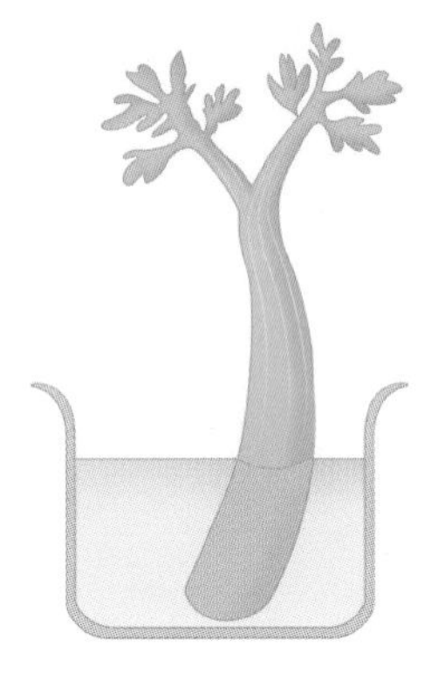

Procedure

1 Cut a piece of celery and stand it in a coloured solution (suitable stains include eosin [red] and methylene blue [blue!]).

2 Leave for a few hours.

3 Carefully (CARE! SHARP SCALPELS) cut off about 5 cm of the celery, and use a hand lens to look for the stain. The coloured solution has been carried up the **xylem**.

Extension: You could also try:

a Carefully (using the scalpel) scrape away the outside tissues of the celery to trace the xylem. What happens as the stem branches to the leaves?

b Use a pale-coloured (e.g. white) flower, such as a carnation. You can change its colour, just as florists do!

Moving vital substances from sources to sinks

The transport tissues are arranged in the stem and root as shown opposite, to offer:

- the most efficient transport of materials from **sources** (where they are taken in or made) to **sinks** (where they are used or stored)
- the most effective support in air (the stem) and soil (the root).

The transport functions of xylem and phloem have been investigated in a number of ways, as shown in the diagram below.

Transport of the products of photosynthesis

Aphids (greenfly) are serious pests of many crops. They can take food meant for the growing regions of plants by inserting their mouthparts (the **stylet**) into the plant tissues.

If feeding aphids are anaesthetised with carbon dioxide their bodies can be 'flicked away' from the plant surface, leaving the stylet in place. The contents of the phloem, the **sap**, will slowly leak out of the stylet and can be analysed. The results show that the phloem transports sucrose (sugar), the main product of photosynthesis.

Application: aphids eat themselves to death!

Many insecticides kill useful insect species as well as pests. **Systemic** insecticides are sprayed onto the plant and absorbed into the phloem tissues, so they only kill aphids.

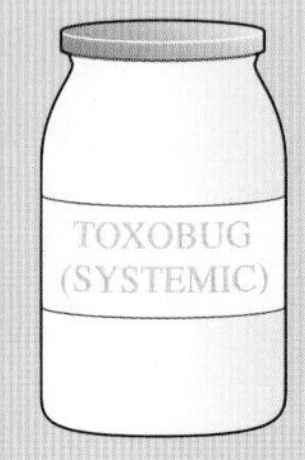

Direction of transport varies with the seasons!
Sucrose is transported **from** stores in the root **to** leaves in spring, but **to** stores in the root **from** photosynthesising leaves in the summer and early autumn. Whatever the time of year the movement of sugars and amino acids (translocation) is from **source** to **sink**. In other words, sucrose and amino acids are translocated from the region where they are made or absorbed to the region where they are stored or used.

Stem – vascular bundles are arranged in a ring with soft cortex in the centre, helping to support the stem.

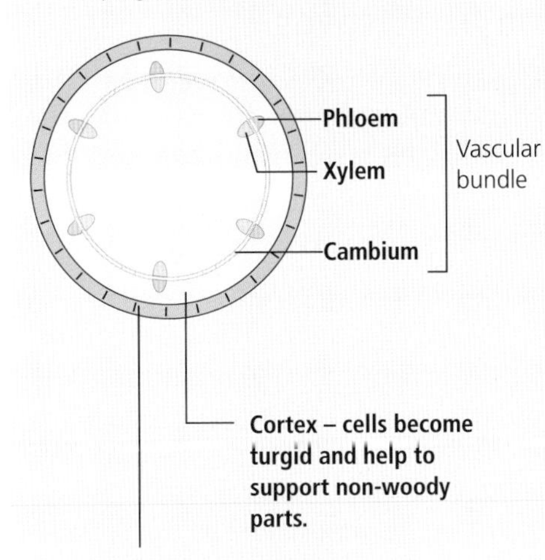

Cortex – cells become turgid and help to support non-woody parts.

Epidermis – protects against infection by viruses and bacteria, and dehydration.

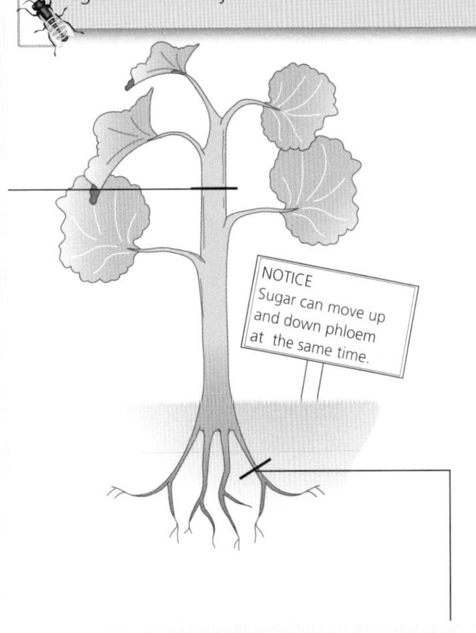

NOTICE
Sugar can move up and down phloem at the same time.

Root – root hairs are extended cells of the epidermis.

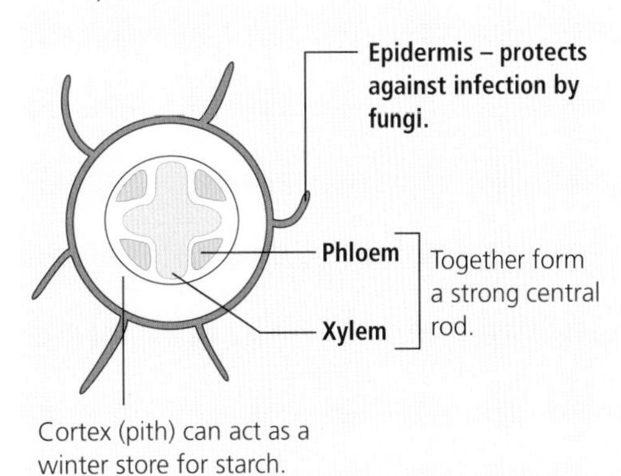

Cortex (pith) can act as a winter store for starch.

▲ The transport tissues xylem and phloem are arranged in vascular bundles. They transport water and dissolved substances around the plant.

Q

1. Name the two vascular tissues in flowering plants. Which tissue divides to form the vascular tissues?
2. What is a **source**? Suggest two examples in a flowering plant.
3. Why does the direction of sugar transport vary from season to season?
4. Why must sugar be transported to **sinks** such as growing points and roots?
5. Many dyes are water soluble. Xylem vessels reach up from roots to flower petals. How could these two observations be useful to a florist?

2.23 Transpiration: water movement through the plant

OBJECTIVES

- To recall that water movement through a plant begins with water loss from the leaves
- To understand that water is lost from leaves via the stomata, through which the exchange of gases between the leaf and the atmosphere also occurs
- To describe how the leaf surface most involved in water loss can be identified
- To understand how environmental conditions can affect water movement through plants

Evaporation from leaves

Water evaporates from the parts of a plant that are exposed to the atmosphere – for example, the whole shoot system of a terrestrial plant and the upper leaf surfaces of a floating aquatic plant. The greatest loss of water takes place through the **stomata** (singular **stoma**), minute pores on the leaf surface (see page 42). There are usually more stomata on the lower surface of leaves than on the upper surface.

Water movement through a plant begins with the diffusion of water vapour out of the leaf and evaporation from the leaf surface (spongy mesophyll). 98% of the water taken up by a plant is lost to the atmosphere by transpiration.

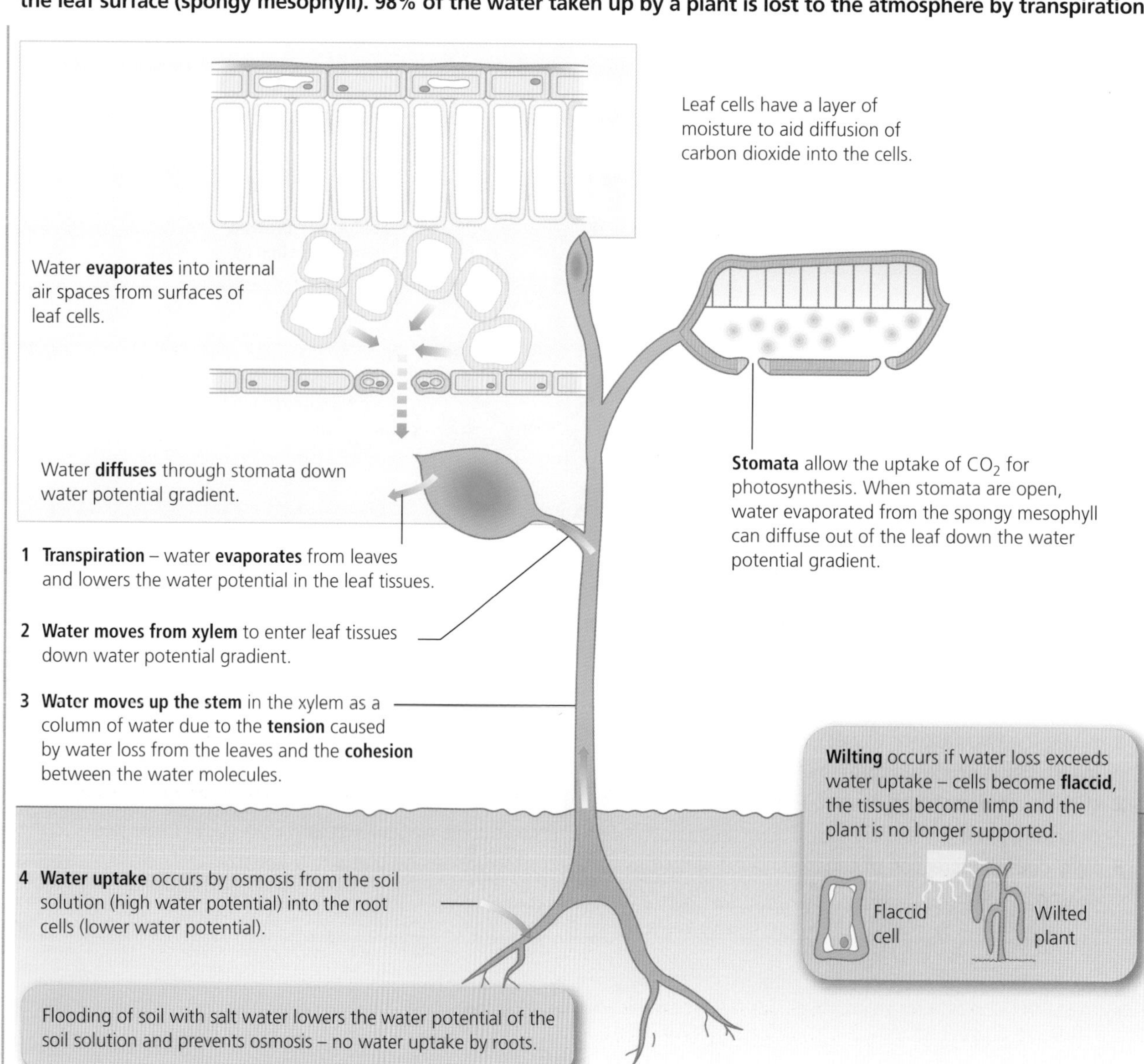

Leaf cells have a layer of moisture to aid diffusion of carbon dioxide into the cells.
Water **evaporates** into internal air spaces from surfaces of leaf cells.
Water **diffuses** through stomata down water potential gradient.
Stomata allow the uptake of CO_2 for photosynthesis. When stomata are open, water evaporated from the spongy mesophyll can diffuse out of the leaf down the water potential gradient.
1 **Transpiration** – water **evaporates** from leaves and lowers the water potential in the leaf tissues.
2 **Water moves from xylem** to enter leaf tissues down water potential gradient.
3 **Water moves up the stem** in the xylem as a column of water due to the **tension** caused by water loss from the leaves and the **cohesion** between the water molecules.
4 **Water uptake** occurs by osmosis from the soil solution (high water potential) into the root cells (lower water potential).
Wilting occurs if water loss exceeds water uptake – cells become **flaccid**, the tissues become limp and the plant is no longer supported.
Flaccid cell
Wilted plant
Flooding of soil with salt water lowers the water potential of the soil solution and prevents osmosis – no water uptake by roots.

The lower surface is less exposed to the warming effects of the Sun's radiation, which would speed up the evaporation rate. Loss of water from the leaf is shown in the diagram on the opposite page.

Water cannot diffuse *into* the leaf through the stomata, because the air spaces inside the leaf are completely saturated. Instead, water must be absorbed from the soil solution and drawn up through the plant (see page 74). This flow of water through the plant to replace the losses by evaporation from the leaf is called the **transpiration stream**, shown in the diagram on the opposite page.

Since water is lost through the stomata to the atmosphere, transpiration is affected by **leaf structure** and by **conditions in the atmosphere** (see opposite). These factors can be investigated using the **potometer** shown below.

Leaf structure may reduce transpiration

- Thick, waxy cuticle reduces evaporation from epidermis
- Stomata may be sunk into pits which trap a pocket of humid air
- Leaves may be rolled with the stomata on the inner surface close to a trapped layer of humid air

- Leaves may be needle shaped to reduce their surface area

Atmospheric conditions may affect transpiration

- **Wind** moves humid air away from the leaf surface and increases transpiration
- **High temperatures** increase the water-holding capacity of the air and increase transpiration
- **Low humidity** increases the water potential gradient between leaf and atmosphere and increases transpiration
- **High light intensity** causes stomata to open (to allow photosynthesis) which allows transpiration to occur

USING A BUBBLE POTOMETER

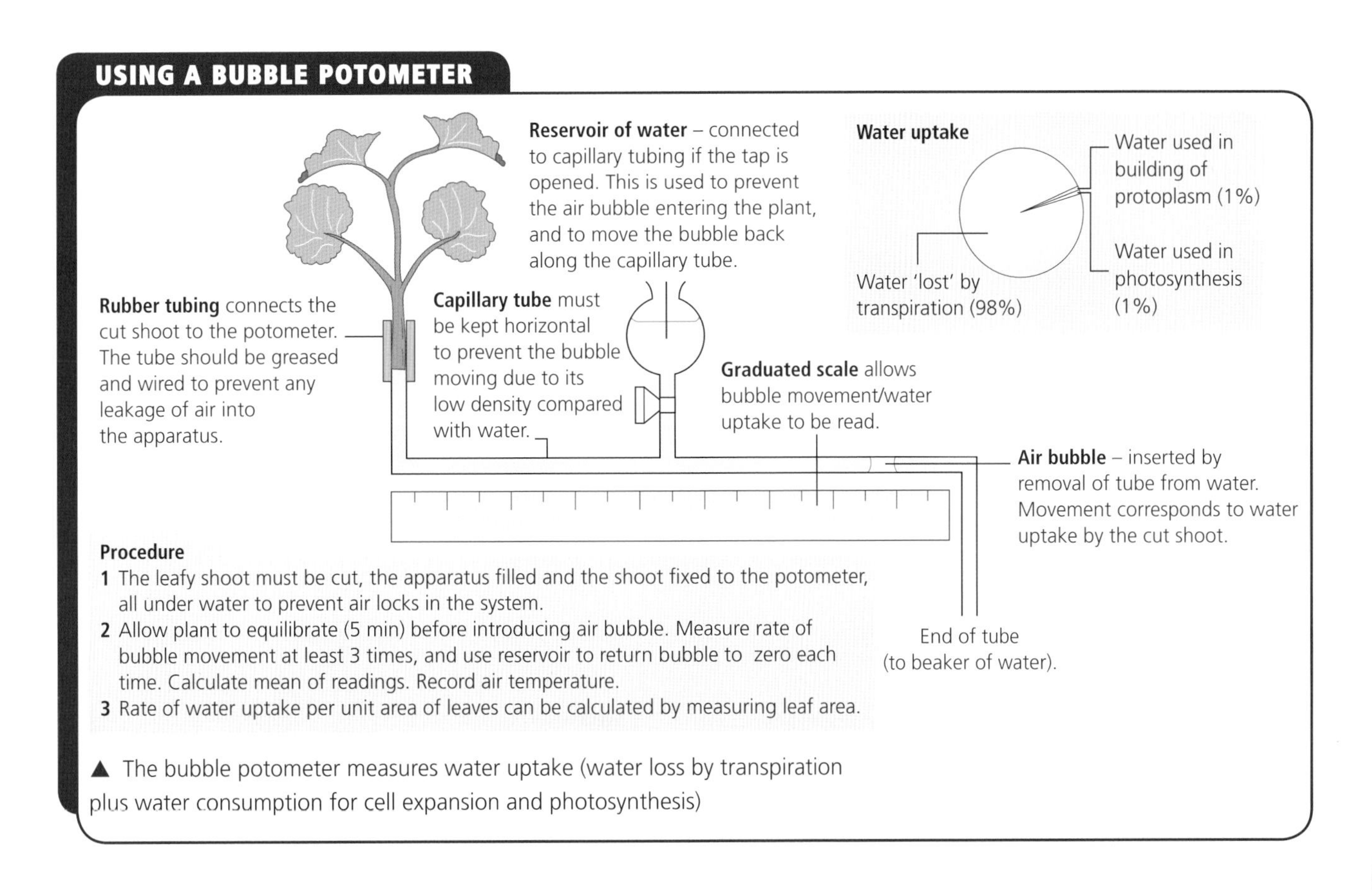

Procedure

1. The leafy shoot must be cut, the apparatus filled and the shoot fixed to the potometer, all under water to prevent air locks in the system.
2. Allow plant to equilibrate (5 min) before introducing air bubble. Measure rate of bubble movement at least 3 times, and use reservoir to return bubble to zero each time. Calculate mean of readings. Record air temperature.
3. Rate of water uptake per unit area of leaves can be calculated by measuring leaf area.

▲ The bubble potometer measures water uptake (water loss by transpiration plus water consumption for cell expansion and photosynthesis)

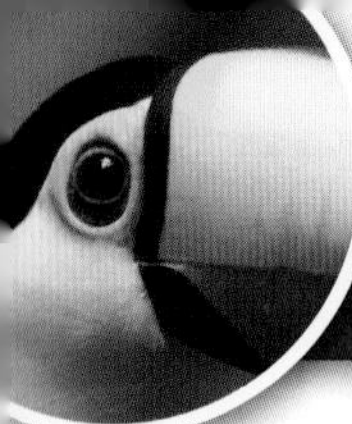

2.24 The leaf and water loss

OBJECTIVES

- To understand that stomata are opened to allow carbon dioxide to enter the leaf, and that this allows water vapour to diffuse out of the leaf
- To describe adaptations of leaves to reduce water losses

Stomata and water loss

Water is lost by evaporation and diffusion from the leaf surface. This water loss happens because the stomata need to open so the leaf can take in carbon dioxide as a raw material for photosynthesis. Plants can open and close the stomata, which helps to minimise water loss whilst allowing photosynthesis to continue. The position and operation of the stomata is explained on page 42, which also shows the adaptations of the leaf to photosynthesis.

Adaptations of plants

The diagrams opposite show how plants may be adapted to the availability of water.

INVESTIGATION OF WATER LOSS FROM LEAF SURFACES

Cobalt chloride paper is blue when dry, and pink when wet. It is handled with forceps to avoid dampness from the fingers affecting its colour. The paper is attached to the upper and lower leaf surfaces using microscope slides. The paper attached to the lower surface of the leaf turns pink, showing that water is lost mainly from the lower surface. This technique gives a *qualitative (non-quantitative)* comparison.

Measuring the mass changes of leaves can be used to give a quantitative comparison of water loss from different leaf surfaces. A number of leaves are smeared with Vaseline as shown in the diagram below. They are weighed and then left in a drying atmosphere for 48 hours, and reweighed at intervals. Some typical results from this investigation are shown in the table.

Leaf number	Initial mass / g	Final mass / g	Percentage change in mass
1	4.2	4.1	
2	4.6	4.4	
3	3.9	2.5	
4	4.1	2.5	

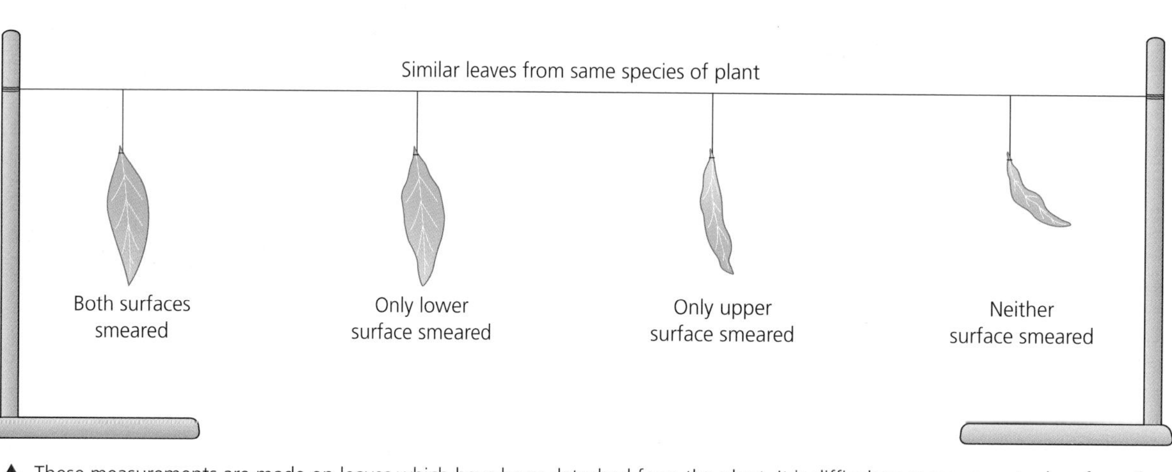

▲ These measurements are made on leaves which have been detached from the plant. It is difficult to measure water loss from the leaves of an intact plant, but relatively simple to measure water uptake using a **potometer** (see page 79).

Q

Questions 1 to 4 refer to the investigation above.

1. Copy the table above. Calculate the percentage change in mass for each leaf.
2. Why did leaf 3 lose a greater proportion of its mass than leaf 2?
3. Why was it important that leaves of the same species were used?
4. How could the results be made more reliable or valid?

Adaptations of plants to reduce water loss in different environments

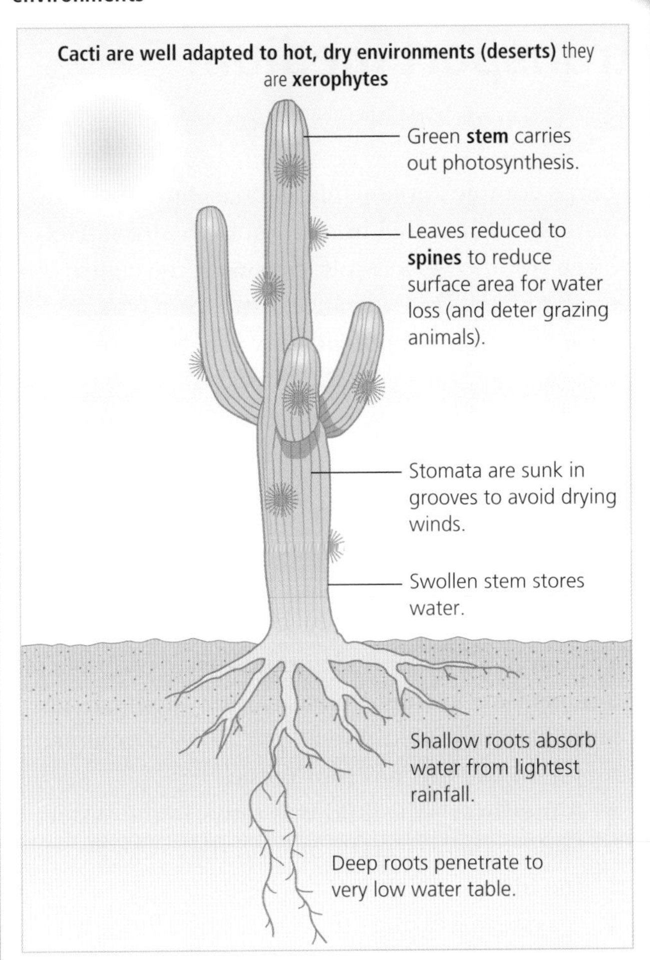
Cacti are well adapted to hot, dry environments (deserts) they are xerophytes
Green stem carries out photosynthesis.
Leaves reduced to spines to reduce surface area for water loss (and deter grazing animals).
Stomata are sunk in grooves to avoid drying winds.
Swollen stem stores water.
Shallow roots absorb water from lightest rainfall.
Deep roots penetrate to very low water table.

Aquatic (pond) plants have leaves with:

- little lignin in the xylem, since the leaf is supported by the water
- a very thin cuticle, since water conservation is not a problem
- stomata on the **upper** surface to allow CO_2 uptake from the atmosphere.

These plants are **hydrophytes**.

Wilting and leaf fall

When water is in short supply, plants may reduce water loss by:

- wilting – leaves collapse are stomata close to reduce heat absorption and evaporation/diffusion of water
- leaf fall – in very severe conditions, e.g. when water is frozen during winter, plants allow the leaves to fall off so that no water loss can occur. No photosynthesis can take place, but the plants can remove chlorophyll from the leaves for storage before allowing leaves to fall.

Modern Christmas trees are bred so that the needles don't fall – a disadvantage in a natural habitat, but good news when clearing up after Christmas!

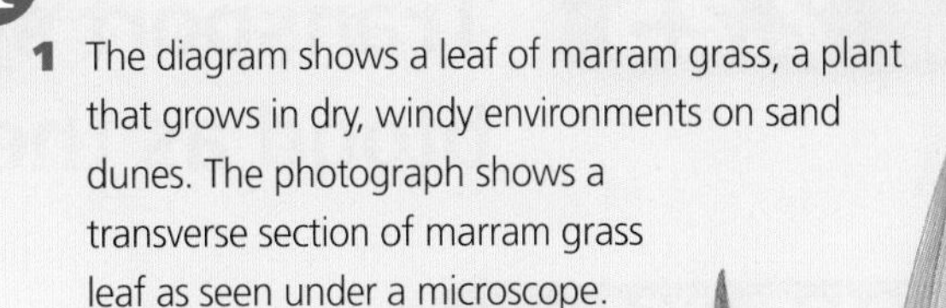

1 The diagram shows a leaf of marram grass, a plant that grows in dry, windy environments on sand dunes. The photograph shows a transverse section of marram grass leaf as seen under a microscope.

a Copy the diagram of the leaf, and draw a line to show the direction in which a transverse section is cut.

b The positions of several stomata are labelled on the photograph. If sufficient light is available, oxygen is released from the leaf through the stomata.

i Name the process that produces oxygen.

ii Which gas is consumed during this process?

iii Name and define the process by which the oxygen moves from leaf to atmosphere.

c When the stomata are open, water vapour will also be lost from the leaf's internal surfaces. Use the diagram and the photograph to suggest three ways in which the structure of the marram grass leaf helps to reduce water loss.

d Give three reasons why plants require water.

e Why is it particularly important that a plant growing in sand dunes should reduce water loss?

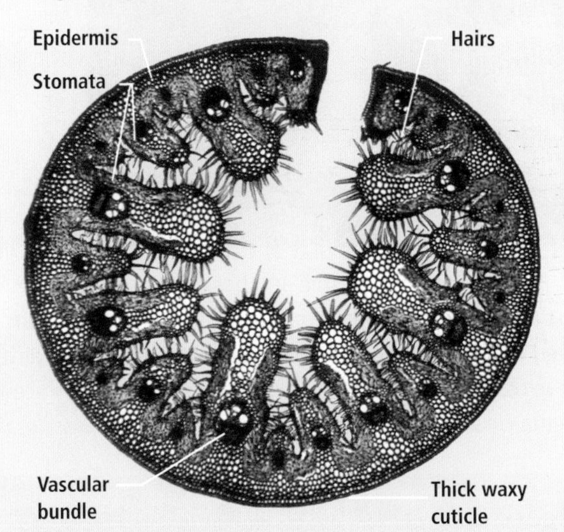
Epidermis
Hairs
Stomata
Vascular bundle
Thick waxy cuticle

2 Copy and complete the following paragraph.
The movement of water through the plant begins at the leaf surface, and a stream of water called the _____ is drawn up through the plant to replace these losses. The gas _____ is required for photosynthesis, and enters the leaves through pores called _____. These pores allow the loss of _____ to the atmosphere, and leaves show many adaptations to reduce this loss. These adaptations include _____, _____ and _____.

2.25 Transport systems in animals use blood as the transport medium

OBJECTIVES

- To understand why animals need a circulatory system
- To know that a transport system has four components
- To know the structure and functions of the components of the blood

All living organisms require energy, which is released from food and oxygen in the process of respiration (page 106). All living cells in the body (for example in the brain, liver, kidney and muscle) need energy, and so glucose and oxygen must be transported throughout the body. In some small animals (especially those with a flat body such as the flatworm in the photograph), substances are transported in and out of the body, and from one tissue to another, by the process of **diffusion**.

▲ The flatworm is very thin and has a high surface area to volume ratio. Its volume is made up of working cells that need various substances including oxygen, which can diffuse in across the large surface. The distance from the surface to all the cells is small.

Large complex organisms, such as humans, have a small surface area in relation to their relatively large volume. As a result, cells near the centre of the body are some distance away from contact with the atmosphere, and may also be some distance from the gut where food is digested. Cells inside such large organisms cannot gain enough oxygen and glucose by diffusion alone.

A mass flow system

To supply oxygen and glucose, as well as other substances, large organisms have a specialised **transport system**. This system, called the **blood system** in all vertebrate animals, is an example of a **mass flow system**. A mass flow system carries large volumes of fluid to all parts of the organism. A system like this has four parts:

- A **medium** – the fluid that flows in the system and carries materials around the body. This is the **blood**.
- A **system of tubes** that carries the fluid from place to place. These are the **arteries** and **veins**.
- A **pump** that supplies pressure to keep the fluid moving through the tubes. This is the **heart**.
- **Sites of exchange** that allow materials delivered by the blood to enter the tissues that need them. These are the **capillaries**.

▼ A whale has a large surface area, but its volume is too large for materials to be moved to and from its body cells by diffusion. The blue whale has a circulatory system made up of 4000 dm^3 of blood, pumped through 50 000 km of blood vessels by a heart the size of a small car.

The blue whale's heart beats about 30000 times a day. What is its heart rate in bpm (beats per minute)?

A shrew has a heart rate of about 900 bpm, a human of about 60 bpm and a horse about 40 bpm. Including the value for the blue whale, what pattern can you see relating size and heart rate?

Blood is the circulatory medium

The average adult human has about 5 dm^3 of blood, which contains a number of blood cells suspended in a watery liquid (**plasma**). If a sample of blood taken from the body is allowed to stand, and a chemical added to prevent it clotting, it will separate into layers as shown opposite.

If a drop of blood is placed on a microscope slide and stained with a special dye, these different types of cell can be seen, as the photograph below shows.

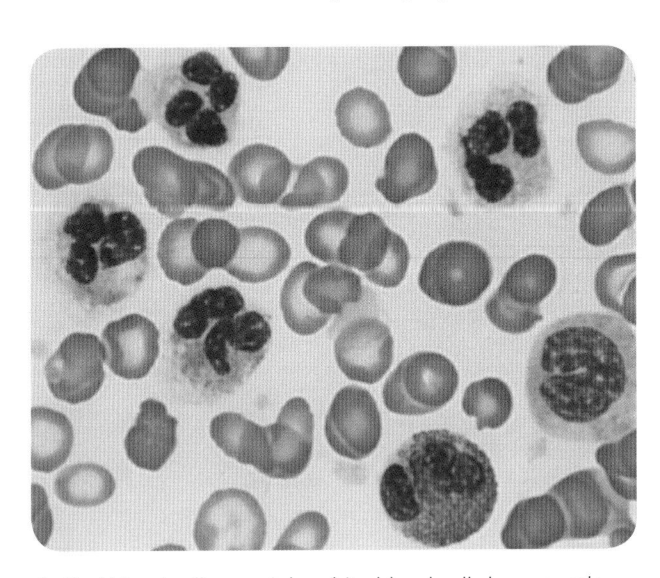

▲ Red blood cells are pink, white blood cells have purple-stained nuclei, of several different shapes (platelets are not visible)

Blood cells are first formed in the bone marrow of long bones such as the femur (thigh bone), although they may go on to other parts of the body before they become fully developed. The structure and function of different types of blood cell are shown in the table below.

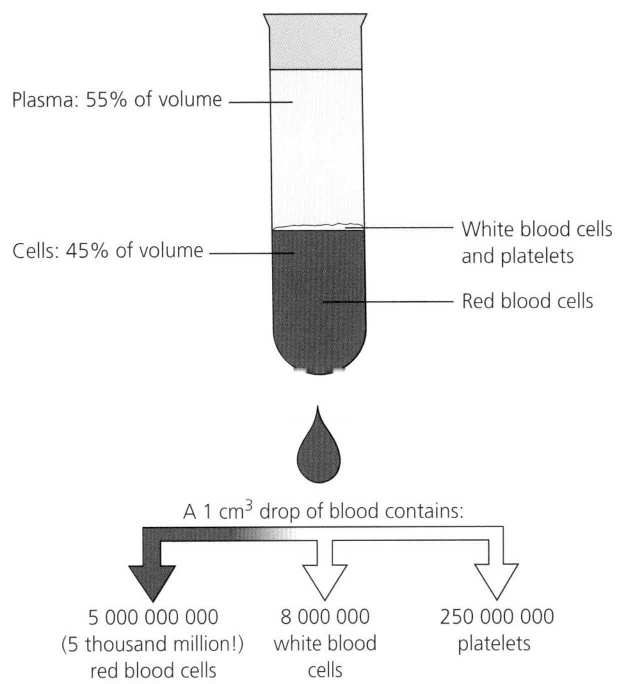

▲ The red blood cells sink to the bottom, and the white blood cells and platelets settle on top of the red blood cells, rather like a layer of dust. The plasma forms a clear, straw-coloured layer at the top.

Cell type	Appearance	Function	How the structure is suited to the function
Red blood cells (erythrocytes)		Transport oxygen from lungs to all respiring tissues. Prepare carbon dioxide for transport from all respiring tissues to lungs.	Contain **haemoglobin**, an iron-containing pigment which picks up oxygen at the lungs and lets go of it at the tissues. Have no nucleus, leaving more space for haemoglobin. Cells are small and flexible, so can squeeze through narrow capillaries.
White blood cells (phagocytes)		Remove any microorganisms that invade the body and might cause infection. The phagocyte engulfs the microorganism (see page 102).	Irregular shaped nucleus allows cell to squeeze through gaps in walls of capillaries. Enzymes in cytoplasm digest microorganisms once engulfed. Sensitive cell surface membrane can detect microorganisms.
White blood cells (lymphocytes)		Produce antibodies – proteins that help in the defence against disease (see page 104).	Large nucleus contains many copies of genes for the control of antibody protein production.
Platelets		Cell fragments involved in blood clotting.	Can release blood-clotting enzymes (see page 102).

Functions of the blood

The table on page 83 shows that one function of red blood cells is the transport of the respiratory gases, oxygen and carbon dioxide, between the lungs and the respiring tissues. The **plasma** also has transport functions. This watery liquid carries dissolved food molecules such as glucose and amino acids, waste materials such as urea, and some control molecules such as hormones. Because the plasma is largely water, it has a very high specific heat capacity (see page 236). This means that the plasma is able to distribute heat around the various parts of the body.

Defence against disease is another important function of the blood. The different roles of the blood are summarised below.

Functions of the blood

Regulatory functions – homeostasis

Blood solutes affect the water potential of the blood, and thus the water potential gradient between the blood and the tissue fluid. The size of this water potential gradient is largely due to sodium ions and plasma proteins. The blood solute level **regulates the movement of water** between blood and tissues.

Water plays a part in the **distribution of heat** between heat-producing areas such as the liver and areas of heat loss such as the skin.

Blood also helps to maintain an **optimum pH** in the tissues.

Too much alcohol in the blood can cause water to leave brain cells, causing the pain and sensation of thirst called a **hangover!**

Protective functions

Platelets, plasma proteins (e.g. fibrinogen) and many other plasma factors (e.g. Ca^{2+} ions) protect against **blood loss** and the **entry of pathogens** by the clotting mechanism.

White blood cells protect against **disease-causing organisms**:

- **phagocytes** engulf them
- **lymphocytes** produce and secrete specific antibodies against them.

Transport functions

Soluble products of digestion/absorption (such as glucose, amino acids, fatty acids, vitamins and minerals) are transported from the gut to the liver and then to the general circulation.

Waste products of metabolism (such as urea, creatinine and lactate) are transported from sites of production to sites of removal, such as the liver and kidney.

Respiratory gases (oxygen and carbon dioxide) are transported from their sites of uptake or production to their site of use or removal.

Hormones (such as insulin) are transported from their sites of production in the glands to the target organs where they have their effects.

Support function

Erection of the penis is achieved by filling large spongy spaces with blood. The penis becomes soft when blood flows out more quickly than it flows in.

The blood detectives

A scientist who specialises in the study of blood is called a **haematologist**. A haematologist can tell a great deal from a tiny sample of blood. For example:

- **Anaemia**, an inability to transport enough oxygen, can be detected by noting a lower than normal number of red blood cells.
- **Sickle cell anaemia** shows some red blood cells shaped like sickles, as the photograph below illustrates.
- **Leukaemia**, a cancer of white blood cells, can be detected by high numbers of oddly shaped white blood cells.
- **AIDS** may be detected in its early stages by the presence of antibodies to the human immunodeficiency virus (HIV) in the plasma – the affected person is HIV positive. In the later stages of the disease, the number of white blood cells is very much reduced.
- **DNA fingerprinting**, which can be important in criminal investigations, can be carried out on nuclei extracted from white blood cells.
- **Diabetes** may be detected by a high glucose concentration in the plasma.
- **Eating disorders** may be detected by higher than normal concentrations of urea in the plasma.

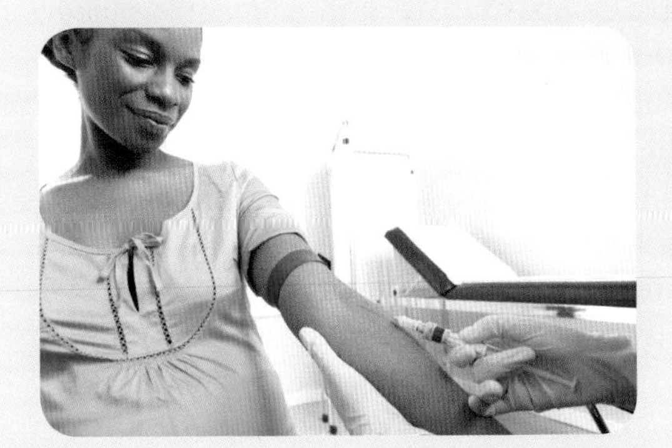

▲ Using modern techniques of biochemistry, a single drop of blood can yield information about all the conditions described above

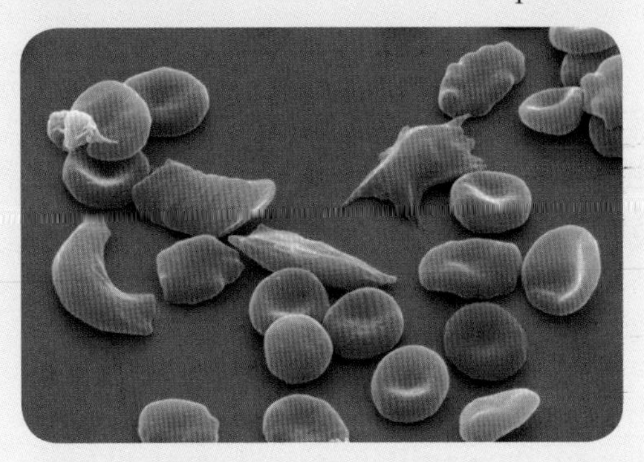

▲ Sickle-shaped red blood cells show the person has sickle cell anaemia. A sickle is a long hooked tool used for cutting crops.

1 Why do large organisms require a transport system?

2 A sample of human blood is collected and placed in a tube and centrifuged. The blood separates into two distinct layers.

a What are the two layers?

b Suggest one dissolved food substance found in the upper layer. Choose a substance for which there is a simple chemical test. Describe the test for this substance, and state what a positive result would be.

c There is a third, rather thin, layer between the two main components. What is this layer made of?

3 'The structure of a cell is closely related to its function.' Is this statement correct for blood cells? Explain your answer.

4 The table shows the cell composition of three samples of blood.

Cell count / number per mm^3	Sample from Jill	Jenny	Jackie
Red blood cells	7 500 000	5 000 000	2 000 000
White blood cells	500	6000	5000
Platelets	250 000	255 000	50

a Which person is most likely to have lived at high altitude recently? Explain your answer.

b Which person would be the most likely to become ill if exposed to a virus? Explain your answer.

c Which person's blood is least likely to clot efficiently? Explain your answer.

d Which person is likely to have an iron deficiency in her diet? Explain your answer.

e These three samples were all taken from 23-year-old women. Explain why this makes comparisons between them valid.

5 Use words from this list to complete the following paragraphs. The words may be used once, more than once or not at all.

epidermis, tissues, specialised, cells, blood, red blood cell, division of labour, organ

Large numbers of ______ that have the same structure and function are grouped together to form ______, for example ______. Several separate tissues may be joined together to form an ______ which is a complex structure capable of performing a particular task with great efficiency.

The structure of cells may be highly adapted to perform one function, i.e. the cells may become ______. One excellent example is the ______ which is highly adapted to carry oxygen in mammalian blood. If the different cells, tissues and organs of a multicellular organism perform different functions they are said to show ______.

2.26 The circulatory system

OBJECTIVES

- To understand that the blood is directed around the body in a set of vessels
- To know the structure and function of arteries and veins
- To understand why humans have a double circulatory system
- To know the names of the main arteries and veins in the human body

Blood vessels – arteries and veins

Blood flows around the body in a system of tube-like **blood vessels**, arranged in such a way that they all eventually lead back to the heart. The blood flows *away* from the heart in vessels called **arteries**, and it flows back *towards* the heart in vessels called **veins**. Joining the arteries and veins are the **capillaries**, and we shall look at these on page 88. In humans (and in many other animals), the main artery is called the **aorta** and the main vein is called the **vena cava**. The structure and functions of arteries and veins are shown in the table below.

The human double circulation

The human circulation is outlined at the top of the page opposite. The arrangement is called a **double circulation** because the blood passes through the heart twice for each complete circuit of the body. The blood flows to the lungs under high pressure (so a large volume of blood flows past the lung surfaces in a short time). Then, having picked up oxygen at the lungs, the blood receives another 'boost' of pressure from the heart to drive it out to the tissues, where the oxygen is needed.

Structure			
Type of vessel	**Transverse section ('cut across')**	**Longitudinal section ('cut along')**	**Adaptations of structure to function**
Artery	Thicker outer wall Narrow central tube (lumen) Thick layer of muscle and elastic fibres	Smooth lining so no obstruction to flow of blood	■ Carries blood away from the heart to the tissues. ■ Blood is at high pressure. ■ Blood is rich in oxygen, low in carbon dioxide (except in the pulmonary artery). ■ Elastic walls expand and relax as blood is forced out of the heart. This causes the pulse that you can feel if you press an artery against a bone, for example in the wrist. ■ Thick walls withstand the high pressure of blood. Rings of muscle can narrow or widen the artery and control the blood flow in it according to the body's needs.
Vein	Thin outer wall Thin layer of muscle and elastic fibres Wide central tube (lumen)	Flap of watch pocket valve	■ Carries blood from the tissues to the heart. ■ Blood is at low pressure. ■ Blood is low in oxygen, high in carbon dioxide (except in the pulmonary vein). ■ Valves prevent the backflow of blood. Blood is at low pressure, but nearby muscles squeeze the veins and help push blood back towards the heart. ■ Large diameter and thin walls reduce resistance to the flow of blood.

▲ The structure and functions of arteries and veins

High pressure – both ventricles pump out blood at high pressure, but the 'body' circuit has a pressure about 5x that in the 'lung' (pulmonary) circuit. This is because blood in the body circuit has so far to travel, and there are many branches in the circuit to take the high pressure blood.

Pulmonary artery – unlike other arteries, this vessel carries **deoxygenated blood** which also has a **high carbon dioxide concentration**.

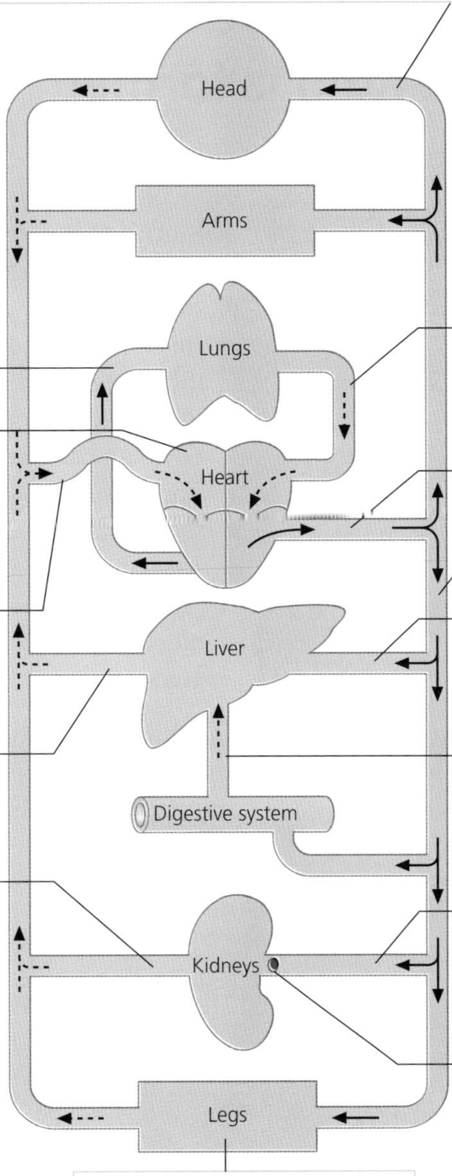

Too high! If pressure in the pulmonary artery is too high, tissue fluid or plasma can leak into the lungs. This sometimes happens to climbers at high altitude.

The heart (see also page 90)

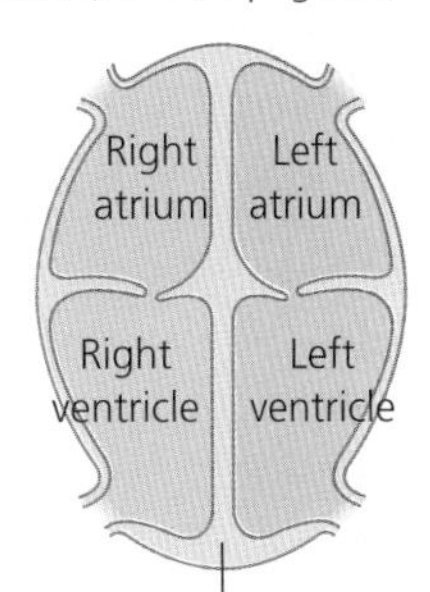

This wall (the septum) separates oxygenated blood from deoxygenated blood.

Head

Arms

Lungs

Heart

Liver

Digestive system

Kidneys

Legs

Pulmonary vein – unlike other veins, this vessel carries **oxygenated blood**. It has the **highest oxygen** and **lowest carbon dioxide** concentration in the circulation.

Aorta – the main artery of the body; supplies oxygenated blood, at high pressure, to the organs and tissues of the body.

Vena cava – the main vein of the body; returns deoxygenated blood at low pressure from organs and tissues to the heart.

Hepatic artery – carries oxygenated blood to the liver.

Hepatic vein – returns blood with a regulated, optimum concentration of food substances to the circulation.

Hepatic portal vein – carries blood containing variable amounts of the absorbed products of digestion from the digestive system to the liver.

Renal vein – blood with reduced urea concentration is returned to the circulation.

Renal artery – carries oxygenated blood with a **high concentration of urea** to the kidneys.

Key

Blood at low pressure ◄--------

Blood at high pressure ◄———

Capillaries link arteries and veins. They are present in all organs and tissues and are the site of exchange of materials between blood and tissue fluid (see next page).

Q

1. State two differences between arteries and veins, and say how these differences are related to the functions of these blood vessels.
2. Look at the diagram above. Follow the path of the red blood cell in the renal artery around the circulation and back to the renal artery. How many times does it pass through the heart? List the blood vessels and chambers of the heart that the cell passes through.

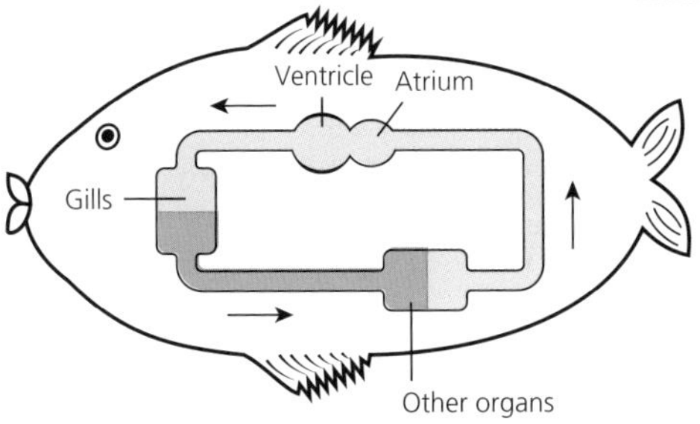

Ventricle

Atrium

Gills

Other organs

Fish have a **single** circulation – blood flows through the heart only once for each complete circuit of the body:
but there is

- rapid fall in velocity and pressure as blood leaves the gills
- pressure too low for efficient kidney function as in mammals.

2.27 Capillaries: materials are exchanged between blood and tissues, and tissue fluid is formed

OBJECTIVES

- To understand that substances carried in the blood must leave the circulation to reach the tissues
- To know that materials are exchanged between tissues and blood in the capillary beds
- To know how the structure of the capillaries is suited to the transfer of materials between blood and tissues

Tissue fluid leaves the capillaries

To reach the cells that need them, dissolved substances carried in the blood must leave the blood vessels and enter the tissues. At the same time, waste materials produced by the tissues need to enter the blood to be carried away. Dissolved substances move between the blood and tissues by **diffusion** across the walls of very fine blood vessels called **capillaries**. Networks or **beds** of capillaries extend through all the tissues, so every body cell is near to a capillary. The capillary beds are adapted to their function of exchange of substances in a number of ways:

- the walls of the capillaries are only one cell thick – substances do not have very far to diffuse through them
- the capillaries are highly branched so they cover an enormous surface area, giving more 'space' for diffusion to occur
- the capillary beds are constantly supplied with fresh blood, keeping up the concentration gradients of dissolved substances between blood and tissues. Without these concentration gradients diffusion could not occur.

The diagram shows how materials are exchanged between the blood and the tissues, and how **tissue fluid** is formed.

Artery – delivers oxygenated blood, rich in nutrients and at high pressure. Can be narrowed or widened to control blood flow (for example, during exercise). Continuous supply of blood keeps up the concentration gradients of substances between the blood plasma and the tissue fluid.

Cells of tissue
Need: oxygen and nutrients such as glucose and amino acids.
Produce: wastes such as carbon dioxide and some useful products such as hormones.

Vein – carries away deoxygenated blood, low in nutrients and at low pressure. Has high concentration of waste products.

Tissue fluid
- formed from plasma
- contains no blood cells or plasma proteins.

Formation of lymph

Useful substances move out from plasma – formation of tissue fluid.

Substances collected from cells

Capillary has:
- wall only one cell thick
- very large surface area.

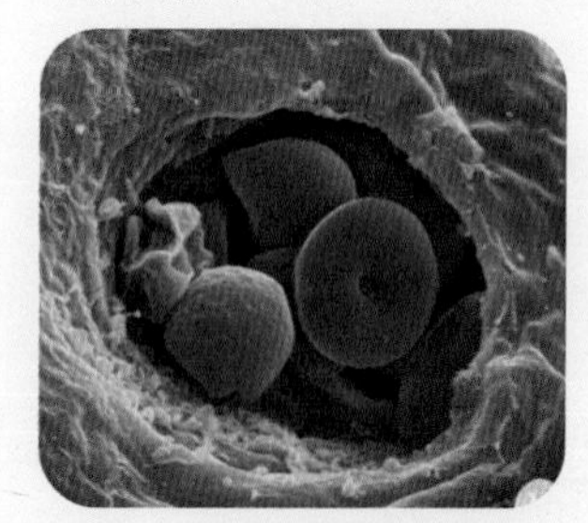

Capillary walls are about 1 μm thick. The central tube (lumen) may only be the same diameter as the red blood cells, so they have to squeeze through the capillary in single file as they unload their oxygen.

Remember how substances cross membranes

Diffusion is:
- the movement of molecules
- down a concentration gradient
- until equilibrium is reached.

Osmosis is:
- the movement of water
- down a water potential gradient
- across a partially permeable membrane.

Problems with the return of tissue fluid

The diagram below shows in detail how tissue fluid is formed and returned to the blood. This tissue fluid is essential for the transport of dissolved substances between blood and cells. If anything goes wrong with this return of tissue fluid, the tissues swell up.

In **elephantiasis**, for example, a parasitic worm lodges in the lymph vessels in the groin and causes fluid to build up in the legs.

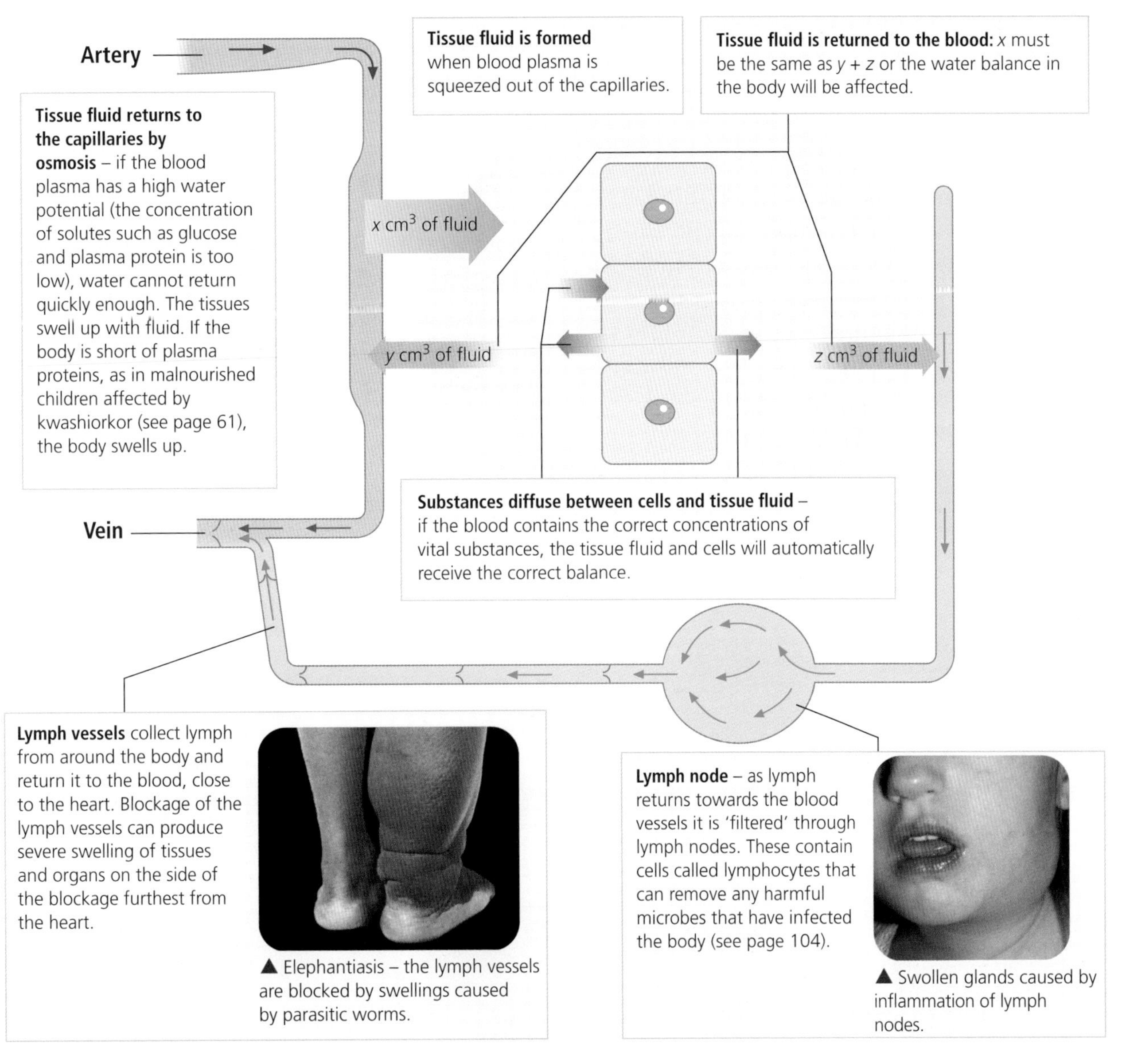

Artery
Tissue fluid is formed when blood plasma is squeezed out of the capillaries.
Tissue fluid is returned to the blood: x must be the same as y + z or the water balance in the body will be affected.
Tissue fluid returns to the capillaries by osmosis – if the blood plasma has a high water potential (the concentration of solutes such as glucose and plasma protein is too low), water cannot return quickly enough. The tissues swell up with fluid. If the body is short of plasma proteins, as in malnourished children affected by kwashiorkor (see page 61), the body swells up.
x cm³ of fluid
y cm³ of fluid
z cm³ of fluid
Vein
Substances diffuse between cells and tissue fluid – if the blood contains the correct concentrations of vital substances, the tissue fluid and cells will automatically receive the correct balance.
Lymph vessels collect lymph from around the body and return it to the blood, close to the heart. Blockage of the lymph vessels can produce severe swelling of tissues and organs on the side of the blockage furthest from the heart.
Lymph node – as lymph returns towards the blood vessels it is 'filtered' through lymph nodes. These contain cells called lymphocytes that can remove any harmful microbes that have infected the body (see page 104).

▲ Elephantiasis – the lymph vessels are blocked by swellings caused by parasitic worms.

▲ Swollen glands caused by inflammation of lymph nodes.

Q

1 This diagram represents a group of body cells and some parts of the circulatory system. The arrows show movement of fluids.

a Name the fluids contained in spaces **A**, **B** and **C**.

b Name two substances that the cells remove from the fluid in **B**. Suggest two substances that the cells might add to the fluid in **B**.

c Describe how the fluid in **C** is returned to the circulation.

d Give a reason why the process shown at **D** might be inefficient. What would be the result of this for the body? How could it be corrected?

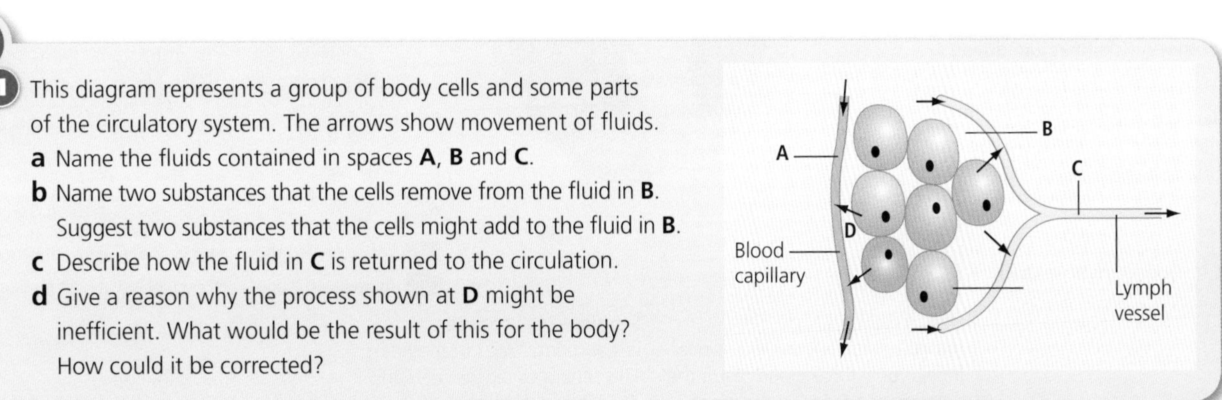

2.28 The heart is the pump for the circulatory system

OBJECTIVES

- To know that the blood is pumped around the circulatory system by the action of the heart
- To know that the heart is a muscular organ with four chambers
- To understand how the flow of blood through the circulation is maintained

The heart of a mammal pumps blood through the circulatory system. It provides the pressure that forces the blood through arteries, capillaries and veins. The pressure is generated by the squeezing of the walls of the heart against the incompressible fluid blood. The heart walls can squeeze the blood because they are made of **muscle**, and the muscle contracts rhythmically.

A double pump

The heart is divided into two sides, each of which acts as a pump. The right side of the heart pumps deoxygenated blood coming from the tissues out to the lungs. The left side pumps oxygenated blood coming from the lungs out to the tissues. A much greater pressure (about five times as much) is needed to force blood out to the extremities of the body than is needed to drive blood to the lungs. Because of this, the left side of the heart is much more muscular than the right side (page 87).

Even though the two sides of the heart generate different pressures, they work in the same way and have the same parts, as shown in the diagram below.

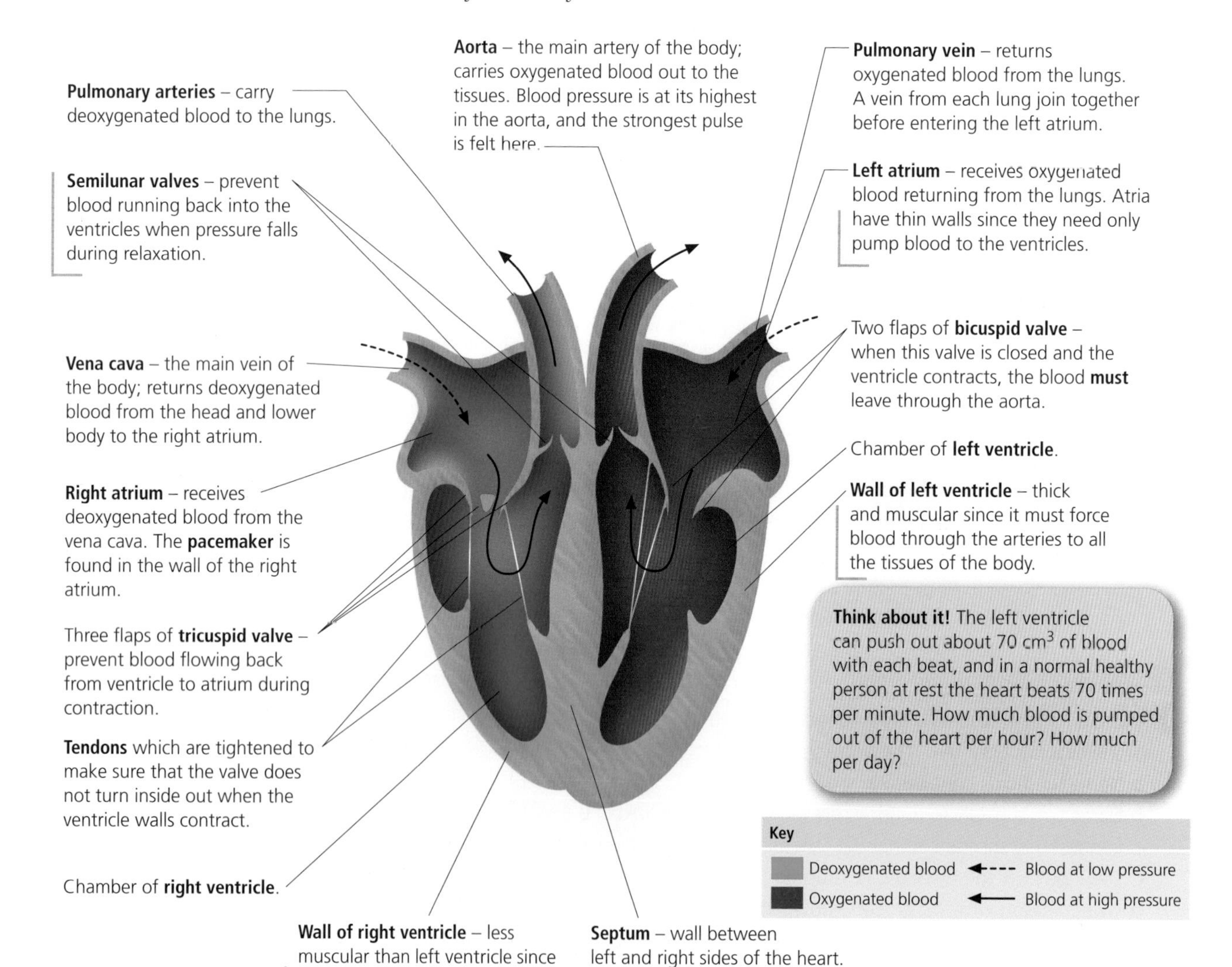

Aorta – the main artery of the body; carries oxygenated blood out to the tissues. Blood pressure is at its highest in the aorta, and the strongest pulse is felt here.
Pulmonary arteries – carry deoxygenated blood to the lungs.
Pulmonary vein – returns oxygenated blood from the lungs. A vein from each lung join together before entering the left atrium.
Semilunar valves – prevent blood running back into the ventricles when pressure falls during relaxation.
Left atrium – receives oxygenated blood returning from the lungs. Atria have thin walls since they need only pump blood to the ventricles.
Vena cava – the main vein of the body; returns deoxygenated blood from the head and lower body to the right atrium.
Two flaps of **bicuspid valve** – when this valve is closed and the ventricle contracts, the blood **must** leave through the aorta.
Chamber of **left ventricle**.
Right atrium – receives deoxygenated blood from the vena cava. The **pacemaker** is found in the wall of the right atrium.
Wall of left ventricle – thick and muscular since it must force blood through the arteries to all the tissues of the body.
Three flaps of **tricuspid valve** – prevent blood flowing back from ventricle to atrium during contraction.
Tendons which are tightened to make sure that the valve does not turn inside out when the ventricle walls contract.
Chamber of **right ventricle**.
Wall of right ventricle – less muscular than left ventricle since it need only force blood along the pulmonary arteries to the lungs.
Septum – wall between left and right sides of the heart. This separates oxygenated and deoxygenated blood.
Key
Deoxygenated blood
Oxygenated blood
Blood at low pressure
Blood at high pressure

Think about it! The left ventricle can push out about 70 cm^3 of blood with each beat, and in a normal healthy person at rest the heart beats 70 times per minute. How much blood is pumped out of the heart per hour? How much per day?

Note that:

- The **atrium** receives blood at low pressure from the veins (coming from the lungs or tissues).
- The **ventricle** pumps blood at high pressure out to the arteries (to the lungs or tissues).
- **Valves** make sure that the blood flows in the right direction.

The beating of the heart is controlled by a pacemaker

In a healthy person the heart beats about 70 times a minute during normal levels of activity. This rate is enough to supply blood containing oxygen and nutrients to tissues.

The muscular walls of the heart differ from other muscles in that they never become tired or **fatigued**, because each contraction of the heart is immediately followed by a relaxation. Even when the heart is beating at its fastest during severe exercise (see page 116), the period of relaxation allows the muscle to recover so it does not fatigue.

The pattern of contraction and relaxation is kept going by electrical signals sent from a region of the heart called the **pacemaker**. This is a specialised piece of tissue in the wall of the right atrium. It is sensitive to the swelling of the heart wall as blood enters the heart from the main veins. The signals from the pacemaker make sure that:

- the atria contract just before the ventricles, so that blood flows from atria to ventricles
- the heartbeat is fast enough to meet the demands of the tissues for oxygen and nutrients, and for the removal of wastes.

If the pacemaker does not work as well as it should, an artificial electronic pacemaker can be fitted inside the chest (see box on the right).

Artificial pacemakers – help for the heart

The beating of the heart is controlled by the natural pacemaker in the wall of the right atrium. If this pacemaker is damaged, pumping goes on automatically at about 30 beats per minute. This is less than half the normal rate, and is only enough to keep a very inactive person alive.

An artificial pacemaker can help people whose natural pacemaker does not work well. This artificial pacemaker is made up of a box containing batteries and an electronic timing device. It is placed in a cavity under the muscle of the upper chest as shown below, and a wire is fed down a vein into the right ventricle. The timing device sends a small electrical charge which triggers the beating of the heart. This is set to give a basic rate of 72 beats per minute. The latest pacemakers can sense changes in breathing, movement and body temperature, and make exactly the right adjustments to heart rate. The battery in the pacemaker is usually replaced every year or so, under local anaesthetic.

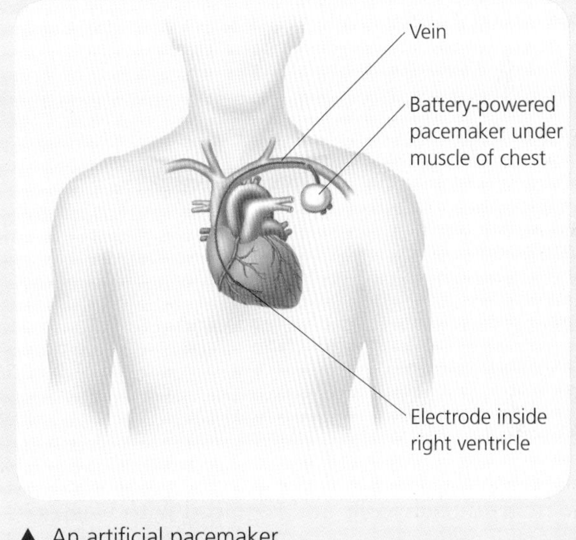

▲ An artificial pacemaker

1 a Name the chamber of the heart that receives blood from the lungs.
  b Explain what happens to the blood in the ventricles when the muscle in the ventricle walls contracts.
  c The muscle around the atria is thinner than the muscle around the ventricles. Suggest a reason for this.

2 A doctor listening to the heartbeat through a stethoscope hears two sounds as the blood flows through the heart: 'lup-dup lup-dup lup-dup'.
  a From your knowledge of the working of the heart, suggest how these two sounds are produced.
  b The doctor records 72 beats per minute. How long a period would there be between two consecutive 'lup' sounds? Explain your answer.

2.29 Coronary heart disease

OBJECTIVES

- To know that the heart muscle requires a supply of oxygen, glucose and other nutrients
- To know that the heart muscle has its own blood supply through the coronary arteries
- To understand that factors such as lifestyle, diet and family history may affect the efficiency of the coronary arteries
- To know that coronary heart disease is one of the major causes of death in the developed world

The supply of blood to the heart muscle

Heart muscle contracts to push blood through the vessels of the circulation. This contraction is work, and it requires energy. The energy is made available by aerobic respiration, so the working cells of the heart need glucose and oxygen for respiration.

The blood passing through the chambers on the left side of the heart carries oxygen and glucose, but the heart cells cannot use these because the muscular walls are very thick and the blood is too far away. The heart muscle has its own blood supply, delivered to capillary beds in the walls of the heart through the **coronary arteries**. The coronary arteries branch off from the aorta, just where the aorta leaves the left ventricle. The heart therefore has a high-pressure supply of blood loaded with oxygen and glucose. Once these useful substances have been removed by the heart muscle cells and replaced with wastes such as carbon dioxide, the blood returns to the circulation through **coronary veins** which pour blood into the vena cava. The coronary circulation is shown below left. The activity of the heart can be checked in a number of ways:

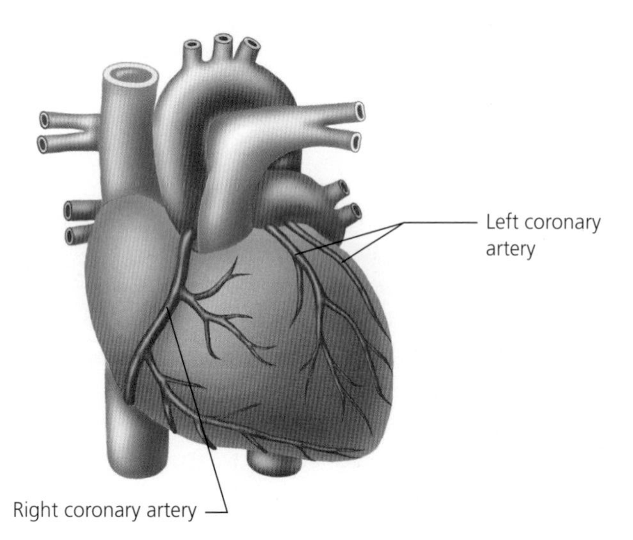

▲ The coronary arteries branch off the aorta, supplying the surface of the heart with oxygenated blood at high pressure

- **making an ECG (electrocardiogram)**. There should be a regular pattern of the flow of electric current through the heart muscle.

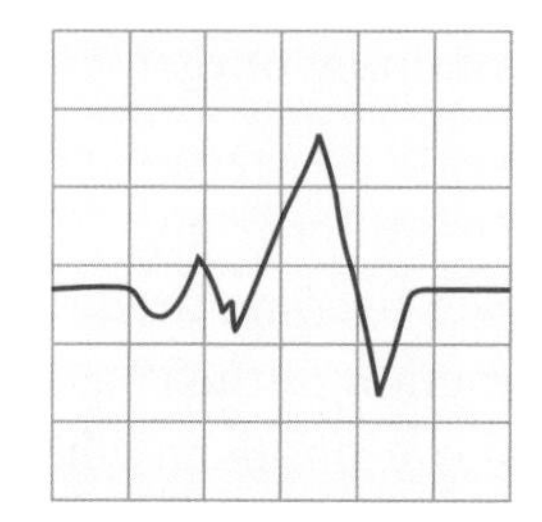

Any irregularity could be a sign of heart disease.

- **measuring pulse rate**. This should be regular, allowing for exercise or activity. Any unusual changes could indicate problems in the heart's nervous system.
- **heart sounds**. The regular 'lub-dup' sound of the heart is caused by the heart valves closing in sequence (cuspid then semilunar). Leaking valves can make the sound less clear.

Coronary heart disease

If any of the coronary arteries become blocked, the supply of blood to the heart muscle may be interrupted. The heart muscle cells are deprived of glucose and oxygen, and poisonous wastes such as lactic acid build up (see page 108). Part of the heart muscle stops contracting, causing a **heart attack**. This can be damaging or even fatal, since other tissues in the body no longer receive their supplies of oxygen and nutrients if the heart stops beating.

What causes the coronary arteries to become blocked? A healthy coronary artery may be narrowed by atheroma so that blood flow through it is restricted. This person has **coronary heart disease (CHD)**.

The risk of developing CHD is increased by:

- **poor diet** – high levels of cholesterol or saturated fatty acids in the blood
- **poor lifestyle** – smoking, lack of exercise, stress
- **genetic factors** – being male; having a family history of heart disease.

Anyone with a genetic risk of developing heart disease should obviously take care that he or she does not have a poor diet or lifestyle. Many middle-aged men, the highest risk group, help to prevent heart disease by taking half an aspirin a day (this seems to help stop small clots forming which could block the arteries) and/or by drinking a small quantity of alcohol (red wine may be the most beneficial).

If a doctor suspects that a patient has CHD, an **angiogram** is carried out. This gives a picture of the state of these arteries (see below). If the coronary arteries are blocked, a **coronary artery bypass** operation may be carried out, as shown below.

Avoid the problem!

- Don't smoke
- Avoid fatty foods
- Take regular aerobic exercise

Aortic arch

Grafted 'bypass'

Blocked part of coronary artery

Coronary artery

Men are more at risk of CHD than women – but men can't avoid CHD by changing their gender!

In a **coronary artery bypass operation** a blood vessel is removed from another part of the body and stitched into place between the aorta and the unblocked part of the coronary artery. Sometimes an artificial vessel is used. The bypass increases blood flow and reduces the likelihood of angina (chest pain).

Angioplasty can also help.

A special cable is passed into the narrowed artery and used to insert a metal 'cage' which forces the artery open

Blood can now flow more easily

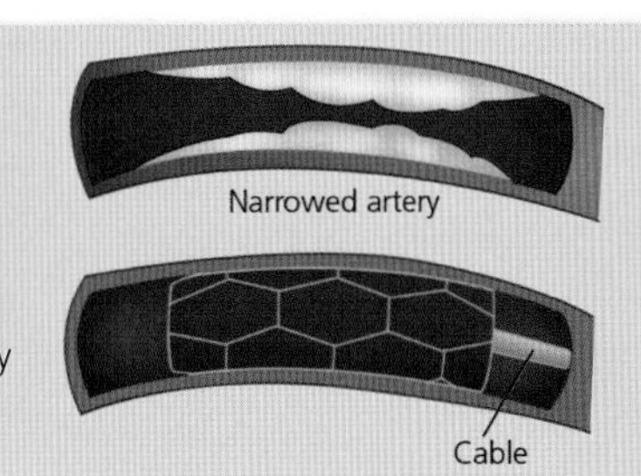

The metal 'cage' (stent) is placed in the artery and the cable is withdrawn.

Q

1 a The heart muscle has its own supply of blood from vessels that run all over its surface. What is the name of these vessels?

b In some people these vessels can become blocked with a fatty substance containing cholesterol. Explain the effects that blocking these blood vessels would have.

2 Look at the graphs. Graph A shows the relationship between the death rate for CHD and the blood cholesterol level. Graph B shows the relationship between cholesterol levels and age, for men and women.

a At which ages do men and women show the same blood cholesterol level?

b Use the information in the graphs to explain why the death rate from CHD is higher for men than for women between the ages of 25 and 45 years.

c Use the graphs to determine the level of blood cholesterol that would keep the death rate from CHD at a minimum.

d What does your answer to **c** suggest about cholesterol as the only cause of CHD?

e What sort of person is most at risk from CHD?

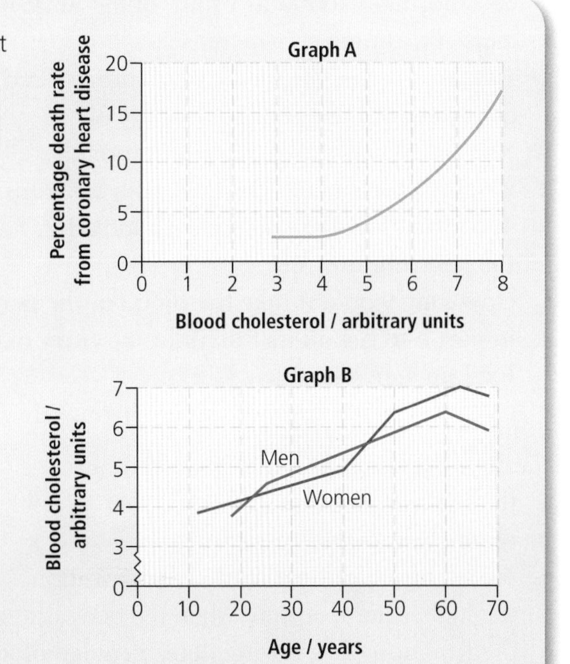

Questions on circulation

1 Read the following article carefully, then answer the questions based on it.

Blood plasma carries red blood cells, white blood cells and platelets to all parts of the body. It also carries many solutes (dissolved substances) – these include oxygen, carbon dioxide, urea, hormones, glucose and amino acids.

In a normal healthy person the number of white blood cells may vary, but is usually no more than 8000 per mm^3 of blood. If a person has an infection, the number of white blood cells may rise to 40 000 per mm^3 of blood.

In a healthy person living at sea level there are about 5 000 000 red blood cells per mm^3 of blood. The cells are regularly replaced from the bone marrow. Old, worn out red blood cells are removed from the blood by the liver, after about 120 days of carrying out their function. Each cell carries oxygen from the lungs to the tissues, combined with a protein called haemoglobin. Haemoglobin will also combine with carbon monoxide, a gas in car exhaust fumes and cigarette smoke. Carbon monoxide combines with haemoglobin about 250 times more readily than oxygen does, and the combination does not break down.

a Name two soluble substances transported in the blood plasma. For each substance you name, suggest where it might be coming from and where it might be going to.

b What is the maximum number of white blood cells normally found in 1 mm^3 of blood? What can cause this number to increase?

c In a healthy person, what is the ratio of red blood cells to white blood cells?

d Name two sources of carbon monoxide.

e What happens to the amount of oxygen transported if a person breathes in carbon monoxide? Explain your answer.

f How long would it take the blood of the person in part **e** to regain its full ability to carry oxygen? Explain your answer.

g Liver is a good dietary source of iron. Why?

2 Name the blood vessel in a human that has:

a the highest pressure

b the highest oxygen concentration

c the highest carbon dioxide concentration

d the highest temperature when the body is at rest

e the function of removing deoxygenated blood from the leg

f the highest concentration of glucose following a meal

g blood at high pressure but with a low oxygen concentration.

3 These diagrams are concerned with the exchange of materials between the blood and tissues.

Part of the circulatory system

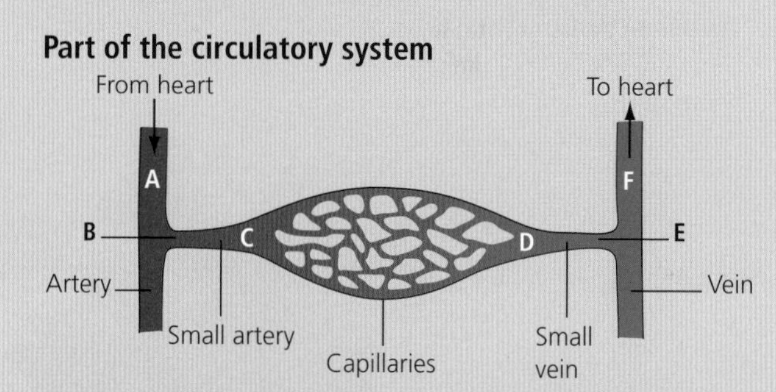

Sections through blood vessels *(not to scale)*

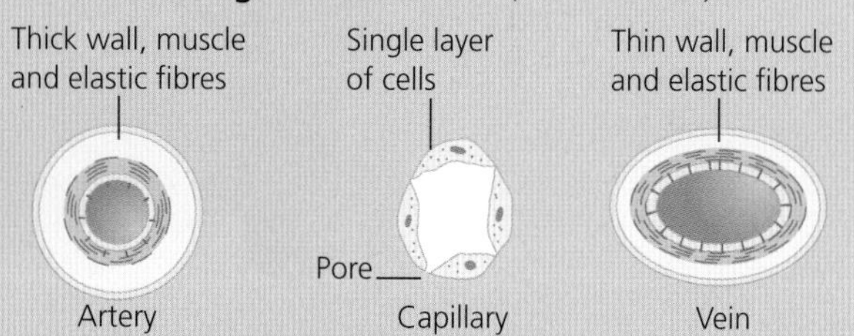

Average blood pressure at different positions in the circulatory system

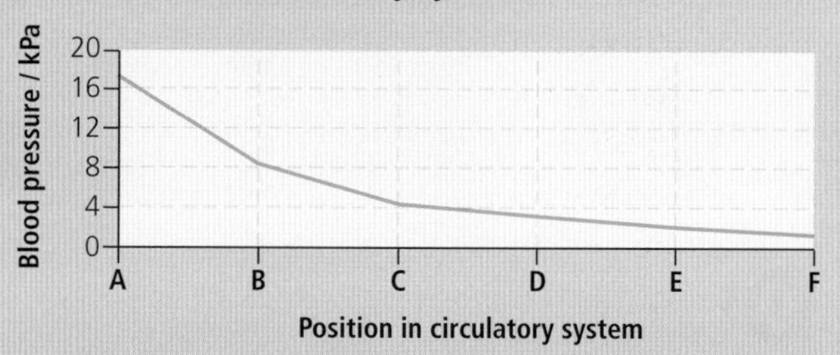

a What happens to the blood pressure as the blood travels from **A** to **F**?

b The highest pressure is found in the artery. Why is this important for the delivery of materials to the cells?

c Veins have structures along them, not shown in the diagram, to help return blood to the heart. What are these structures called?

d How can the small arteries help control the distribution of blood to tissues which have a high demand for oxygen and nutrients?

e Use information shown in the diagrams to explain how the structure of capillaries allows substances to pass from them to surrounding cells.

4 List the blood vessels and chambers of the heart through which:

a a molecule of glucose absorbed through the gut passes on its way to an active muscle in the leg

b a red blood cell passes as it delivers oxygen from the lungs to the brain and returns to be oxygenated again.

5 On a visit to a sports physiology laboratory, a student underwent a series of tests. He was made to exercise on a rolling road, and the following information was collected.

heart rate / beats per minute	total heart output / dm^3 per minute	output per beat / cm^3
55	4.0	
70	4.8	
80	5.2	
90	5.6	
120	6.0	
140	6.0	
150	5.8	
170	4.6	

a Copy the table and complete it by calculating the output per beat for each heart rate value.

b Plot a graph of total output (vertical axis) against heart rate (horizontal axis). On the same graph plot output per beat (vertical axis) against heart rate.

c Describe the relationship between:

i heart rate and total output

ii heart rate and output per beat.

d i Even during severe exercise, the heart rate seldom rises above 140 beats per minute. Use the data in the table to explain why this is so.

ii Well trained athletes can keep up a heart rate of 170 beats per minute. Suggest how their output per beat may be different from that of an untrained person such as the student.

e Distance runners often have low resting heart rates. How might this be of advantage to them?

f Increased total output means that more blood can be delivered to active tissues. As heart rate increases, so does breathing rate. Suggest why.

6 The bar chart below shows how the risk of CHD varies in different parts of the world.

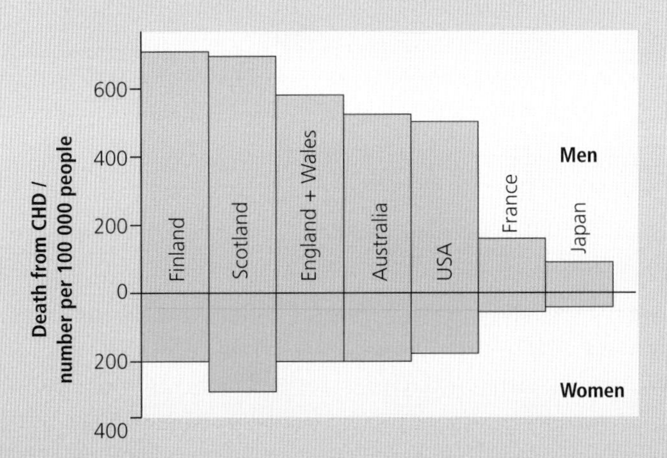

A US citizen is almost three times as likely to develop CHD as a French citizen. Medical scientists suggested that this difference could be due to either diet or inheritance. They set up a study of 1000 French nationals who had emigrated to the US, and who had taken up the American lifestyle, including diet.

a How many of these people would you expect to develop CHD if:

i diet is responsible

ii inheritance is responsible?

b All the people in this study were male, and non-smokers. Why was this important to the validity of the results?

7 The following tables show the effects of smoking and blood cholesterol levels on the risk of developing CHD.

cigarettes smoked per day	relative risk of CHD
0	1
5	1.2
10	1.5
15	2.0
20	2.5
25	2.9
30	3.2

▲ Effect of smoking

cholesterol level	relative risk of CHD
Male, normal	1
Female, normal	0.3
Male, 30% above normal	1.4
Male, 45% above normal	2.0
Male, 75% above normal	3.1

▲ Effect of blood cholesterol level

a Plot the two sets of results in an appropriate way.

b i Which appears to carry greater risk, smoking 10 cigarettes per day or having a blood cholesterol level 45% above normal?

ii How much more likely is a man who smokes 30 cigarettes per day and has a 75% higher than normal blood cholesterol level to develop CHD than a non-smoking woman with normal blood cholesterol level?

8 Suggest how:

a lifestyle

b diet

c inheritance

may cause problems with the circulatory system.

2.30 Health and disease

OBJECTIVES

- To understand what is meant by disease
- To appreciate that disease may have a number of causes
- To understand that some microorganisms may cause disease

What is disease?

The process of homeostasis maintains optimum conditions for body function (see page 128). However, sometimes the mechanisms of homeostasis cannot cope with changes in the internal environment of the body. A person in this situation will show **signs** (such as a raised body temperatures) and experience **symptoms** (such as feeling tired). The person is no longer 'at ease' but is 'dis-eased'. Disease, then, is the state of the body when it cannot cope with changes by the normal homeostatic methods.

Classification of diseases

There are two main classes of disease – **infectious** and **non-infectious**. Infectious diseases are caused by some other living organism, usually a microorganism. They can be 'caught', or passed on from one individual to another and so are also called transmissible diseases. These diseases are caused by a living organism, called a **pathogen**. Infectious diseases can be classified according to how the disease-causing organism is passed from one individual to another, as illustrated below.

Non-infectious diseases are not 'caught' from another individual. These diseases may have a number of causes, illustrated opposite.

Patterns of disease vary in time and space

In the eighteenth century many British children died from bacterial infections. With the development of antibiotics, immunisation programmes and

Infectious diseases

These are caused by organisms, called pathogens, which may spread in a number of ways:

In infected water e.g. *Cholera* bacteria

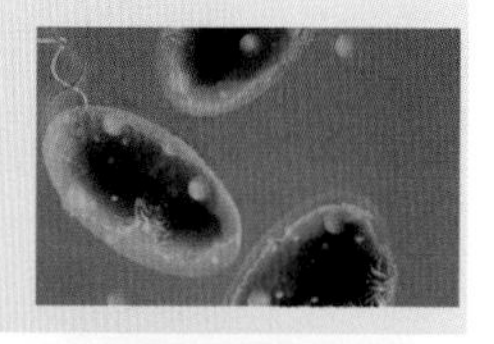

In droplets in the air e.g. influenza virus

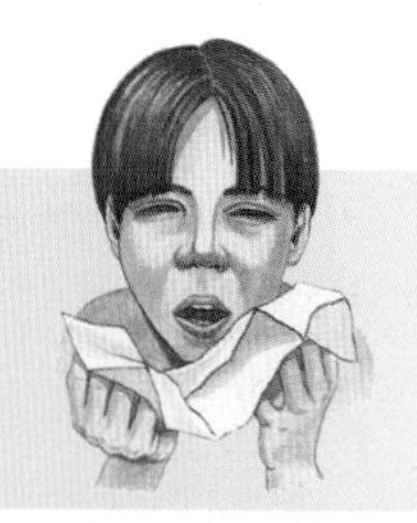

By direct contact (contagious) e.g. athlete's foot-fungus or head lice – insects

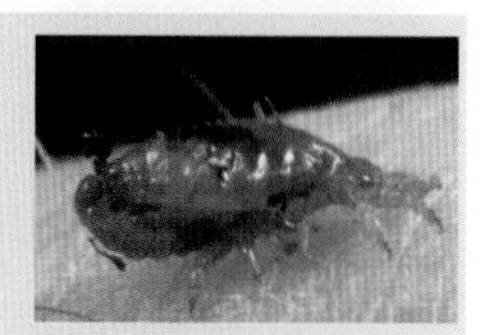

In contaminated food (causes food poisoning) e.g. *Salmonella* bacteria or pork tapeworm

By animal vectors e.g. *Plasmodium* protoctist (causes malaria) via *Anopheles* mosquito

Via body fluids e.g. hepatitis B virus or human immunodeficiency virus (HIV) (causes AIDS)

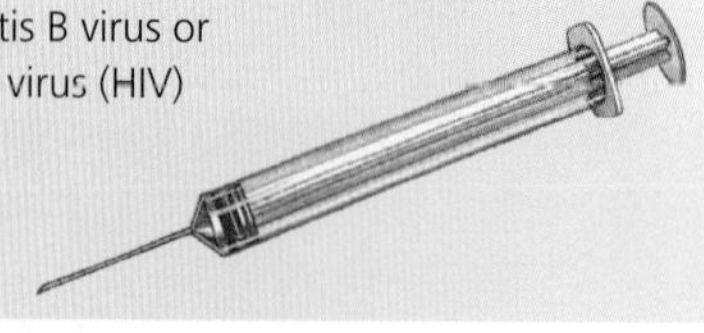

Typical symptoms of disease include:

- **sweating/fever** due to resetting of body's thermostat
- **vomiting/diarrhoea** due to body's attempts to clear gut of irritants
- **pain** due to release of toxins by pathogens.

A doctor might use a thermometer to recognise a sign such as raised body temperature.

▲ Infectious diseases. Parasites such as tapeworms and head lice are not considered to cause 'diseases' by some people, but they do cause distress to the person affected by them and the body is less 'at ease' when it has been colonised by such a parasite. Parasites may be thought of as causing long-term disease.

Non-Infectious diseases

Degenerative – organs and tissues work less well as they age. This is thought to be due to changes in body chemicals caused by **free radicals** such as the peroxide ion e.g. heart attacks, cataracts, hardening of the arteries.

Deficiency – poor diet may deprive the body of some essential substance e.g. scurvy is caused by lack of vitamin C – see page 54.

Allergy – sensitivity to some antigen in the environment, e.g. hay fever (pollen is the antigen) – see page 104.

Environmental – some factor in the environment may trigger a dangerous or abnormal reaction e.g. overexposure to ultraviolet radiation may cause abnormal cell division leading to skin cancer – see page 212.

Inherited/metabolic – some failure in the body's normal set of chemical reactions, e.g.
- sickle cell anaemia (abnormal haemoglobin, page 204)
- cystic fibrosis (production of thick mucus, page 204)
- diabetes (failure to produce enough insulin, page 133).

These conditions are due to alterations in the genes.

Psychological/mental – changes in the working of the brain may lead to abnormal behaviour, e.g. schizophrenia, depression.

Self-induced – some abuse of the body may affect its function, e.g. lung cancer caused by cigarette smoking, cirrhosis of the liver caused by alcohol abuse, see page 155.

improved public hygiene, deaths from infectious diseases in the UK are now much less common. Smallpox was a major killer worldwide until the 1960s, but effective vaccination has now eliminated this disease (see page 105).

In the Western world the major killers are now 'diseases of affluence', caused by our relatively wealthy lifestyle. Along with accidents, coronary heart disease and cancer cause most deaths in Britain, largely the result of smoking, eating too many sugary and fatty foods, and lack of exercise.

Infectious diseases that are spread by vectors (e.g. malaria spread by mosquitoes) are naturally confined to those parts of the world in which the vector can live and breed. In the same way, diseases spread by the number of people likely to meet an infected individual. In recent years there has been concern over the possible spread of infectious diseases because:

- Easier travel by air means that diseases can be carried from one country to another before the infected person develops any symptoms.
- Global warming has increased the range of some insect vectors.
- A greater dependence on communal eating and fast food has led to the easier transmission of organisms that cause food poisoning.

1. Suggest three ways in which microbes might be harmful to humans, and three ways in which they might be helpful. Give examples to support your suggestions.
2. Suggest three cause of non-infectious diseases, and give one example of each.
3. How can infectious diseases spread? Give examples to support your answer.
4. What is the difference between the signs and the symptoms of a disease? What are the causes of 'typical' disease symptoms?
5. Suggest why measles is very rarely fatal in Britain yet still ranks among the top five killer diseases in the developing world.
6. How have changes in human lifestyle contributed to the spread of infectious diseases?

2.31 Pathogens are organisms that cause disease

OBJECTIVES

- To recall that a pathogen is an organism that causes disease
- To give examples of different types of pathogen and the diseases they cause

Pathogens are parasites

Any organism that affects the body to cause disease is a pathogen (or pathogenic organism). Pathogens are parasites, that is, they live on the body of a host and cause it some harm:

- **by secreting poisons (toxins)** – this is especially common from bacteria. The toxins have different effects, for example the organism *Clostridium botulinum* produces a deadly nerve poison, whereas bacteria of the *salmonella* group release a toxin which irritates the lining of the gut.
- **as a result of multiplication** – the organism may reproduce quickly and produce such a large colony that it damages cells directly, as in malaria, or it uses up compounds which should be used by the host cell, as in polio. Many viruses multiply and cause host cells to burst.
- **as a result of the immune response** – when the host detects pathogens it directs more blood to the site of infection. This can cause swelling and soreness, and usually causes a rise in body temperature.

There are many pathogens that are parasites on humans, causing diseases. The following table illustrates the range of pathogenic organisms.

Pathogen type	Size	Disease in humans
Virus	About nm (1/1000 μm)	Influenza AIDS
Bacterium (Prokaryote)	About 1μm	Cholera Food poisoning
Protoctist	Up to 1 mm	Dysentery Malaria
Fungus	May be extensive	Athlete's foot Ringworm
'Worms'	Up to several meters	Tapeworm *Toxocara*

Viruses

Viruses are responsible for many of the most serious human diseases. Viruses differ from the true living organisms in that they cannot survive and reproduce outside the cells of their host – every virus is a parasite. The structure of viruses was described on page 6. Remember that viruses are so small that they can pass through filters and screens which will trap any other organism. We use the electron microscope to study viruses.

Bacterial diseases

The structure of bacteria and their importance to humans are outlined on page 244. Some bacteria cause serious human diseases, TB for example.

TB is now the infectious disease that causes most deaths worldwide, due to:

- poor disease control programmes
- resistance to antibiotics (see page 248)
- co-infection with HIV (see page 189)
- a rapid increase in the population of young adults, the age group most at risk from TB.

Any bacterial disease can be treated with antibiotics.

Controlling the spread of disease

Transmissable diseases are caused by pathogens so controlling these diseases often involves preventing the spread of these organisms. Important techniques include:

- hygienic food preparation (see opposite)
- good personal hygiene
- waste disposal (page 280)
- sewage treatment (page 277).

Individuals and communities all contribute to the control of disease.

2.32 Preventing disease: safe food

OBJECTIVES

- To understand how microorganisms can affect human food supplies
- To understand the dangers of food poisoning
- To understand how the risks of food poisoning can be minimised

Human foods need to be protected from microorganisms for two reasons:

- the microorganisms might be decomposers and spoil the food
- the microorganisms might be pathogenic and cause disease.

Food poisoning

Food poisoning is caused by eating contaminated food which contains harmful numbers of food-poisoning microbes. The symptoms of food poisoning may be due to:

- the microorganism 'feeding' on the host tissues as it reproduces
- toxins released onto the food by the microorganism or released inside the host as the microbe reproduces.

A number of bacteria can cause food poisoning, including *Clostridium botulinum* (found in soil, fish and meat – it makes the most deadly nerve toxin produced by a living organism), *Listeria monocytogenes* (found in soft cheeses and pâtés, and becoming more common as we eat more 'convenience foods') and the *Salmonella* group. The diagram below shows some questions and answers about bacteria and food poisoning. Thorough cooking kills any bacteria in food, but food may become contaminated after being cooked, or may not be cooked well enough.

The following principles of good food hygiene help to prevent food poisoning:

- Avoid contamination of food by bacteria – wash hands, cooking utensils and surfaces carefully during food preparation, and package food carefully during transport and storage.
- Prevent any bacteria that do gain entry to food from multiplying – see page 160.
- Destroy any remaining bacteria – cook the food thoroughly.

The causes and symptoms of food poisoning

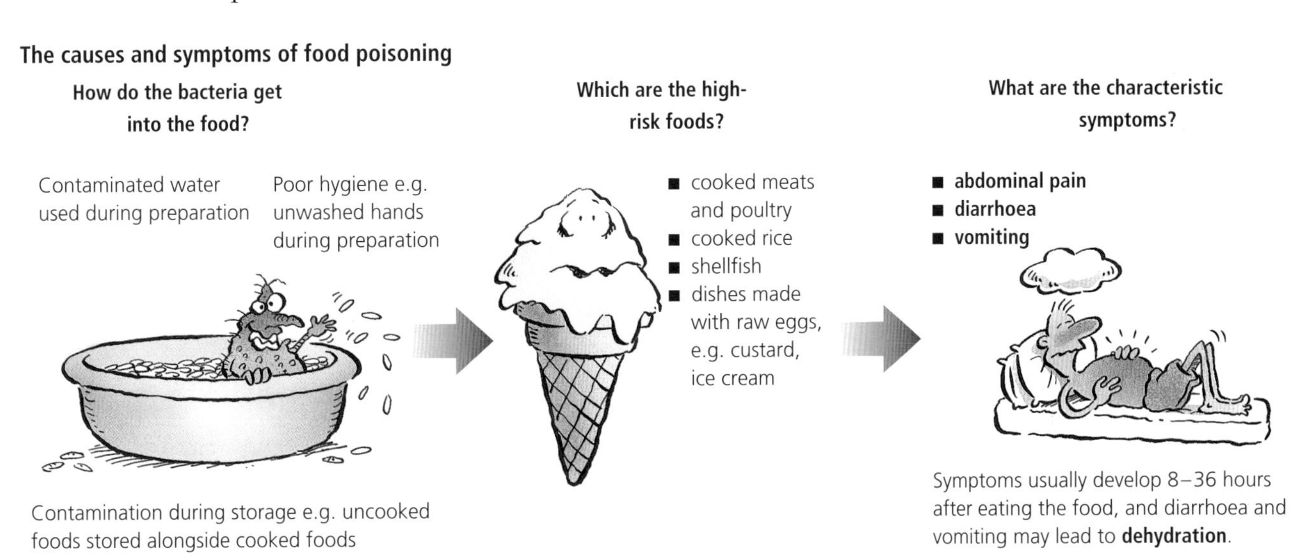

2.33 Individuals and the community can fight disease together

OBJECTIVES

- To understand that the fight against disease involves several levels of responsibility
- To provide examples of individual, community and worldwide responsibilities

The fight against disease has three levels:

- personal – for example, each individual can take responsibility for his or her own social habits
- communities – for example, local health services must be correctly managed and financed
- worldwide – for example, many nations carry out vaccination programmes (see page 105).

Personal responsibility

The individual can reduce his or her chances of contracting some diseases by:

- personal hygiene
- balanced diet
- regular exercise
- not smoking
- controlling alcohol intake.

Community responsibility

Living close together in towns and cities means we share many facilities which affect our health.

Community health responsibilities include:

- Providing safe drinking water and treating sewage (see page 277)
- removing refuse (see diagrams below and right)
- providing medical care for the unwell
- monitoring standards of health and hygiene.

Worldwide responsibility

The **World Health Organisation (WHO)** aims to raise the level of health of all the citizens of the world so that they can lead socially productive lives. They have had some successes:

- reduce infant mortality, by providing a better diet for mothers and their infants
- elimination of smallpox, by a well coordinated vaccination programme
- reduction in malaria, by a variety of control measures such as draining swamps
- improved provision of safe water, by the construction of water-treatment plants.

Refuse disposal in land-fill sites

Advantages

- Can help **land reclamation** e.g. in filling old quarry workings.
- Can be made economical in terms of space:
  - lorries use rams to compress rubbish
  - very deep pits can be used.
- Can be situated well away from residential areas, reducing impact of smells and unsightly rubbish.

Disadvantages

- Sites can attract **pests**, such as flies, rats and gulls. These might spread disease (e.g. flies contaminate food) or leave a mess (e.g. gulls leave droppings).
- Ecologically important areas such as marshes and heathland may be used as dumps.

Refuse may be buried under 0.5m of soil. This

- reduces access to pests
- allows bacteria and fungi to decompose organic compounds
- generates **biogas** which can be used to fuel machinery.

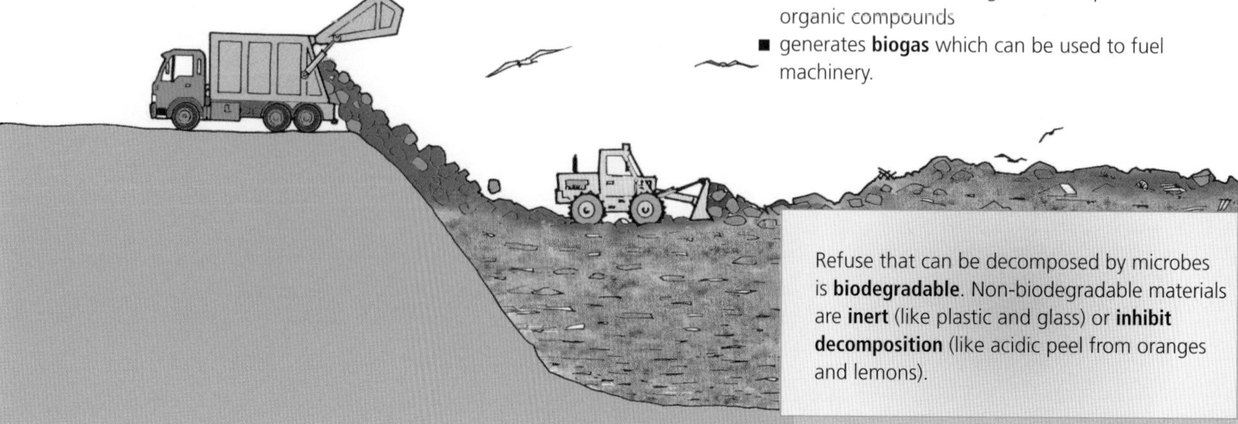

Refuse that can be decomposed by microbes is biodegradable. Non-biodegradable materials are inert (like plastic and glass) or inhibit decomposition (like acidic peel from oranges and lemons).

Refuse disposal by incineration

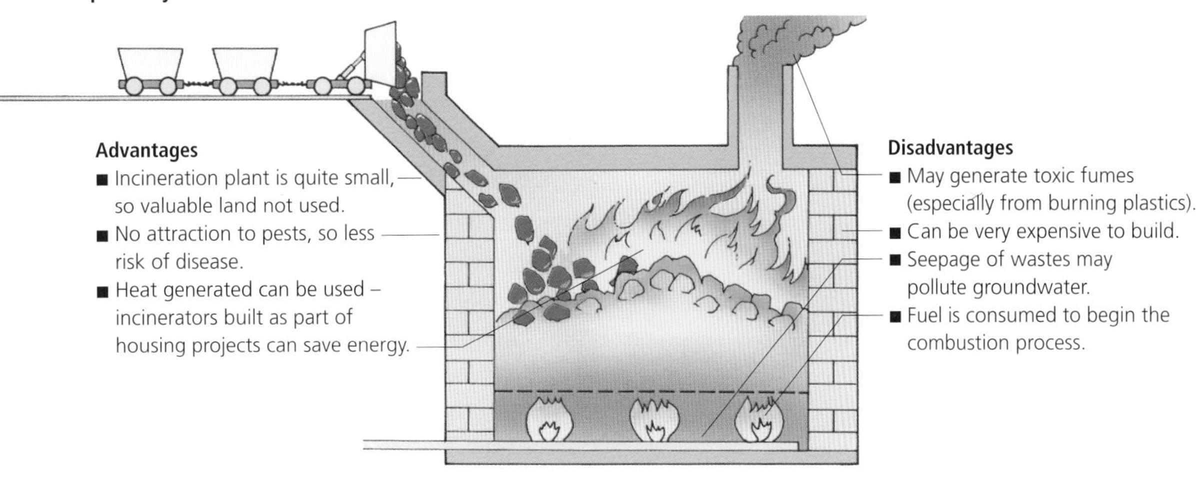

Advantages
■ Incineration plant is quite small, so valuable land not used.
■ No attraction to pests, so less risk of disease.
■ Heat generated can be used – incinerators built as part of housing projects can save energy.
Disadvantages
■ May generate toxic fumes (especially from burning plastics).
■ Can be very expensive to build.
■ Seepage of wastes may pollute groundwater.
■ Fuel is consumed to begin the combustion process.

▲ Open rubbish dumps are a source of infection. Waste food is infected with harmful microbes, which may be spread by rats, birds such as the Marabou stork, and humans.

1 Suggest two steps an individual can take to reduce the risk of named disease.
2 What are the responsibilities of a community health service?
3 Name one viral and one bacterial sexually transmitted infection (STI).
4 For any one named STI suggest how individuals, local communities and scientists worldwide might be involved in its control.

2.34 Combating infection: blood and defence against disease

OBJECTIVES

- To recall what is meant by disease
- To recall that disease can be caused by pathogens, which must first invade the body
- To understand that the body may be able to defend itself against pathogens

Disease is often caused by the invasion of the body by another organism. Organisms that cause disease in this way are called **pathogens** and their attacks on the body result in **infections**.

The skin and defence against disease

The outer layer of the skin, the epidermis, is waxy and impermeable to water and to pathogens (although microorganisms can live on its surface). Natural 'gaps' in the skin may be protected by **chemical** secretions, for example:

- the mouth leads to the gut which is protected by hydrochloric acid in the stomach
- the eyes are protected by lysozyme, an enzyme that destroys bacterial cell walls, in the tears
- the ears are protected by bactericidal ('bacteria-killing') wax.

Physical defences against the entry of microorganisms include the cilia and mucus-secreting cells of the respiratory pathways (see page 112). If the potential pathogens do penetrate these first lines of defence, they might reproduce quickly in the warm, moist, nutrient-filled tissues. Further defence depends upon the **blood**.

Bleeding and clot formation

Blood clotting seals wounds. The blood clot limits the loss of blood and also prevents entry of any pathogens. Clotting depends on **platelets** and **blood proteins**, as outlined in the diagram below.

White blood cells and defence

Organisms that gain entry to the tissues are removed or destroyed by **white blood cells**. These white blood cells must attack only invading organisms and not the body's own cells (although this does happen sometimes, see page 105). The white blood cells recognise foreign particles such as bacteria, or perhaps large molecules such as proteins in snake venom, and react against them. These foreign particles are called **antigens**. Potential pathogens have antigens on their

Blood clotting reduces loss of blood and seals the wound against pathogens

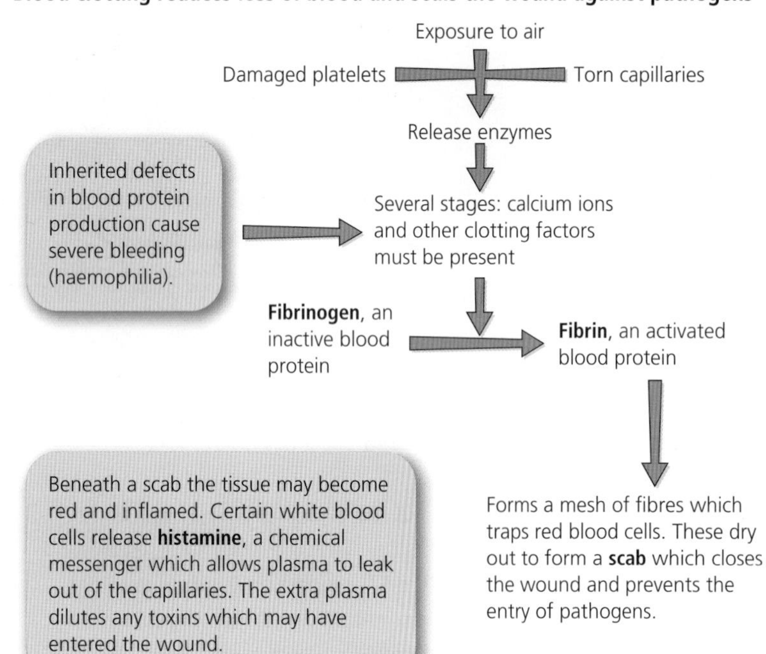

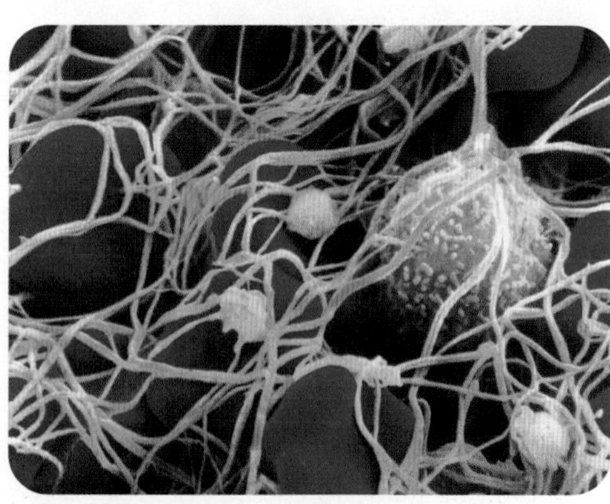

▲ Fibrin fibres trap blood cells to form a scab

Phagocyte action

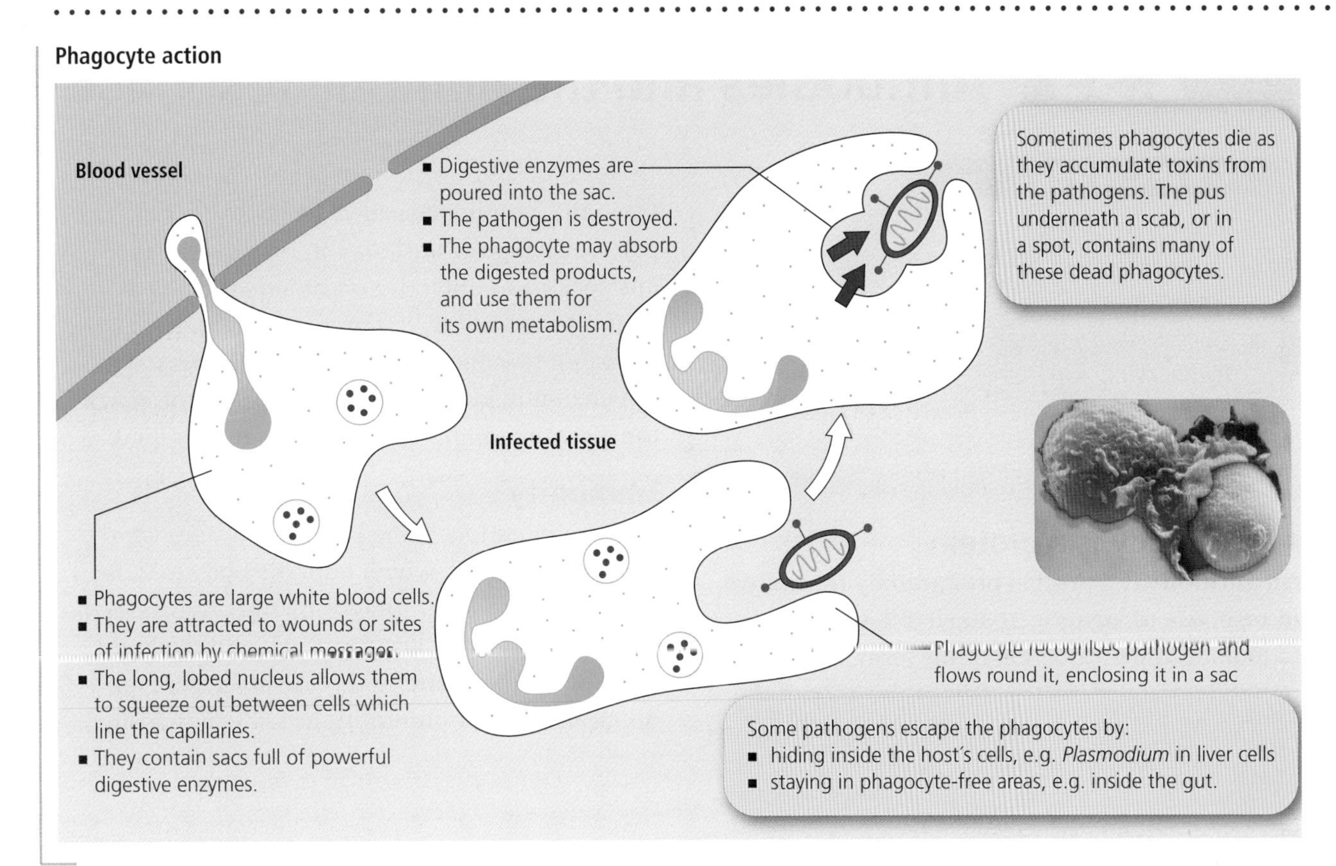

surface, and they are recognised and destroyed by **phagocytes** and by **antibodies**. Phagocytes engulf and then destroy the pathogens with digestive enzymes, as outlined above. Antibodies are described on page 104.

There are many phagocytes present in areas of the body likely to suffer infection. The exposed surfaces of the lungs, for example, are patrolled by phagocytes. If the lungs are regularly attacked by the free radicals in tobacco smoke, large numbers of phagocytes collect and may become disorganised. They can destroy lung tissue rather than foreign particles, leading to the disease **emphysema** (see page 121).

1 Suggest how the skin may limit the entry of pathogens to the body. Why is it necessary to prevent the entry of pathogens?

2 This question concerns the process of blood clotting.

a Why might blood clotting be necessary?

b When could blood clotting be a disadvantage?

c Blood clotting occurs in a number of stages. This is quite common in biological processes, since it allows **amplification**. Each step produces a product which can trigger many repeats of the next step – for example, each enzyme molecule released from a platelet can catalyse the conversion of 100 inactive protein molecules to their active form.

i Suggest how this amplification might be an advantage in a rapid response to wounding.

ii If there were five steps, each allowing an amplification of 100, how many 'product' molecules would be present at the end of the complete process for each 'signal' molecule released at the start of the process?

3 Describe the process of phagocytosis. Suggest two ways in which the structure of a phagocyte is related to its function. Suggest two ways in which a pathogen might avoid phagocytosis.

First aid can save lives

Severe blood loss, or haemorrhage, can cause a number of problems. Following an accident there may be so much blood loss that the blood pressure falls to dangerously low levels. This may affect the function of vital organs such as the heart, brain and kidney.

If the victim has no pulse or is not breathing, steps are taken to restart these essential functions.

If there is considerable bleeding, the first-aider applies firm pressure to the site of the injury. The victim should not be moved unless he or she is in danger. No attempt should be made to clean wounds – this may force a foreign body such as a glass fragment deeper into the wound.

Organisations such as the St John's Ambulance Brigade in the UK provide training in first aid.

2.35 Antibodies and the immune response

OBJECTIVES

- To know what an antibody is
- To understand how antibodies are involved in defence against disease
- To understand how memory cells protect against infections
- To know how the immune response can be enhanced

An antibody is a protein

An **antibody** is a protein produced by the body in response to an antigen. Each different antigen stimulates the production of the particular type of antibody that will destroy that antigen. Antibodies are made by white blood cells called **lymphocytes**. They defend the body as shown below.

Once the lymphocytes have learnt to make a particular type of antibody in response to the antigens on an infective organism, the body begins to recover as the organisms are destroyed. It takes a few days to produce antibodies, so the infected individual will show some symptoms of the disease (a high temperature, for example).

Immunity

After an infection, some lymphocytes are kept as 'memory cells', which help the body to defend itself against further attacks by the same antigen. This 'memory' may last for years, and the body is said to be **immune** to the disease. There are different types of immunity, as shown opposite.

The action of white blood cells – lymphocytes

Scientists estimate that humans can make up to 1 000 000 different antibodies – enough to account for every pathogen or 'foreign' substance we might ever meet!

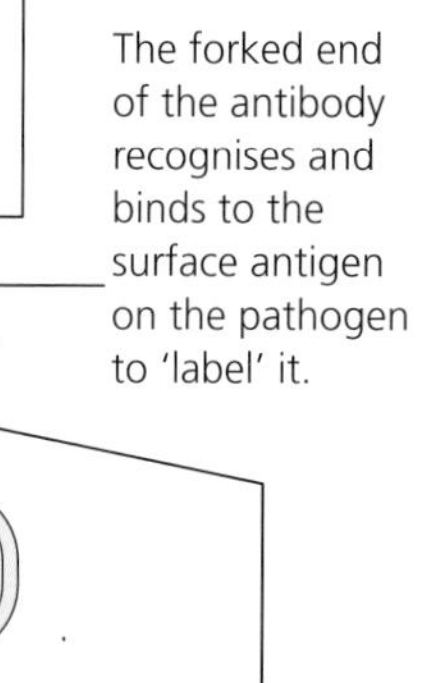

This end of the antibody acts as a signal to destroy the pathogen.
The forked end of the antibody recognises and binds to the surface antigen on the pathogen to 'label' it.

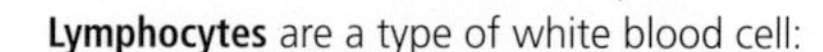

Lymphocytes are a type of white blood cell:
- found in circulating blood and in the lymph nodes (see page 89)
- have a large nucleus and no granules in the cytoplasm
- stimulated by contact with pathogens to produce **antibodies**.

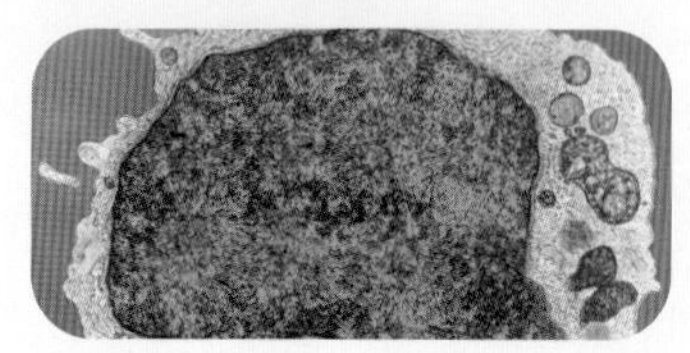

Antibodies are:
- proteins produced by lymphocytes
- able to recognise, bind to and help to destroy pathogens
- always Y-shaped.

'Labelled' pathogens may be destroyed by:
- sticking together in clumps so they can be **ingested by phagocytes**
- T-lymphocytes, which **burst membranes around the pathogen**
- antibodies directly – a few antibodies may actually destroy the pathogen's cell walls or membranes.

An **antigen** is:
- a protein or carbohydrate on the surface of the pathogen
- able to provoke the immune system of the host.

Pathogens may evade the immune system by mutation – they change and produce different antigens which the host has not learned to recognise and has no antibodies for.

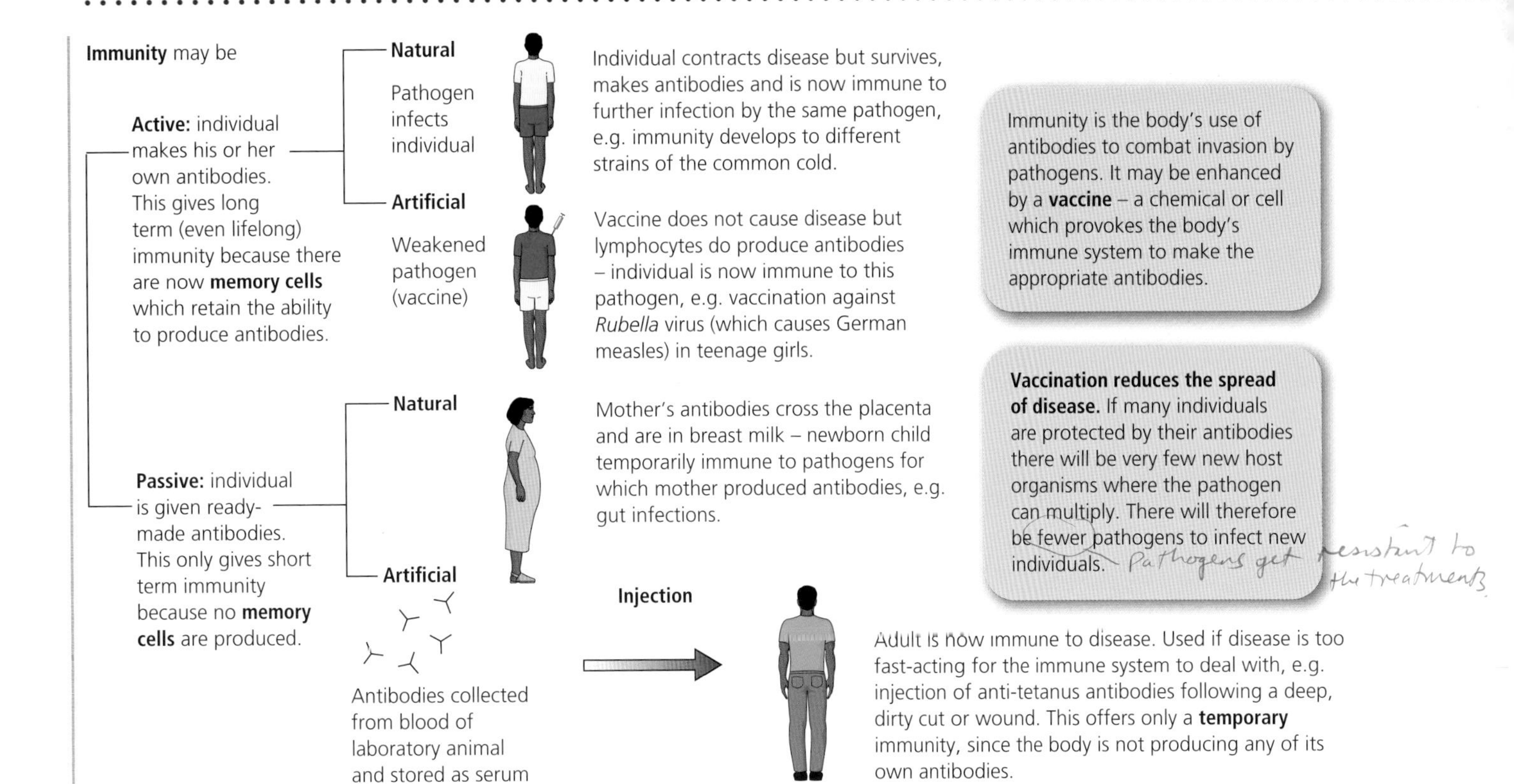

Immunity is the body's use of antibodies to combat invasion by pathogens. It may be enhanced by a **vaccine** – a chemical or cell which provokes the body's immune system to make the appropriate antibodies.

Vaccination reduces the spread of disease. If many individuals are protected by their antibodies there will be very few new host organisms where the pathogen can multiply. There will therefore be fewer pathogens to infect new individuals.

Some problems with the immune response

The activity of the immune system saves all our lives, many times over. There are occasions, however, when it may actually reduce the likelihood of survival. These are described below.

Autoimmune diseases are caused by the body producing antibodies which destroy its own cells. Why this should happen is not known. An example is Type I diabetes where the body destroys its own insulin-producing cells by an immune reaction.

Other problems*

Allergies are conditions in which the body becomes sensitive to a substance and over-reacts to it. This may cause swelling and tissue damage. Examples include hay fever and sensitivity to bee and wasp stings. Allergies are treated with drugs such as antihistamines to reduce the immune response.

Transplant rejection – the most common organ transplant in Britain is the kidney transplant (see page 127), but heart, intestine, lung, liver and pancreas transplants are becoming more common. The recipient's lymphocytes may recognise antigens on the surface of the donor organ as foreign and slowly destroy it. This problem of rejection is being overcome by:

- drugs that suppress the immune system of the recipient long enough to allow the transplanted organ to become established
- matching tissues wherever possible, for example by seeking out relatives of people needing bone marrow transplants, since relatives are more likely to have similar antigens to the recipient.

Q

1. How do antibodies recognise pathogens?
2. State one difference in structure between a lymphocyte and a phagocyte.
3. Explain how a single infection by a pathogen can provide lifelong protection against a disease.

2.36 Respiration provides the energy for life

OBJECTIVES

- To understand that energy is needed to carry out work
- To appreciate that different forms of energy can be interconverted
- To be able to list some of the energy-demanding processes in living organisms
- To describe how the process of respiration releases energy from chemical foods

Energy conversions in living things

Energy can be defined as 'the capacity for doing work'. The processes that keep organisms alive (for example, pumping ions from one side of a membrane to another) usually require work to be done. It is clear, therefore, that life requires energy.

Life also depends on energy **conversions**. The first and most important energy conversion in living things is **photosynthesis**. This process is described in greater detail on page 38, but in energy terms it can be described simply as:

light energy from the Sun
⇓
converted in green plants
⇓
chemical energy in organic molecules

A second energy-conversion process releases the chemical energy stored in organic food molecules and converts it to energy forms that organisms use to stay alive. This process is called **respiration**. Respiration is the set of chemical reactions that break down nutrient molecules in living cells to release energy. Respiration involves the action of enzymes in cells (see page 32). The diagram at the bottom of the page shows some of the uses of energy by living organisms.

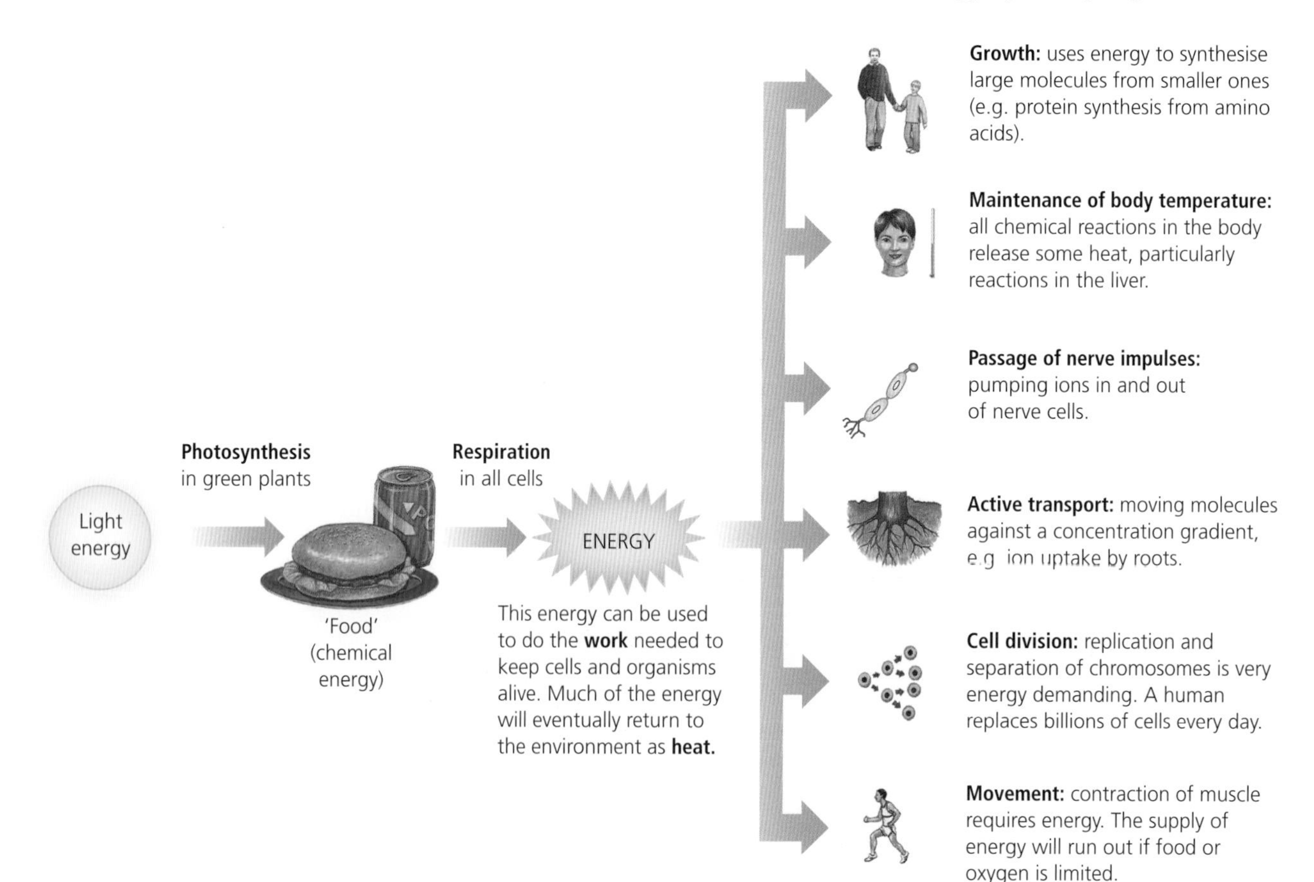

Light energy
Photosynthesis in green plants
'Food' (chemical energy)
Respiration in all cells
ENERGY
This energy can be used to do the work needed to keep cells and organisms alive. Much of the energy will eventually return to the environment as heat.
Growth: uses energy to synthesise large molecules from smaller ones (e.g. protein synthesis from amino acids).
Maintenance of body temperature: all chemical reactions in the body release some heat, particularly reactions in the liver.
Passage of nerve impulses: pumping ions in and out of nerve cells.
Active transport: moving molecules against a concentration gradient, e.g. ion uptake by roots.
Cell division: replication and separation of chromosomes is very energy demanding. A human replaces billions of cells every day.
Movement: contraction of muscle requires energy. The supply of energy will run out if food or oxygen is limited.

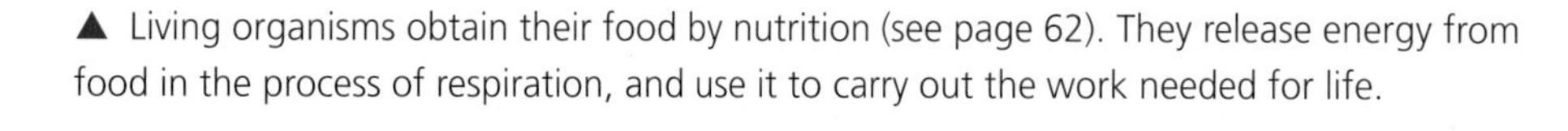

▲ Living organisms obtain their food by nutrition (see page 62). They release energy from food in the process of respiration, and use it to carry out the work needed for life.

What is respiration?

Respiration involves the **oxidation** of food molecules, and energy is released during the process. These food molecules all contain carbon, hydrogen and oxygen and the complete oxidation process converts these to carbon dioxide and water. Complete oxidation occurs only if oxygen is present. For example:

words: glucose + oxygen → energy + carbon dioxide + water

symbols: $C_6H_{12}O_6 + 6O_2 \rightarrow \text{energy} + 6CO_2 + 6H_2O$

Definition: respiration is the release of energy from food substances, and goes on in all living cells. So much energy is released in this process that, if it was all released at once, the cell might be damaged.

The energy is released in small 'packets' which can then drive the reactions that keep the cell alive. These energy packets act as a short-term store of energy. The one used most commonly by cells is a molecule called **adenosine triphosphate (ATP)**.

The reactions are summarised in the following equation.

glucose + oxygen → ATP → 'work' in cells
↓
carbon dioxide + water

This process, which releases energy in the presence of oxygen, is called **aerobic respiration** ('aerobic' means 'in air'), and takes place in the **mitochondria** (see page 17). The reactions involve enzymes, and some energy is 'lost' as heat.

VACUUM FLASK EXPERIMENT

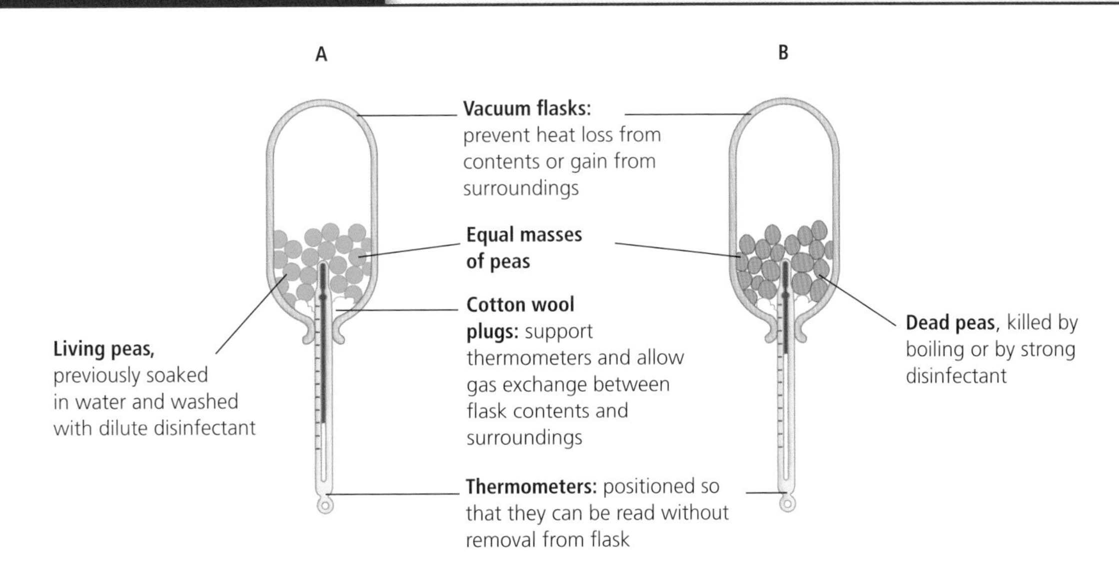

- Apparatus is assembled as shown.
- Temperature readings are taken immediately, and then every day for 4–5 days.

In an experiment like this one, the following results were obtained:

Time / days	Temperature / °C	
	A	B
0	20	20
1	23	20
2	25	20
3	27	21
4	28	21
5	28	21

a Explain these results.
b Why is it important to include flask B?
c Disinfectant kills bacteria – why is this important?

2.37 Contraction of muscles requires energy supplied by respiration

OBJECTIVES

- To know that respiration is the source of energy for muscular work
- To understand that anaerobic respiration is less efficient than aerobic respiration and produces a toxic product
- To understand that exercise is limited by the build-up of lactic acid

How muscles contract

Muscles contract and pull on bones to move the skeleton. Muscles are collections of very long **muscle fibres**. When lots of fibres shorten at the same time, the muscle contracts.

Muscular work requires energy

Aerobic respiration

Work must be done to contract muscle. **The energy** for this comes from **aerobic respiration**. The equation is a reminder of how this energy is released:

glucose + oxygen → energy + carbon dioxide + water

The oxygen comes from the air. It is taken in at the lungs and carried around the body in the blood, pumped by the heart. Glucose comes from food digested in the gut, and is also carried in the blood. Muscles have extensive capillary beds to supply glucose and oxygen for respiration, and to carry away carbon dioxide. The blood also carries away heat that is produced during respiration.

Anaerobic respiration

When we work very hard our muscles use up a lot of energy. The heart and lungs, even working flat out, cannot supply enough oxygen to provide this energy by aerobic respiration. Muscles can release energy from food without using oxygen, by a process called **anaerobic respiration**:

glucose → energy + lactic acid

Anaerobic respiration has two drawbacks:

- It gives only about one-twentieth of the energy per glucose molecule that aerobic respiration yields.
- Lactic acid is poisonous – if it builds up in the cells it inhibits muscular contraction, which leads to fatigue and, eventually, death.

This harmful lactic acid is carried out of the muscles in the blood. It is transported around the body to the heart, liver and kidneys where it is oxidised to **pyruvate**, which can be used to release energy by aerobic respiration. The heart, the liver and the kidneys will need extra oxygen to get rid of this lactic acid, provided by the deep fast breathing that follows hard exercise. This extra oxygen is the **oxygen debt**.

1 One of the comparisons between aerobic and anaerobic respiration shown below is incorrect. Which one?

	Respiration	
	Aerobic	**Anaerobic**
A	Uses oxygen gas	Does not use oxygen gas
B	Produces ethanol or lactic acid	Produces no ethanol or lactic acid
C	Large amount of energy released	Small amount of energy released
D	Mitochondria involved	Mitochondria not involved
E	Carbon dioxide always produced	Carbon dioxide sometimes produced

2 What is meant by energy transformation? Briefly describe one energy transformation that is important to living organisms. What eventually happens to all of the energy taken in by living organisms?

The diagram below shows the part played by aerobic and anaerobic respiration during rest, exercise and recovery.

Rest – all respiration is aerobic. Normal breathing and heart rates can supply the tissues with all the oxygen they need,
Glucose + oxygen → energy + carbon dioxide + water

Heart rate
70 beats per minute

Breathing
15 breaths per minute

Hard exercise – respiration is mainly anerobic. Very high breathing and heart rates still cannot provide the muscles with enough oxygen for aerobic respiration.
Glucose → energy + lactic acid

Heart Rate
140 beats per minute

Breathing
50 breaths per minute

The muscles are getting energy without 'paying' for it with oxygen. They are running up an oxygen debt.

Recovery – paying off the oxygen debt. The breathing and heart rates remain high, even though the muscles are at rest. The extra oxygen is used to convert the lactic acid into carbon dioxide and water, paying off the oxygen debt.

Heart rate
140 beats per minute falling to normal after some minutes

Breathing
50 breaths per minute falling to normal after some minutes

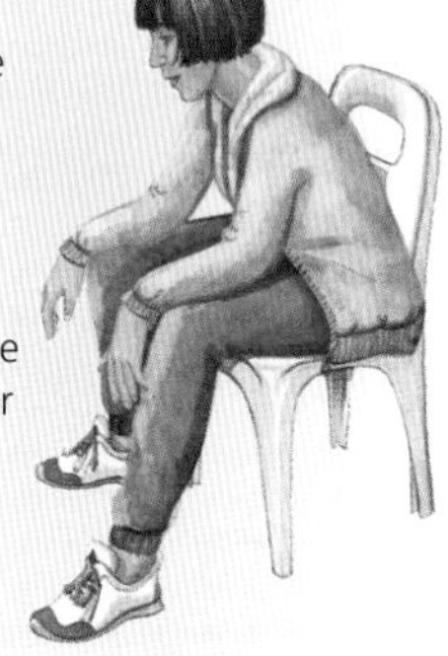

Panting and rapid heartbeat continue until the lactic acid has been romoved. Physically fit people recover faster.

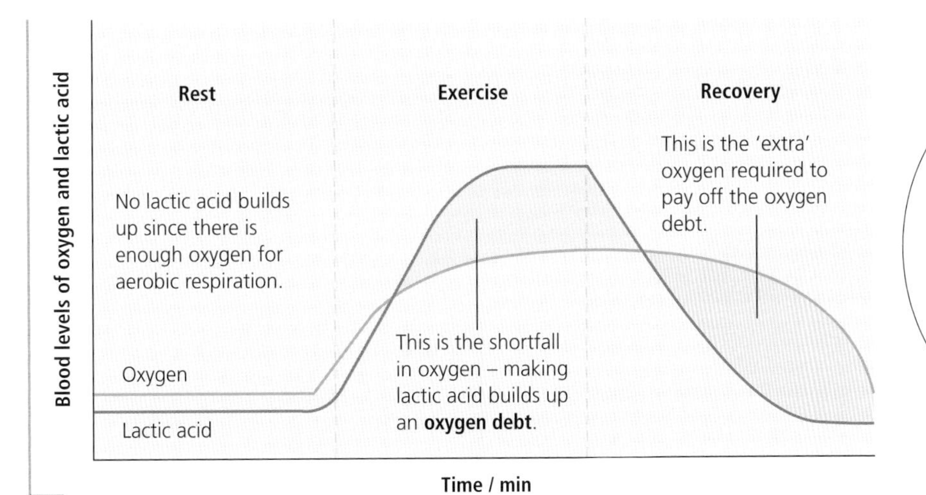

Anaerobic respiration is important in other organisms such as

- yeast, for brewing and baking (page 250)

Glucose ⟶ alcohol + carbon dioxide

$C_6H_{12}O_6 \longrightarrow 2C_2H_5OH + 2CO_2$

▲ Trained athletes can exercise for longer than people who do not usually exercise before they build up an oxygen debt

Q

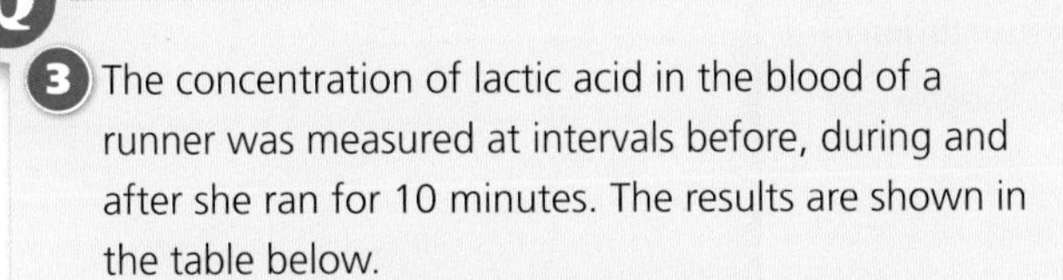

3 The concentration of lactic acid in the blood of a runner was measured at intervals before, during and after she ran for 10 minutes. The results are shown in the table below.

Time / minutes	Concentration of lactic acid / arbitrary units
0	18
10	18
15	56
25	88
35	42
50	21
65	18

a Plot this information in the form of a graph.
b What was the lactic acid concentration at the end of the run?
c For how long did the concentration of lactic acid increase after the end of the run?
d Why did the blood still contain lactic acid after the run?
e In which tissues was the lactic acid produced?
f How long after the run was it before the oxygen debt was paid off?

2.38 The measurement of respiration

OBJECTIVE

- To explain how it is possible to detect the process of respiration

Since respiration is essential for life, an obvious 'sign of life' is that respiration is taking place. We can demonstrate that respiration is happening, and we can even measure how quickly it is going on.

The equation for aerobic respiration suggests methods that we might use to demonstrate this process.

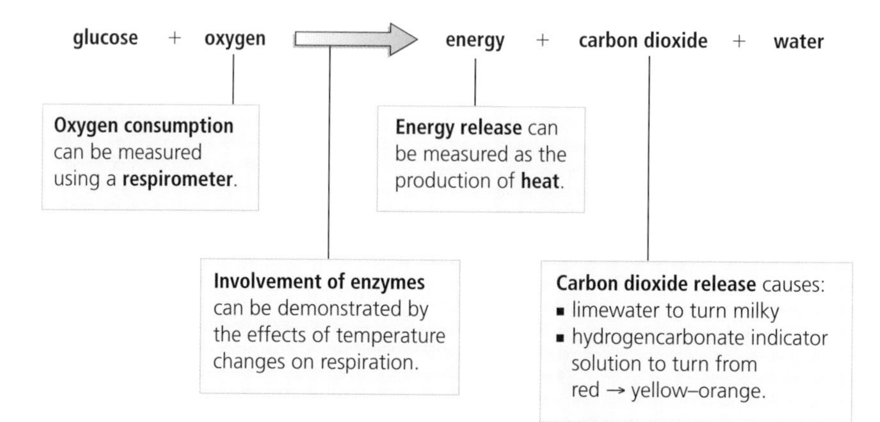

glucose + oxygen → energy + carbon dioxide + water
Oxygen consumption can be measured using a **respirometer**.
Energy release can be measured as the production of **heat**.
Involvement of enzymes can be demonstrated by the effects of temperature changes on respiration.
Carbon dioxide release causes:
- limewater to turn milky
- hydrogencarbonate indicator solution to turn from red → yellow–orange.

To demonstrate that respiration is going on, we need to use some living organisms. Germinating seeds are useful for this experiment since they are chemically very active (the fast growth during germination requires a great deal of energy). Animals such as blowfly larvae (maggots) or woodlice can also be used, although they are rather more difficult to handle.

1 MEASUREMENT OF OXYGEN CONSUMPTION USING A RESPIROMETER

The apparatus shown in the diagram on the next page is used.

a A measured mass of living organisms is put in the chamber.

b First the spring clip is open, so that an equilibrium of temperature and pressure is set up between the chamber and the surroundings.

c After five minutes or so, the spring clip is closed and the movement of the coloured liquid along the capillary tube is observed. The time taken for it to move a measured length along the tube is noted.

d The living organisms are removed and the experiment is repeated.

Using this apparatus, a group of students obtained the following data.

Experiment	Distance moved / mm	Time taken / s	Relative oxygen consumption / mm per s
1 Seeds	39	100	
2 Seeds	42	100	
3 Seeds	24	60	
4 Maggots	46	90	
5 Maggots	55	100	
6 Maggots	30	60	

The effect of temperature on respiration can be investigated using this apparatus. The respiratory chamber can be placed in a thermostatically controlled water bath and the measurements made at a series of temperatures (see Q.3 on page 111 opposite).

Principle: Carbon dioxide produced during respiration is absorbed by potassium hydroxide solution. If the system is closed to the atmosphere, a change in volume of the gas within the chamber must be due to the consumption of oxygen. The change in volume, i.e. the oxygen consumption, is measured as the movement of a drop of coloured liquid along a capillary tube.

Spring clip – when open, contents of chamber are in equilibrium with atmosphere.

Rubber stopper – can be made more airtight by smearing Vaseline along seal between chamber and stopper.

Filter paper wick ensures that maximum surface area of potassium hydroxide solution is available to contents of chamber.

Potassium hydroxide solution absorbs carbon dioxide evolved during respiration.

Graduated scale against which movement of coloured liquid can be measured.

Coloured liquid – narrow-bore capillary tube means that any small change in volume of gas in chamber produces a large movement of liquid.

Gauze basket to hold respiring material must be porous to allow free exchange of gases.

Respiratory chamber has relatively low volume, so that changes in volume due to respiration are significant enough to be measured.

Weighed amount of respiring material – the mass, m, of the respiring material should be known so that results can be compared with other experiments.

Glass rod to keep respiring material out of direct contact with potassium hydroxide solution.

2 MEASUREMENT OF CARBON DIOXIDE RELEASE

To detect carbon dioxide released by respiring organisms, we can use an **indicator solution**. For example **hydrogencarbonate indicator** is purple at high pH (alkaline), red around neutral pH and orange-yellow at low pH (acidic). The diagram shows the apparatus.

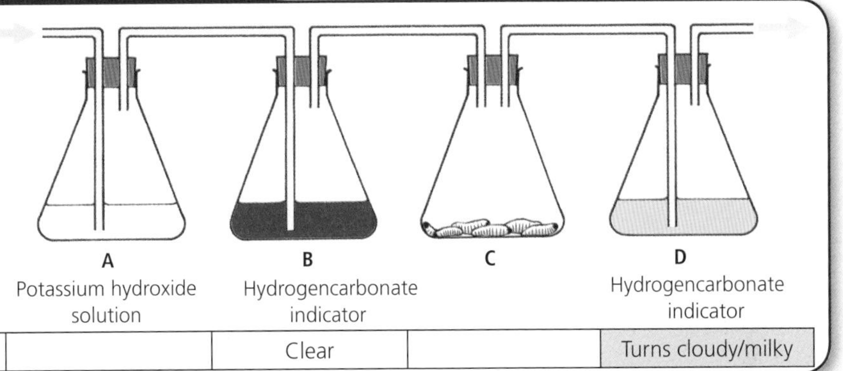

Q

1 **a** In experiment **1** why did the students take more than one set of readings for each group of organisms?

b Calculate the mean relative oxygen consumption for the seeds and for the maggots. Suggest a reason for the difference.

2 The students repeated the experiment with no living organisms in the chamber. The coloured liquid did not move at all. Why was this important?

3 In an extension of this investigation, the students measured the effect of temperature on the rate of oxygen consumption by the maggots. They obtained the following results:

Temperature / °C	Relative oxygen consumption / mm per s
15	0.3
25	0.6
35	1.1
45	0.8
55	0.2
65	0.0

a Plot these results on a graph, and explain the shape of the curve.

b In this investigation, identify the **independent** (input) **variable** and the **dependent** (outcome) **variable**.

c Suggest any **fixed variables**. (Refer to page 34 if you are unsure about these terms.)

In experiment **2**:

4 What is the purpose of the potassium hydroxide solution?

5 What does the indicator solution in flask **B** show?

6 How can you explain the change in flask **D**?

7 Suggest a control for this investigation, and explain why it is a suitable control.

8 Suggest any visible change that might happen in flask **C**. Explain your answer.

2.39 Gas exchange supplies oxygen for respiration

OBJECTIVES

- To understand why living organisms must obtain oxygen from their environment, and why they must release carbon dioxide to their environment
- To know the properties of an ideal gas exchange surface
- To be able to identify the parts of the human gas exchange system

Exchanging oxygen and carbon dioxide

Respiration uses oxygen to 'burn' (oxidise) food and so release the energy that cells need to stay alive. Respiration produces carbon dioxide and water vapour as waste products:

glucose + oxygen → energy + carbon dioxide + water

Living organisms must be able to take oxygen from the air and get rid of carbon dioxide to the air. Swapping oxygen for carbon dioxide in this way is called **gas exchange** (or **gaseous exchange**).

Gas exchange takes place through a **gas exchange surface**, also known as a **respiratory surface**. The diagram below shows how the process happens. The properties of an ideal respiratory surface are given in the table on the right.

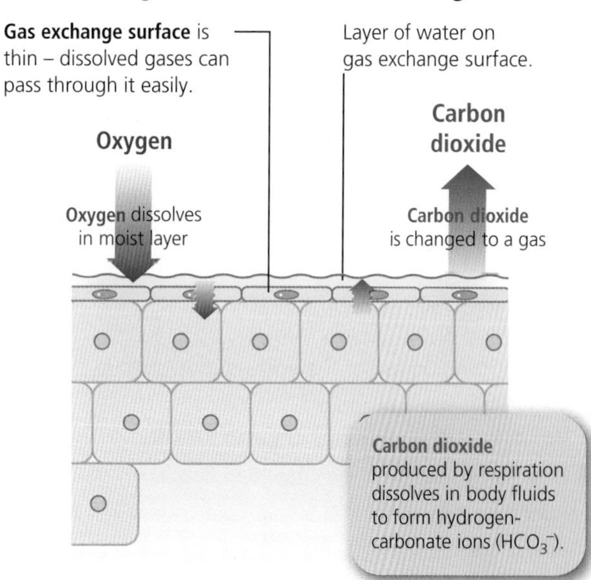

▲ A gas exchange surface allows cells to obtain the oxygen they need for respiration, and get rid of the carbon dioxide they produce

Property of surface	Reason
Thin (ideally one cell thick)	Gases have a short distance over which to diffuse.
Large surface area	Many molecules of gas can diffuse across at the same time.
Moist	Cells die if not kept moist.
Well ventilated	Concentration gradients for oxygen and carbon dioxide are kept up by regular fresh supplies of air.
Close to a blood supply	Gases can be carried to and from the cells that need or produce them.

Gas exchange in humans

Like other mammals, humans are active and maintain a constant body temperature. This means they use up a great deal of energy. Mammals must have a very efficient gas exchange system.

The gas exchange system in humans is shown opposite and is made up of:

- a **respiratory surface** – membranes lining the alveoli (air sacs) in the lungs
- a **set of tubes** to allow air from the outside to reach the respiratory surface. This set of tubes has many branches, and is sometimes called the 'bronchial tree'
- a **blood supply** (carried by the pulmonary artery and pulmonary vein) to carry dissolved gases to and from the respiratory surface
- a **ventilation system** (the intercostal muscles and the diaphragm) to keep a good flow of air over the respiratory surface.

Q

1. What are the properties of an ideal gas exchange surface?
2. List the structures through which a molecule of oxygen passes to get from the atmosphere to the cytoplasm of a named working cell.
3.  Most larger animals transport oxygen in red blood cells. What are the advantages of transporting oxygen in this way? How is a red blood cell adapted to its function of oxygen transport?
4. What is the difference between respiration and gas exchange?

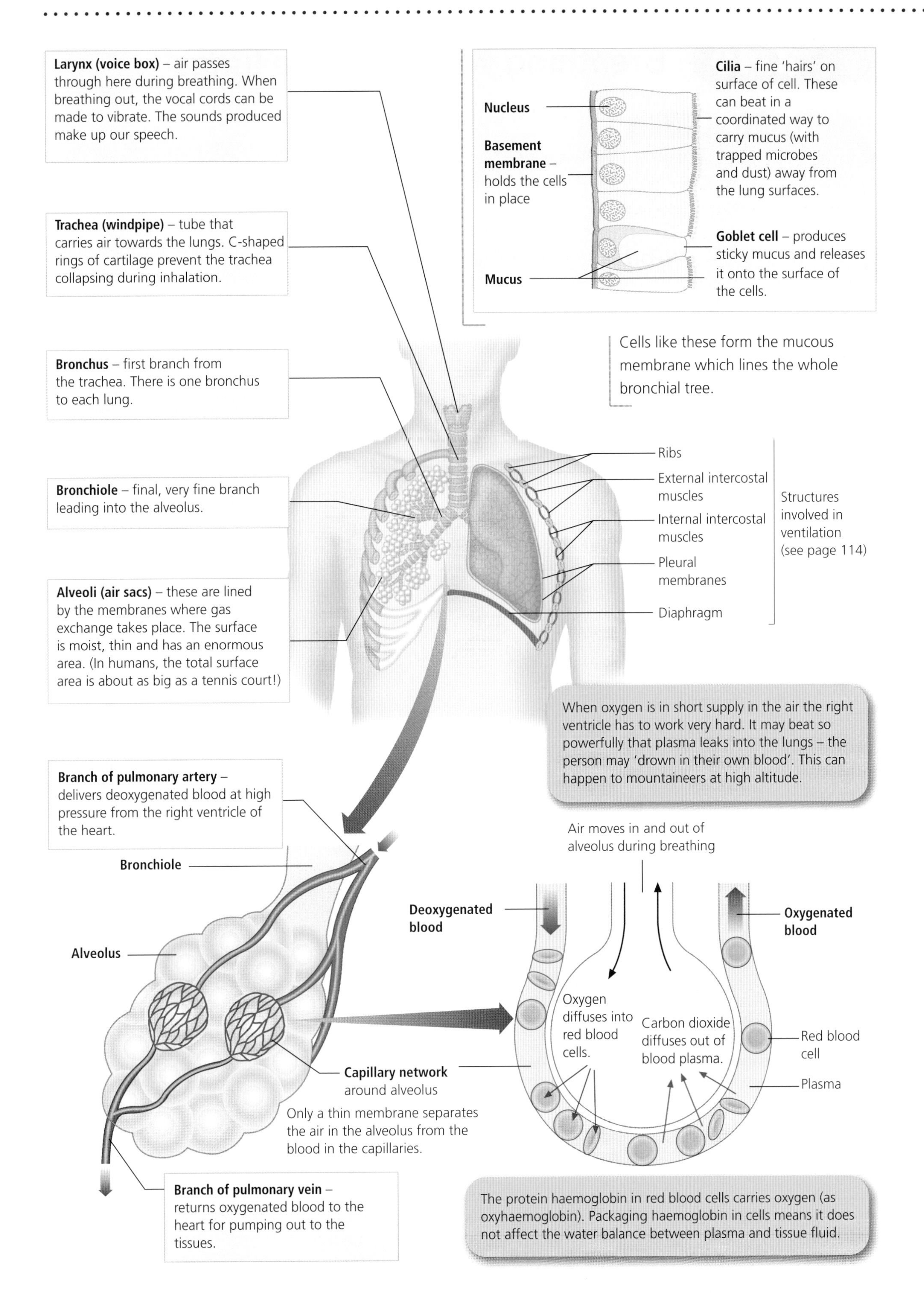

Larynx (voice box) – air passes through here during breathing. When breathing out, the vocal cords can be made to vibrate. The sounds produced make up our speech.
Trachea (windpipe) – tube that carries air towards the lungs. C-shaped rings of cartilage prevent the trachea collapsing during inhalation.
Bronchus – first branch from the trachea. There is one bronchus to each lung.
Bronchiole – final, very fine branch leading into the alveolus.
Alveoli (air sacs) – these are lined by the membranes where gas exchange takes place. The surface is moist, thin and has an enormous area. (In humans, the total surface area is about as big as a tennis court!)
Nucleus
Basement membrane – holds the cells in place
Mucus
Cilia – fine 'hairs' on surface of cell. These can beat in a coordinated way to carry mucus (with trapped microbes and dust) away from the lung surfaces.
Goblet cell – produces sticky mucus and releases it onto the surface of the cells.
Cells like these form the mucous membrane which lines the whole bronchial tree.
Ribs
External intercostal muscles
Internal intercostal muscles
Pleural membranes
Diaphragm
Structures involved in ventilation (see page 114)
When oxygen is in short supply in the air the right ventricle has to work very hard. It may beat so powerfully that plasma leaks into the lungs – the person may 'drown in their own blood'. This can happen to mountaineers at high altitude.
Branch of pulmonary artery – delivers deoxygenated blood at high pressure from the right ventricle of the heart.
Bronchiole
Alveolus
Capillary network around alveolus
Only a thin membrane separates the air in the alveolus from the blood in the capillaries.
Branch of pulmonary vein – returns oxygenated blood to the heart for pumping out to the tissues.
Air moves in and out of alveolus during breathing
Deoxygenated blood
Oxygenated blood
Oxygen diffuses into red blood cells.
Carbon dioxide diffuses out of blood plasma.
Red blood cell
Plasma
The protein haemoglobin in red blood cells carries oxygen (as oxyhaemoglobin). Packaging haemoglobin in cells means it does not affect the water balance between plasma and tissue fluid.

2.40 Breathing ventilates the lungs

OBJECTIVES

- To understand the muscular movements involved in the ventilation of the lungs
- To know how the efficiency of breathing can be measured
- To understand how breathing is affected by exercise
- To appreciate that the function of the lungs may sometimes need to be supported

Breathing is the set of muscular movements that gives the respiratory surface a constant supply of fresh air. This means there is always a concentration gradient between the blood and the air in the alveoli for both oxygen and carbon dioxide. As shown below, breathing is brought about by:

- the action of two groups of muscles – the **intercostal muscles** and the **diaphragm**
- the properties of the **pleural membranes** that surround the lungs.

The pleural membranes 'stick' the outside of the lungs to the inside of the chest cavity. The lungs themselves do not have any muscles, but the 'stickiness' of the pleural membranes means that the lungs will automatically follow the movements of the chest wall. If the volume of the chest cavity increases, the volume of the lungs will increase at the same time. The pressure inside the lungs will decrease as the volume gets bigger.

Air always moves down a pressure gradient, from a region of higher air pressure to a region of lower air pressure. If the air pressure in the lungs is less than the pressure of the atmosphere, air will move into the lungs along a pressure gradient. In the same way, if the air pressure in the lungs is greater than the pressure of the atmosphere, air will move out of the lungs along a pressure gradient.

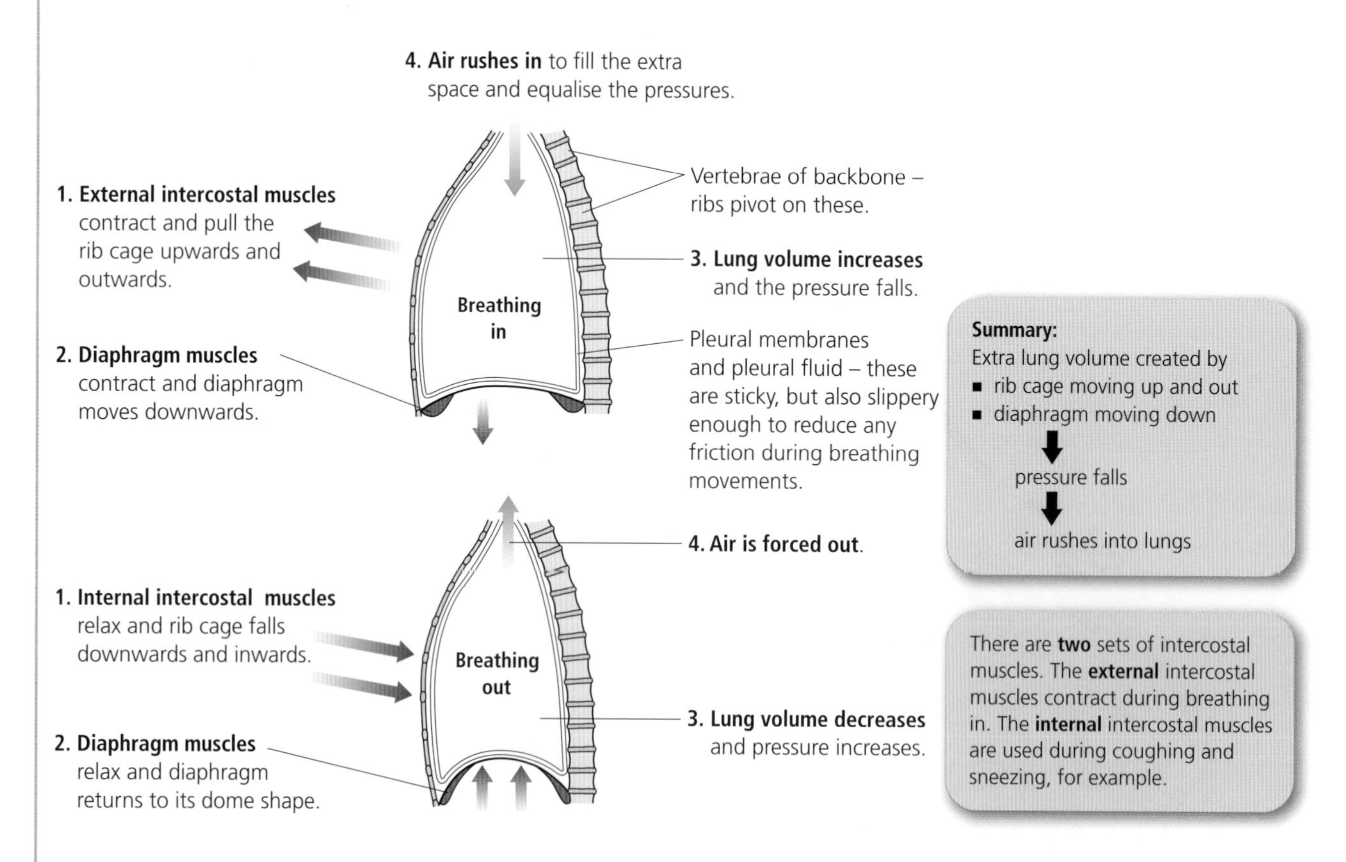

▲ The intercostal muscles ('intercostal' means 'between the ribs') and the diaphragm work together to alter the volume of the chest cavity. Changing the volume of the chest cavity will automatically change the pressure of air inside it. (It is a law of physics that pressure × volume is a constant – in other words, if pressure increases then volume must decrease, and vice versa.)

Air is breathed in, gas exchange happens in the alveoli, and the air is breathed out again. The composition of the inspired (breathed-in) air is therefore different from the composition of the expired (breathed-out) air, as shown in the table.

Component of air	Inspired (inhaled) air / %	Expired (exhaled) air / %	Reason
Oxygen	21	18	Oxygen has diffused from the air in the alveoli into the blood
Carbon dioxide	0.04	3*	Carbon dioxide has diffused from the blood into the air in the alveoli
Nitrogen	78	78	Nitrogen gas is not used by the body
Water vapour	Very variable	Saturated	Water evaporates from surfaces in the alveoli
Temperature	Very variable	37 °C	Heat is lost to the air from the lung surfaces

▲ Composition of inhaled and exhaled air

* The **increase in carbon dioxide concentration** can be detected with **limewater** (turns milky/cloudy) or **hydrogen carbonate indicator** (turns from red to yellow-orange), as shown on page 111.

Asthma*

Breathing is difficult, and an asthma sufferer may become very distressed. Air cannot easily move along the pressure gradient because:

- the muscles in the wall of the bronchi contract
- the lining of the bronchi 'leaks' a sticky mucus.

An asthma attack can be brought on by various factors including allergy to pollen or dust (or fur), emotion, breathing in cold air, smoke (page 120) and air pollution (page 264) and exercise. Treatment involves:

- removal of the factor causing the asthma
- use of a **bronchodilator** – usually a spray containing a drug which relaxes the bronchial muscles.

Measuring the efficiency of the lungs

The amount of air that enters and leaves the lungs is measured using a **spirometer**. A person breathes in and out of a mouthpiece connected to a chamber. Inside the chamber a piston moves up and down, and its movements are measured electronically. The changes in volume during breathing are plotted on a graph called a **spirogram**, shown below.

▼ A spirogram gives a great deal of information about someone's breathing and the efficiency of their lungs. The lung volumes are expressed in the SI unit dm^3. 1 dm^3 is the same as 1 litre.

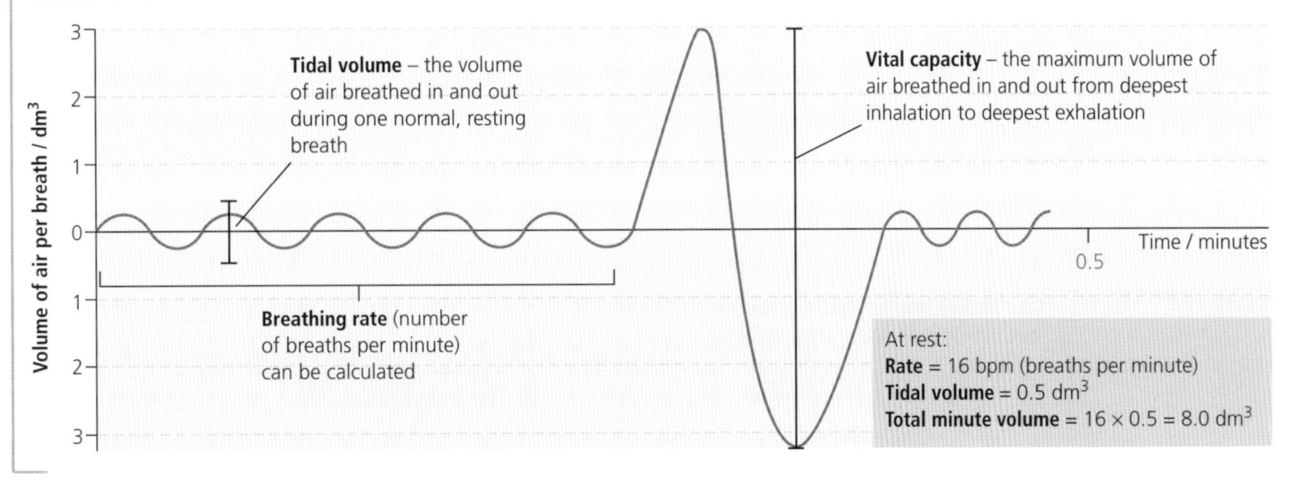

1 Explain the part played by the intercostal muscles and diaphragm in breathing.
2 What can be learned from a spirogram?
3 Why are there differences in the oxygen and carbon dioxide compositions of inhaled air and exhaled air?

Exercise and breathing

During exercise the muscles work hard, and need to release more energy by respiration. Greater volumes of air must therefore be breathed in and out, by:

- increasing the **breathing rate** – more breaths per minute
- increasing the **tidal volume** – more air per breath.

These two changes can increase the volume of air passing in and out of the lungs from the typical 8 dm^3 per minute at rest to 50–60 dm^3 per minute during strenuous exercise.

▲ The effect of exercise on breathing can be measured after a race

Any alterations in blood CO_2 concentration is detected by sensors in the brain. Nerve impulses are then sent to muscles in the chest and the heart.

At rest: CO_2 concentration in blood, and blood pH, are kept at safe levels (see page 128 – homeostasis)

Exercise:
- deeper, more rapid **breathing** (so more CO_2 lost from lungs)
- deeper, more rapid **heartbeat** (so more blood pumped to tissues to carry CO_2 away)

Changes in breathing and bloodflow keep CO_2 concentration and blood pH at safe levels.

Other factors affecting breathing rate

As well as exercise (see also page 109) some other factors affect breathing rate:

Factor	Effect
Smoking	Increase, due to the effect of **carbon monoxide** (see page 121)
Anxiety	Increase, due to the effect of **adrenaline** (see page 150)
Drugs	Some cause an increase e.g. **amphetamines** because they are stimulants (see page 152) Some cause a decrease e.g. **alcohol** and **barbiturates** because they are depressants (sedatives) (see page 152)
Environmental	Increased by **high CO_2 concentration** (see page 265) Sometimes increased by **high temperature or humidity**, as an attempt to lose body heat by panting (see page 130)
Altitude	Increased by **low O_2 concentration** in the atmosphere. Climbers at high altitude breathe quickly and with a low tidal volume – this can cause problems of dehydration.
Weight	Can increase because fat makes lung ventilation harder (i.e. tidal volume falls) Can decrease if excess body weight is a symptom of low activity of the thyroid gland (see page 149)

Exercise: too much or too little?*

Humans are well-adapted to exercise. Some evolutionary biologists believe that one of the reasons that humans have been so successful is that we have evolved to run: from our large *gluteus maximus* muscles to our almost hairless skin, we can exercise for longer periods than most other land animals. Our brains are able to cope with lengthy exercise, and have allowed us to develop the social skills necessary to become successful hunters.

Recent generations of humans in highly economically developed countries rarely have to chase game to feed themselves. However, our physiology and anatomy have not become adapted *not* to exercise, and we have developed many artificial ways to make sure that we are not harmed physically or mentally by our change in feeding habits. Some of the benefits of exercise – to heart and breathing system – have already been described on page 93. When we consider the benefits or risks associated with exercise we should take into account:

- Whether the effect is short term (a runner's 'high' for example), medium term (management of body mass for example), long term (benefits to the circulatory system) or catastrophic (heart attack, for example!)
- Whether the effect is related to exercise generally (e.g. changes in muscle structure) or to a specific exercise (e.g. shoulder injuries in swimmers).
- What we actually mean by **physical activity** and by **exercise**. **Physical activity** is any movement that uses energy. **Exercise** is physical activity that is structured and is done at a certain intensity for a certain length of time. We engage in physical activity and exercise for fitness benefits in order to improve some components of physical fitness (heart-lung condition, muscle strength, muscle endurance, flexibility, and body composition).

Hair on head allows upright posture without overheating of brain from solar radiation.

Large **gluteus maximus** muscles allow upper body to lean forward without individual falling over. They also allow legs to alternately swing forward and swing backward without overbalancing.

Hairless **skin** allows evaporation of sweat – essential for cooling during prolonged periods of exercise.

Sweat glands secrete sweat, containing water which can lose latent heat of evaporation as it is converted to water vapour.

▲ Adapted to run

Is too much exercise bad for you?*

Menstrual cycle may stop
Results from low oestrogen levels, and can cause delayed puberty. It is particularly common in athletes involved in endurance events (e.g. distance runners), and in women who participate in sports where appearance is considered important (e.g. gymnasts, ballet dancers).

Cardiac myopathy
A disease of the heart muscle that causes the heart to become enlarged and to pump less strongly. As a result the muscle of the heart becomes weak, thin, or floppy and is unable to pump blood efficiently around the body. This causes fluid to build up in the lungs, which therefore become congested, and results in a feeling of breathlessness: this is called left heart failure. It is probably partly genetic (inherited), but is only likely to be seen if exercise is very severe.

Stress fracture
A stress fracture occurs when the forces are much lower than in a car crash or fall, but happen repetitively for a long period of time. Stress fractures are commonly seen in athletes who run and jump on hard surfaces, such as distance runners, basketball players, and ballet dancers. A stress fracture can occur in any bone, but is commonly seen in the foot and shin bones. They rarely occur in the upper extremity because the weight of your body is not supported by your arms as it is by your legs.

Compulsive exercise
An exercise addict no longer chooses to exercise but feels compelled to do so and struggles with guilt and anxiety if he or she doesn't complete their exercise. Any of the above injuries can result from compulsive exercise. Compulsive exercise is more common in girls and young women, and is often associated with eating disorders.

Torn Achilles tendon
The Achilles tendon joins the *gastrocnemius* (calf) and the muscles of the lower leg to the heel of the foot. If the muscles are weak and become fatigued, they may tighten and shorten. Overuse can also be a problem by leading to muscle fatigue. The more fatigued the calf muscles are, the shorter and tighter they will become. This tightness can increase the stress on the Achilles tendon and result in a rupture.

Torn muscle
Pain that occurs in the calf muscle on the lower part of the leg often is the result of a pulled or torn calf muscle. This is called a calf strain. It occurs when part of the muscle of the lower leg (*gastrocnemius* or *soleus*) is torn away from the Achilles tendon.

Despite these problems, doctors believe that moderate exercise is one of the most important methods we have for maintaining our health and mental well-being.

Questions on gas exchange

1 Two boys were asked to take part in an investigation into the effect of exercise on breathing. The number of breaths they took in each half minute was measured and recorded, first of all while sitting still, then when recovering from two minutes of hard exercise. The results are shown in the table below.

time / minutes	activity	number of breaths in each half minute	
		Tom	Alan
0.0		7	8
0.5		7	8
1.0	Sitting still	7	8
1.5		7	8
2.0			
2.5			
3.0	Exercise (step-ups)		
3.5			
4.0		24	24
4.5		23	17
5.0		18	13
5.5		15	10
6.0	Recovery (sitting)	12	10
6.5		12	9
7.0		10	8
7.5		8	8
8.0		8	8
8.5		7	8

a Draw a graph to show the changes in breathing rate over the time period of this investigation. Plot both lines on the same axes.

b Which boy appears to be fitter? Explain your answer.

The teacher of the class was interested in the changes in breathing during the exercise period. She used a sensor, computer interface, monitor and printer to obtain the following information on another member of the class.

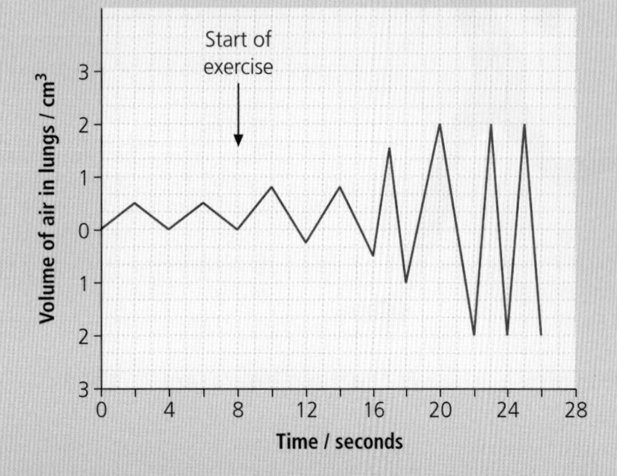

c What is the ratio of the volume of a breath during exercise to the volume of a breath at rest?

d Calculate the rate of breathing, in breaths per minute, during strenuous exercise.

e Using the data gathered, describe two effects of exercise on breathing.

f The computer could also measure the effects of exercise on heart rate. Suggest what these effects might be.

g What is the benefit to the body of the effects described in **e** and **f**?

2 These diagrams show apparatus that can be used to explain the mechanism of breathing.

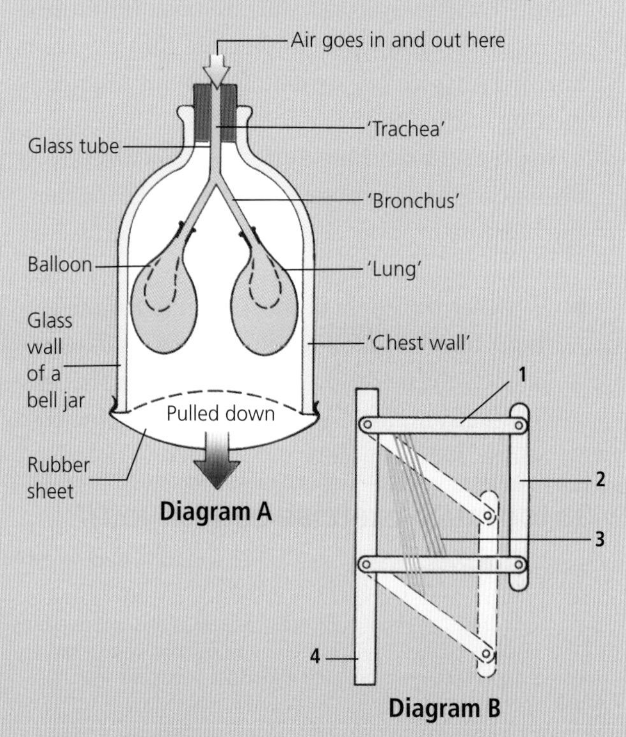

Look at diagram A.

a What does the rubber sheet represent?

b What will happen to the balloons when the rubber sheet is pulled downwards?

c What phase of the breathing cycle is represented when the rubber sheet is allowed to return to its resting position?

d In what way is this model an incomplete demonstration of the mechanism of breathing?

Look at diagram B.

e What parts of the human ventilation system do the labels **1**–**4** represent?

f Use the information in your answers to **a**–**e** to comment on the statement: 'Breathing in is an active process, but breathing out is completely passive.'

3 This table shows the energy reserves for skeletal muscles in an athlete.

			time the reserve would last / min	
energy reserve	**mass / g**	**energy / kJ**	**walking**	**marathon running**
blood glucose	3	48	4	1
liver glycogen	100	1660	86	20
muscle glycogen	350	5800	288	71
fat in skin	9000	337 500	15 500	4018

a **i** Compare the effect of walking and marathon running on energy reserves.

ii Suggest which two energy reserves would be most readily available to muscles during exercise.

iii Choose **two** food groups to which the energy reserves in the table belong:

protein mineral fibre fat carbohydrate

iv Calculate the energy per gram of glycogen. Show your working.

b Suggest why athletes eat foods high in

i proteins, during training

ii carbohydrates, for three days before a marathon race.

c During a fast race (a 100 metre sprint), 95% of the energy comes from anaerobic respiration. During a marathon, only 2% of the energy comes from anaerobic respiration.

i State the equation, in symbols, for anaerobic respiration in muscles.

ii Suggest and explain why a sprinter can use mainly anaerobic respiration during the race, while a marathon runner needs to use aerobic respiration.

iii Explain how, during a marathon race, the blood glucose concentration stays fairly constant, but the mass of glycogen in the liver decreases.

Cambridge IGCSE Biology 0610
Paper 3 Variant 1 Q5 November 2008

4 Respiration is one of the characteristics of living things.

a List four other characteristics of living things **not** including respiration.

b Describe the difference between *respiration* and *breathing*.

Cambridge IGCSE Biology 0610
Paper 2 Q1 May 2008

5 a This table shows the percentage of haemoglobin that is inactivated by carbon monoxide present in the blood of taxi drivers in a city.

city taxi drivers		percentage of haemoglobin inactivated by carbon monoxide
day time drivers	smokers	5.7
	non-smokers	2.3
night time drivers	smokers	4.4
	non-smokers	1.0

i The carbon monoxide in the blood of these taxi drivers comes from two sources. One source is from vehicle exhaust fumes. Name the other source of carbon monoxide that may be inhaled by drivers.

ii Using data from the table, suggest which of these two sources contributes most to the inactivation of the haemoglobin. Explain your choice.

iii Calculate the difference in the percentage of haemoglobin inactivated by carbon monoxide in day and night time taxi drivers and suggest a reason for the difference.

b **i** Name two other harmful components of cigarette smoke, apart from carbon monoxide. For each, describe an effect it can have on the body of a person who smokes.

ii Suggest a possible effect that might happen to the fetus of a pregnant woman who smokes.

Cambridge IGCSE Biology 0610
Paper 2 Q2 November 2008

2.41 Smoking and disease

OBJECTIVES

- To understand that smoking tobacco is harmful to health
- To list some of the harmful components of tobacco smoke and the damage they cause
- To understand why it is so difficult to give up smoking

The risks of smoking

Many national advertising campaigns stress that smoking is harmful. At the same time, the manufacturers of cigarettes emphasise the 'glamorous' side of smoking. However, manufacturers have to include, by law, a statement on their advertisements and cigarette packets that 'smoking can seriously damage your health'.

Life insurance companies routinely ask 'Do you smoke?' because they are aware that smokers are more likely to die younger. Whether or not to take up smoking is the most important health decision that many of us will ever make. For this reason everyone should know about the possible effects of smoking. Nobody who starts smoking now can say 'But I didn't know the risks' when they suffer the effects of their smoking habit later in life.

How is smoking harmful?

Smoking is inhaling the smoke from burning tobacco (and paper). This smoke can harm the lungs and respiratory passages for a number of reasons:

- it is hot
- it has a drying effect
- it contains many harmful chemicals.

The heat and dryness irritate the lungs, but the main dangers of smoking come from the chemicals in the burning tobacco. There are over 1000 known chemicals in tobacco smoke. These include tars, carbon monoxide, sulfur dioxide, nicotine and even small quantities of arsenic and plutonium! When doctors treat lung diseases with medicine, the molecules of the medicine are delivered in a spray; the droplets of water in the spray carry the medicine down through the respiratory tubes and deep into the lungs. Burning tobacco produces tiny droplets of water too, and these carry the harmful chemicals deep into the lungs in just the same way as a medicine spray. It would be hard to find a more efficient way of delivering harmful chemicals to the lungs than smoking! Some of these dangerous chemicals, and the effects they have on the body, are shown on the opposite page.

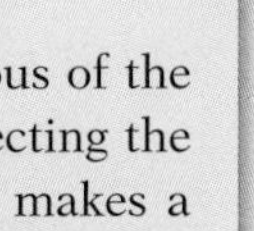

Why is it so difficult to give up smoking?

In many ways nicotine is the most dangerous of the chemicals in tobacco smoke. As well as affecting the heart and blood pressure directly, nicotine makes a person become **addicted** to smoking. Addiction comes in two forms:

- in **physical addiction**, the body cannot function properly in the absence of the chemical because it has partly replaced a natural body chemical
- in **psychological addiction**, the addicted person links smoking with comfort or lack of stress – when they feel stressed they may automatically reach for a cigarette.

Q

1. How is smoke harmful to the lungs?
2. What is the difference between physical and psychological addiction? How can smokers be helped to overcome their addiction?
3. Suggest three harmful effects of smoking other than damage to the lungs and breathing passages.
4. Why are smokers more likely to develop infections of the lungs than non-smokers?
5. Draw a single cube with a side of 10 cm, and then the same cube divided into smaller cubes each with a side of 1 cm.
   a. How many small cubes fit into the large cube?
   b. What is the surface area of each small cube?
   c. What is the total surface area of all of the small cubes?
   d. What is the surface area of the large cube?
   e. Use your answers to explain why emphysema sufferers are often very breathless.

Free radicals are extremely dangerous chemicals that damage proteins and DNA. They cause ageing of cells (smokers are often 'wrinkled', inside and out!) but smokers can reduce their effects with high doses of vitamin C.

Nicotine is the chemical that causes **addiction**. It is also a **stimulant** which makes the heart beat faster and at the same time makes blood vessels narrow. Together these two effects **raise blood pressure** (causing long-term damage to the circulation). The increased heart rate increases the demand for oxygen, but carbon monoxide (see below) reduces oxygen availability – so heart muscle is more likely to be damaged.

Cilia are destroyed which means that mucus accumulates in the respiratory tubes. Dust and microbes, trapped in the mucus, slide down towards the lungs making the person cough. This coughing inflames the lining of the bronchi, causing **bronchitis**. A smoker is 20 times more likely to develop bronchitis than a non-smoker.

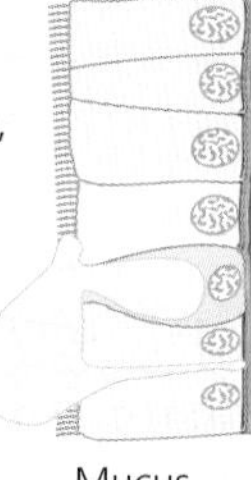

Mucus, microbes and cell fragments build up

Emphysema results when the walls of the air sacs are destroyed. This happens because smoke affects white blood cells which then destroy lung tissue. When the walls break down there is less surface for gas exchange, and breathing becomes very difficult. Somebody with emphysema may only be able to walk 2 or 3 metres before becoming breathless. Emphysema is almost unknown in non-smokers.

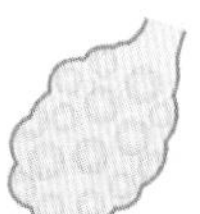

Healthy alveoli

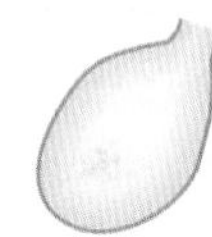

Alveoli in person with emphysema

Chronic Obstructive Pulmonary Disease (COPD) can be a result of emphysema

- very poor airflow
- worsens over time
- does not respond to anti-asthma drugs.

Carbon monoxide reduces the oxygen supply
Oxygen combines with haemoglobin to form **oxyhaemoglobin** in red blood cells. Carbon monoxide (CO) reduces oxyhaemoglobin formation because it binds very tightly to haemoglobin, and the effect is permanent since **carboxyhaemoglobin is very stable**. This:

- reduces aerobic respiration (bad for sport)
- reduces oxygen transport across the placenta (babies born to smokers have low birth-weight).

Carbon monoxide poisoning is especially likely when car engines are allowed to run in enclosed spaces (such as garages) and when the atmosphere is of poor quality (e.g. in smog).

Tar causes cancer, which is uncontrolled division of cells. These cells, usually those lining the lower part of the bronchus, grow through the basement membrane and invade other tissues.
Tar is also an **irritant** which makes coughing more likely. This causes physical damage to the lungs, and makes the effects of emphysema even worse. Other irritants in tobacco smoke include **smoke particles**, **ammonia** and **sulfur dioxide**.

Lung cancer is 10 times more likely in a smoker than a non-smoker. The tumour invades other tissues. This causes pain and loss of function of other tissues, often resulting in death.

▲ The lung of a smoker destroyed by cancer (right), and a normal lung (left)

Smoking causes other diseases including:

- cancer of stomach, pancreas and bladder
- loss of limbs – amputated because of poor circulation
- coronary heart disease
- lower sperm counts in men.

2.42 How do we know that smoking causes disease?

OBJECTIVES

- To understand the statistical evidence linking smoking to disease
- To be able to design an epidemiological investigation

Sir Richard Doll's research into smoking

We know today that smoking causes disease, but how have we found this out? The diseases could result from living in a polluted environment, or exposure to chemicals at work, or diet, or any number of other factors.

Sir Richard Doll was an **epidemiologist** – he studied patterns in the distribution of diseases. He was particularly interested in comparing the habits and environment of people who developed lung disease with people who did not. His results were very valuable because he collected data in a scientific way, removing many variables from his studies. For example:

- he carried out many of his studies on doctors, so he could rule out profession as a cause of lung disease
- he separated data for people living in cities from those living in the countryside, so environment was not causing large differences between people in one study group.

Some examples of epidemiological studies that related lung disease and early death to smoking tobacco are shown on the opposite page. The box shows some data about smoking in Britain and other countries.

◀ Sir Richard Doll carried out epidemiological research on lung diseases.

Some disturbing data

Fewer boys are taking up smoking in the UK than 20 years ago (good news) but more girls are now smoking (bad news). Can you suggest reasons for these changes?

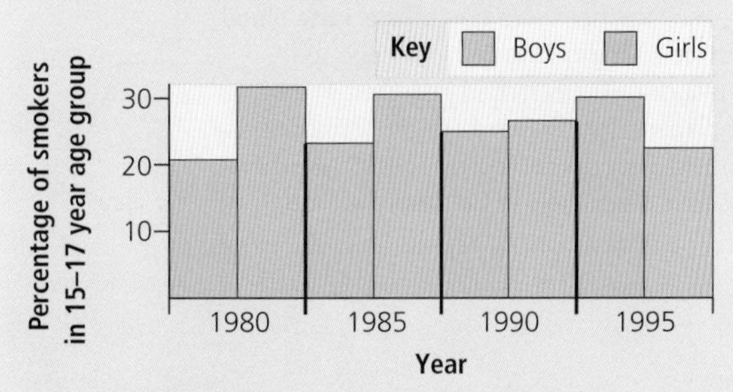

Cigarette sales in the UK have fallen because people are becoming more aware of the health risks of smoking. Advertising campaigns emphasise the dangers, and smoking is banned in many public places to reduce 'passive smoking' (breathing in someone else's cigarette smoke). Tobacco companies are focusing on selling to developing countries where people are less aware of the health risks. The rise in lung disease in Britain in the 1980s and 1990s will be repeated in 20 years' time in these countries.

The number of cigarettes sold in the UK and in Mexico, 1960–1995.

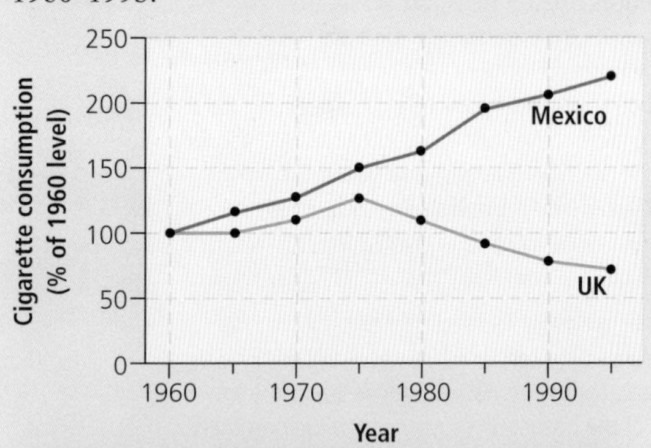

The cigarette companies are now targeting countries such as Africa to increase their profits.

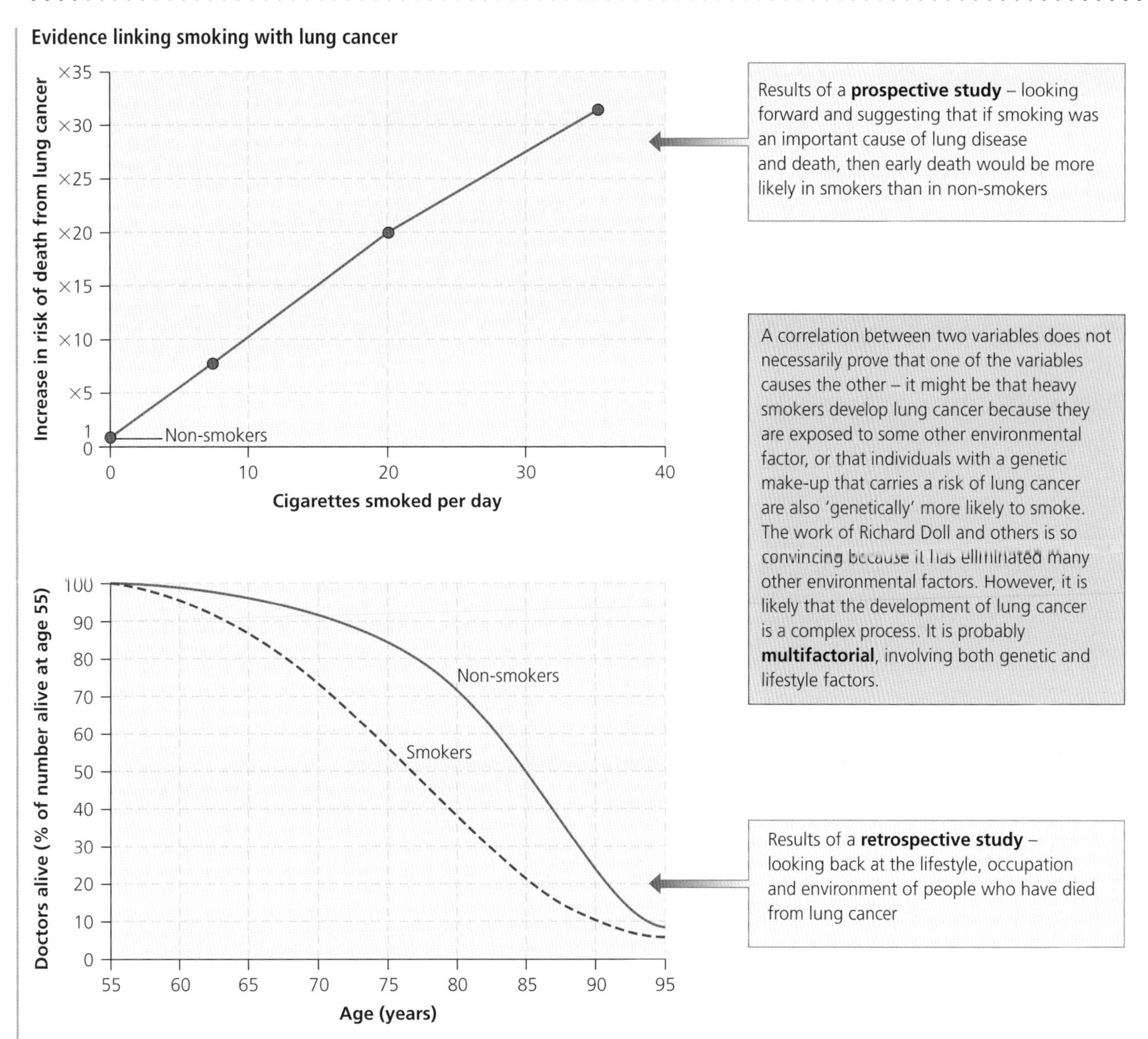
Evidence linking smoking with lung cancer
Increase in risk of death from lung cancer
×35
×30
×25
×20
×15
×10
×5
1
0
Non-smokers
0
10
20
30
40
Cigarettes smoked per day
Results of a **prospective study** – looking forward and suggesting that if smoking was an important cause of lung disease and death, then early death would be more likely in smokers than in non-smokers
A correlation between two variables does not necessarily prove that one of the variables causes the other – it might be that heavy smokers develop lung cancer because they are exposed to some other environmental factor, or that individuals with a genetic make-up that carries a risk of lung cancer are also 'genetically' more likely to smoke. The work of Richard Doll and others is so convincing because it has eliminated many other environmental factors. However, it is likely that the development of lung cancer is a complex process. It is probably **multifactorial**, involving both genetic and lifestyle factors.
Doctors alive (% of number alive at age 55)
100
90
80
70
60
50
40
30
20
10
0
Non-smokers
Smokers
55
60
65
70
75
80
85
90
95
Age (years)
Results of a **retrospective study** – looking back at the lifestyle, occupation and environment of people who have died from lung cancer

▲ In 1962 the Royal College of Physicians published a report on smoking and health, which suggested a clear link between cancer and smoking. For example, among a sample of doctors living in similar-sized cities, those who smoked regularly were more likely to develop cancer of the lung.

1 What is meant by the term epidemiology?

2 Suggest two factors other than cigarette smoking that might increase the risk of developing lung cancer.

3 How much more likely to die of lung cancer is a person who smokes 25 cigarettes a day than someone who does not smoke at all?

4 One epidemiological study has suggested that living close to power lines can cause leukaemia. How would you try to prove this link?
- Which populations would you study?
- How old would they be?
- Which sex?
- Which occupation?
- How about their diet?

In what way do you think the results of your study might be useful?

5 In a typical laboratory experiment, data are collected by manipulating one variable and measuring the responding change in another, with all other identifiable variables kept constant (see page 34). Why can this approach not be used to investigate the effect of smoking on the development of lung disease in humans?

2.43 Excretion: removal of the waste products of metabolism

OBJECTIVES

- To understand that living cells produce wastes
- To name some human waste products
- To name the organs involved in excretion
- To understand the functions of the kidney

Homeostasis (page 128) means keeping a constant environment around the cells of the body. This involves providing cells with essential raw materials, but also means removing waste products. These waste products can be very toxic (poisonous), for example:

- **Carbon dioxide**, produced during respiration, dissolves in plasma and tissue fluid to form a weak acid (carbonic acid, H_2CO_3) which can denature enzymes and other proteins at high concentrations.
- **Urea**, produced in the liver during deamination of excess amino acids, can denature enzymes.
- **Salts**, which can be in excess in the diet, can have an effect on the water potential of the blood.

Excretion is the removal of toxic materials, the waste products of metabolism and substances in excess of requirements.

The role of the kidney

The kidneys are specialised organs that:

- remove the toxic waste product **urea** from the circulating blood – they carry out **excretion**
- regulate the **water content** of the blood – they carry out **osmoregulation**.

The structure of the kidneys is well adapted to enable them to carry out these processes.

- Each kidney receives a good supply of blood at high pressure through the **renal artery**.
- Each kidney contains hundreds of thousands of tubes, the **nephrons** (kidney tubules), that can filter substances from the blood.
- Each kidney has an exit tube, **the ureter**, to carry away the **urine** (a solution of wastes dissolved in water).
- The kidney is under close control by a **feedback system** so that water saving is always exactly balanced to the body's needs.

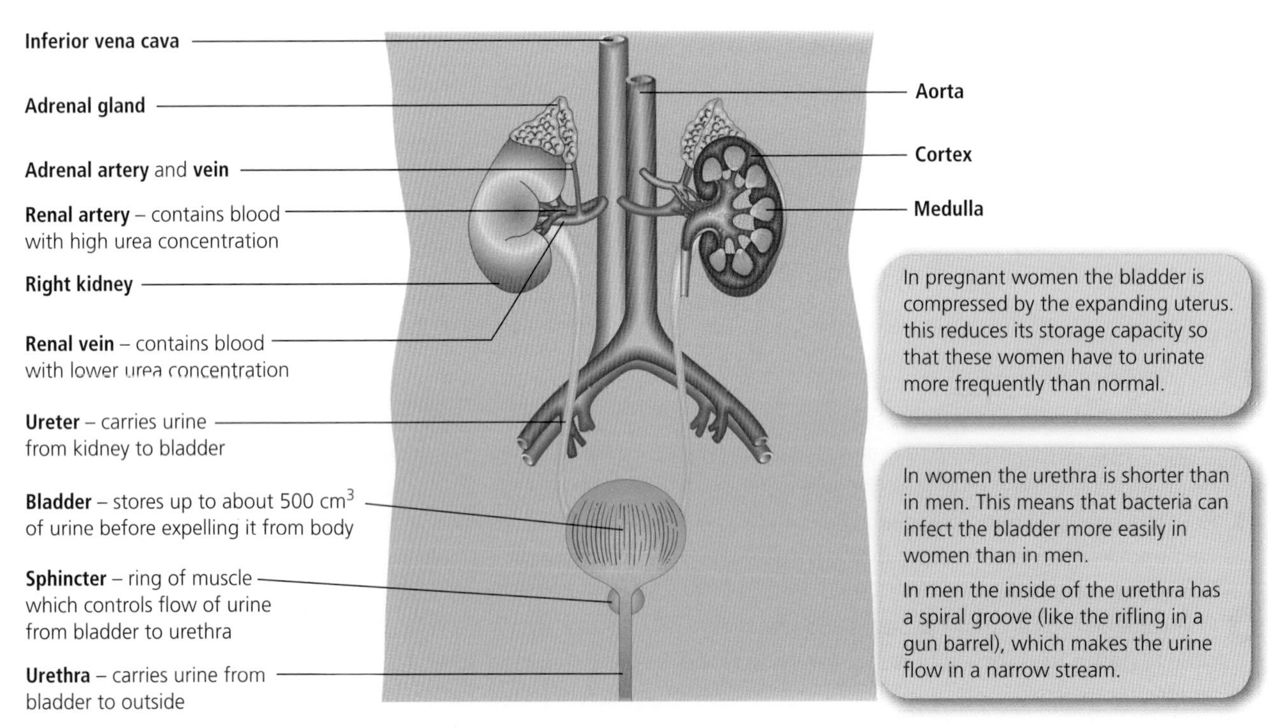

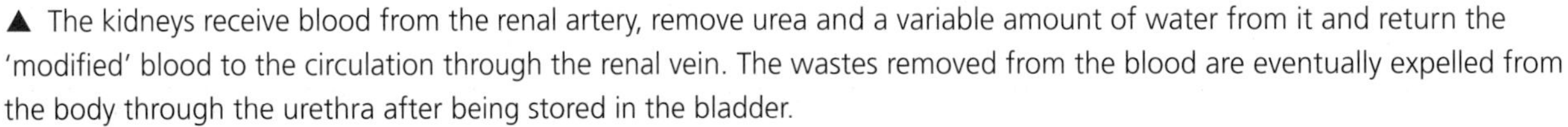
▲ The kidneys receive blood from the renal artery, remove urea and a variable amount of water from it and return the 'modified' blood to the circulation through the renal vein. The wastes removed from the blood are eventually expelled from the body through the urethra after being stored in the bladder.

The structure of the kidney

The kidneys and their blood vessels are located in the abdomen, as shown in the diagram opposite. The kidneys constantly produce urine which then passes to the **bladder**.

The functional unit – the kidney tubule (nephron)

Each kidney contains hundreds of thousands of long tubes called **nephrons**, each with its own branch of the renal artery and vein. Each nephron works in exactly the same way, so the function of the kidney can be explained by considering the working of just one nephron, as shown below.

▼ The kidney is made up of many nephrons (kidney tubules). Substances are filtered out of the blood into the nephron. Useful molecules and most of the water are reabsorbed into the blood.

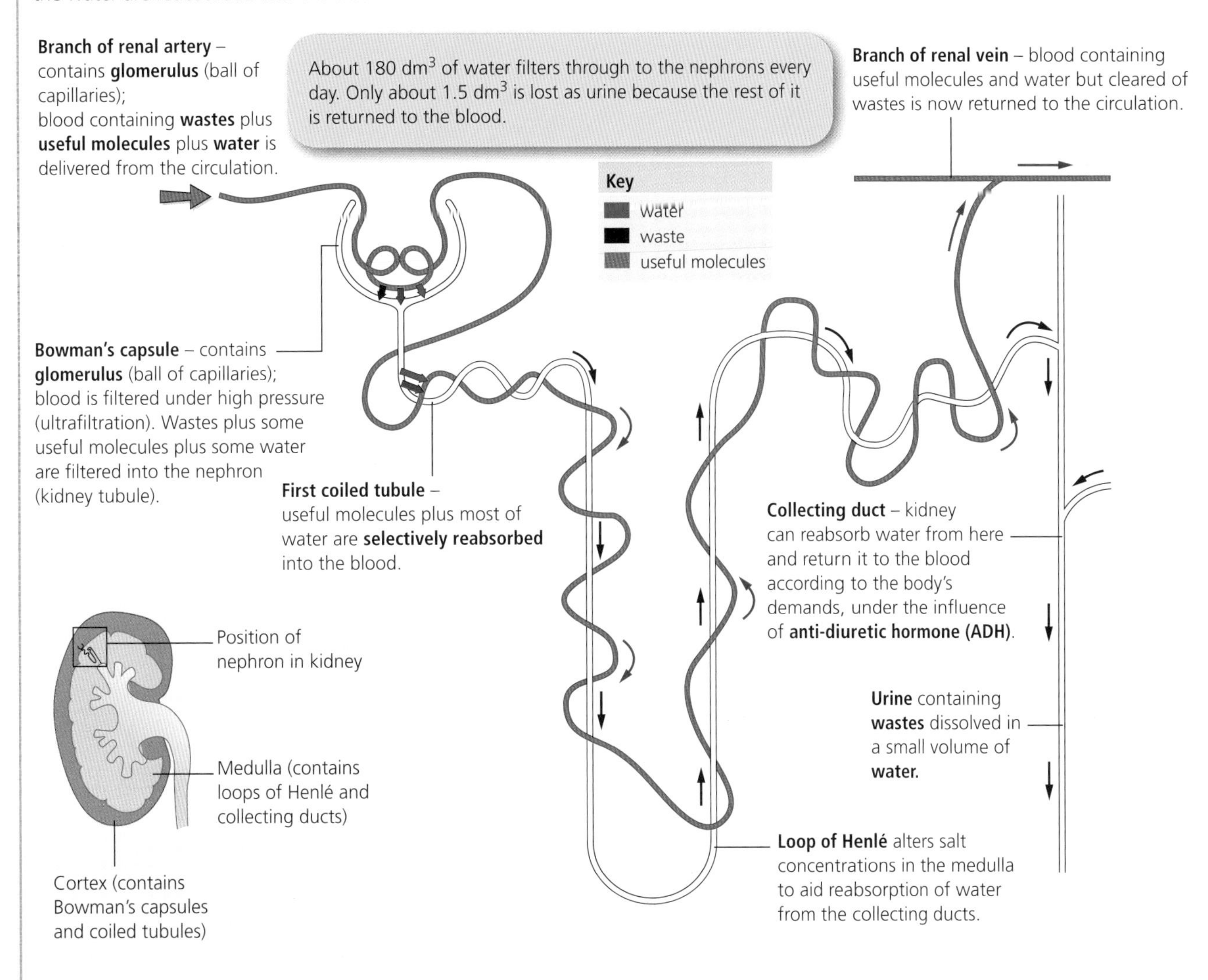

Branch of renal artery – contains **glomerulus** (ball of capillaries); blood containing **wastes** plus **useful molecules** plus **water** is delivered from the circulation.
About 180 dm^3 of water filters through to the nephrons every day. Only about 1.5 dm^3 is lost as urine because the rest of it is returned to the blood.
Branch of renal vein – blood containing useful molecules and water but cleared of wastes is now returned to the circulation.
Key
water
waste
useful molecules
Bowman's capsule – contains **glomerulus** (ball of capillaries); blood is filtered under high pressure (ultrafiltration). Wastes plus some useful molecules plus some water are filtered into the nephron (kidney tubule).
First coiled tubule – useful molecules plus most of water are **selectively reabsorbed** into the blood.
Collecting duct – kidney can reabsorb water from here and return it to the blood according to the body's demands, under the influence of **anti-diuretic hormone (ADH)**.
Urine containing **wastes** dissolved in a small volume of **water.**
Loop of Henlé alters salt concentrations in the medulla to aid reabsorption of water from the collecting ducts.
Position of nephron in kidney
Medulla (contains loops of Henlé and collecting ducts)
Cortex (contains Bowman's capsules and coiled tubules)

Q

1. Define the terms excretion and osmoregulation.
2. Name two waste products of metabolism. State their source and how they are removed from the blood.
3. Copy and complete the following paragraph.
   The main excretory organ in the mammal is the ____. There are two of these, each supplied with blood through the ____ and each composed of many thousands of tubules called ____. Each of these tubules receives materials from the blood after filtration in the ____ capsule. The tubules then remove useful substances such as ____ from this filtrate by the process of ____. The remaining waste or excess materials pass down the tubule and leave the excretory organ via the ____. They are collected for temporary storage in the ____. Eventually a dilute solution of these wastes, the urine, leaves the body through the ____.

2.44 Dialysis and the treatment of kidney failure

OBJECTIVES

- To explain how body water is regulated
- To explain how kidney disease can be treated

Osmoregulation

The final part of each nephron, the **collecting duct**, removes water from the filtered solution passing down inside the tubule and returns it to the blood. The body can control the amount of water that is returned to the blood in this way, and balance it with the amount of water taken in in the diet and the amount of water lost from the body by other means. This vital control of water balance is called **osmoregulation**. Water is the most common substance in the body, and has many functions (see page 236). The average daily intake and loss of water by different methods is outlined in the diagram below.

Kidney failure

Damage to the kidneys, perhaps through infection or following an accident, can stop the nephrons working efficiently. The body can no longer control the composition or amount of urine formed, so the content of the blood plasma and tissue fluid is not kept at its optimum. Death may follow quite quickly if the kidney failure is not corrected. Two types of treatment are available:

- **dialysis** using a kidney machine (an artificial kidney)
- **kidney transplant**.

Dialysis and the artificial kidney

A kidney machine takes a patient's blood, 'cleans' it and returns the blood to the circulation. This process is called **dialysis**. Wastes diffuse out of the blood, across a partially permeable membrane, into a fluid that is constantly renewed. In this way urea is removed from the blood without altering any of its other features. The diagram on the opposite page shows the workings of a kidney dialysis machine.

▼ The water excreted in the urine is adjusted so that total water intake and total water loss are balanced

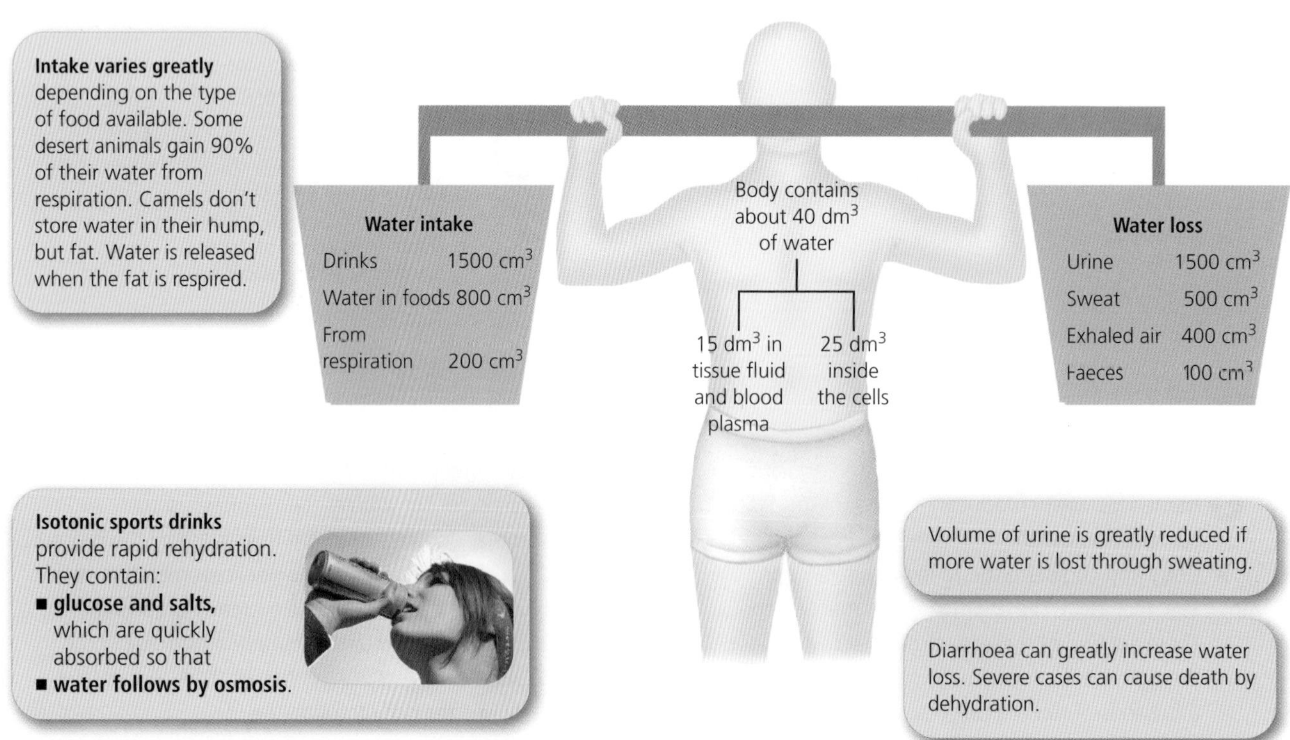

Intake varies greatly depending on the type of food available. Some desert animals gain 90% of their water from respiration. Camels don't store water in their hump, but fat. Water is released when the fat is respired.
Water intake
Drinks 1500 cm3
Water in foods 800 cm3
From respiration 200 cm3
Body contains about 40 dm3 of water
15 dm3 in tissue fluid and blood plasma
25 dm3 inside the cells
Water loss
Urine 1500 cm3
Sweat 500 cm3
Exhaled air 400 cm3
Faeces 100 cm3
Isotonic sports drinks provide rapid rehydration. They contain:
■ glucose and salts, which are quickly absorbed so that
■ water follows by osmosis.
Volume of urine is greatly reduced if more water is lost through sweating.
Diarrhoea can greatly increase water loss. Severe cases can cause death by dehydration.

Kidney transplants

A **kidney transplant** involves surgically transferring a healthy kidney from one person (the **donor**) to a person with kidney failure (the **recipient**). It is relatively simple to connect up the donor kidney in the recipient's body, but a problem arises with **tissue rejection**. The recipient's immune system will attack the donor kidney and slowly destroy it (see page 105) unless the recipient takes drugs to stop this happening. Blood groups and tissue types of donors and recipients are carefully matched to reduce the likelihood of rejection. Successful kidney transplants have advantages over dialysis treatment:

- In the long term, a transplant is much cheaper.
- The patient's life is less disrupted once they have recovered from the operation.

1 The body must maintain a water balance.
   a Why does the body need water?
   b How is water gained by the body?
   c How is water lost by the body?
2 The water balance of the body is maintained by negative feedback. Explain what this term means.
3 Why is a kidney transplant considered better than dialysis? What problems are associated with kidney transplantation?

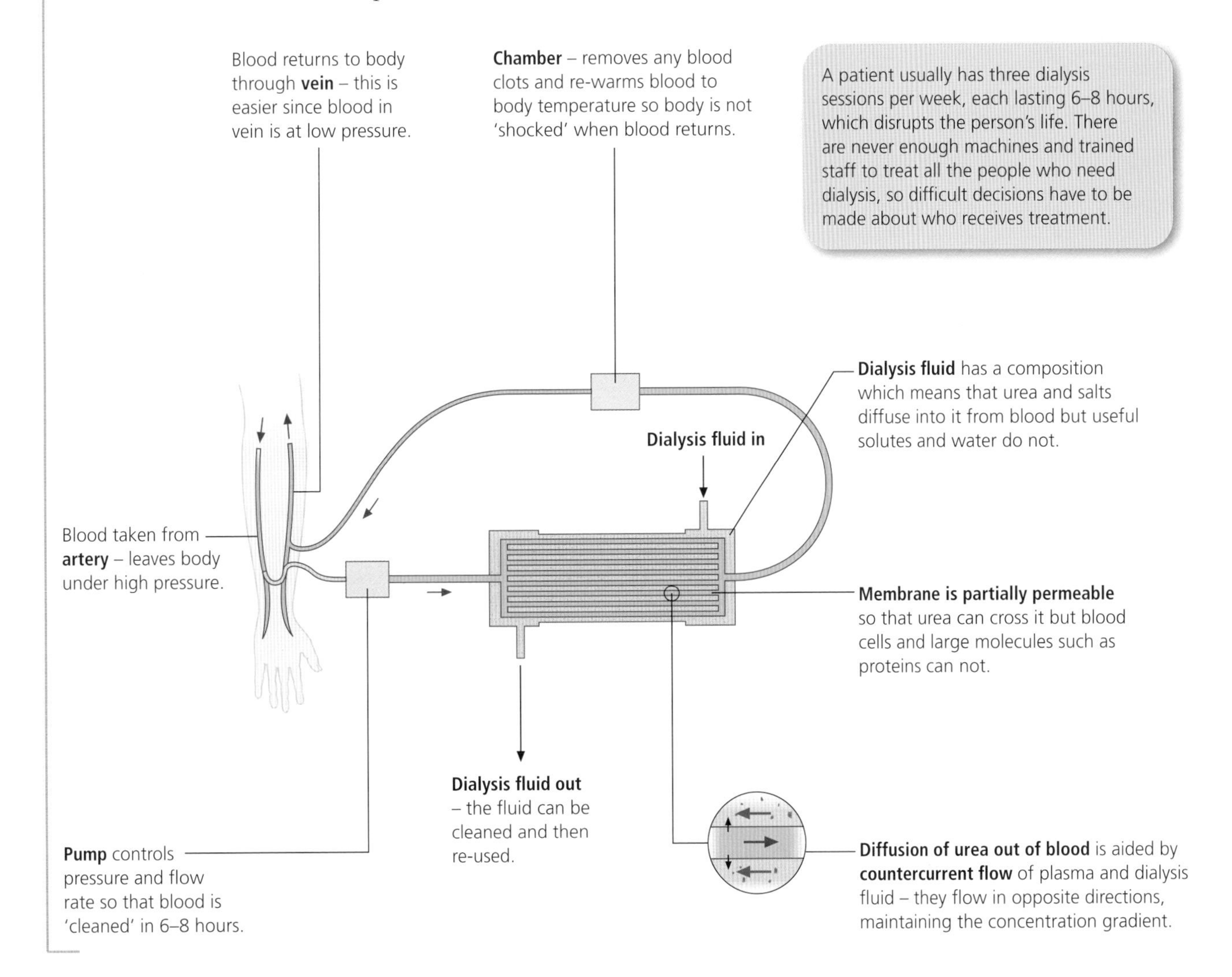

Blood returns to body through **vein** – this is easier since blood in vein is at low pressure.
Chamber – removes any blood clots and re-warms blood to body temperature so body is not 'shocked' when blood returns.
A patient usually has three dialysis sessions per week, each lasting 6–8 hours, which disrupts the person's life. There are never enough machines and trained staff to treat all the people who need dialysis, so difficult decisions have to be made about who receives treatment.
Dialysis fluid has a composition which means that urea and salts diffuse into it from blood but useful solutes and water do not.
Dialysis fluid in
Blood taken from **artery** – leaves body under high pressure.
Membrane is partially permeable so that urea can cross it but blood cells and large molecules such as proteins can not.
Dialysis fluid out – the fluid can be cleaned and then re-used.
Pump controls pressure and flow rate so that blood is 'cleaned' in 6–8 hours.
Diffusion of urea out of blood is aided by **countercurrent flow** of plasma and dialysis fluid – they flow in opposite directions, maintaining the concentration gradient.

2.45 Homeostasis: maintaining a steady state

OBJECTIVES

- To be able to define the term homeostasis
- To understand why the body must keep constant conditions around its cells
- To remember that tissue fluid surrounds all cells
- To understand the principle of negative feedback
- To know the organs involved in homeostasis

The cells in any living organism will only function properly in the correct conditions. The conditions *outside* the body (the **external environment**) are continuously changing – for example, temperature on a cool day might vary from 0°C out of doors to 20°C indoors. The body has mechanisms for adjusting conditions *within* the body (the **internal environment**) so that conditions around the cells remain constant.

Keeping the internal environment constant

The blood cells lie in the blood plasma, and other cells are surrounded by tissue fluid (see page 88). Conditions in the blood (and therefore in the tissue fluid) are maintained at an **optimum** – the best values for the cells to function. Keeping constant conditions in the tissue fluid around the cells is called **homeostasis**.

Homeostasis involves several organs, but the basic principle is always the same, as shown below. The diagram on the right shows some of the organs involved in homeostasis, and the particular conditions in the tissue fluid which they regulate.

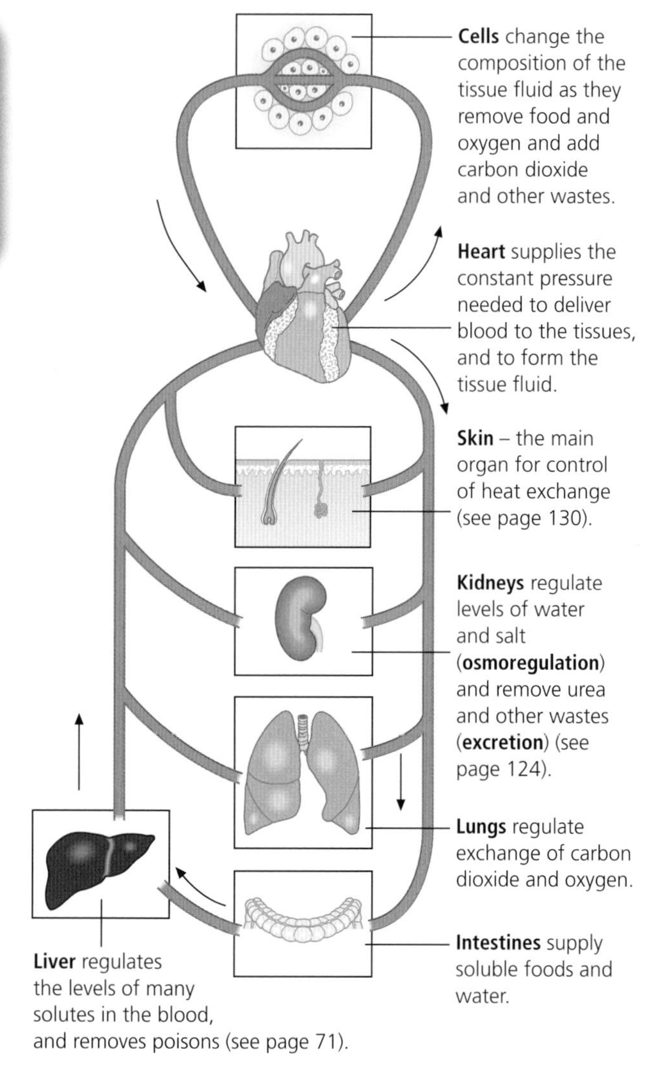

▲ Many organs play a part in homeostasis. (A more accurate diagram of the human circulation is given on page 86.)

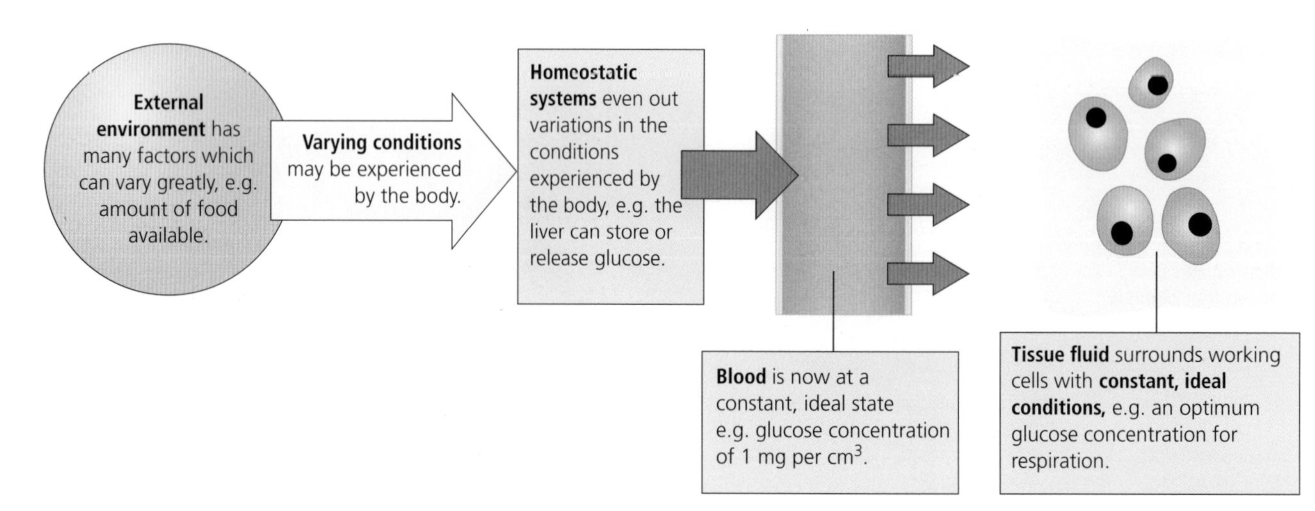

External environment has many factors which can vary greatly, e.g. amount of food available.
Varying conditions may be experienced by the body.
Homeostatic systems even out variations in the conditions experienced by the body, e.g. the liver can store or release glucose.
Blood is now at a constant, ideal state e.g. glucose concentration of 1 mg per cm^3.
Tissue fluid surrounds working cells with **constant, ideal conditions,** e.g. an optimum glucose concentration for respiration.

How is homeostasis brought about?

The work of these organs must be coordinated to achieve homeostasis. Information about the conditions in the body is continuously fed to the brain from **sensory receptors** around the body.

For example, if the body temperature rises, temperature receptors in the skin send information to the brain. In response, the brain starts off mechanisms that will lower the body temperature again. The same temperature receptors 'inform' the brain when the temperature is back to normal. Homeostasis depends on this continual **feedback** of information, as explained in the diagram below.

To summarise:

- Homeostasis is the maintenance of a constant internal environment.
- Homeostasis involves control by negative feedback.
- In negative feedback, a change sets off a response that cancels out the change.

Flying with 'George'

An aeroplane may be flown by an automatic pilot, sometimes called 'George'. This involves a series of feedback controls that correct any slight deviation from the aircraft's flight path. For example, if the plane tips to one side, a gyroscope detects the deviation and sends an electronic signal to the computer. This computer then sends out another signal to the plane's ailerons (levelling flaps) and the flight path is corrected.

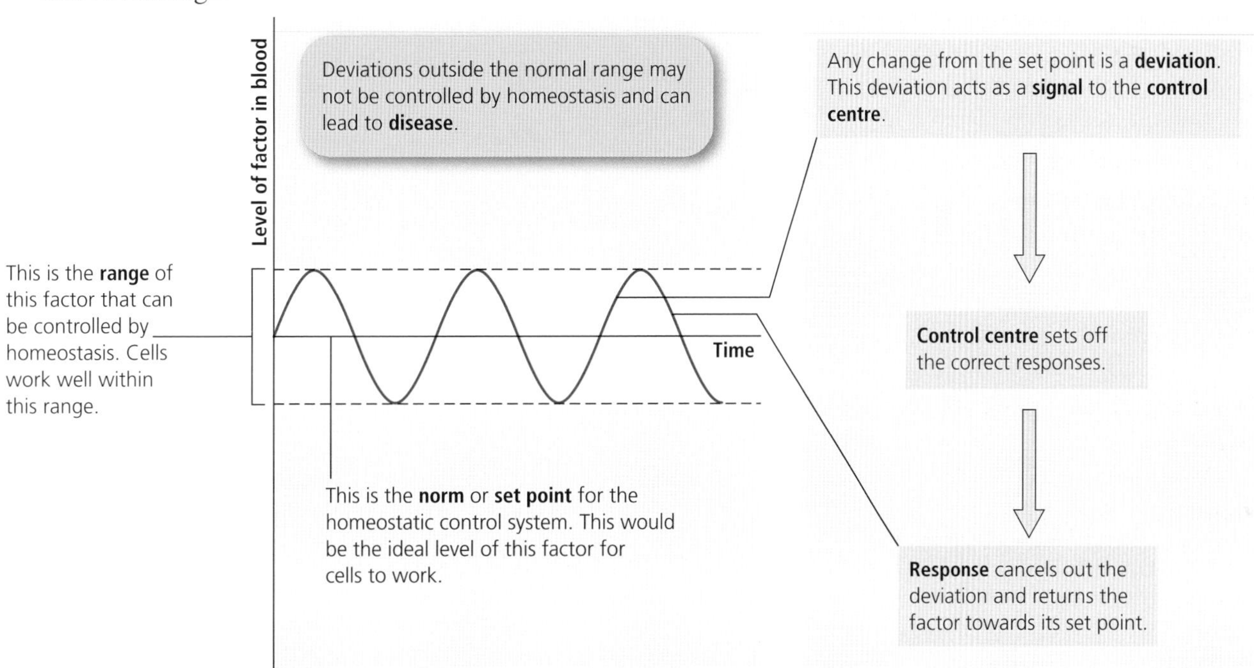

▲ In homeostasis, deviation from a set point acts as the signal that sets off the correction mechanism. This negative feedback keeps variable factors within the narrow range suitable for life.

Q

1 Copy and complete the following paragraph.
Cells work best when conditions around them remain constant. Cells are bathed in _____ fluid, and many systems work to keep the composition of this fluid constant. Each system has _____, which detect any changes from the _____ (the ideal conditions for the cells). The changes are then communicated to the _____, usually along _____ neurones. The central control area then sets off the correct responses. These responses cancel out the original change and so this sort of control is often called _____.

2 Copy and complete the following table.

Homeostatic organ	Factor controlled
Liver	
Lungs	
	Water content of blood
	Heat loss or gain
Intestines	

3 Use the principle of feedback control to compare the regulation of blood glucose level in humans (see page 132) with the maintenance of a stable temperature in an aircraft cabin.

2.46 Control of body temperature

OBJECTIVES

- To understand why body temperature must be controlled
- To know the difference between heat and temperature
- To understand the part played by the skin in control of temperature
- To appreciate that control of body temperature is an example of negative feedback

Heat and temperature

Heat is a type of energy, so it can be 'held' by an object or passed from one object to another. Temperature is a measure of how concentrated the heat energy is in an object, such as the human body. A body will rise in temperature if it gains heat energy.

The importance of a constant temperature

Many biological and physical processes are affected by temperature. For example:

- enzymes work best at their optimum temperature, and are denatured by wide deviations from this
- cell membranes become more fragile as temperature rises
- diffusion rates are increased by higher temperatures, and decreased by lower ones
- liquids such as blood become more viscous (thicker) as the temperature falls.

Birds and mammals are **endotherms** ('inside heat') – they can maintain a constant body temperature by generating heat internally. Humans have several mechanisms which work non-stop to balance heat production against heat loss, as shown in the diagram below. This balance is achieved by a **temperature control centre** in the **hypothalamus**, a region of the brain. The diagram opposite shows the negative feedback systems that control body temperature in an endotherm.

The role of the skin

As the barrier between the body and its environment, the skin is also the main organ concerned with heat loss and heat conservation.

▼ An endothermic animal maintains an ideal body temperature by balancing heat losses and heat gains

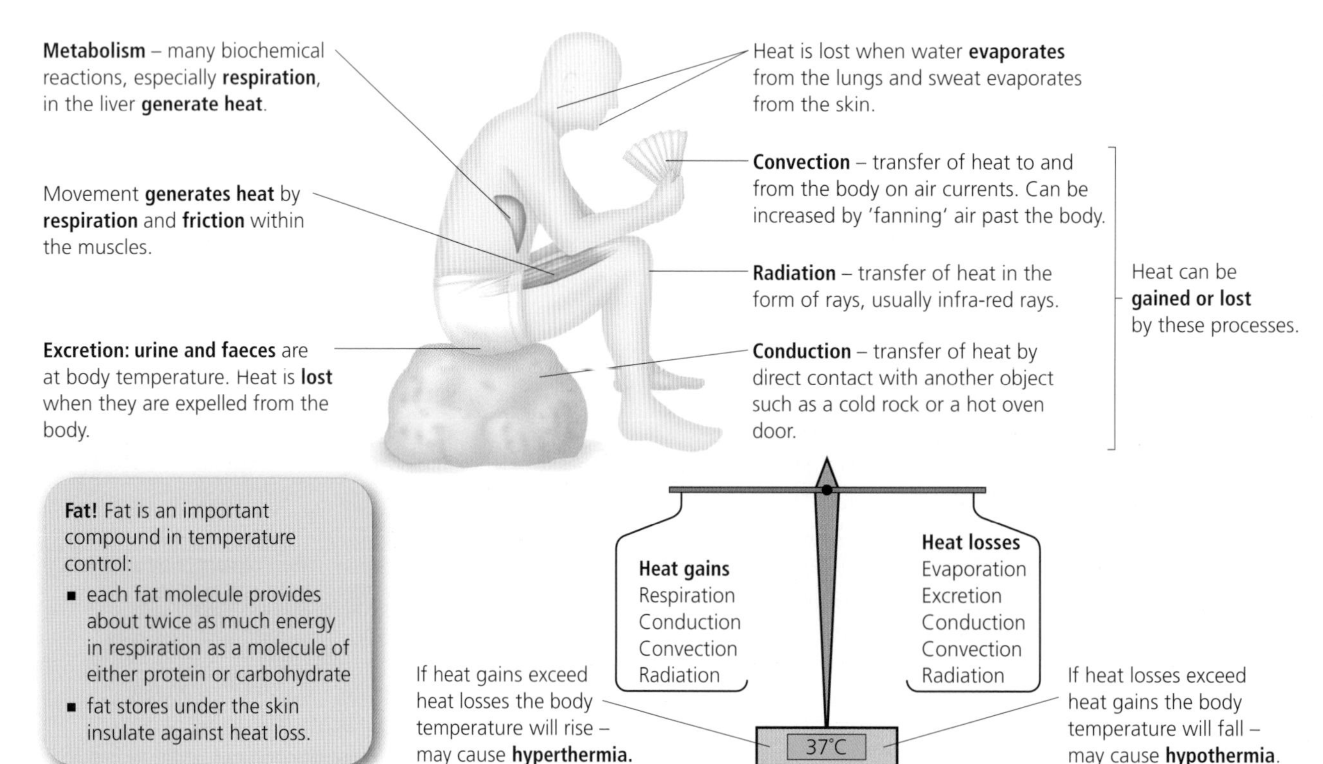

Fat! Fat is an important compound in temperature control:

- each fat molecule provides about twice as much energy in respiration as a molecule of either protein or carbohydrate
- fat stores under the skin insulate against heat loss.

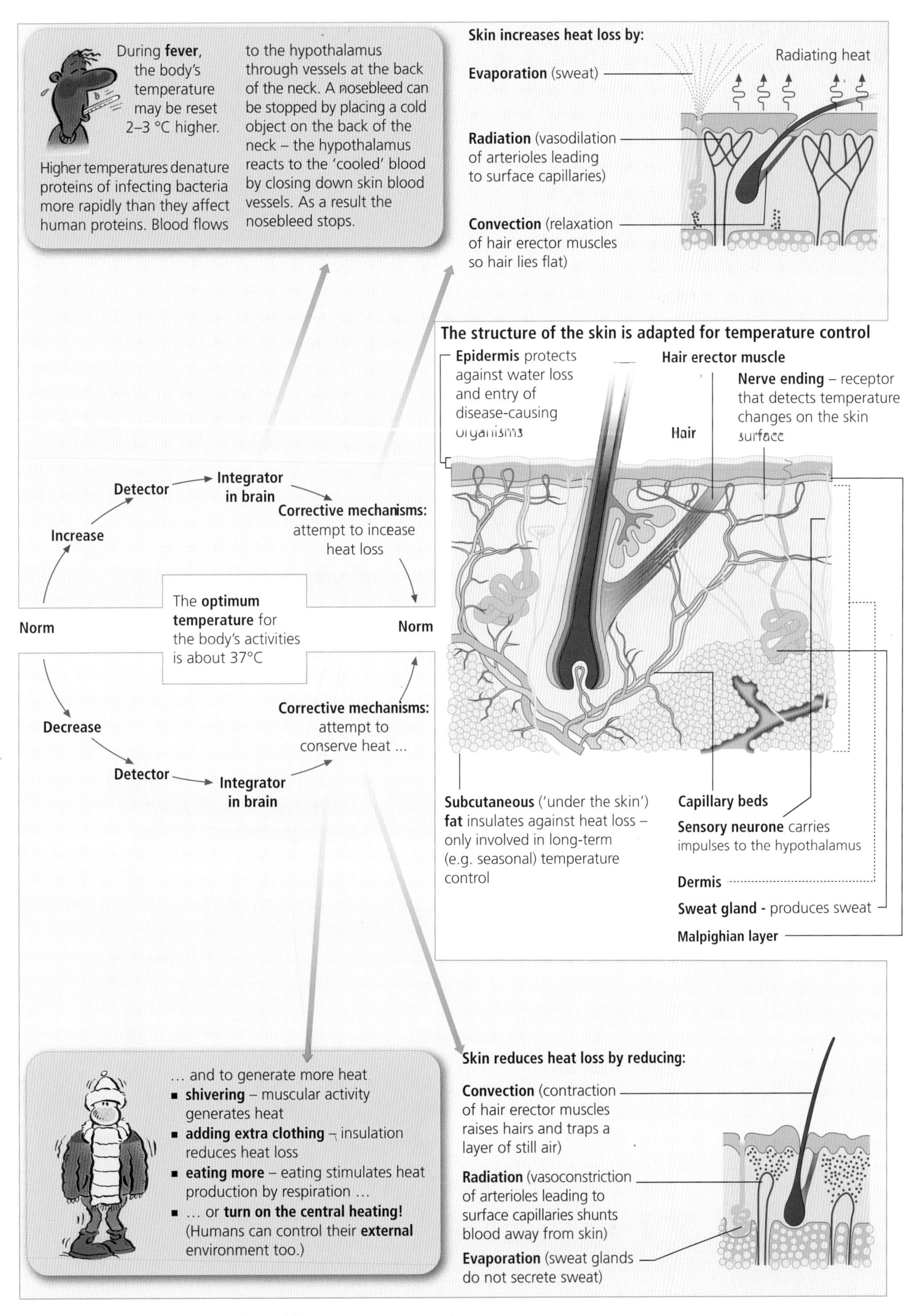

▲ The structure of the skin is adapted for temperature control

Control of blood glucose level

One of the functions of adrenaline is to increase the concentration of glucose in the blood for respiration. However, constant high concentrations of glucose in the blood are harmful.

The ideal concentration of glucose in the blood is normally maintained by two further hormones, **insulin** and **glucagon**. These are secreted by cells in the pancreas in response to changes in blood glucose concentration. They affect liver, fat tissue and muscle.

- **Insulin** is released when blood sugar is too high. It stimulates the removal of glucose from the blood.
- **Glucagon** is released when blood sugar is too low. It stimulates the release of glucose into the blood.

Insulin controls the conversion of **glucose** to **glycogen**; glucagon controls the conversion of **glycogen** to **glucose**. Glucose is a simple sugar and is soluble in blood plasma and cell cytoplasm. Glycogen is a polysaccharide (see page 28) and is insoluble. Glucose is therefore the usable form of carbohydrate and glycogen is the storage form of carbohydrate. The way in which the blood glucose level is kept within safe limits is shown in the diagram below.

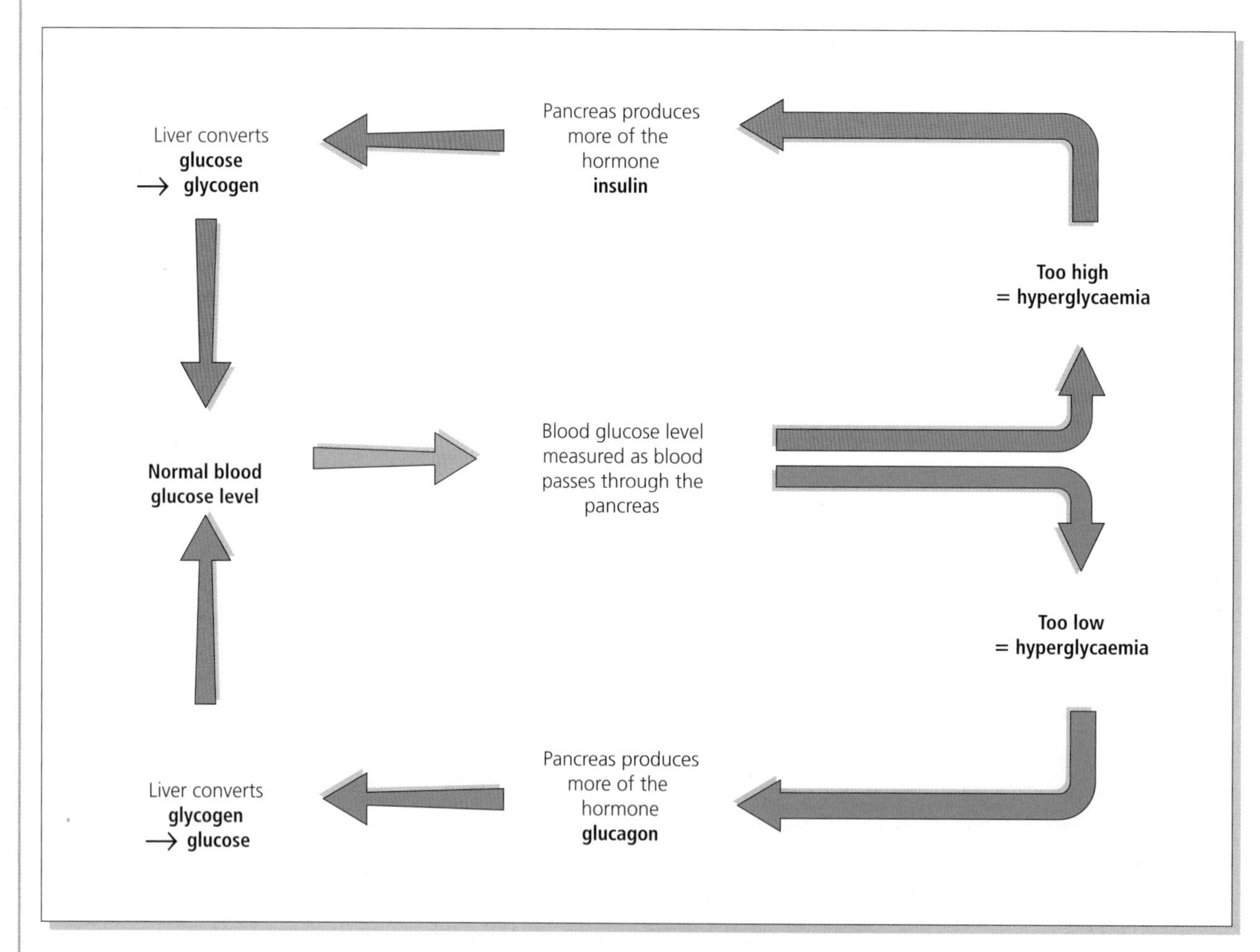

▲ Blood glucose level is under feedback control by the hormones insulin and glucagon

Blood glucose level

Glucose is the cells' main source of energy, and it must always be available to them for respiration …

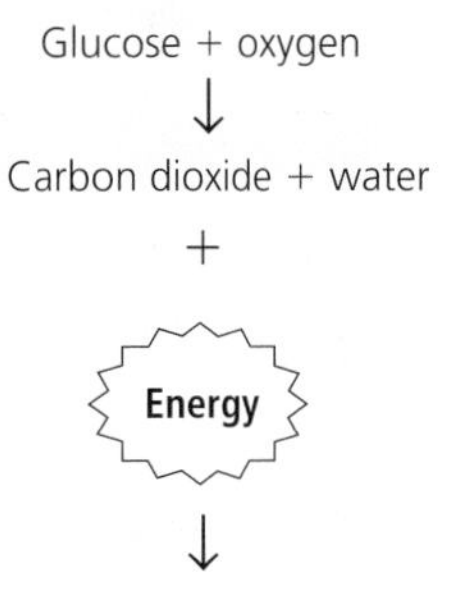

… so the body keeps a constant amount of glucose in the blood. This blood glucose level is usually maintained at about 1 mg of glucose per cm^3 of blood.

What is diabetes?

Diabetes is a condition in which the blood glucose concentration is higher than normal.

- **Type I diabetes** is usually the result of the pancreas failing to secrete enough insulin.
- Symptoms include
  - excessive thirst, hunger or urine production
  - sweet smelling breath
  - high 'overflow' of glucose into urine (test with Clinistix).
- Long-term effects if untreated include
  - premature ageing
  - cataract formation
  - hardening of arteries
  - heart disease.
- Treatment is by regular injection of pure insulin – much of this is now manufactured by genetic engineering (see page 245).
- A diet that contains too much fat and too much sugar can also cause a form of diabetes. This **Type II diabetes** can be controlled by adjusting the diet to limit fat and sugar, and does not need injection of insulin. This 'non-insulin-dependent' diabetes is a common problem for obese people.

Clinistix are thin strips of plastic with a small pad at the bottom. The pad contains an enzyme and a dye. If glucose is present the enzyme uses it to change the colour of the dye. A Clinistix dipped into a urine sample from a diabetic person will give a positive result within seconds. Clinistix are an excellent example of the medical uses of enzymes (see page 246).

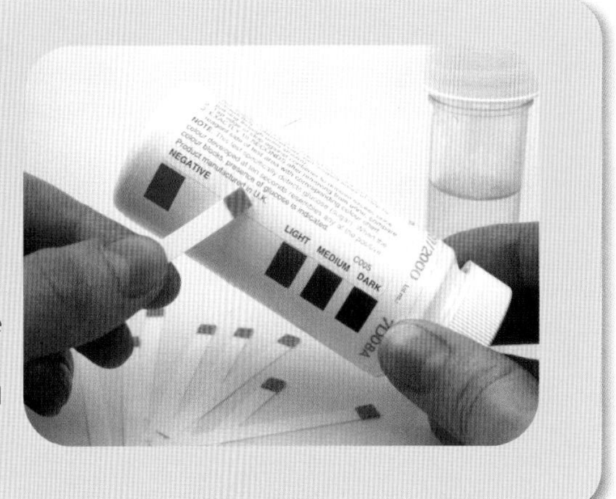

Questions on excretion and homeostasis

1 **a** What is hypothermia?
b Why are old people particularly at risk from hypothermia?
c Why do you think children lose heat very quickly?

2 This diagram shows how body temperature is controlled.

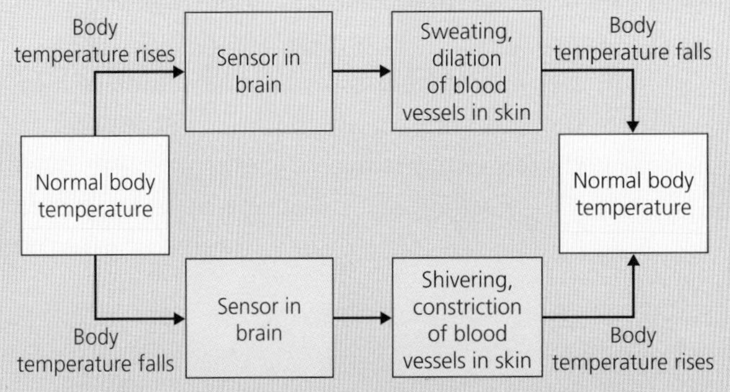

a Use this information to explain how humans control body temperature.
b Explain how shivering affects body temperature.
c How does the liver contribute to body heat?
d Use the diagram to explain the meaning of 'negative feedback'.

3 If a person's kidneys are diseased he or she may have a kidney transplant. The transplanted kidney is connected to blood vessels which go into and out of the leg.
a Copy and complete the diagram on the next page to show clearly how the artery, vein and tube carrying urine should be connected in a transplant operation.

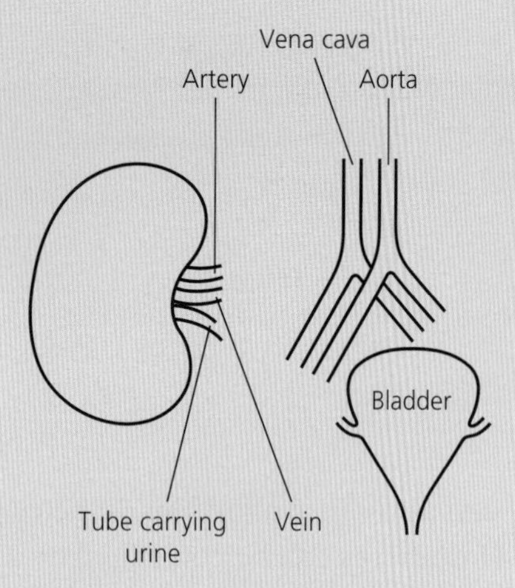

b Sometimes a transplanted kidney is rejected. Explain why rejection may occur.

c Explain how the kidneys prevent sugar and useful ions being lost from the blood.

4 A person who suffers kidney failure may be treated every few days by dialysis. This uses an artificial kidney machine. The diagram shows the working of a kidney machine. In this machine a special solution flows around the outside of an inner tube which carries the patient's blood.

a How is the waste substance urea removed from the blood by dialysis?

b i It is better for the body if a kidney transplant operation is performed instead of having to undergo dialysis every few days. Apart from the improved convenience, why is a kidney transplant better than dialysis treatment?

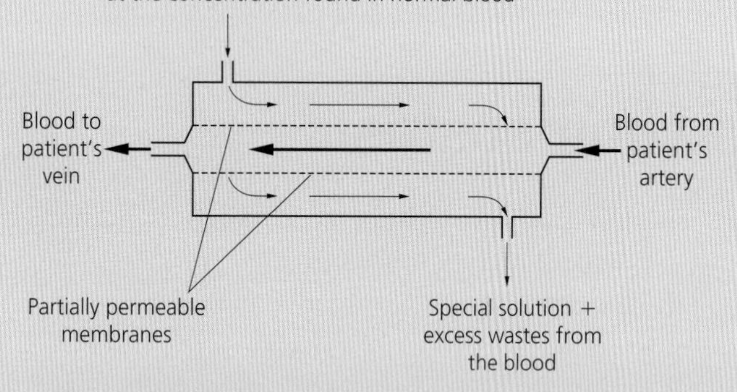

ii It is important to match carefully the tissue type of the kidney donor and of the patient before doing a kidney transplant. What would happen after the transplant if this matching were not done?

5 Humans and other mammals are able to maintain a relatively constant body temperature, despite widely ranging environmental temperatures. Mammals, unless adapted to living in water, seem to prefer not to get wet.

Three flasks were set up. Each flask represents a hot mammal cooling down. Flask **A** had nothing around the flask. This represents a hairless mammal. Flask **B** had a dry covering of cotton cool around the flask. This represents a mammal with dry fur. Flask **C** had a covering of cotton wool soaked in water around the flask. This represents a mammal with wet fur.

Each flask was covered with a lid through which a thermometer was suspended. The bulb of the thermometer was immersed in the water, but did not touch the sides of the flask. Each flask was filled with an equal volume of hot water. The temperature of the water in each flask was measured as it cooled. Readings were taken every 2 minutes and recorded in the table below. A laboratory clock was used to check the time.

	temperature / °C		
time / min	flask A	flask B	flask C
0	70	70	70
2	66	68	64
4	61	67	58
6	58	65	52
8	50	61	42
10	45	60	40

a i On the same axes plot a graph of the three sets of results.

ii Compare cooling of the water in the three flasks (flask **A** compared with flask **B**, flask **B** compared with flask **C** and flask **C** compared with flask **A**).

iii Explain what has happened to produce these results.

b i Describe **three** ways in which this investigation was a fair test.

ii Describe **two** improvements which would increase the accuracy and reliability of this investigation.

Cambridge IGCSE Biology 0610
Paper 6 Q1 May 2008

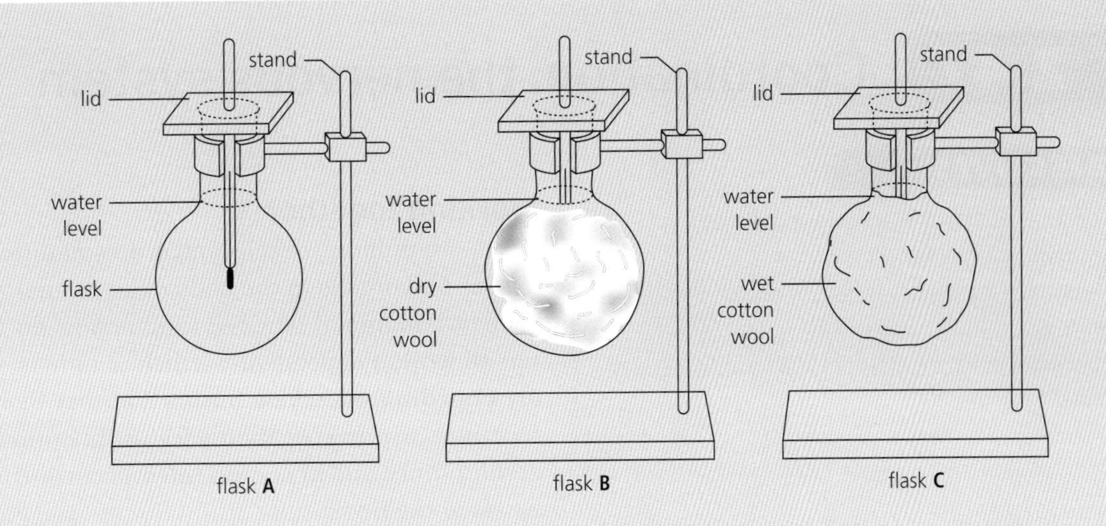

6 a The diagram below shows the urinary system and its blood supply.

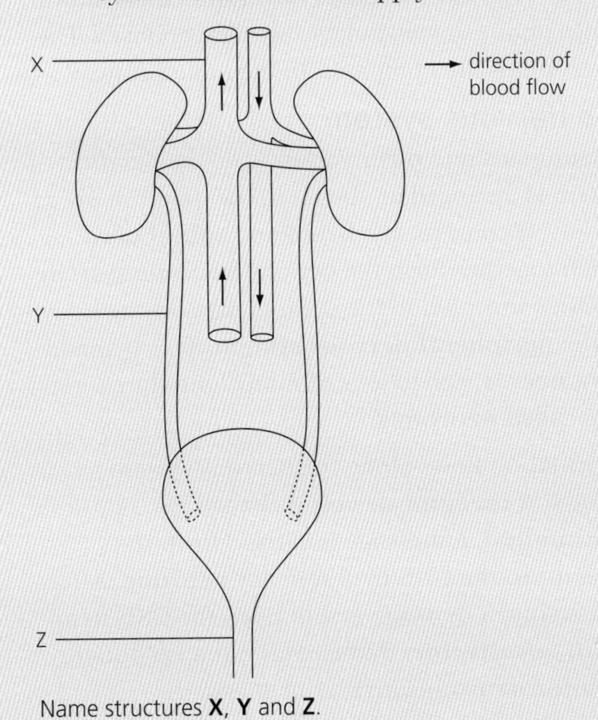

Name structures **X**, **Y** and **Z**.

b The table shows the relative quantities of several substances in the blood in the renal artery and renal vein.

substance	relative quantities in blood in renal artery / arbitrary units	relative quantities in blood in renal vein / arbitrary units
glucose	10.0	9.7
oxygen	100.0	35.0
sodium salts	32.0	29.0
urea	3.0	1.5
water	180.0	178.0

Explain what is happening in the kidney to bring about **three** of the differences between the blood in the renal artery and renal vein, shown in the table.

Cambridge IGCSE Biology 0610
Paper 2 Q11 May 2008

2.47 Coordination: the nervous system

OBJECTIVES

- To understand that different cells and tissues must work in a coordinated way
- To know that there are two systems of coordination in mammals
- To understand that the structure of a neurone is highly adapted to its function

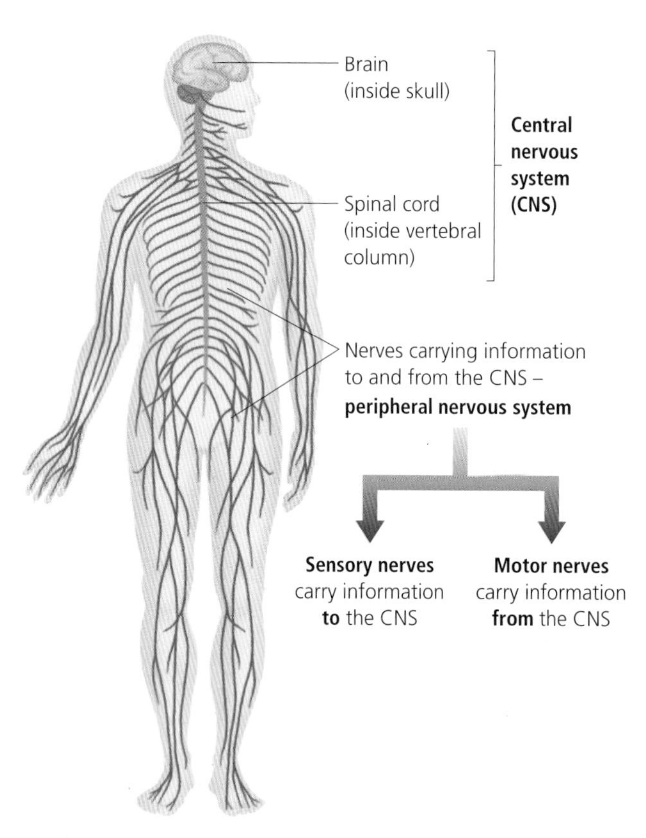

▲ The arrangement of the human nervous system

Working together

There are millions of cells and scores of different tissues and organs in the body of an animal such as a mammal. The cells and organs do not all work independently – their activities are **coordinated**, which means that they work together, carrying out their various functions at certain times and at certain rates, according to the needs of the body.

Coordination in mammals is achieved through two systems, each with its own particular role. The **nervous system** deals with rapid but short-lasting responses, whereas the **endocrine system** brings about slower, longer lasting responses. The two systems are compared in the table below.

The nervous system

In mammals and other vertebrates, the nervous system is arranged as shown in the diagram on the left. It consists of a **brain** and **spinal cord**, which together form the **central nervous system** (**CNS**), connected to the various parts of the body by the **peripheral nervous system**. This is made up of **nerves**, collections of many long thin nerve cells called **neurones**.

Information flows along the nervous system as follows. A **receptor** detects a change in conditions (a **stimulus**). A message is carried from the receptor to the CNS by a **sensory neurone**. After processing, a message is sent from the CNS to an organ (an **effector**) that carries out a **response**. A **motor neurone** carries this message.

The endocrine and nervous systems compared

Comparison	Nervous system	Endocrine system
Speed of action	Very rapid	Can be slow
Nature of message	Electrical impulses, travelling along nerves	Chemical messengers, travelling in the bloodstream
Duration of response	Usually completed within seconds	May take years before completed
Area of response	Often confined to one area of the body – the response is **localised**	Usually noticed in many organs – the response is **widespread**
Examples of processes controlled	Reflexes such as blinking; movement of limbs	Growth; development of reproductive system

Nerves and neurones

All the information carried by the nervous system travels along specialised cells called neurones (sometimes just called nerve cells). The structure of a single neurone, is well adapted to its function of carrying information, as shown on the right.

Nerve impulses

Messages pass along neurones in the form of **electrical impulses**, called **action potentials,** which travel very quickly from one end of a nerve cell to the other. In a living mammal the impulses always travel along a neurone in a certain direction. They are then passed on to another neurone, to a muscle cell or to a gland cell. The end of the neurone is separated from the next cell by a tiny gap, visible under a microscope, and the impulses can only cross this gap in one direction. This gap, called a **synapse**, acts like a valve as explained in the diagram below.

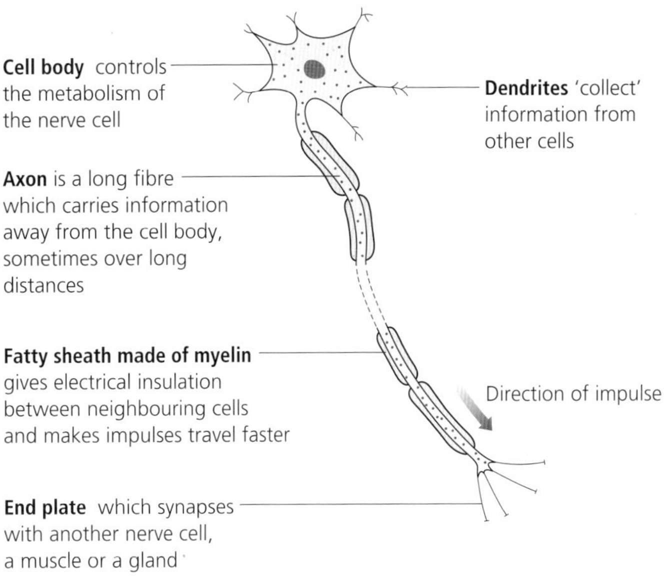

Cell body controls the metabolism of the nerve cell

Dendrites 'collect' information from other cells

Axon is a long fibre which carries information away from the cell body, sometimes over long distances

Fatty sheath made of myelin gives electrical insulation between neighbouring cells and makes impulses travel faster

Direction of impulse

End plate which synapses with another nerve cell, a muscle or a gland

▲ A nerve is made of many neurones. This **motor neurone** carries information from the CNS to an effector.

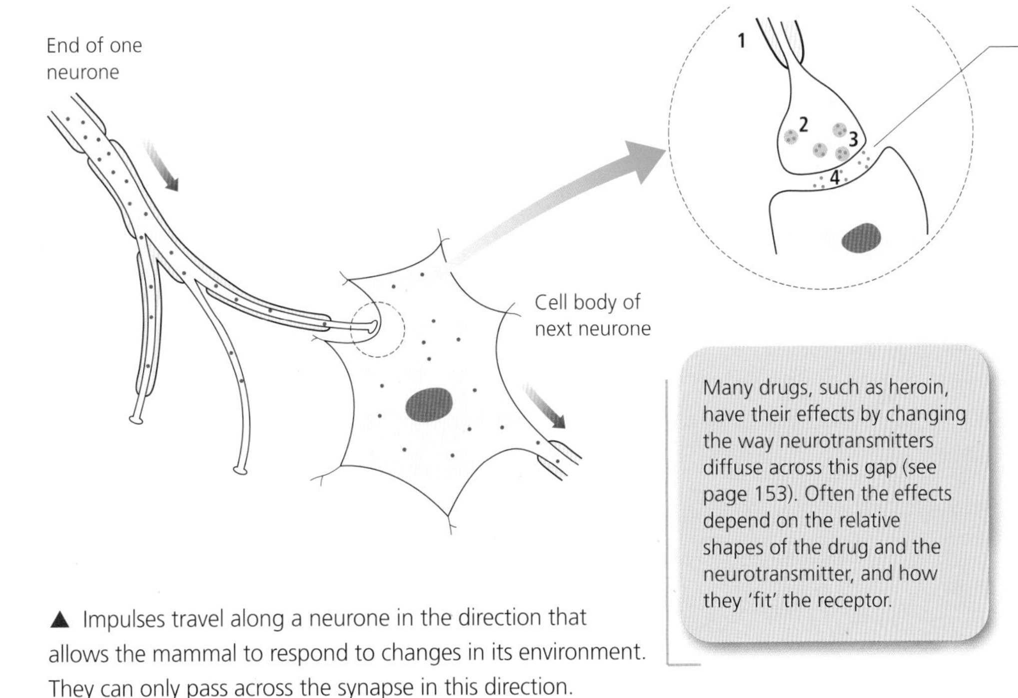

Many drugs, such as heroin, have their effects by changing the way neurotransmitters diffuse across this gap (see page 153). Often the effects depend on the relative shapes of the drug and the neurotransmitter, and how they 'fit' the receptor.

The **synapse** is the junction between the end plate of one neurone and the dendrite of the next.

1. An impulse arrives at the synapse.
2. At the end plate are tiny vesicles containing a chemical (**neurotransmitter**).
3. The chemical is released into the gap.
4. The chemical diffuses across the gap and the impulse restarts on the other side when the neurotransmitter binds to the correct receptor molecule.

▲ Impulses travel along a neurone in the direction that allows the mammal to respond to changes in its environment. They can only pass across the synapse in this direction.

Q

1. Suggest two similarities and two differences between the endocrine system and the nervous system. What is the importance of these differences?
2. Explain how the structure of a neurone is related to its function.
3. Explain the difference between:
   a. motor and sensory neurones
   b. central and peripheral nervous systems.
4. How does a nerve impulse:
   a. pass along a neurone
   b. cross a synapse?

2.48 Neurones can work together in reflex arcs

OBJECTIVES

- To understand that neurones work together in a reflex arc
- To understand that all reflex arcs are important for survival

Neurones act together in many complex ways to bring about the correct response to a stimulus. The simplest type of response is called a **reflex action**. A reflex action is a rapid automatic response to a stimulus, for example jerking your hand away from a sharp or hot object. The nerve pathway involved in the reflex action is called a **reflex arc**, shown in the diagram on the opposite page.

Reflexes and survival

All reflex actions have evolved to help us survive. The table below lists four reflexes, and shows how each helps us survive.

The size of the pupil can change quickly and automatically in response to changes in the intensity of light. This reflex action, described in more detail on page 145, prevents damage to the retina.

A reflex is an involuntary action. Voluntary actions only take place if the brain is involved in initiating the action (see page 140).

Name of reflex	Stimulus	Response	Survival value
Coughing	Particles making contact with the lining of the respiratory tree	Violent contraction of the diaphragm and internal intercostal muscles	Prevents lungs being damaged or infected, so that gas exchange remains efficient
Pupil reflex	Bright light falling on the retina	Contraction of the circular muscles of the iris	Prevents bleaching of the retina so that vision remains clear
Knee jerk	Stretching of the tendon just under the knee, holding the kneecap in place (a doctor may tap this tendon to test the reflex)	Contraction of the muscles of the upper thigh so that the leg straightens	The leg can support the body's weight during walking
Swallowing	Food particles making contact with the back of the throat	Contraction of the muscle of the epiglottis, which closes off the entrance to the trachea	Prevents food entering the respiratory pathway, so that the lungs are not damaged

Q

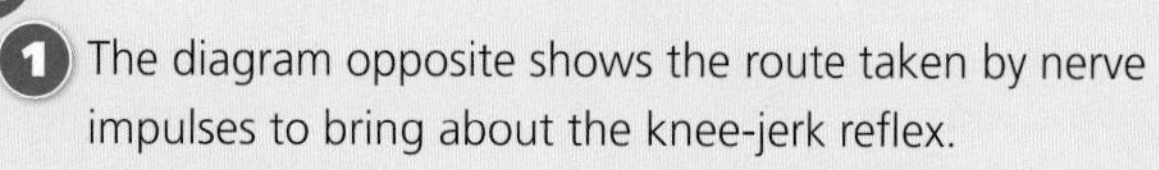

1 The diagram opposite shows the route taken by nerve impulses to bring about the knee-jerk reflex.
   a Name the structure tapped by the hammer. How does this set off the reflex action?
   b Name the structure that carries impulses towards the spinal cord.
   c Which structure **A**–**E** is responsible for the response in this reflex action?
   d The distance between structure B and the spinal cord is about 30 cm. Assuming that the impulses travel at 100 m per s, how long should it take for an impulse to travel through this reflex arc?
   e Careful measurement suggests that the actual time taken for the impulse to travel through this arc is 2 or 3 times this value. Can you explain why?

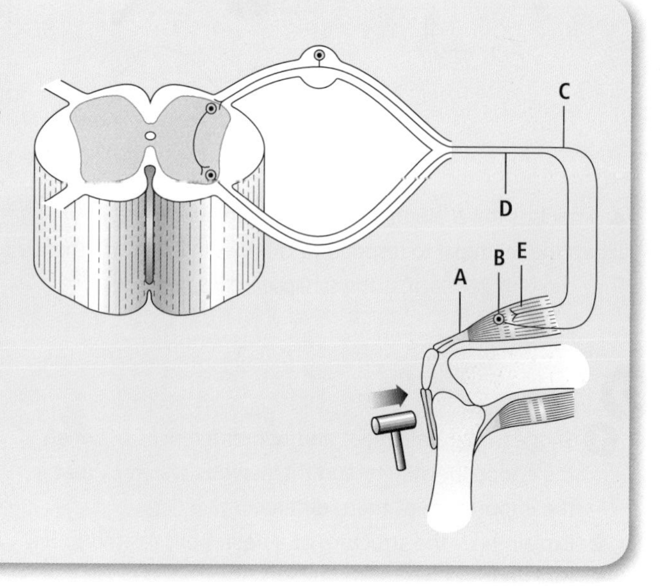

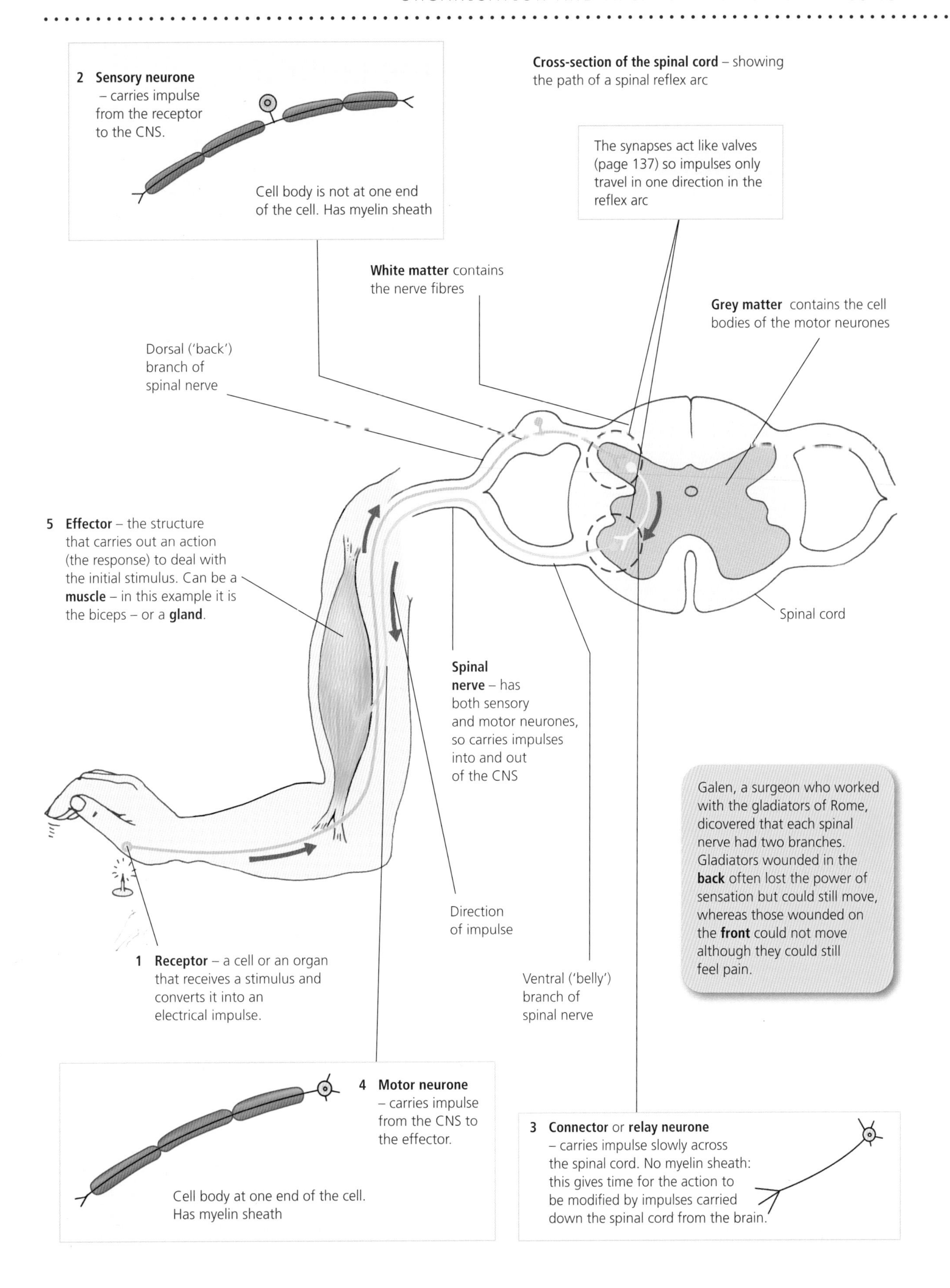

Galen, a surgeon who worked with the gladiators of Rome, dicovered that each spinal nerve had two branches. Gladiators wounded in the **back** often lost the power of sensation but could still move, whereas those wounded on the **front** could not move although they could still feel pain.

2.49 Integration by the central nervous system

OBJECTIVES

- To understand that the central nervous system integrates and coordinates the responses of the body

The central nervous system (CNS) processes information from receptors and passes instructions to effectors to tell the organism how to respond. This complex series of operations, referred to as **integration**, is outlined below.

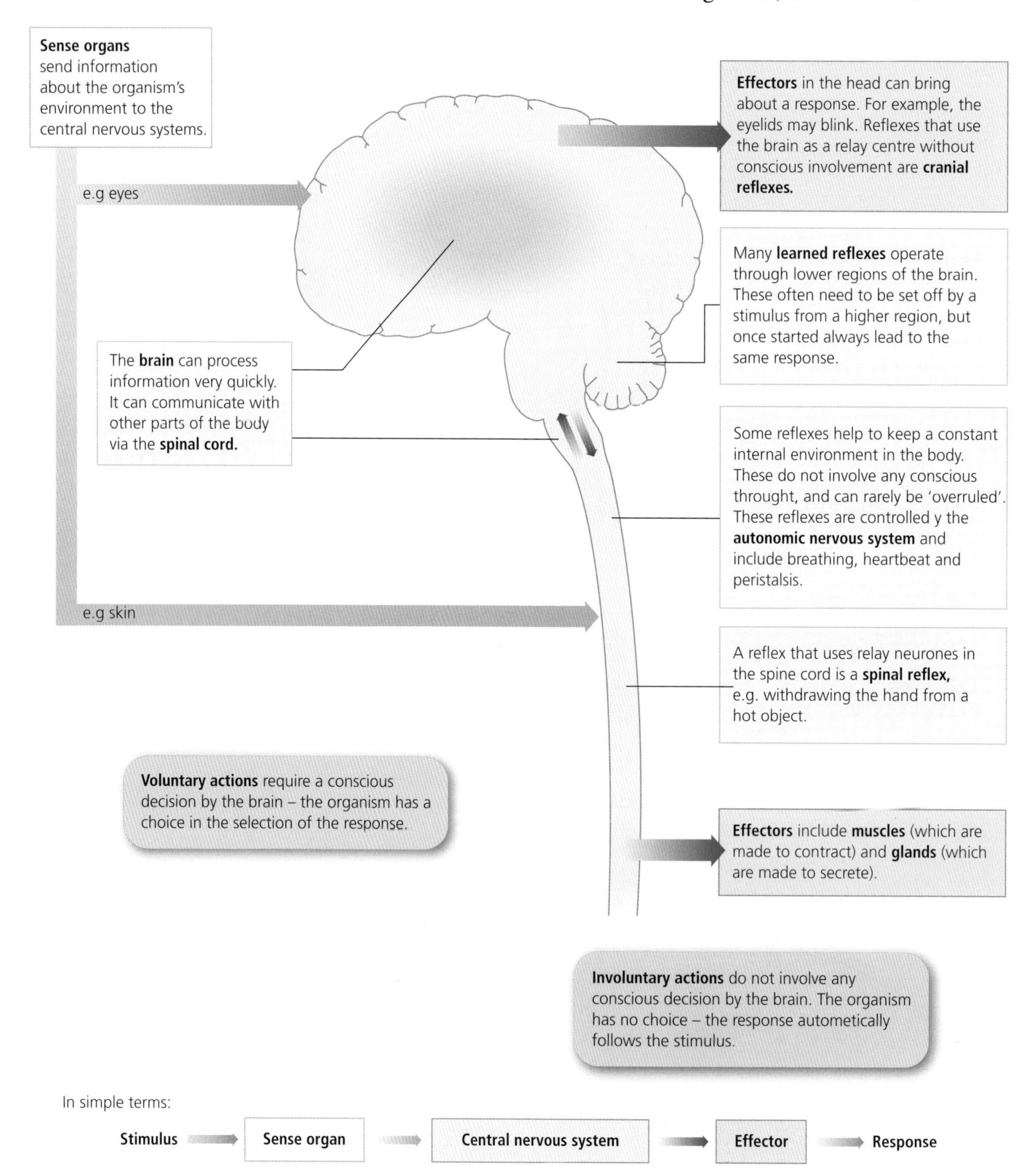

Sense organs send information about the organism's environment to the central nervous systems.
e.g eyes
e.g skin
The **brain** can process information very quickly. It can communicate with other parts of the body via the **spinal cord.**
Effectors in the head can bring about a response. For example, the eyelids may blink. Reflexes that use the brain as a relay centre without conscious involvement are **cranial reflexes.**
Many **learned reflexes** operate through lower regions of the brain. These often need to be set off by a stimulus from a higher region, but once started always lead to the same response.
Some reflexes help to keep a constant internal environment in the body. These do not involve any conscious throught, and can rarely be 'overruled'. These reflexes are controlled y the **autonomic nervous system** and include breathing, heartbeat and peristalsis.
A reflex that uses relay neurones in the spine cord is a **spinal reflex,** e.g. withdrawing the hand from a hot object.
Voluntary actions require a conscious decision by the brain – the organism has a choice in the selection of the response.
Effectors include **muscles** (which are made to contract) and **glands** (which are made to secrete).
Involuntary actions do not involve any conscious decision by the brain. The organism has no choice – the response autometically follows the stimulus.

In simple terms:

Stimulus → Sense organ → Central nervous system → Effector → Response

Involuntary actions

Reflexes concerned with the 'housekeeping' tasks of the body, such as breathing, do not reach the conscious level of the brain. They are dealt with by the **autonomic** branch of the nervous system.

Responses may be more complex than a simple reflex arc. For example, the CNS may store information as **memory** and then compare an incoming stimulus with a previous one. It chooses the correct response for this particular situation, and sends information out to the effectors to bring about the appropriate action. Each time a particular stimulus leads to a certain response, the impulse passes along the same route, so that reflex actions become **learned reflexes**. Talking and cycling are examples of learned reflexes. Whether or not a reflex has been learned, it is an **involuntary action** – a particular stimulus always leads to the same response.

Voluntary actions

During evolution, the front of the spinal cord became highly developed to form the brain. The advanced development of the brain, particularly those parts that deal with learning, sets mammals (and especially humans) apart from 'lower' animals. The brain is involved in **voluntary actions,** in which a conscious choice is made about the response to a particular stimulus.

Conditioned reflexes

Conditioned reflexes are learned reflexes in which the final response has no natural relationship to the stimulus. In an experiment, a Russian scientist, Ivan Pavlov, rang a bell when he fed dogs. The dogs then salivated in response to the bell, even when no food was given. The natural stimulus (the food) had been replaced by an unnatural one (the sound of the bell). Conditioned reflexes can be 'unlearned' if the unnatural stimulus is not repeated with the natural one – if the food was produced without the bell over a period of time, the dogs would no longer salivate at the sound of the bell.

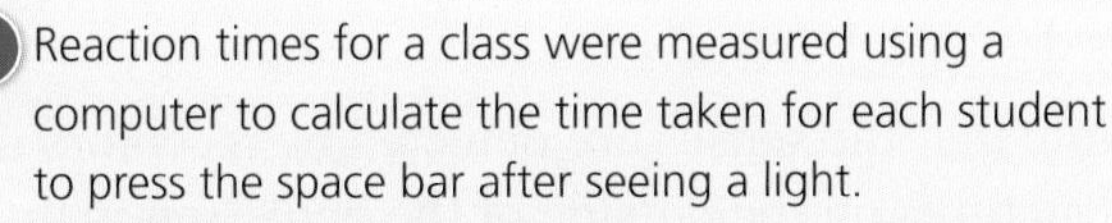

Q

1 Reaction times for a class were measured using a computer to calculate the time taken for each student to press the space bar after seeing a light.

The table opposite shows the results for student 1.

a Calculate the mean reaction time of this student. Why is the mean value useful?

Attempt	1	2	3	4	5	6	7	8	9	10
Reaction time / ms	330	340	290	320	320	280	270	290	400	260

The mean class results are shown in the table below. (You have just calculated the result for student 1.)

Student number	Reaction time / ms	Student number	Reaction time / ms
1		16	150
2	220	17	300
3	130	18	140
4	220	19	230
5	210	20	150
6	250	21	190
7	190	22	180
8	200	23	240
9	220	24	120
10	240	25	170
11	140	26	160
12	280	27	190
13	210	28	210
14	330	29	270
15	270	30	200

b Place the students' reaction times into groups by copying and completing the table below. Draw a bar chart of the results.

Reaction time / ms	Number of students in group
110–150	
160–200	
210–250	
260–300	
310–350	

c The teacher suggested that this reaction time could affect driving ability – a motorcycle travelling at 55 km per h would cover about 15 m in a second. How far would the motorcycle travel before:

i the student with the shortest reaction time pulled the brake lever

ii the student with the longest reaction time pulled the brake lever?

d In a further experiment a loud noise was made at the same time as the light was shown. Eventually the student began to respond when just the noise was made. How does this result explain the meaning of the term 'conditioned reflex'?

2.50 Receptors and senses: the eye as a sense organ

OBJECTIVES

- To understand that receptors are the first stage in reflex arcs
- To know the different types of stimulus to which a mammal is sensitive
- To know that a sense organ combines receptors with other cells
- To know the structure and function of the eye

Receptors detect changes in the environment

A **stimulus** is a change in the environment that affects an organism. All living organisms are **sensitive** – they can respond to stimuli. Animals, including mammals, have a nervous system which receives information from the environment, decides how to respond and then tells the body. A **receptor** is a part of the nervous system that is adapted to receive stimuli. Receptors can be classified according to the type of stimulus they respond to, as shown in the table.

Receptor type	Responds to stimulus	Example in humans
Photoreceptor	Light	Rod cells in retina of eye
Chemoreceptor	Chemicals	Taste buds
Thermoreceptor	Changes in temperature	Thermoreceptors in skin
Mechanoreceptor	Mechanical changes such as changes in length	Hair cells in ear (hearing and balance)

▲ Classification of receptors

Receptors are transducers

All receptors are **transducers**, which means they convert one form of energy into another. They convert the energy of the stimulus (such as light energy) into the kind of energy that the nervous system can deal with (electrical impulses).

The general principle of receptor action is outlined in the diagram opposite.

The senses

Our **senses** are our ability to be aware of different aspects of the environment. For example, the sense of sight allows us to be aware of light stimuli, detected by photoreceptors. The photograph opposite shows the different human senses, and the stimuli to which they are sensitive.

The receptor cells that provide our senses do not work on their own. They need a supply of blood to deliver nutrients and oxygen and remove wastes. Receptors may need help in receiving the stimulus, and receptor cells are often grouped together with other tissues to form a **sense organ**. The other tissues allow the receptor cells to work efficiently.

The working of the eye illustrates the involvement of other tissues in the operation of a sense organ, as described opposite.

Note that the senses are detected by receptors in structures on the outside of the body, and mainly around the head. This is because the stimuli come from outside the body, and the head is often the first part of the body that goes into a new environment. There are also many **internal** receptors inside the body. These detect blood temperature and pH, for example, and are vital in the process of homeostasis (see page 128).

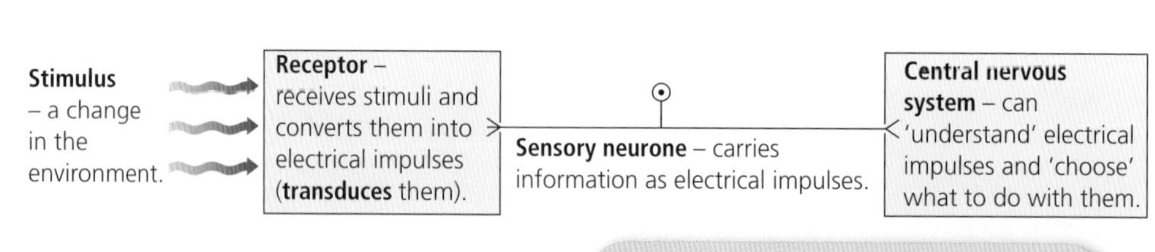

Think about it! We are not aware of any stimulus until the impulse reaches the correct area of the brain. The receptor can be working perfectly but unless the sensory nerve and brain are working, the 'sense' will be incomplete – we 'see' with our brain as much as our eyes.

▲ All receptors work in the same way – they convert one form of energy (the stimulus) into another form that the nervous system can understand

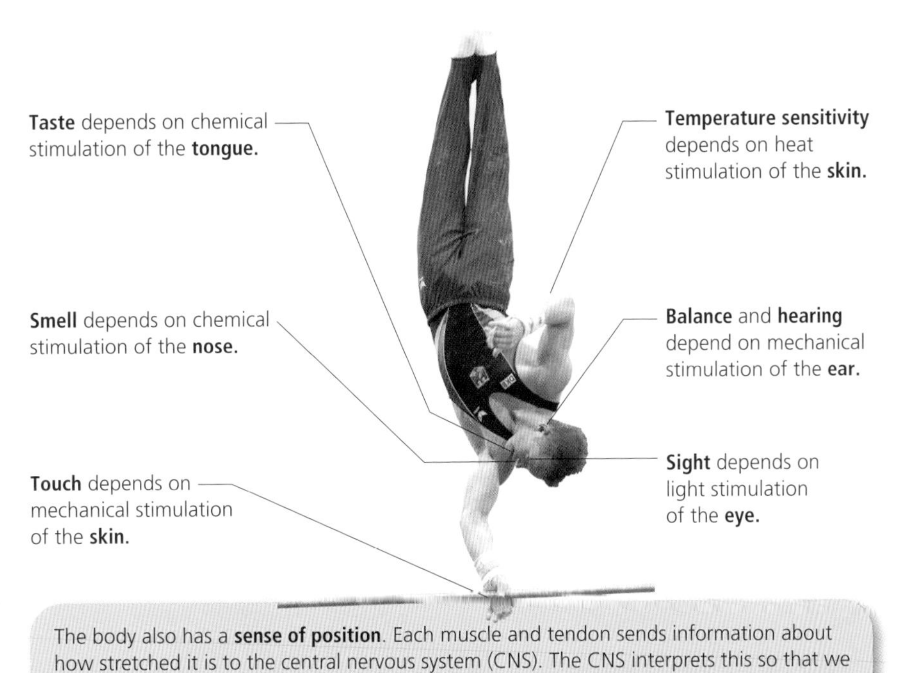
Taste depends on chemical stimulation of the **tongue.**
Smell depends on chemical stimulation of the **nose.**
Touch depends on mechanical stimulation of the **skin.**
Temperature sensitivity depends on heat stimulation of the **skin.**
Balance and **hearing** depend on mechanical stimulation of the **ear.**
Sight depends on light stimulation of the **eye.**

The body also has a **sense of position**. Each muscle and tendon sends information about how stretched it is to the central nervous system (CNS). The CNS interprets this so that we 'know' where each part of the body is in relation to the other parts (see how easy it is to clasp your hands behind your back, or to touch your nose with your eyes shut).

The eye as a sense organ

The eye is an example of a sense organ. It contains:

- **receptors** – the rod and cone cells on the retina
- **systems for making the most of the light stimulus**, including the lens and the iris
- **its own blood supply** and **physical protection** via the choroid and the sclera.

The diagram below shows the arrangement of the structures in the eye.

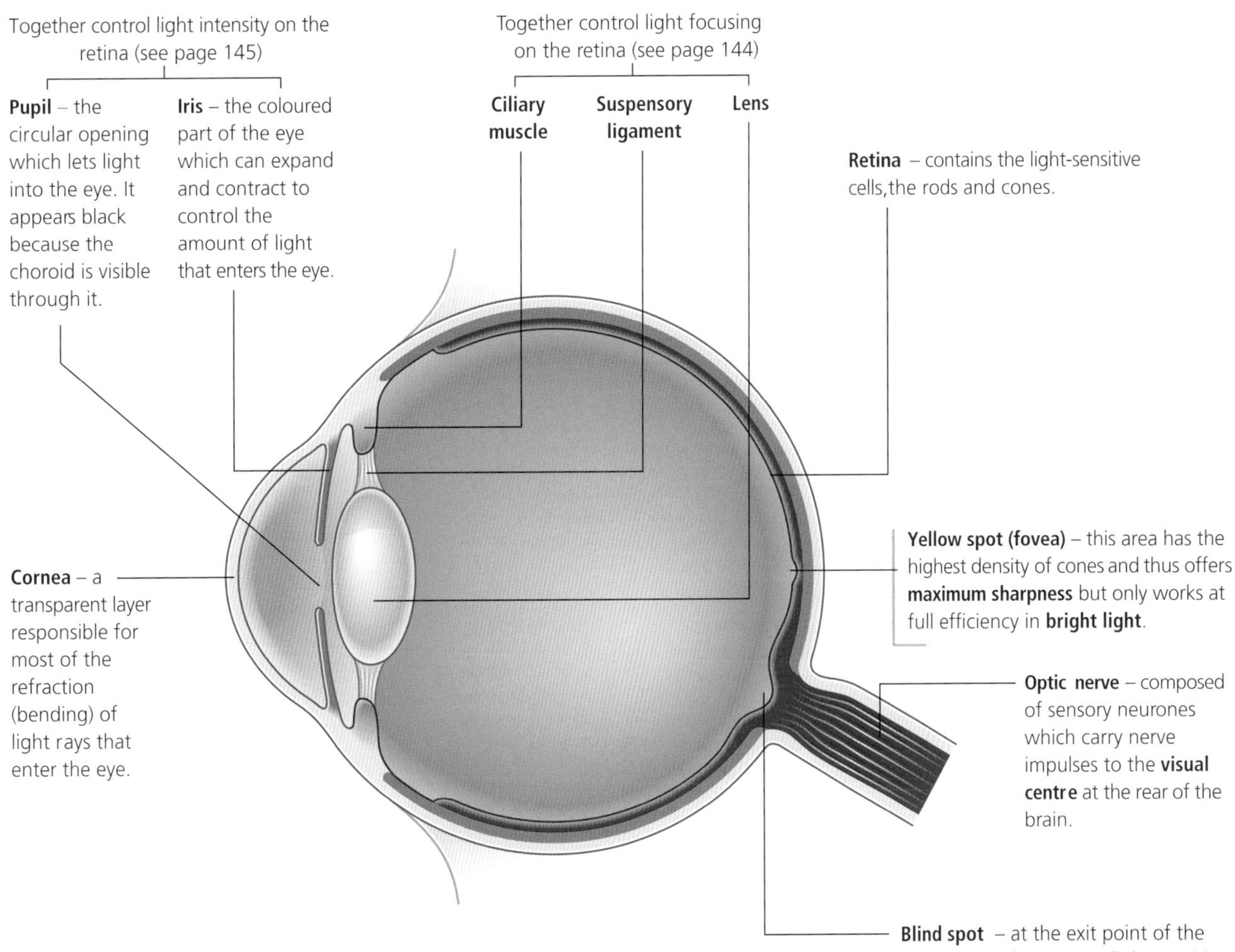
Together control light intensity on the retina (see page 145)
Pupil – the circular opening which lets light into the eye. It appears black because the choroid is visible through it.
Iris – the coloured part of the eye which can expand and contract to control the amount of light that enters the eye.
Together control light focusing on the retina (see page 144)
Ciliary muscle
Suspensory ligament
Lens
Retina – contains the light-sensitive cells,the rods and cones.
Cornea – a transparent layer responsible for most of the refraction (bending) of light rays that enter the eye.
Yellow spot (fovea) – this area has the highest density of cones and thus offers **maximum sharpness** but only works at full efficiency in **bright light**.
Optic nerve – composed of sensory neurones which carry nerve impulses to the **visual centre** at the rear of the brain.
Blind spot – at the exit point of the optic nerve. There are no light-sensitive cells here so light falling on this region cannot be detected.

▲ A horizontal section through the eye. The eye is a sense organ, containing different tissues working together to perform one function.

Rods and cones are photoreceptors on the retina

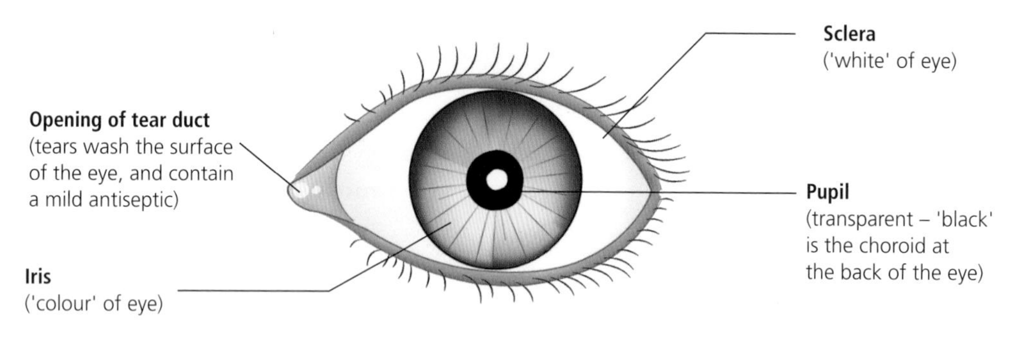

▲ Eye: front/surface view

The retina contains two types of light-sensitive cell, **rods** and **cones**, as shown in the diagram below.

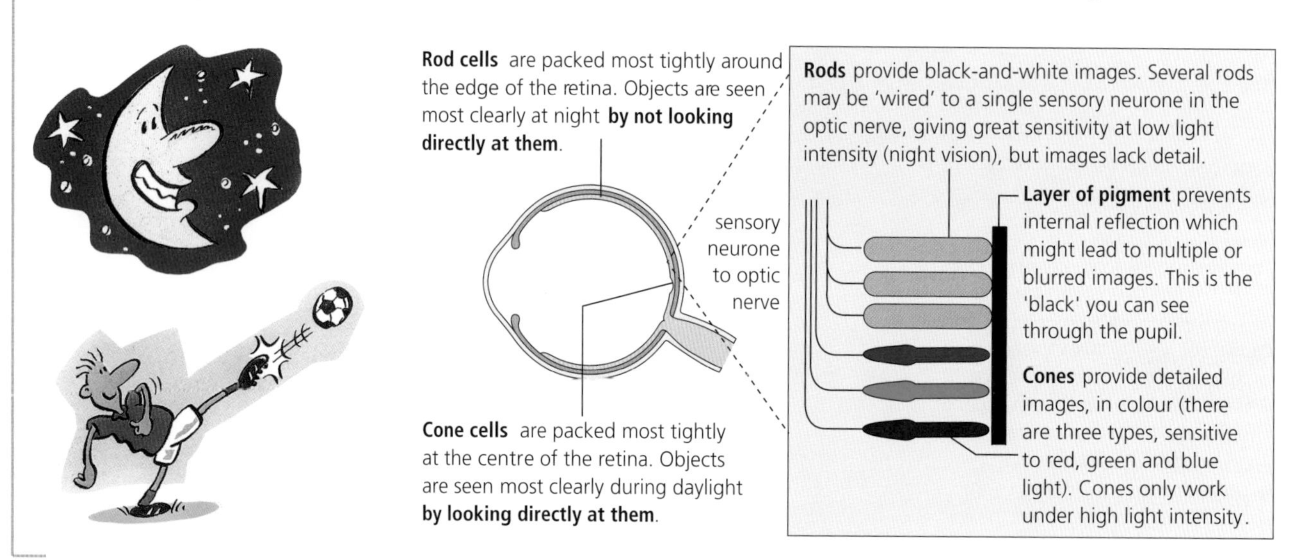

How a sharp image is formed on the retina

An **image** is formed when rays of light from an object are brought together (**focused**) onto the retina, as shown here.

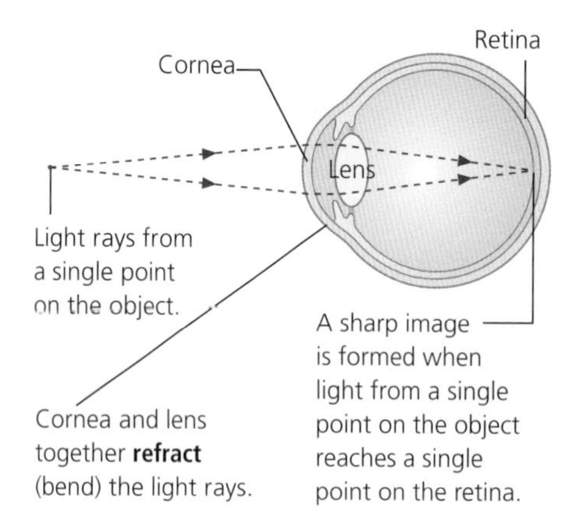

Cornea
Retina
Lens
Light rays from a single point on the object.
Cornea and lens together **refract** (bend) the light rays.
A sharp image is formed when light from a single point on the object reaches a single point on the retina.

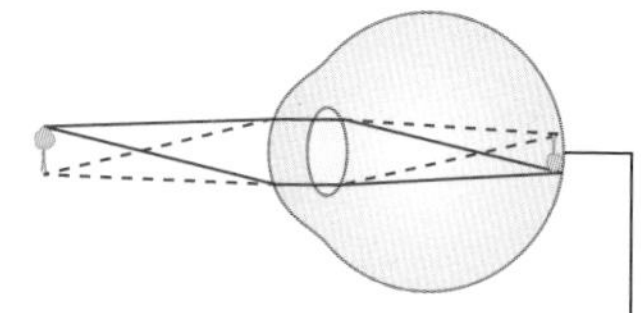

The image formed on the retina is **inverted** (upside down) and **diminished** (smaller than the object). The brain 'corrects' this inversion and reduction in size. This process is called **integration**.

Accommodation – adjusting for near and distant objects

The amount of **refraction** (bending) of the light is adjusted depending on the distance between the object and the eye. The light rays coming from very distant objects are parallel. They only need to be refracted a little to form a single, sharp, focused image on the retina. However, the lens does not need to be very powerful for this. The light rays coming from close objects are diverging. They need to be refracted more to form a sharp image on the retina, and the lens needs to be more powerful. The ability of the lens system to produce a sharp image of objects at different distances is called **accommodation**. As people get older the lens becomes less elastic and loses its ability to change shape. This makes it harder to refocus quickly on objects at different distances. Smokers may experience this problem earlier than non-smokers because smoke 'ages' molecules in the lens.

The iris controls the light intensity at the retina

Light falls on the retina and stimulates the rods and cones to produce nerve impulses. These travel to the brain along the optic nerve. It is important that the rod and cone receptor cells are not over-stimulated. If too much light fell on them they would not recover in time to allow continued clear vision. The iris contains muscles that alter the size of the pupil, thereby controlling the amount of light that falls on the retina. This control is automatic, and is a good example of a reflex action. This **pupil reflex** is explained in the diagram opposite.

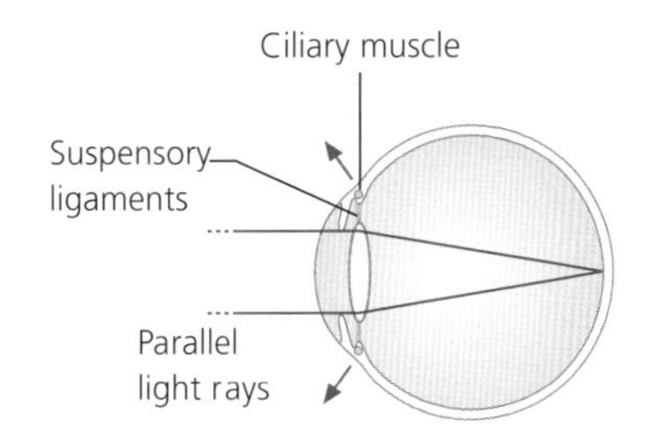

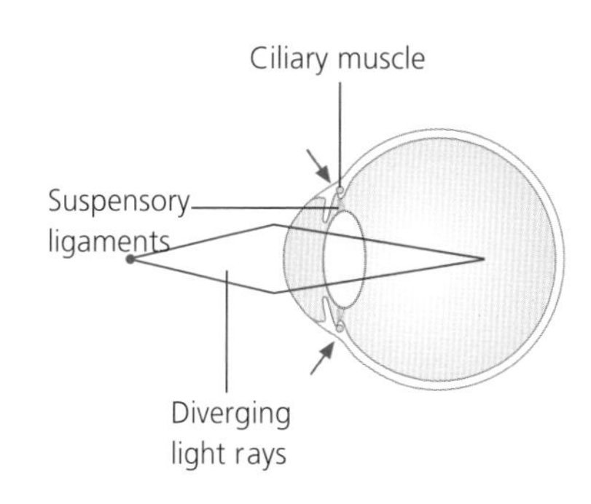

Distant object
- Light needs to be **refracted** (bent) **less**.
- Ciliary muscles **relax**, eyeball becomes spherical.
- Ligaments are **tight**.
- Lens is pulled **long and thin**.

Close object
- Light must be **greatly refracted** (bent).
- Ciliary muscles **contract**, pull eyeball inwards (eyeball 'bulges' forward).
- Ligaments **relax**.
- Lens becomes **short and fat**.

Relaxed or waking eyes are set for viewing distant objects so that images of close objects (such as alarm clocks!) are blurred.

Eyestrain is caused by long periods of close work. The ciliary muscles are contracting against the pressure of fluid in the eyeball.

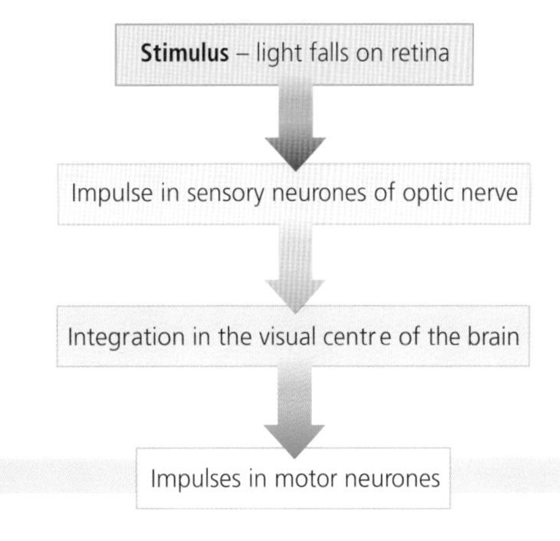

Low light intensity – radial muscles of iris contract and the pupil is opened wider, so more light can enter and reach the retina.

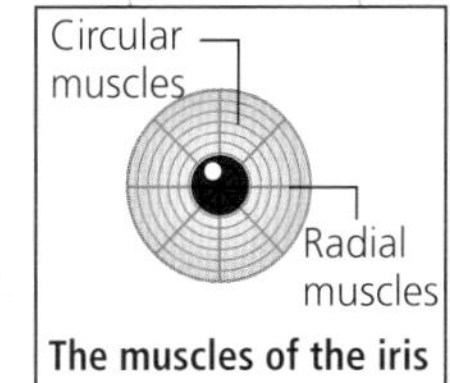

High light intensity – circular muscles of iris contract and the pupil is reduced in size, so less light can enter and the retina is protected from bleaching.

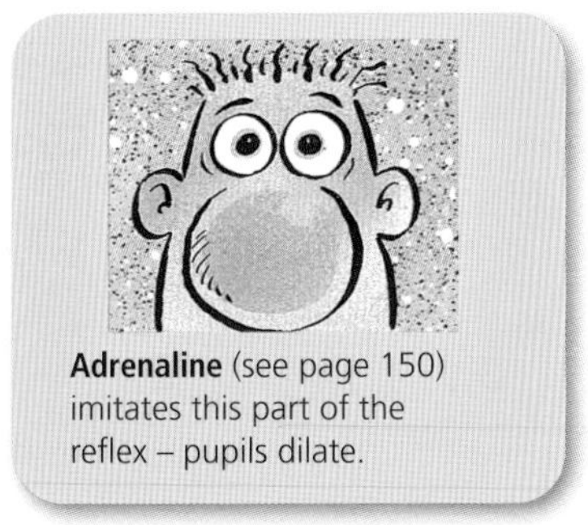

Adrenaline (see page 150) imitates this part of the reflex – pupils dilate.

Heroin (see page 152) imitates this part of the reflex – pupils constrict.

▲ The pupil reflex prevents bleaching of the retina by regulating the amount of light that enters the eye

Questions on receptors and senses

1 What is a transducer? Explain how:
a taste buds **b** rod cells
can be called transducers.

2 Copy and complete this paragraph.
Light rays from an object are refracted by the ____ and the ____ which focus the light rays onto the ____. The amount of light that reaches this light-sensitive layer is controlled by the ____ which is able to adjust the size of the ____ (the black 'hole' in the front of the eye). There are two types of light-sensitive cells, the ____, which are responsible for ____ vision in low light intensity, and the ____, which are responsible for ____ vision in ____ light intensity. The image formed on the ____ is ____ and ____ than the object, but the nerve impulses that pass along the ____ nerve to the brain are interpreted so that they make sense. The ability of the brain to compare incoming information with previous experience, and to set off the correct response, is called ____.

3 Bimla was sitting in a well-lit room. She covered one eye with an eye patch. A pencil was held in front of her eye for 10 seconds. Bimla focused on it and at the same time the thickness of her eye lens was measured using an optical instrument. The pencil was then moved a different distance from the eye. This was repeated over a short period. The results are shown in the table.

distance from eye / cm	thickness of lens / mm
10	4.0
20	3.6
30	3.2
50	2.9
100	2.7
150	2.6
200	2.6

a Name the structures in the eye that bring about the change in the thickness of the lens.
b In this investigation, which is the independent variable and which is the dependent variable?
c Suggest two important fixed variables. Explain why they must be fixed.
d How could the experiment be improved to make the data more reliable?

4 a Copy the diagram of the eye. Use arrows and the letters listed below to label the diagram.
A Layer containing rods and cones

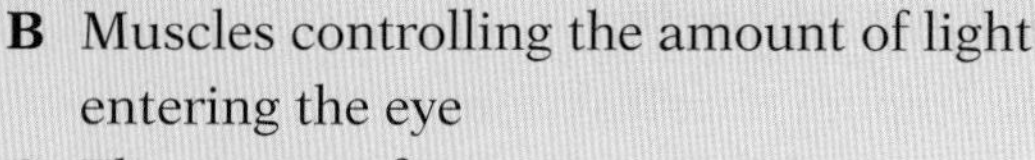

B Muscles controlling the amount of light entering the eye
C The source of tears
D A black layer containing blood vessels
E A very thin layer which protects the surface of the eye from bacteria
F A tough white protective layer

b In bright light the iris changes shape to reduce the size of the pupil.
i What is the advantage of this?
ii For this purpose, name the stimulus, receptor, coordinator and effector. Copy and complete the diagram below.

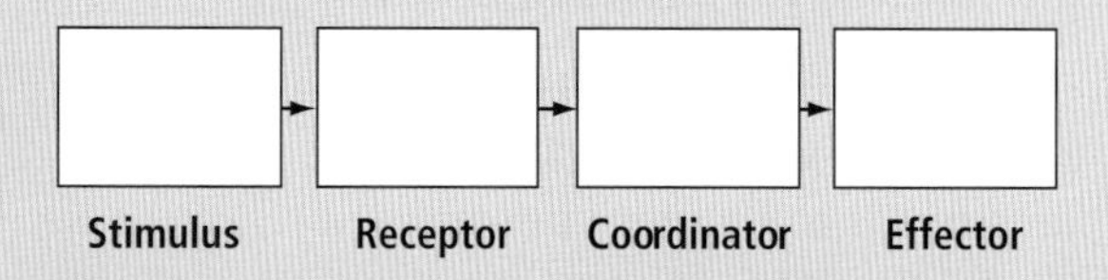

c Rosie went from bright light into a dark room. The graph below shows the dimmest light in which she could see during the first half hour after entering the room.
i How long was it before Rosie's eyes were completely 'used to the dark'?
ii Rosie's friend, Matilda, suggests that Rosie's eyes become 'used to the dark' because her iris changes shape. From the evidence in the graph, do you think that this is likely? Explain your answer.

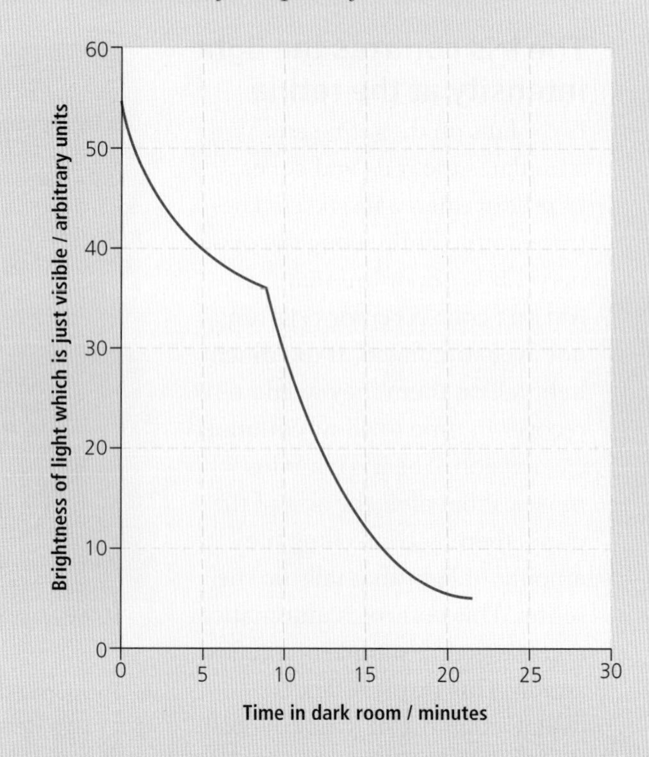

5 The figure below shows the bones and muscles of a human leg.

a Muscles in the leg work antagonistically.

i State which muscle is antagonistic to muscle **A**.

ii Explain what is meant by *antagonistic*.

b In the figure, the person is sitting with the foot clear of the ground. If a sharp tap is given at **X** then the lower leg swings forwards. This is a reflex action.

i Describe the general features of any reflex action.

ii If the spinal cord is cut through near the chest, this reflex action still takes place. Suggest where in the central nervous system this reflex response is coordinated.

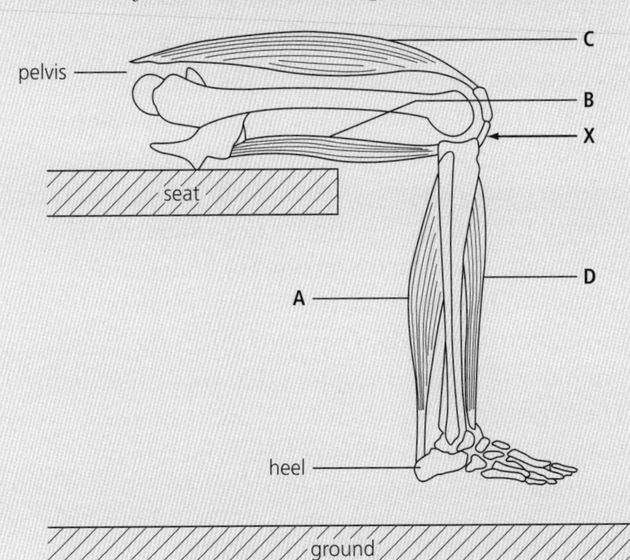

c In an emergency, a person might have to run suddenly and very quickly.

i Name the hormone that the body releases in such an emergency.

ii Describe three changes that occur in the body when this hormone is released in such an emergency.

Cambridge IGCSE Biology 0610
Paper 2 Q8 November 2008

6 a Complete the following paragraph using appropriate words.

Sense organs are composed of groups of __________ cells that respond to specific __________. The sense organs that respond to chemicals are the __________ and the __________

b The eye is a sense organ that focuses light rays by changing the shape of its lens. It does this by contracting its ciliary muscles.

i What links the ciliary muscles to the lens?

ii Describe the change in shape of the lens when a person looks from a near object to a distant object.

c The graph shows changes in the contraction of the ciliary muscles as a person watches a humming bird move from flower to flower while feeding on nectar.

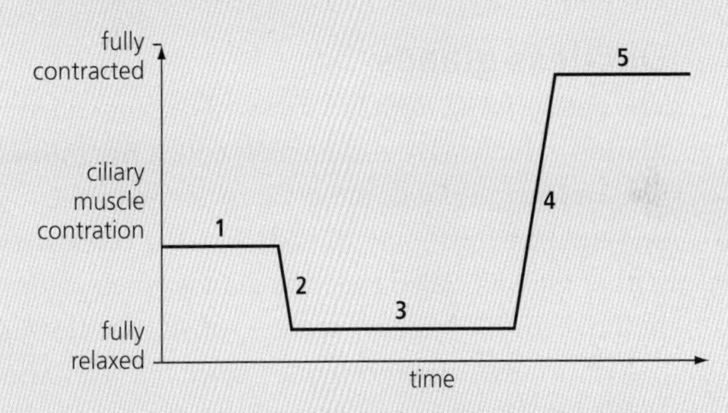

In which period of time, **1**, **2**, **3**, **4** or **5**, was the bird

i feeding from a flower very near to the person,

ii flying away from the person,

iii flying towards the person.

Cambridge IGCSE Biology 0610
Paper 2 Q7 May 2009

2.51 The endocrine system

OBJECTIVES

- To understand the need for a chemical system of coordination
- To define the terms hormone and endocrine organ
- To give examples of hormones in the human body

A second control system

The responses controlled by the nervous system happen quickly, but there are some responses that go on over a long period of time. Growth and development, for example, continue for years. Animals have a second coordination system, the **endocrine system**, which carries out this sort of control.

Ductless glands

The endocrine system is a series of organs called **glands**, which secrete chemicals called **hormones**. The endocrine glands are ductless glands – they secrete their hormones directly into the bloodstream. (Other glands, called exocrine glands (such as those in the digestive system), secrete substances through a duct or tube.) The hormones, once released, travel in the blood to any part of the body that is supplied with blood. The hormones affect only their **target organs,** as outlined in the diagram below.

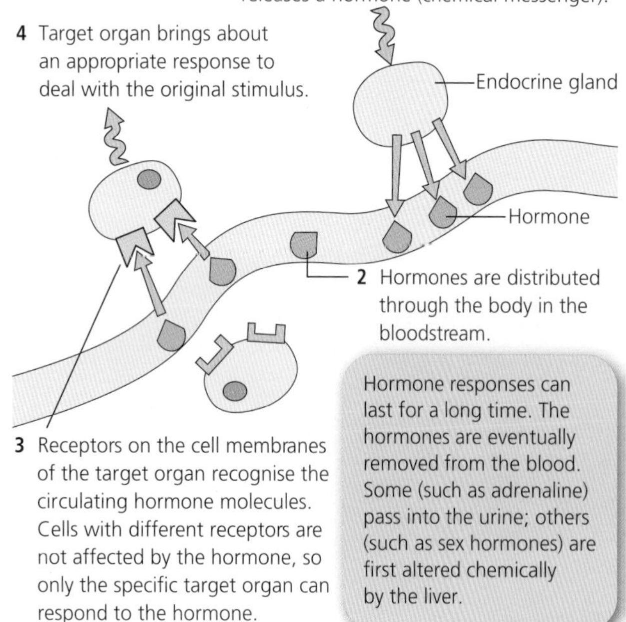

▲ The mechanism of hormone action

Hormones in humans

The main hormone-producing glands, and the hormones they produce, are shown in the diagram on the opposite page.

A comparison between hormonal and nervous co-ordination is shown on page 136.

Hormones control puberty in humans

Hormones control long-term processes, and often have widespread effects on the body. **At puberty** a person becomes physically able to reproduce. The development of sexual maturity is a good example of a hormone-controlled process (see page 173).

As a young person develops physically, certain signals are processed by the brain, which then instructs the pituitary gland to stimulate the **primary sex organs** – the testes in males and the ovaries in females. Sex hormones – **oestrogen** (in females) and **testosterone** (in males) – are released into the bloodstream and circulate throughout the body. They only affect the target organs which have receptors that recognise them. These target organs then carry out responses, such as the growth of body hair, which may continue for many years.

Hormone production is under feedback control

The production and secretion of hormones is accurately controlled by **feedback** – the hormones regulate their own production. As the level of hormone in the blood rises, it switches off (**inhibits**) its own production so that the level never gets too high. As the level of hormone in the blood falls, it switches on (**stimulates**) its own production so that the level never gets too low. **Feedback control** is very important in biology, particularly in homeostasis (see page 129). It is outlined in the diagram opposite.

Hormones and food: growth hormones are widely used to increase meat production in domestic animals (see page 150).

Endocrine organs and their secretions

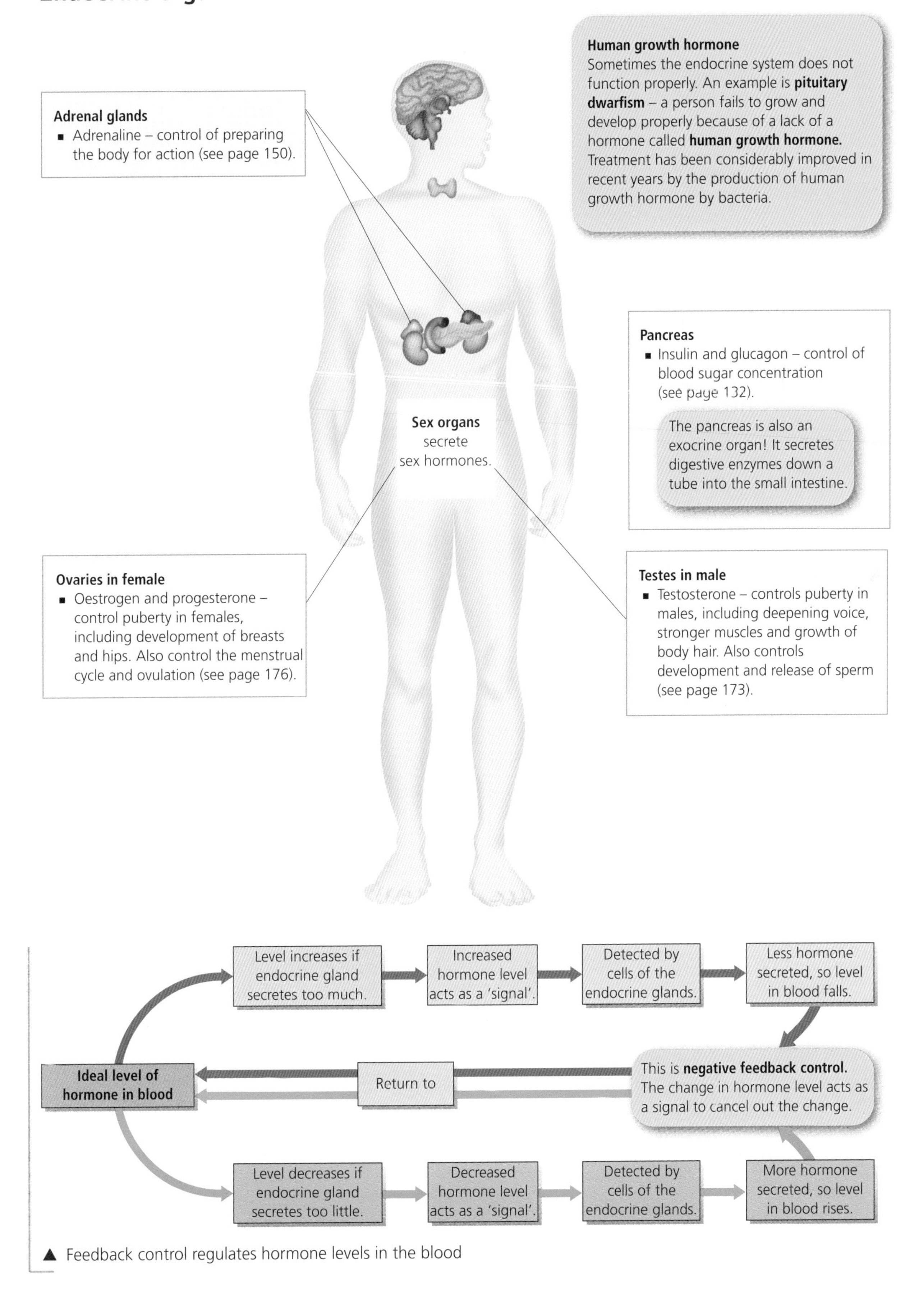

Human growth hormone
Sometimes the endocrine system does not function properly. An example is pituitary dwarfism – a person fails to grow and develop properly because of a lack of a hormone called human growth hormone. Treatment has been considerably improved in recent years by the production of human growth hormone by bacteria.
Adrenal glands
Adrenaline – control of preparing the body for action (see page 150).
Pancreas
Insulin and glucagon – control of blood sugar concentration (see page 132).
The pancreas is also an exocrine organ! It secretes digestive enzymes down a tube into the small intestine.
Sex organs secrete sex hormones.
Ovaries in female
Oestrogen and progesterone – control puberty in females, including development of breasts and hips. Also control the menstrual cycle and ovulation (see page 176).
Testes in male
Testosterone – controls puberty in males, including deepening voice, stronger muscles and growth of body hair. Also controls development and release of sperm (see page 173).
Level increases if endocrine gland secretes too much.
Increased hormone level acts as a 'signal'.
Detected by cells of the endocrine glands.
Less hormone secreted, so level in blood falls.
Ideal level of hormone in blood
Return to
This is negative feedback control. The change in hormone level acts as a signal to cancel out the change.
Level decreases if endocrine gland secretes too little.
Decreased hormone level acts as a 'signal'.
Detected by cells of the endocrine glands.
More hormone secreted, so level in blood rises.

▲ Feedback control regulates hormone levels in the blood

Adrenaline

One hormone that has been widely studied is **adrenaline**. This substance seems to bridge the gap between nervous and endocrine control. It is definitely a chemical messenger, and is released directly into the bloodstream, yet its actions are often very rapid indeed and may only last for a very short time. The widespread and instant effects of adrenaline are described below.

Skin becomes pale as blood is diverted away.

Deeper, more rapid breathing and airways become wider.

Heart beats more rapidly.

Blood is diverted away from digestive system to muscles by using sphincters (see page 88).

Adrenal glands (on top of the kidneys) release the hormone adrenaline.

Glycogen in muscles is converted to **glucose**, and released into the blood.

▲ Adrenaline is known as the 'flight or fight' hormone, released when the body is given a shock

The overall effect is to provide more glucose and more oxygen for working muscles – preparation for action!

Hormones and food production*

Hormones are important in the beef, poultry and dairy industries: they mimic the effects of hormones produced by the animals concerned. Protein hormones must be injected into the animal (as proteins in food would be digested), steroid hormones are supplied in food or from pellets implanted in a part of the body not consumed by humans. Many people are worried about the use of these hormones.

steroid pellets in flesh of ear: ears are not usually eaten by humans

Hormone	Use	Concerns for consumers
rBGH (recombinant bovine growth hormone) (protein)	Rapid growth of beef cattle: better conversion ratio of feed in to muscle increased period of milk production	May cause cattle to produce other growth factors which could cause cancer in humans
Synthetic oestrogen (steroid)	Rapid growth in beef and poultry	Early onset of puberty in girls; feminisation in males e.g. breast development

Worries about steroid levels have led to a ban on their use in the US, UK and much of Europe.

Hormones can be used as drugs in sport

The effects of hormones on the body's activities can be used to improve performance in sport. Some hormones which have been abused in this way are listed in the table below.

Hormone	Effect on body	Effect on sporting performance
Anabolic steroids	Increase growth of muscle. Reduce fat content of body.	Increase strength and power: weight ratio useful in 'explosive' sports such as sprinting and shot putting.
Cortisone	Repair of damaged tissues.	Allow rapid recovery after intensive training.
Testosterone	Stimulates male's aggressive behaviour.	Aggression can be important in contact sports like rugby.

BUT BEWARE!

- Drug abuse is illegal and can lead to lengthy bans for sportsmen and women.
- 'Artificial' hormones can switch off natural hormone production by feedback inhibition. Men may lose their sexual potency, gain body fat and develop 'squeaky' voices, and women may cease menstruation and ovulation (and sometimes develop excessive body hair).

▲ Ben Johnson, 100 m sprinter who was disqualified from the 1988 Summer Olympics and 1987 World Championships after testing positive for banned drugs.

Questions on hormones

1 Copy and complete the following paragraph. A human exposed to a severe shock responds by producing the hormone ____. This hormone causes the storage polysaccharide ____ to be converted to ____, a soluble sugar used to release energy via respiration. Aerobic respiration requires ____ as well as this sugar, and more of this gas is made available because the hormone causes ____ and ____ breathing. The body makes the most of its resources by adjusting blood flow to different organs – less blood flows to the ____, for example, and more flows to the ____. The face of a shocked person shows three effects of this hormone – the skin ____, the pupils ____ and the hair ____. Because of these effects this hormone is often called the ____ or ____ hormone.

2 A dangerously aggressive animal is unlikely to fit into society. Aggression may be an important part of puberty in male animals, and may help to win females. Explain how negative feedback (feedback control) would keep aggression within acceptable limits in a young male animal.

2.52 Drugs and disorders of the nervous system

OBJECTIVES

- To know what is meant by a drug
- To be able to list the main types of drug
- To understand the effects of some drugs on the nervous system
- To appreciate that the nervous system may not always work perfectly

Types of drug

A **drug** is a chemical substance that can modify or affect chemical reactions in the body. A **medicine** is a chemical treatment for an illness or disorder. **Antibiotics** are drugs which are medicines (see page 248). All medicines contain drugs, for example, aspirin tablets contain the drug acetylsalicylic acid, but not all drugs are medicines. Nicotine and alcohol are drugs, for instance, but are not usually thought of as medicines.

There are many drugs, affecting different parts of the body. Some drugs act directly on the nervous system. These are often grouped according to the effect they have, for example:

- **Stimulants** promote (speed up) the action of the nervous system. Stimulants usually make the drug user feel more confident and alert, and include **amphetamines**, **caffeine** and **nicotine**. (Nicotine is taken in tobacco smoke. Other effects of smoking are described on page 120.)
- **Depressants** inhibit (slow down) the action of some part of the nervous system. The user feels sleepy and less anxious, but may become dependent on the drug. Examples of depressants are **barbiturates**, **alcohol** and **cannabis**.
- **Narcotics** act like depressants but particularly target the brain. They work as pain-killers and may bring about a feeling of drowsy well-being or **euphoria**. Narcotics such as **heroin** are very likely to bring about drug dependence in the user.
- **Analgesics** are mild pain-killers. **Aspirin** and **paracetamol** are widely used analgesics.

Some of these drugs and the effects that they have are described on the opposite page.

Social drugs may be abused

A drug that is taken for non-medical reasons can be described as a **social** or **recreational drug**. Examples include legal drugs, such as nicotine in cigarette smoke, and alcohol, and illegal drugs, such as amphetamines and LSD. These drugs are taken for the pleasurable sensations that they give the user. Users may become **dependent** on the drugs because they are unwilling to give up these pleasurable sensations. This **psychological** or **emotional addiction** may then be followed by a **physical addiction**. A person becomes physically addicted to a drug when the drug is necessary for the normal working of the body. If the person cannot get the drug, then he or she will get **withdrawal symptoms**. Someone suffering withdrawal from heroin, for example, may vomit, tremble, sweat profusely and have severe abdominal pain after as little as four hours without the drug.

The dangers of abusing social drugs

Many casual users of social drugs soon become dependent upon them (physically and/or psychologically), so that the drug becomes a dominant feature of their everyday life. The user will do almost anything to satisfy the desire for the effects of the drug, and this may lead to problems such as:

- malnourishment as the drug depresses the appetite
- financial problems – drugs can be expensive and users often resort to stealing
- infections from shared needles used to inject drugs, including HIV and hepatitis
- dangers from other substances mixed with the drugs.

The drug may also lead to dangerous behaviour such as poorly coordinated driving after drinking alcohol or erratic behaviour when using a hallucinogen.

Drugs may affect the central and peripheral nervous system

Alcohol affects emotional centres in the forebrain. It acts as a **depressant** and overrules normal social restraints. At low concentrations alcohol therefore 'lifts' social inhibitions.

The brain is affected by several groups of drugs:

- **Heroin** (an example of a narcotic) mimics the action of the body's natural pain-killers. This gives a pleasurable sense of well-being (a 'high').
- **Alcohol** can upset normal sleep patterns by reducing the levels of a 'calming agent' in the brain.

Alcohol slows down impulses in peripheral nerves, causing slower reactions. It also affects nerves that control blood flow to the skin, causing flushing of the skin.

At high concentrations, **alcohol** depresses the life-support centres in the medulla. Breathing may stop, causing brain damage or even death.

Many drugs have their effects at synapses
Synapses transfer nerve impulses as chemicals called **neurotransmitters** (see page 139). Many drugs have their effects by changing the concentration of neurotransmitters, or by mimicking what they do. Heroin has its effects by interacting with receptors normally sensitive to natural neurotransmitters.

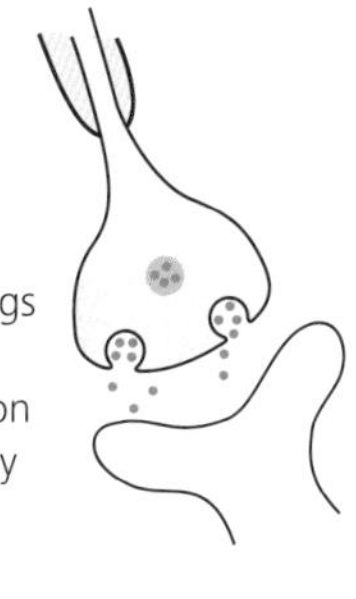

Nicotine in tobacco smoke is a **stimulant** which mimics the natural neurotransmitters in the part of the nervous system concerned with control of heartbeat and blood pressure (see page 91).

Aspirin does not affect the nervous system directly but inhibits an enzyme involved in the response that leads to inflammation.

Ecstasy

- Ecstasy is known by many other names, including 'Disco burgers', 'Doves', 'Dennis the Menace' and 'Fantasy'.
- It provides a 'feeling of energy' which can enable people to dance for long periods. This causes **dehydration** – users need to drink a pint of non-alcoholic liquid each hour.
- Death or disability can result from high body temperatures, **but** rapid rehydration has also caused some deaths.
- Ecstasy is one of the most commonly 'cut' drugs. This means it is mixed with other substances, and its quality is extremely variable.
- Long-term use has been shown to reduce the number of nerve cell connections in the brain. This causes memory loss and inability to perform simple tasks.
- Is a powerful stimulant and so is especially dangerous for anyone with high blood pressure, heart problems or epilepsy.
- Affects coordination so users should never drive or use machinery.

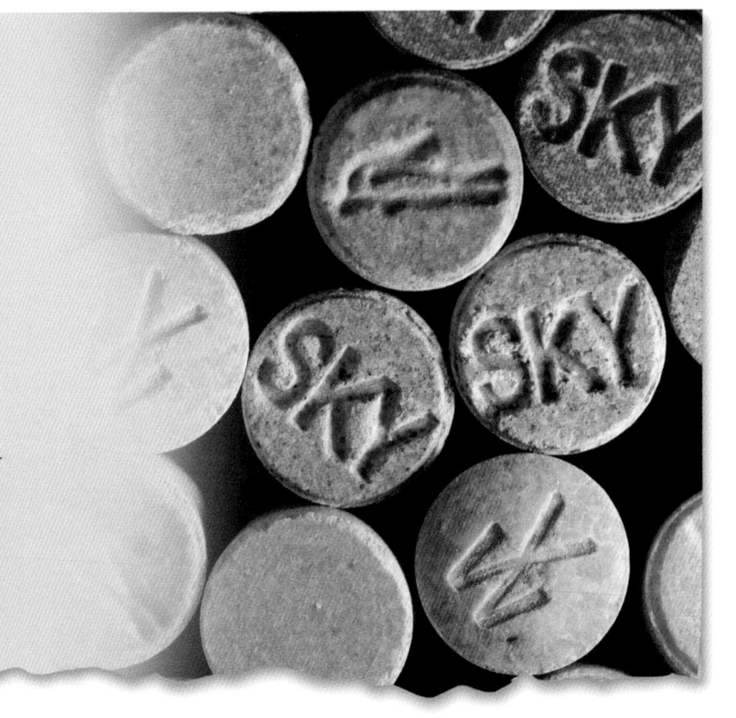

Alcohol is the most widely used drug

Alcohol is an unusual drug in that it is widely available legally in many countries. Even young people that cannot buy alcohol legally may still be able to obtain it quite easily. Apart from some cultures which ban alcohol, very few people have never tried the drug, and most young people would not consider it a 'drug of abuse'. However, alcohol can be an addictive drug, and causes great harm when used in excess. Most people greatly underestimate how much alcohol they drink. The effects of alcohol on the body extend beyond the nervous system – these are described in the diagram below.

Alcohol has widespread effects throughout the body

Cardiovascular effects include **anaemia** (alcohol poisons bone marrow) and increased **deposits of fat** in the coronary arteries.

Skin blood vessels dilate so warm blood flows to the skin. The body feels warm and comfortable, but **hypothermia** may result at low environmental temperatures.

One unit of alcohol is the amount that can be processed by the liver in one hour in the average person (see page 71). This is equivalent to one small glass of wine, half a pint of beer, or one measure of spirits.

Sex organs are stimulated but do not work well. Sperm count may be reduced.

Intestines are irritated, causing indigestion, nausea, diarrhoea and ulcers.

In pregnant women, alcohol may cross the placenta to the unborn child. The child may develop slowly, especially its nervous system (this is called **fetal alcohol syndrome**).

Liver cells work harder to detoxify the alcohol. Cells lining blood vessels are damaged causing **cirrhosis** – liver function begins to fail.

Cancers of the tongue and oesophagus are much more likely in heavy drinkers (especially if they also smoke).

Many effects on the nervous system – see page 153.

Kidneys cannot reabsorb water very well. Too much water is excreted and the body becomes dehydrated (this is responsible for the headache following a drinking bout).

Because alcohol is a depressant, **withdrawal** leads to tremors, high pulse rate, sweating and visual hallucination. This is called **delirium tremens** (the DTs) and is treated with sedatives and multivitamins.

Disorders of the nervous system

We have seen that many drugs have their effects by altering the activity of the nervous system. The nervous system may also be damaged by medical conditions. Some of these are more likely in old age (such as Alzheimer's disease), some may strike at any time (such as multiple sclerosis), some are caused by infection (such as meningitis) while others may result from an injury (such as a broken back). Such problems may cause great hardship to the affected person, and to those who care for them.

Q

1 **a** Several drug types affect the brain. These include stimulants, depressants, and narcotics. Give one example of each, and describe how it affects the brain.

b What is the difference between psychological and physical addiction to a drug?

c What is meant by drug tolerance?

Some drug users suggest that ordinary people are drug dependent. They claim that people who drink six or more cups of coffee per day may experience withdrawal symptoms if they stop. These symptoms are said to include headaches and poor concentration.

d Design an experiment that would allow you to investigate the validity of this claim. State clearly the independent (input) and dependent (outcome) variables, and say how you would manipulate or record them. Suggest some fixed variables that might improve the validity of your data.

2 Alcohol is a powerful and widely used drug. It affects the central nervous system, slows down reflex actions and causes small blood vessels in the skin to dilate. Most alcohol is absorbed in the small intestine, although its rapid effects are partly explained by the observation that up to 30% of it may be absorbed from the stomach. Alcohol is broken down in the liver, and most of it appears as carbon dioxide and water. A medical student was given six units of alcohol to drink, then blood samples were taken over the next five hours. The table shows the **blood alcohol level (BAL)** over the time of the experiment.

Time since drinking / hours	BAL / mg of alcohol per 100 cm^3 of blood
0.0	0.0
0.5	70
1.0	135
1.5	140
2.0	115
3.0	75
4.0	50
5.0	40

a Use these data to plot a graph.

b **i** What would be the likely BAL after 2.5 hours?

ii The legal limit for driving is a BAL of 80 mg per 100 cm^3 of blood. For how long would this student be above this limit?

iii How would driving be affected by a BAL of 100 mg per 100 cm^3 of blood?

c Suggest one visible effect of the dilation of blood vessels in the skin.

d The liver detoxifies (breaks down) alcohol, and much of it is lost as carbon dioxide and water. How would these substances be lost from the body? Suggest one long-term effect of alcohol consumption on the liver.

e People who drink large amounts of alcohol over long periods often become malnourished. In particular they may be deficient in calcium and vitamin C. What symptoms might be seen in such people?

f What is delirium tremens? Does this condition suggest that alcohol is a stimulant or a depressant? Explain your answer.

3 Illegal or controlled drugs are classified in the UK into three groups – A, B and C. Group A contains the most harmful drugs, including the opiates such as heroin. Drugs seized by police and customs are put into these categories for the purposes of keeping records. The table shows the number of drug seizures by the police in the UK in 1980, 1985 and 1990.

a **i** Suggest one reason for classifying a drug as a group A drug.

ii Calculate the percentage increase in the total quantity of cocaine seized by the police between 1980 and 1990. Show your working.

iii Identify three main trends in the information shown in this table.

b Drug dependence is characterised by personal neglect – poor nutrition, lack of care in personal hygiene and being willing to share needles during injections. Infections such as hepatitis are common, as are vitamin deficiencies. Heroin users are 50 times more likely to commit suicide than non-users. A female user who becomes pregnant may pass on the dependence to her child. Users who are denied access to the drug show violent withdrawal symptoms that include convulsions, diarrhoea and vomiting.

i What factors contribute to the early death of heroin users?

ii Following withdrawal, users are treated with methadone (a drug similar to heroin but less likely to case dependence) and multivitamin supplements. Why?

Seizures of group A drugs in the UK

Drug	Number of seizures			Quantity seized / kg		
	1980	1985	1990	1980	1985	1990
Cocaine (Group A)	365	510	1410	4.2	6.7	49.6
LSD (Group A)	244	448	1772	0.003	0.006	0.020
Heroin (Group A)	612	3003	2321	1.8	32.2	26.9
Cannabis resin	6218	13734	43474	312.2	469.8	57164
Amphetamines	706	3401	4490	5.0	50.2	222.7

2.53 Sensitivity and movement in plants: tropisms

OBJECTIVES

- To recall that plants, like all living things, are sensitive to their environment
- To know that plants respond to stimuli by changes in their growth patterns
- To know that plant growth responses are controlled by hormones
- To appreciate the importance of plant growth responses
- To understand how plant hormones may be used commercially

Plants respond to their environment

Plants respond to their environment – they show **sensitivity** (**irritability**). Plant responses are rather slow compared with those of animals – plants respond to **stimuli** (to changes in their environment) by changing their growth patterns. These growth responses enable a plant to make the most of the resources available in its environment.

Plants respond to many stimuli, but two are of particular importance: **light** (the photo-stimulus) and **gravity** (the gravi-stimulus). A growth response carried out by a plant in response to the direction of a stimulus is called a **tropism**.

A **positive** response is a growth movement towards the stimulus, and a **negative** response is a growth movement away from the stimulus. For example:

- a stem growing towards light is a **positive phototropism**
- a stem growing upwards (away from gravity) is a **negative gravitropism**
- a root growing downwards (away from light but towards gravity) is a positive gravitropism but a negative phototropism.

Roots are **positively gravitropic**:

- They grow into the soil, which provides a source of water and mineral ions.
- They provide an extensive system of support and anchorage for the plant.

Shoots are **positively phototropic**:

- Leaves are in the optimum position to absorb light energy for photosynthesis.
- Flowers are lifted into the position where they are most likely to receive pollen. They will be held out into the wind, or may be more visible to pollinating insects.

EXPERIMENTS PROVIDE CLUES ABOUT PHOTOTROPISM

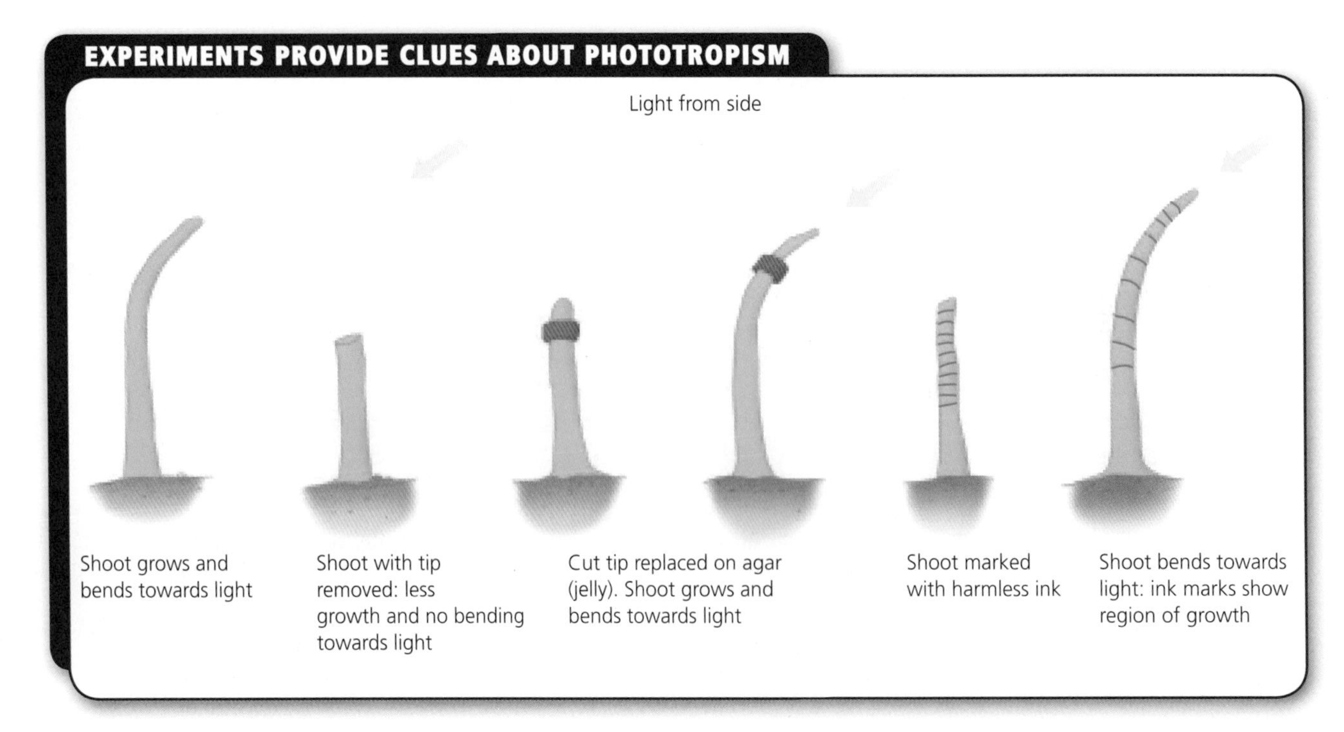
Light from side
Shoot grows and bends towards light
Shoot with tip removed: less growth and no bending towards light
Cut tip replaced on agar (jelly). Shoot grows and bends towards light
Shoot marked with harmless ink
Shoot bends towards light: ink marks show region of growth

INVESTIGATION OF THE LIGHT-SENSITIVE REGION

Lightproof boxes, allowing light in from one side only

Gap to allow entry of light

A — Coleoptiles with tips removed

B — Coleoptiles with tips covered

C — Untreated coleoptiles

Procedure

1 Three groups of coleoptiles (oat shoots) are treated as shown in A, B and C.
2 The coleoptiles are measured, and the lengths recorded.
3 The three groups are placed in the lightproof boxes and exposed to lateral (sideways) light for 2 or 3 days.
4 The coleoptiles are measured again, and the new lengths recorded.
5 Results can be recorded and analysed:

Box	A	B	C
Treatment of coleoptiles			
Mean length / mm at start			
Mean length / mm at end			
Change in length			
Direction of growth			

What conclusions can you draw? Explain your answers.

Phototropism is controlled by auxin

Growth is a relatively slow response to a stimulus in animals (see page 148) and plants. Growth in plants is controlled by **plant hormones** or **plant growth substances**. These are sometimes grouped together as **'auxin'**, which means 'growth substance'. The following observations have been made about auxin:

- If the tip of a young shoot is cut off, the shoot can no longer respond to stimuli. This suggests that the tip produces the auxin.
- A shoot responding to a stimulus always bends *just behind* the tip. The auxin appears to travel from the tip (where it is made) to a region behind the tip (where it has its action).
- Auxin can diffuse back from the tip and can be collected in blocks of agar jelly. These blocks can then allow a decapitated tip to respond to light.
- When shoot tips are exposed to light from one side, auxin accumulates on the 'dark' side of the shoot. The auxin is somehow affecting the growth of the 'dark' side of the shoot.

The role of auxin in phototropism is shown below.

More light means less growth!
Plants grown in the dark grow much taller and thinner than those grown in the light. This is called **etiolation**. Light reduces the amount of auxin and so reduces growth.

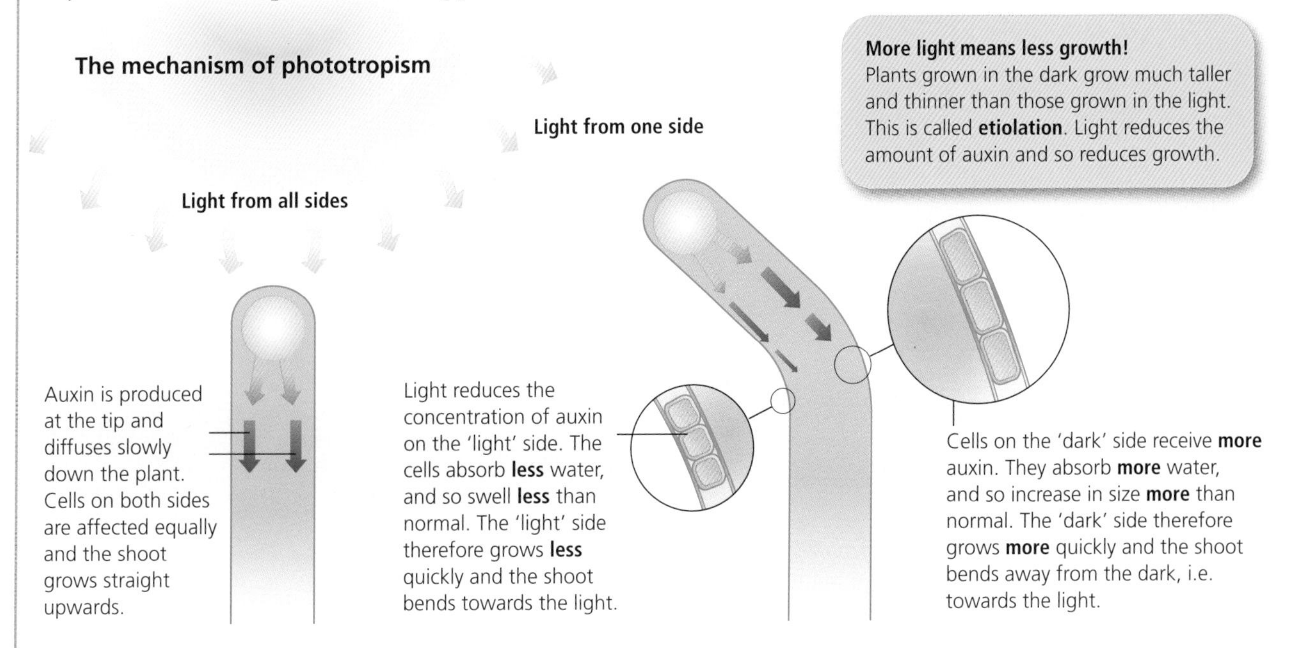

The mechanism of phototropism
Light from all sides
Light from one side
Auxin is produced at the tip and diffuses slowly down the plant. Cells on both sides are affected equally and the shoot grows straight upwards.
Light reduces the concentration of auxin on the 'light' side. The cells absorb less water, and so swell less than normal. The 'light' side therefore grows less quickly and the shoot bends towards the light.
Cells on the 'dark' side receive more auxin. They absorb more water, and so increase in size more than normal. The 'dark' side therefore grows more quickly and the shoot bends away from the dark, i.e. towards the light.

Plants respond to gravity: the mechanism of gravitropism

Many experiments can be carried out to investigate a plant's response to gravity. A useful organism for this type of experiment is the broad bean. Soaked beans germinate (see page 168) to provide obvious roots and shoots. These can be fixed inside a clear glass jar and their growth observed over a series of days. A simple experiment of this type is shown in the figure below.

It is obvious *why* the shoot grows away from gravity (i.e. is negatively gravi tropic) because this means the leaves will be exposed to light energy for photosynthesis. It is more difficult to explain *how* the shoot grows in this way whilst the root is positively gravitropic. One suggestion is that the root and the shoot are responding to different growth substances, but scientists believe that the same growth substance (auxin) affects both roots and shoots. It seems likely that the root and shoot cells have **different levels of sensitivity** to auxin, so that the same concentration of this substance reduces cell expansion in roots but stimulates cell expansion in shoots.

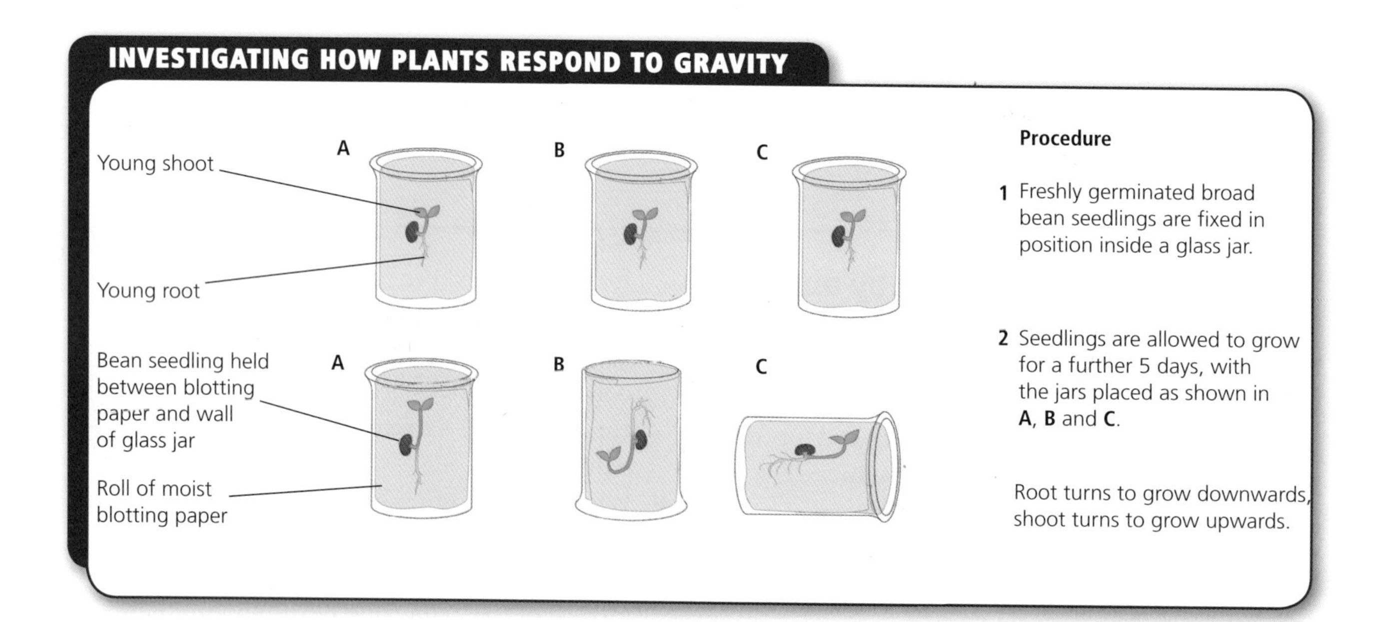

The mechanism of gravitropism

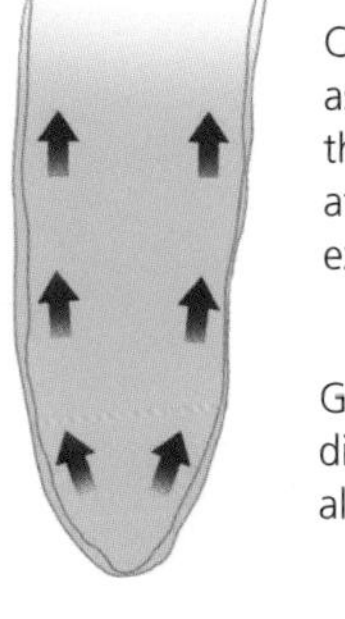

Cells expand less as hormone reaches them: both sides affected to the same extent

Growth substance diffuses back along root

Tip produces growth substance

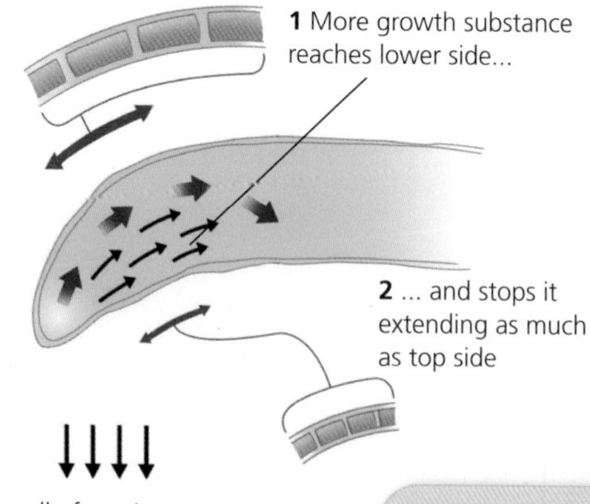

Take note: auxin inhibits (slows down) cell growth in roots, but stimulates (speeds up) cell growth in shoots!

Plant hormones have commercial uses

Humans put their knowledge of how hormones work in plants to good use. Being able to control the growth of plants is valuable, since plants are at the base of all human food chains. Some examples of the commercial use of plant hormones are illustrated above and in the diagram below.

Synchronised fruiting – spraying hormone onto fruits can make them develop at the same rate. This allows efficient picking of the crop by machine.

This happens naturally!
One ripe fruit in a bowl releases a hormone gas which makes the other fruit ripen.

Weeds can be killed – spraying with high concentrations of the synthetic hormone 2,4–D upsets normal growth patterns. Different plant species are sensitive to different extents, so this weed killing can be selective, e.g. grasses may be killed when shrubs survive.

Agent Orange worked like this!
During the Vietnam war, hormone sprays were used to clear areas of vegetation and make bombing of bridges and roads easier. These sprays can also be used to clear vegetation from overhead power lines, where removal by hand could be expensive and dangerous.

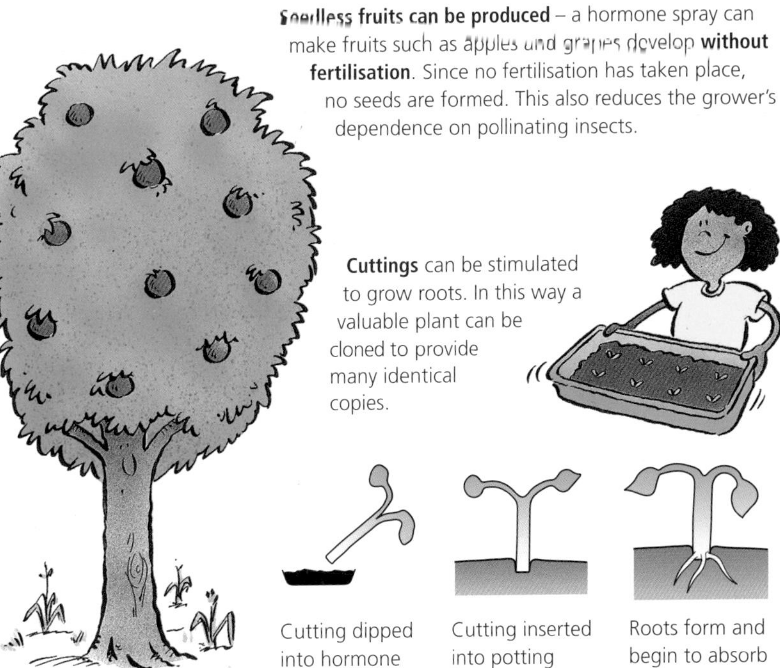

Seedless fruits can be produced – a hormone spray can make fruits such as apples and grapes develop **without fertilisation**. Since no fertilisation has taken place, no seeds are formed. This also reduces the grower's dependence on pollinating insects.
Cuttings can be stimulated to grow roots. In this way a valuable plant can be cloned to provide many identical copies.
Cutting dipped into hormone rooting powder
Cutting inserted into potting compost
Roots form and begin to absorb mineral ions and water

Q

Copy and complete the following paragraphs.

1 The _____ of a growing shoot produces a plant hormone or _____. This substance causes cells behind the tip to _____ by the absorption of _____. When the shoot is lit from one side, more _____ accumulates on the dark side. As a result the cells _____ more and the shoot bends towards the light. This response is called _____ and offers several advantages to the plant, including greater access to light energy to drive the process of _____.

2 Plant hormones have many commercial uses. These include:

- the stimulation of _____ on cuttings, which allows growers to produce many _____ of valuable plants
- the control of _____, which allows growers to harvest economically with machinery
- the destruction of _____, which could otherwise compete with crops for _____, _____ and _____. Careful selection of hormone concentrations allows this destruction to be _____: only pest plants are killed.
- the production of _____ fruits, since the hormone can make the plant develop a fruit without _____ taking place.

3.1 Reproduction is an important characteristic of living organisms

OBJECTIVES

- To understand the importance of reproduction
- To know that there are different methods of reproduction

It is possible to tell whether an organism is living or not by testing for certain characteristics (see page 2). Among these characteristics is included **reproduction**, although reproduction is *not* necessary for an *individual* organism to survive.

However, no individual lives forever and if the *species* is to survive the individuals must replace themselves before they die. This generation of new individuals is called **reproduction**.

Reproduction: a summary

Living organisms can pass on their characteristics to the next generation (**reproduce**) in two ways.

Asexual reproduction:

- involves only one parent organism
- all the characteristics of this one parent are passed on to all of the offspring
- asexual reproduction produces genetically identical offspring from one parent
- many organisms reproduce asexually when conditions are favourable (e.g. when there is plenty of food), and build up their numbers quickly.

Sexual reproduction:

- requires two organisms of the same species, one male and one female
- each individual produces sex cells (**gametes**)
- sexual reproduction always involves **fertilisation** – the fusion of the gametes to produce a zygote
- offspring receives some genes from each parent, so shows a mixture of parental characteristics
- each of the offspring is different to other offspring and to the parents.

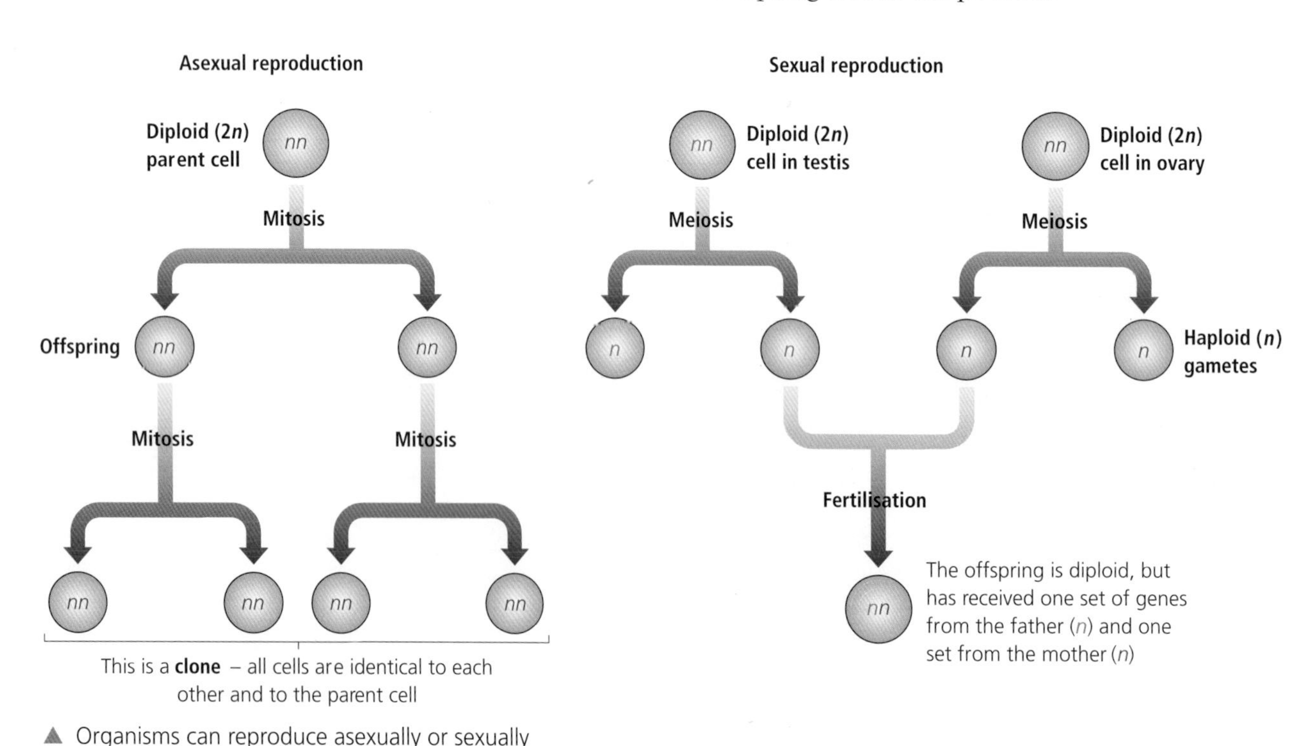

▲ Organisms can reproduce asexually or sexually

Sexual reproduction provides variation

The plant life cycle (page 162) and the human life cycle (page 172) both show sexual reproduction, which depends on the fusion of male and female **gametes**. Sexual reproduction consumes energy that otherwise could be used by the parent. However, it also leads to **variation** (see page 210). Sexual reproduction occurs in all advanced organisms.

Asexual reproduction can be advantageous

Plants arrive in new locations by the dispersal of seeds. There are occasions, however, when a plant would benefit from simply producing many copies of itself. For example:

- When a single plant arrives in a new habitat it can occupy this habitat if many copies can be produced quickly.
- When a plant is well suited to its habitat, any variation might be a disadvantage.

Sexual and asexual reproduction compared

Sexual reproduction and vegetative propagation both have advantages and disadvantages. Many plants make the best of both worlds, and reproduce both sexually and asexually.

	Advantage	Disadvantage
Asexual	■ Only one parent needed ■ Rapid colonisation of favourable environments	■ No variation, so any change in environmental conditions will affect all individuals
Sexual	■ Variation, so new features of organisms may allow adaptation to new environments	■ Two parents needed ■ Fertilisation is random, so harmful variations can occur

Sexual: provide new varieties
Asexual: make many copies of useful variety

.... and what's best for growing crops?

Sexual: new varieties of a plant can be developed. These might give a better yield, or tolerate dry conditions for example.

Asexual: varieties with useful features can be cloned to produce large numbers of identical crop plants.

BUT BE CAREFUL:
A clone of plants could all be susceptible to the same disease. For example, great hardship occurred in the Irish Potato Famine of 1845 when almost all of the potato crop was infected by the potato blight fungus.

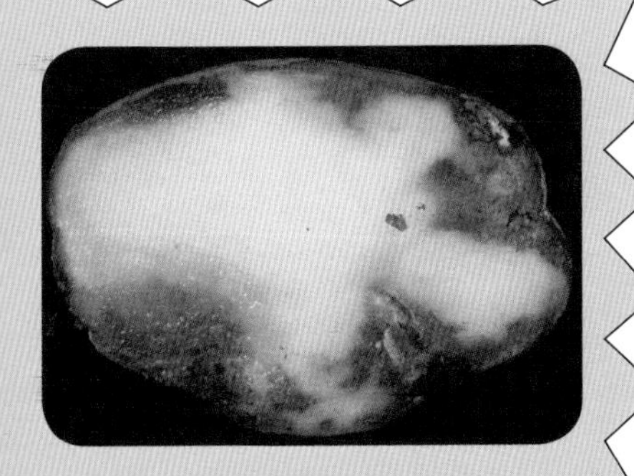

3.2 Reproduction in flowering plants: flowers

OBJECTIVES

- To understand the part played by flowers in the life of a flowering plant
- To be able to identify the parts of a typical flower
- To be able to state the functions of the parts of a flower

An individual plant, like any other living organism, eventually dies. For a *species* of plant to survive, the *individual* plants must be able to replace themselves. This is the process of **reproduction**, an essential part of the life cycle of the plant.

Sexual reproduction in plants

Flowering plants, as their name suggests, are able to reproduce using highly adapted structures called **flowers**. The life cycle of a flowering plant is outlined below.

Flowering plants reproduce sexually. The following list shows the stages that can be recognised in the reproduction of flowering plants.

1. The young plant develops reproductive organs.
2. Sex cells (gametes) develop inside the reproductive organs.
3. The male sex cells are transferred to the female sex cells.
4. Fusion of male and female sex cells (fertilisation) occurs, and a zygote is produced.
5. The zygote develops into an embryo.
6. The embryo grows into a new young plant, and the cycle starts all over again.

Whereas animals move around freely (they are **motile**), plants live in a fixed position. So, in plants:

- Male gametes may have to be carried some distance to meet female gametes.
- Young plant embryos may have to be carried some distance to get away from their parents!

Most plants are **hermaphrodite**, that is, they have male and female sexual parts on the same individual. This means that the male gametes only have to travel a short distance to the female gametes.

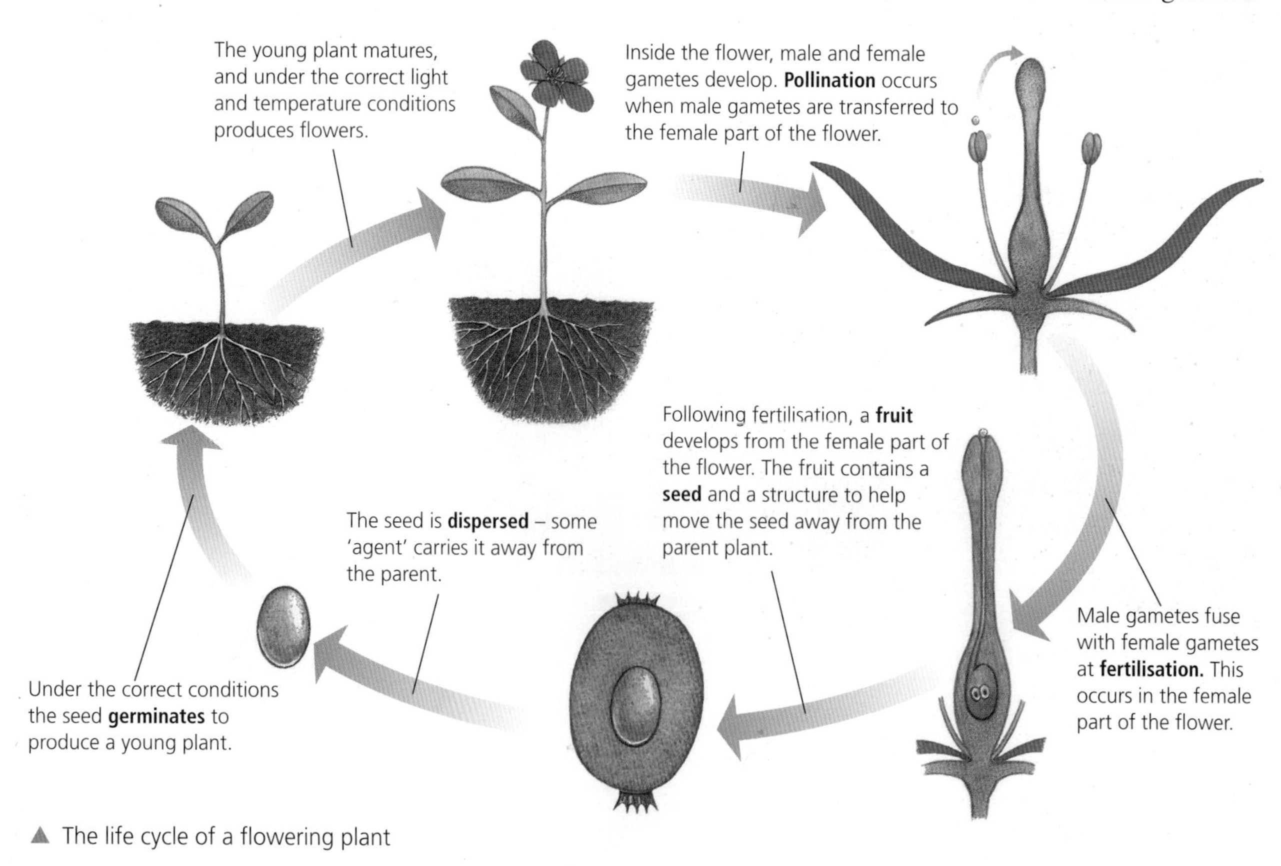

▲ The life cycle of a flowering plant

The formation of flowers

A flower is formed from a bud, which is a collection of cells at the end of a flower stalk. The cells receive hormone messages from the main plant body, and gradually develop into four rings of specialised leaves – the flower. These have the sole function of forming sex cells and making sure that fertilisation occurs.

The structure of a typical insect-pollinated flower is shown in the diagram below.

Each plant species has its own uniquely shaped pollen grains. Scientists can find out about the vegetation that once existed in a particular area by collecting and studying pollen grains released thousands of years ago.

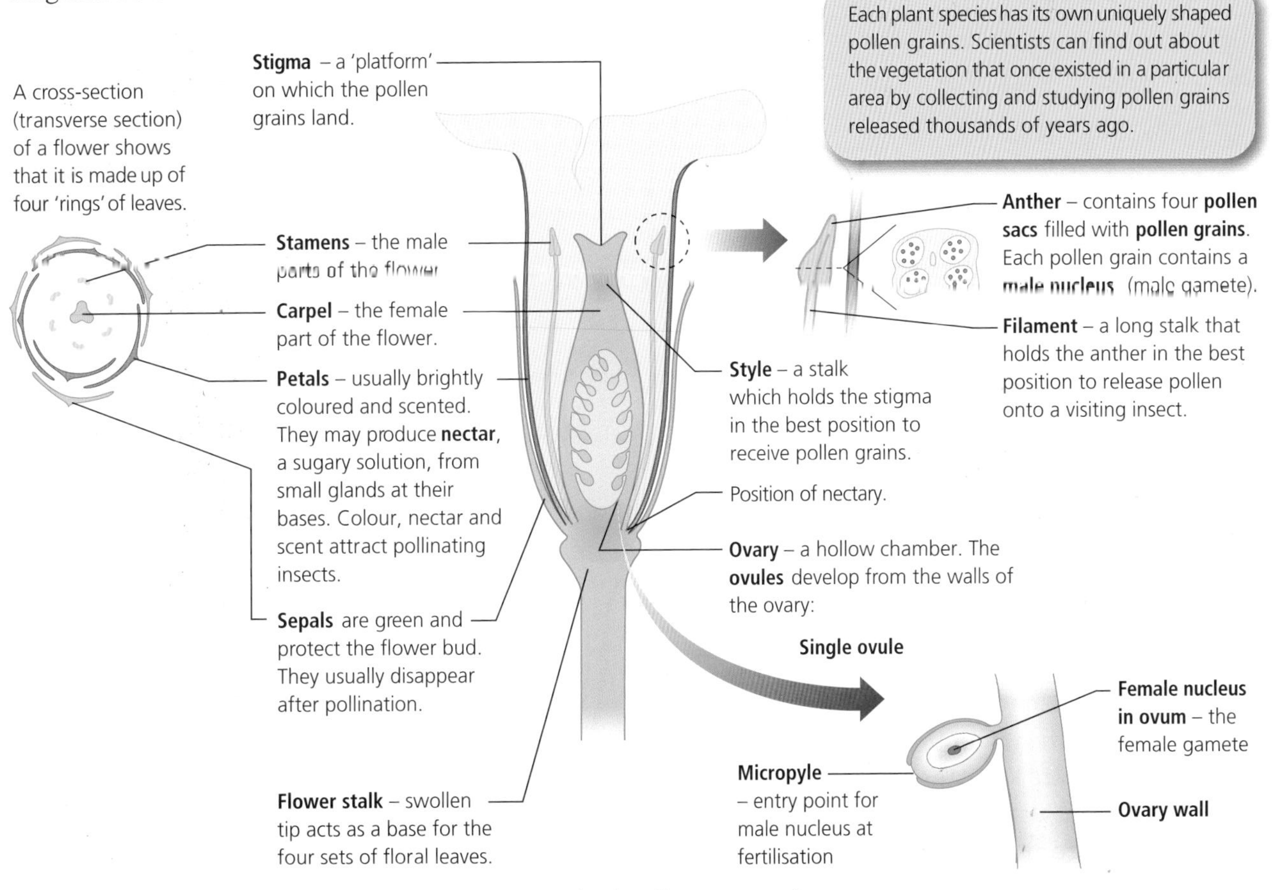

▲ The structure of a flower: the convolvulus (morning glory). A flower grows from a flower bud and is specialised to produce and release the male and female gametes.

Q

1 Look at the table on the right. Match each part of a flower to its function. Write the letter and number to show your answer, for example, **a–5**.

Flower parts	Functions
a pollen	**1** to support the anther
b flower stalk	**2** to secrete a sugary solution
c style	**3** to contain the female gametes
d filament	**4** to protect the flower in bud
e anther	**5** to deliver the male gamete
f sepal	**6** to form a base for the flower
g petal	**7** to hold up the stigma
h nectary	**8** to attract insects
i ovary	**9** to produce pollen
j stigma	**10** to receive pollen

2 Copy and complete the following paragraph.
The life cycle of a flowering plant has a number of stages. A young plant develops when a seed _____. The plant matures until it produces a _____ which is a collection of leaves specialised for _____. Male gametes are transferred to the female part of the flower by _____ and fuse with female gametes at _____. Following this process the ovary develops into a _____ which contains a _____ and a structure to help move it away from the parent plant. This process, which is called _____, requires some agent or vector to remove the seed from the parent plant.

3.3 Pollination: the transfer of male sex cells to female flower parts

OBJECTIVES

- To define the term pollination
- To understand the difference between self-pollination and cross-pollination
- To describe how flowers may be adapted to pollination by insects or wind
- To describe how honey bees are adapted as insect pollinators

Self-pollination and cross-pollination

For sexual reproduction to occur, the male gametes must be transferred to the female part of the flower – this transfer is the process of **pollination**. The transfer may come about within the same flower (self-pollination) or from one flower to another of the same species (cross-pollination), as shown in the diagram on the right.

Cross-pollination offers many advantages, and some species make sure that it happens.

- Some have special proteins on the surface of the stigma that prevent pollen tubes forming if the pollen comes from the anthers of the same plant. These are **self-sterile** plants.
- In some plants, the anther and the stigma are so far away from one another in the flower that it is not very easy for pollen to travel from the anther to the stigma of the same flower.
- A few plants, such as ash, willow and holly, have separate male and female plants.

Pollination by wind and insects

Whether self- or cross-pollination occurs, some agent or vector is needed to carry the pollen grains from the anthers to the stigma. This agent is most often an insect or the wind. Flowers show many adaptations to successful **insect pollination** or **wind pollination**. Some of the insects that act as pollinators have also become adapted to make the most of their relationship with flowers. The table on the opposite page shows how some plants are adapted for pollination by insects or plants. The honey bee feeds on nectar and pollen.

Pollination

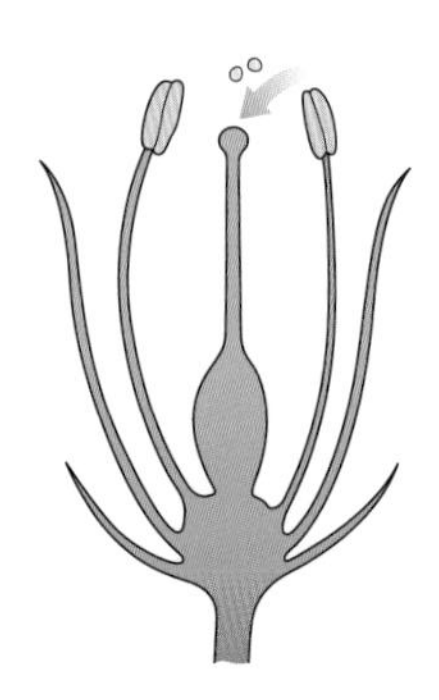

In **self-pollination** pollen is transferred from anther to stigma **of the same plant**. This is very efficient (the pollen doesn't have to travel very far) but does not offer much chance of genetic variation (see page 212).

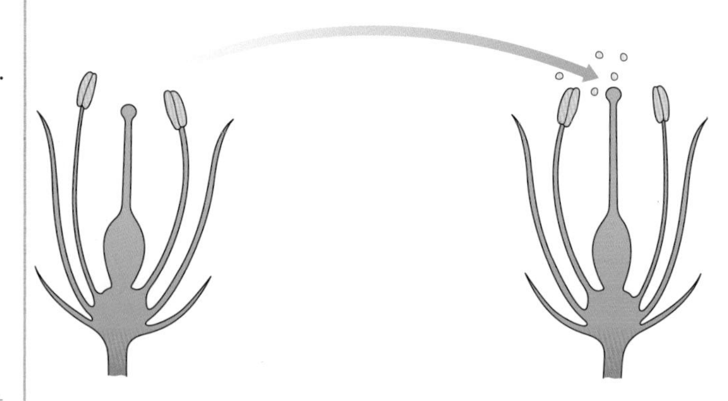

In **cross-pollination** pollen is transferred from anther to stigma **of another plant of the same species**. This is risky (pollen may never reach the other plant) but offers a greater chance of genetic variation than self-pollination.

Cross-pollination is the only possibility for flowers that are not hermaphrodite. Sometimes a plant has only male or only female flowers. This is **very** risky, since plants of the opposite sex may not be nearby, but offers a very great chance of genetic variation (page 212).

▲ Wind-pollinated flowers such as this grass have small, inconspicuous petals. The anthers and stigmas hang outside the flower.

Wind-pollinated flower e.g. maize

Wind-pollinated flowers are often produced at colder times of the year, when very few insects are flying.

Bract: leaf-like structure which holds the parts of the flower.

Tiny petal: helps to push the bracts apart to expose stigma and stamens.

Pollen: light and produced in huge quantities. Has smooth coat and tiny 'wings' to help transfer by the wind.

Stigma: long and feathery to give a large surface area for pollen to land on. Often hang out into the wind.

Filaments: long and flexible so that anthers are held out into the wind.

No nectar or scent – no need to attract insects.

Anthers: held at the middle so that they can shake to release the pollen.

Part of flower	Insect-pollinated (e.g. convolvulus)	Wind-pollinated (e.g. grass)	Reason
Petals	Usually large, brightly coloured, scented, often with nectaries. Guide lines may be present.	Small, green or dull in colour, no scent or nectaries	Insects are attracted to colour and scent. Guide lines direct insects to nectaries, past anthers.
Anthers	Stiff, firmly attached and positioned where insects must brush against them	Hang loosely on long thin filaments	Wind is more likely to dislodge pollen from exposed, dangling anthers than from enclosed ones.
Pollen	Small amounts of large, sticky grains	Enormous quantities of light, smooth pollen grains	Sticky grains attach to hairs on insect's body. Larger amounts from wind-pollinated flowers mean pollination is more likely.
Stigma	Usually flat or lobe-shaped, and positioned where insect must brush against them	Long and feathery, and hanging outside flower	Feathery stigmas form a large network to catch pollen being blown past the flower.

EXPERIMENTAL DESIGN: USING MODELS

Scientists use **models** of living organisms to study one or two features of the organism. Living organisms may have features that other organisms can detect, but that humans cannot. These might affect the results of an experiment, so a model is easier to study.

Two students were interested in why honey bees visit wallflowers. One of them believed that the *colour* of the flower attracted the bees but the other was convinced that it was the *scent*.

1. Design an experiment to determine whether colour or scent is the more important stimulus to the honey bees. In your design:
   - devise suitable models for the wallflowers
   - clearly state the independent (input) and the dependent (outcome) variables
   - suggest any variables that should be fixed, and say how you would fix them
   - consider whether there are any controls that you might use
   - draw a table that would be suitable for the presentation of your data.

3.4 Fertilisation and the formation of seed and fruit

OBJECTIVES

- To define the process of fertilisation
- To understand the role of the pollen tube
- To recall which parts of a flower form the fruit
- To understand that a seed contains a food store as well as an embryo

Fertilisation follows pollination

Pollination is complete once the pollen released from an anther has landed on the stigma of the same or another flower. The next step in the plant's reproductive process is **fertilisation**, the fusion of the male and female gametes. The diagram on the right shows how the nucleus from the pollen grain travels down a **pollen tube** to combine with the nucleus of the ovum.

The formation of fruit and seed

After the male nucleus has fused with the ovum, the resulting **zygote** divides many times to produce an **embryo**. The development of a seed and the structure of the embryo are described in the diagram on the opposite page.

Once fertilisation is complete, the developing seed sends hormone messages to the flower, and a number of changes take place:

- The sepals and petals wither away, and may fall off.
- The stamens, stigma and style wither away. These structures have now completed their function, and would use up valuable food compounds if they remained.
- The wall of the ovary changes. It may become hardened and dry, or fleshy and succulent, and in the wallflower it forms a leathery pouch.

The ovary is now called a **fruit**. A fruit is a fertilised ovary, and has the function of dispersing the seeds away from the parent plant, which helps to reduce competition for light, water and minerals.

Pollen grain – chemical signals released by the stigma ensure that a pollen tube is only produced when pollen lands on a stigma of the same species.

Pollen tube grows down through the style and acts as a channel to deliver the male gamete from the pollen grain to the female gamete in the ovule.

Ovary wall – ovules are attached to the inside of the ovary wall by a short stalk.

Ovule contains the female gamete, and some other cells which may develop into food reserves.

Stigma

Style

Fertilisation occurs when the haploid male and female gametes fuse to form a diploid zygote.

Fertilisation also triggers some other cells in the ovule to divide rapidly and form a food store inside the seed.

Micropyle – a gap in the covering of the ovule. The tip of the pollen tube locates this gap and the male gamete enters the ovule through it.

Note that only one fertilisation is shown here. Each ovule needs its own pollen grain and pollen tube to be fertilised. A plum has only one ovule in each ovary, a wallflower has a few tens and a poppy may have thousands of ovules in one ovary!

Stages in the development of a fruit

▲ Tomato flowers – the petals are still obvious

▲ After fertilisation, the petals have fallen off, the stigma and style have withered and the carpel is beginning to swell

▲ In the ripe fruit, the ovary wall is swollen and succulent. What do you think is the purpose of the bright red colour?

The formation of seed and fruit following fertilisation.

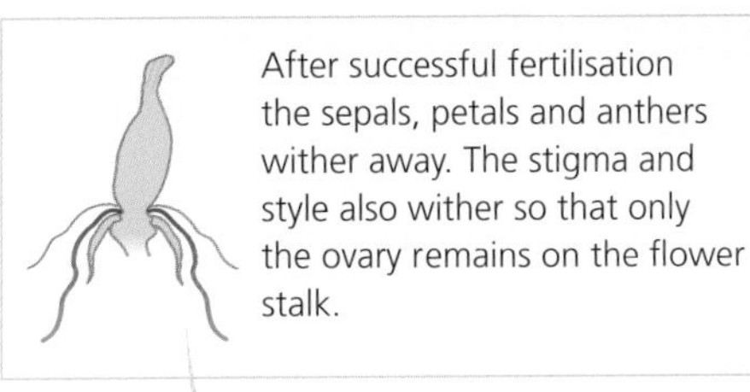

After successful fertilisation the sepals, petals and anthers wither away. The stigma and style also wither so that only the ovary remains on the flower stalk.

Ovary wall may become dry and hard (in a Brazil nut, for example), very soft and fleshy (in a plum, for example) or form a leathery pouch (as in the wallflower).

Fertilised ovule develops into a seed, with the zygote forming the embryo.

Seed coat (testa) prevents drying out of the embryo.

Cotyledons (seed leaves) may form the food store for the embryo. There is **one** seed leaf in monocotyledons such as grasses, or **two** in dicotyledons such as peas and beans.

Endosperm – tissue which forms the food store in cereal crops.

Plumule (young shoot)
Radicle (young root)
together with the cotyledons these make up the **embryo**. The embryo develops into a new young plant after germination.

Micropyle – a small hole in the testa which allows the entry of water and oxygen as the seed germinates.

Remember!

- The **carpel** becomes the **fruit**.
- The **ovule** becomes the **seed**.

Fruit or vegetable?

If it's been formed following **F**ertilisation it's a **F**ruit; if there's no sex in**V**olved it's a **V**egetable. Potatoes and carrots are vegetables but tomatoes are fruits.

Not all seeds are edible

Seeds contain stores of carbohydrate, fat, protein, minerals and other food compounds. These are used up by the embryo as it develops into a young plant, but also form an excellent food source for animals (including humans). But note …

- Uncooked castor oil seeds contain a deadly poison, **ricin**, which was used to kill the Romanian dissident Georgi Markov.
- Almonds contain **cyanide**. Don't eat too much marzipan since this is made from almond paste!

Q

1 In one sentence, explain the difference between pollination and fertilisation.

2 Draw a simple diagram of a typical hermaphrodite flower. On your diagram label:
   a the parts that fall off after fertilisation
   b the parts that develop into a fruit.

3 Define the word 'seed'.

4 Classify each of the following as:
   a fruit
   b seed or
   c neither fruit nor seed.
   Tomato, cucumber, Brussels sprout, baked bean, runner bean, celery, pea, grape

5 Two students suggested that the wallflower cannot produce fruit unless pollination has taken place. Their teacher showed them how to prevent bees reaching the flowers by covering the flowers in a fine mesh bag, and how to transfer pollen with a fine paintbrush.
   a Describe how the students could carry out an experiment to test their hypothesis.
   b Suggest how they could modify their experiment to test whether self- or cross-pollination produced more seeds in the wallflower.
   In your answer be sure to describe any controls which they could include, and any steps they could take to ensure that their results were valid.

3.5 Germination of seeds*

OBJECTIVES

- To define the term germination
- To understand the conditions required for germination
- To describe the structural changes that accompany germination of the broad bean

Dispersal allows plants to spread their seeds so they can develop without competition from their parents. A seed is made up of an embryo and a food store, all enclosed within a seed coat. If environmental conditions are suitable, the embryo will begin to use the food store in the seed and grow into a new young plant. This development of a seed to a new young plant is called **germination**.

Conditions for germination

A seed needs the following to germinate:

- a **supply of water**
- **oxygen** for **aerobic respiration** (see page 106)
- a **temperature suitable** for the enzymes involved in germination (see page 33).

These requirements are explained more fully in the diagram below.

Some seeds have other requirements as well as the ones shown in the diagram. A few need particular conditions of light (for example, lettuce).

Start here!

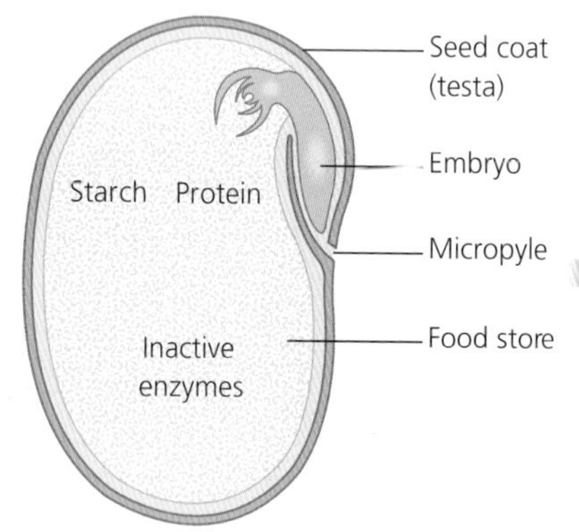

Dormant seed – embryo and food stores are surrounded by an impermeable seed coat. The micropyle is the only gap in the seed coat.

Environmental factors that affect germination are very similar to those that affect enzyme activity. This indicates that **germination is a process controlled by enzymes** (page 32).

Start of germination

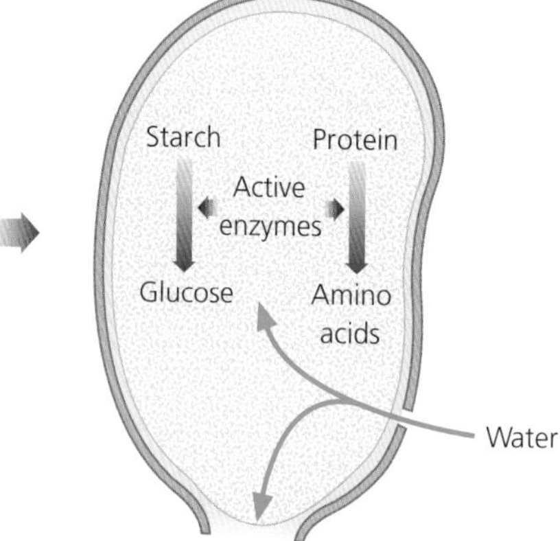

Water enters through the micropyle and:

- activates enzymes to convert insoluble stores to soluble foods
- makes tissues swell so that the testa is split open.

Enzymes work at their **optimum temperature**.

Germination proceeds

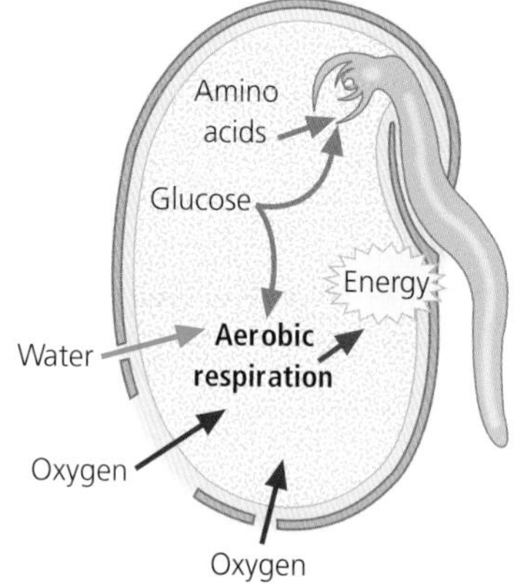

Water and oxygen enter through gaps in the testa. Oxygen and glucose enable aerobic respiration, which releases energy. The embryo is able to grow as it receives raw materials and energy.

Dormancy will continue if the embryo in the seed does not experience the right conditions:

- if kept in anaerobic conditions
- if kept dry
- if kept cool.

Oxygen and water cannot reach the embryo if the testa remains impermeable. Some seeds must pass through an animal's gut (where digestive juices are present) before the testa is weakened enough for the seeds to germinate.

Seeds kept dry in a vacuum, as in seed packets, can be stored for long periods

Dormancy

If a seed does not experience ideal conditions for germination immediately, it will not die. Most seeds can survive long periods of poor conditions, only germinating when conditions improve. Seeds survive in a resting state called **dormancy** during which they use hardly any food. The very low water content of seeds allows them to remain dormant, and the availability of water is one of the conditions that allows a seed to escape dormancy and germinate.

Structural changes during germination

Many biochemical processes go on inside the seed during germination. There are also external changes that can easily be seen as a seed begins to develop into a young plant. Different types of seed germinate in different ways. We shall take the broad bean as an example.

The broad bean shows **hypogeal germination**. 'Hypogeal' means 'below the ground' ('hypo' sounds like 'low' and 'geo' refers to 'earth'). The seed remains below the ground throughout the whole process. The stages involved in hypogeal germination are described in the diagram below.

The seedling must start to photosynthesise before all the food stores are used up. For this reason each seed type has an optimum depth for planting. If a seed is planted too deep, the shoot can't get above the ground before all the reserves are used up. Very tiny seeds (lettuce for example) often require a light signal to trigger germination. They will not germinate unless they are close to the surface of the soil, where they are in less danger of using up their food stores too quickly.

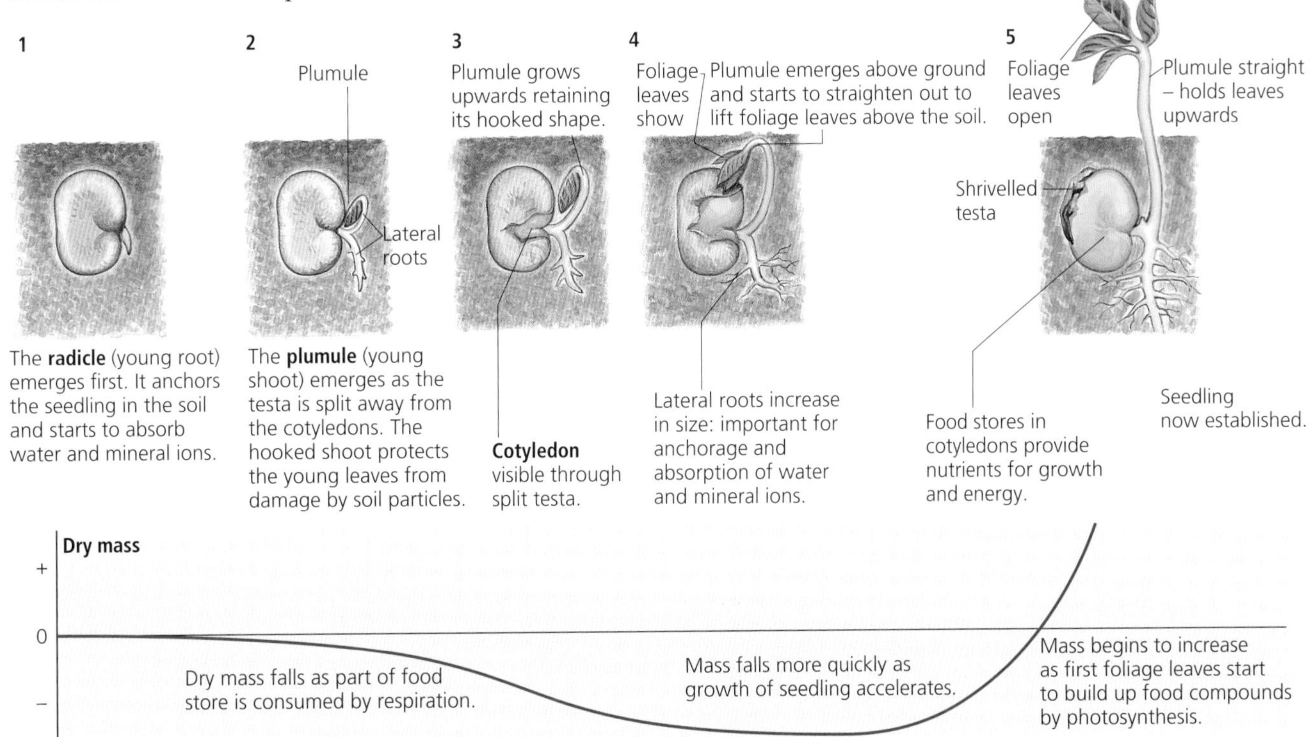

▲ Germination of the broad bean. Dry mass is used as a measure of the food stores because wet mass would include water absorbed from the soil, and the amount of water absorbed and lost can vary greatly.

Q

1 Copy and complete the following paragraph.
Before germination, a seed must absorb ______ from the soil. The seed coat is impermeable to this substance, and it enters through a hole called the ______. This absorption causes the seed to ______ and split the ______. The gas ______ can now enter, which is necessary for ______. The energy from this process, together with soluble food compounds, allows the ______ to grow. The young root or ______ appears first and grows downwards by ______, providing ______ and absorbing water and minerals for the seedling. The young shoot or ______ appears next. This grows upwards and eventually bursts through the soil. The first ______ leaves develop and the seedling is able to produce its own food by ______.

2 How would you attempt to prove that germination is controlled by enzymes?

Questions on plant reproduction

1 Choose words from the list to complete each of the spaces in the paragraph. Each word may be used once only and some words are not used at all.

bright	**dry**	**dull**	**heavy**
large	**light**	**sepals**	**small**
stamens	**sticky**	**style**	

Flowers of plants that rely on the wind to bring about pollination tend to have ____ petals that have a ____ colour.
Their pollen is normally ____ and ____.
In these flowers, the ____ and the ____ both tend to be long.

Cambridge IGCSE Biology 0610
Paper 2 Q2 May 2008

2 a Match the names of the flower parts with their functions. One has been completed for you.

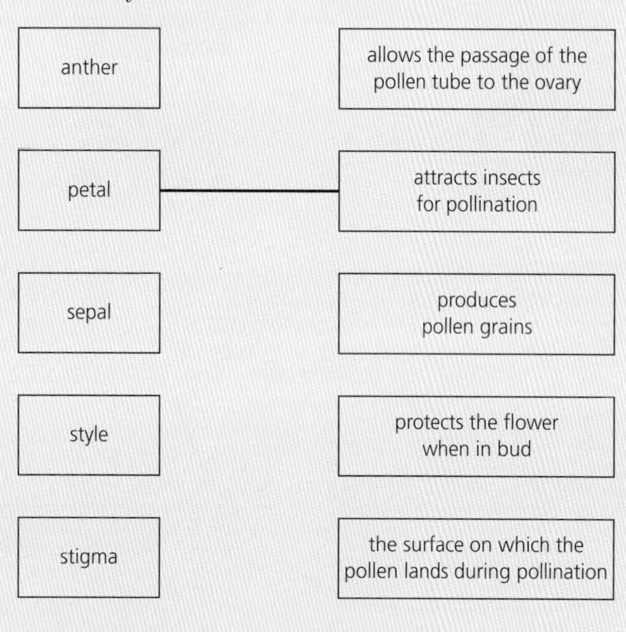

b Describe how the stigmas of wind-pollinated flowers differ from the stigmas of insect-pollinated flowers. Relate these differences to the use of wind as the pollinating agent.

c Discuss the implications to a species of self-pollination.

Cambridge IGCSE Biology 0610
Paper 3 Variant 1 Q1 May 2008

3 Below you can see photographs of a tomato and an apple. Both are cut in half longitudinally through the middle.

a Make a large, labelled drawing of the cut surface of the tomato fruit.

b i List four **differences** between the two fruits visible in the photographs.

ii Describe two **similarities** between the two fruits visible in the photographs.

c Describe an investigation you could carry out to compare the reducing sugar content of these two fruits.
Include any safety precautions you will need to consider.

Cambridge IGCSE Biology 0610
Paper 6 Q2 May 2008

4 a Sexual reproduction in flowering plants involves both pollination and fertilisation.

i Explain the difference between pollination and fertilisation.

ii Name the part of a flower where pollination happens.

iii Name the part of a flower where fertilisation happens.

b Sexual reproduction in flowers results in the production of seeds and fruits. From which part of a flower is each of these formed?

c Describe the role of the wind in the life cycle of some flowering plants.

Cambridge IGCSE Biology 0610
Paper 2 Q3 May 2009

5 a All seeds need oxygen, water and a suitable temperature to germinate. 22°C is a suitable temperature for the germination of pea seeds. Light and dark conditions have no effect on pea seed germination. The diagram shows an experiment on germination of pea seeds.

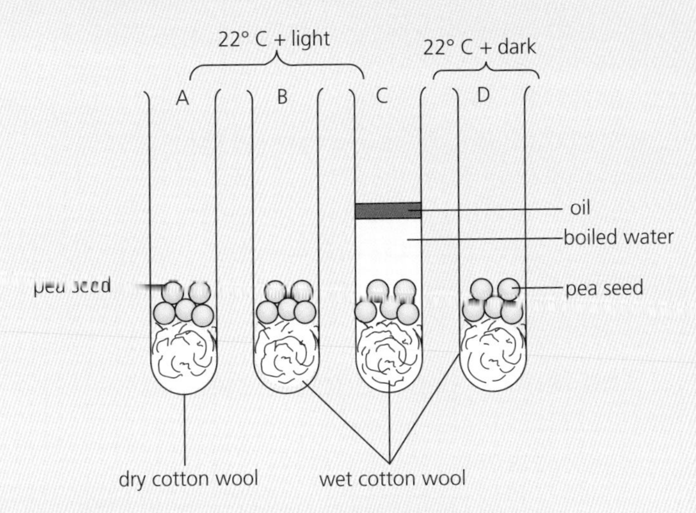

i Complete **Table 1** below.

Tube	Would seeds germinate? (Write YES or NO)
A	
B	
C	
D	

▲ Table 1

ii Germination is quicker if the temperature is raised to 30°C. Explain why.

b **Table 2** show how the dry mass of barley seedings changed over the first 35 days after sowing.

Time after sowing/days	0	7	14	21	28	35
Dry mass/g	4.0	2.8	2.8	4.4	9.6	17.8

▲ Table 2

i Plot these results in a line graph.

ii How many days after sowing did the barley seedings regain their original dry mass?

iii How many days after sowing did the barley seedlings TREBLE their original dry mass?

iv Explain why dry mass falls in the early stages of germination.

3.6 Reproduction in humans

OBJECTIVES

- To define the term 'sexual reproduction'
- To know the steps involved in sexual reproduction
- To describe the human reproductive systems

Reproduction may be sexual or asexual. In **sexual reproduction**, genetic information from two parents combines to produce a new individual. Humans, like all other mammals, only use sexual reproduction. Sexual reproduction produces individuals that are different from each other. The process involves a number of stages, as shown in the diagram below.

The male reproductive system

The male reproductive system has two functions:

- to manufacture the male gametes
- to deliver them to the site of fertilisation.

The male gametes, called **spermatozoa**, or **sperm** for short, are manufactured in the **testes**. These are enclosed in a sac of skin, the **scrotum**, which hangs outside the body between the legs. This position helps protect the testes from physical damage, and more importantly keeps them at a temperature 2–3 °C lower than body temperature, ideal for development of the sperm. The sperm are delivered inside the female body through a series of tubes that eventually release the sperm from the tip of the **penis**. The male reproductive system lies very close to the part of the excretory system that removes urine from the body; indeed the urethra is a tube used to expel urine and seminal fluid from the body. A valve prevents this happening at the same time! The two systems together are called the **urinogenital system**. The male reproductive system is shown in the diagram on page 174.

The female reproductive system

In addition to producing female gametes, the female reproductive system receives male gametes. It provides a site for fertilisation and for the development of the zygote. The female gametes, called **ova**, are produced one at a time by the two **ovaries**. The ovum travels along the **oviduct** or **Fallopian tube** towards the **uterus**. It may be fertilised while in the oviduct. The zygote grows and develops into a baby in the uterus. The female reproductive system is shown in the diagram on page 174.

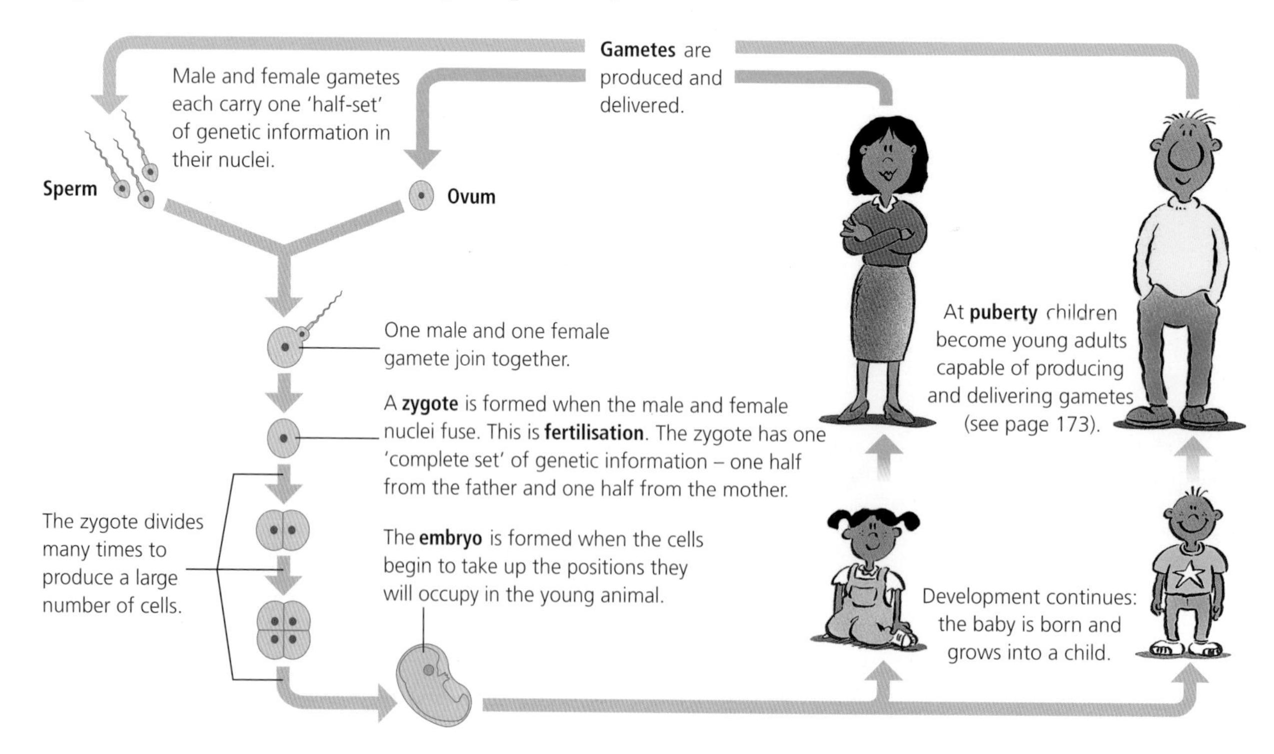

Hormones control puberty in humans

Hormones control long-term processes, and often have widespread effects on the body. **At puberty** a person becomes physically able to reproduce. The development of sexual maturity is a good example of a hormone-controlled process.

As a young person develops physically, certain signals are processed by the brain, which then instructs the pituitary gland to stimulate the **primary sex organs** – the testes in males and the ovaries in females. Sex hormones – **oestrogen** (in females) and **testosterone** (in males) – are released into the bloodstream and circulate throughout the body. They only affect the target organs which have receptors that recognise them. These target organs then carry out responses, such as the growth of body hair, which may continue for many years. The effects of these hormones at puberty are explained further in the diagram below.

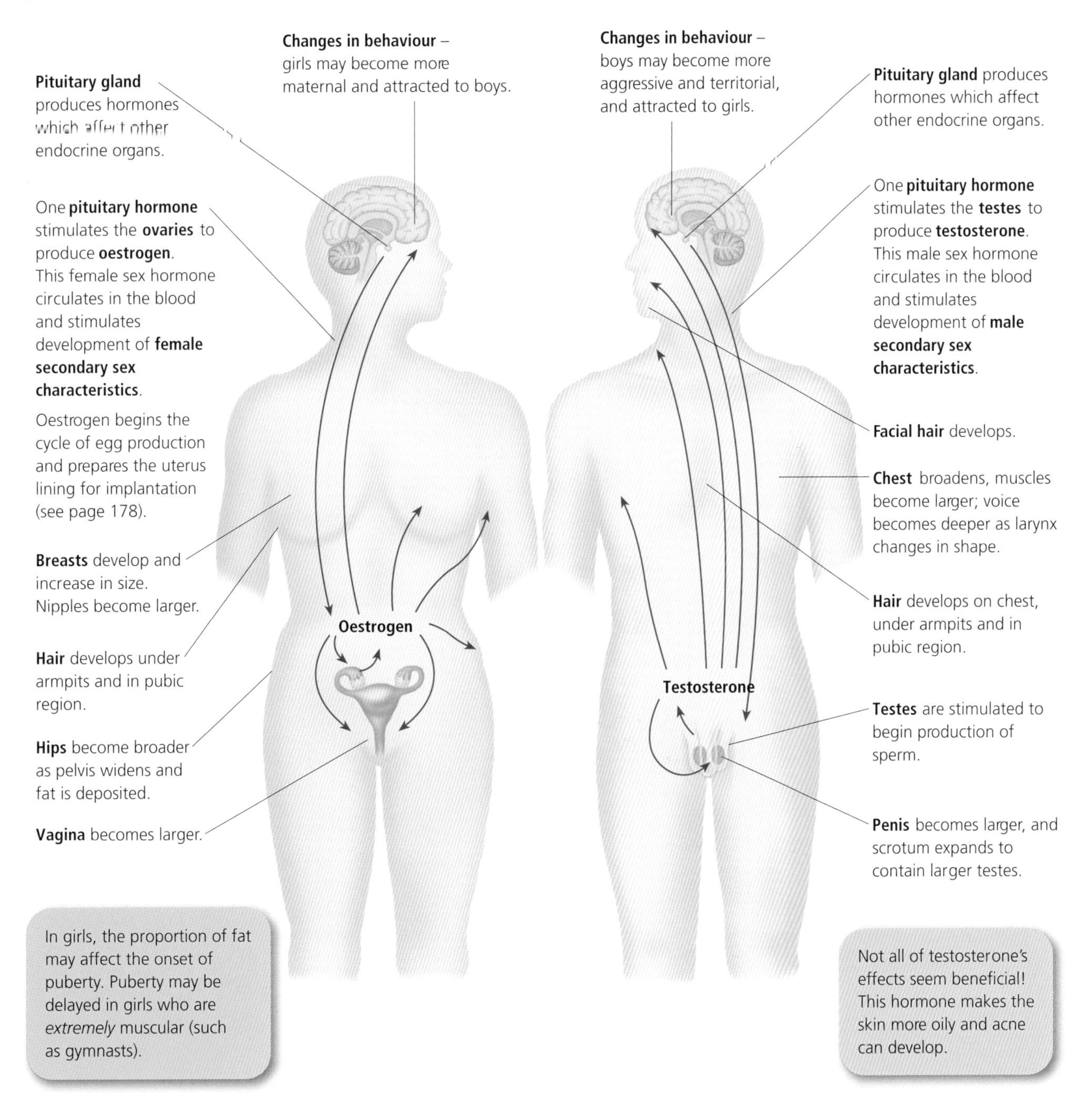

Changes in behaviour – girls may become more maternal and attracted to boys.
Pituitary gland produces hormones which affect other endocrine organs.
One pituitary hormone stimulates the ovaries to produce oestrogen. This female sex hormone circulates in the blood and stimulates development of female secondary sex characteristics.
Oestrogen begins the cycle of egg production and prepares the uterus lining for implantation (see page 178).
Breasts develop and increase in size. Nipples become larger.
Hair develops under armpits and in pubic region.
Hips become broader as pelvis widens and fat is deposited.
Vagina becomes larger.
Oestrogen
Changes in behaviour – boys may become more aggressive and territorial, and attracted to girls.
Pituitary gland produces hormones which affect other endocrine organs.
One pituitary hormone stimulates the testes to produce testosterone. This male sex hormone circulates in the blood and stimulates development of male secondary sex characteristics.
Facial hair develops.
Chest broadens, muscles become larger; voice becomes deeper as larynx changes in shape.
Hair develops on chest, under armpits and in pubic region.
Testes are stimulated to begin production of sperm.
Penis becomes larger, and scrotum expands to contain larger testes.
Testosterone

In girls, the proportion of fat may affect the onset of puberty. Puberty may be delayed in girls who are *extremely* muscular (such as gymnasts).

Not all of testosterone's effects seem beneficial! This hormone makes the skin more oily and acne can develop.

▲ Oestrogen and testosterone are hormones – they are chemical messengers secreted directly into the bloodstream and bring about widespread and long-lasting effects

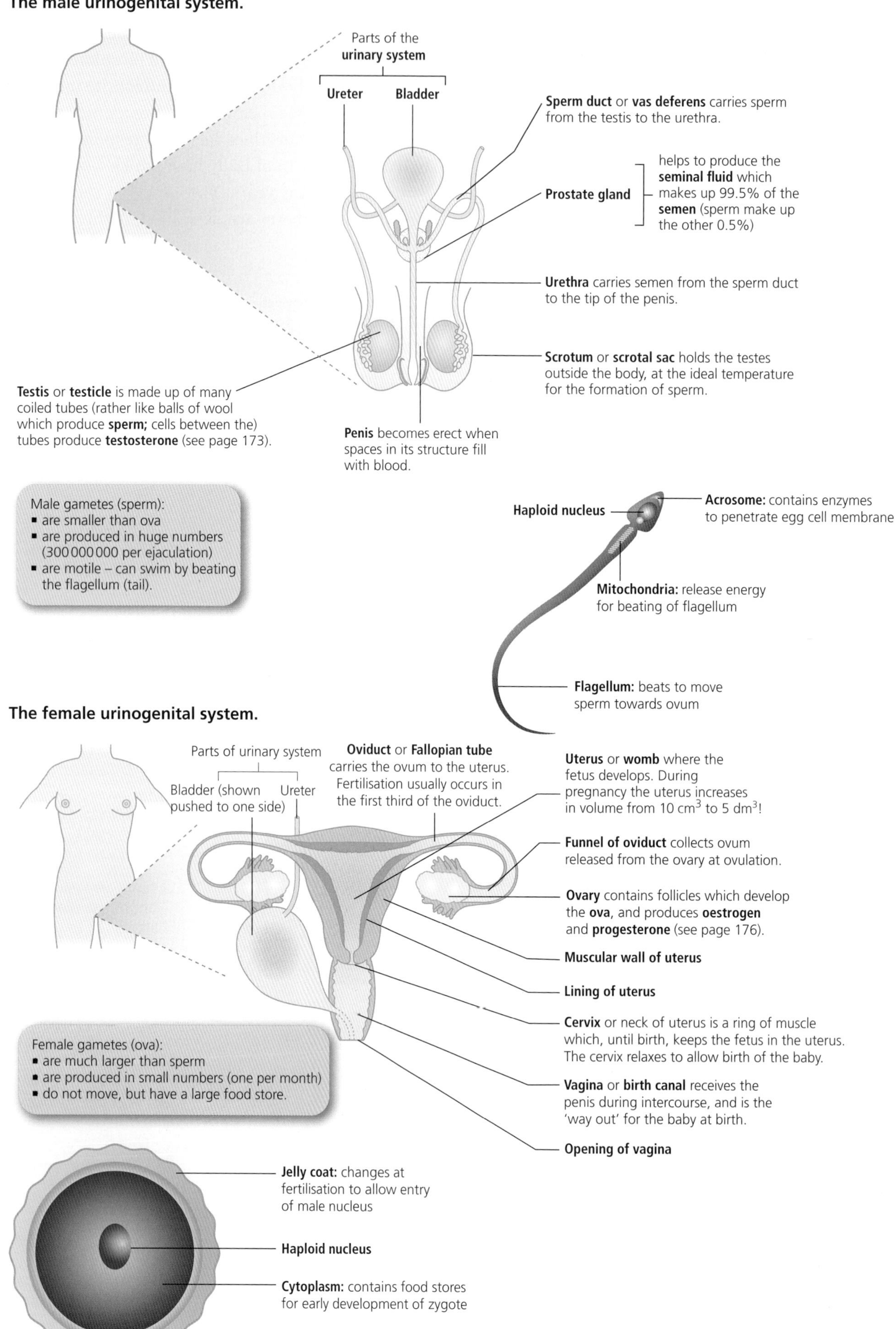
The male urinogenital system.
Parts of the urinary system
Ureter
Bladder
Sperm duct or vas deferens carries sperm from the testis to the urethra.
Prostate gland
helps to produce the seminal fluid which makes up 99.5% of the semen (sperm make up the other 0.5%)
Urethra carries semen from the sperm duct to the tip of the penis.
Scrotum or scrotal sac holds the testes outside the body, at the ideal temperature for the formation of sperm.
Testis or testicle is made up of many coiled tubes (rather like balls of wool which produce sperm; cells between the) tubes produce testosterone (see page 173).
Penis becomes erect when spaces in its structure fill with blood.
Male gametes (sperm):
▪ are smaller than ova
▪ are produced in huge numbers (300 000 000 per ejaculation)
▪ are motile – can swim by beating the flagellum (tail).
Haploid nucleus
Acrosome: contains enzymes to penetrate egg cell membrane
Mitochondria: release energy for beating of flagellum
Flagellum: beats to move sperm towards ovum
The female urinogenital system.
Parts of urinary system
Bladder (shown pushed to one side)
Ureter
Oviduct or Fallopian tube carries the ovum to the uterus. Fertilisation usually occurs in the first third of the oviduct.
Uterus or womb where the fetus develops. During pregnancy the uterus increases in volume from 10 cm3 to 5 dm3!
Funnel of oviduct collects ovum released from the ovary at ovulation.
Ovary contains follicles which develop the ova, and produces oestrogen and progesterone (see page 176).
Muscular wall of uterus
Lining of uterus
Cervix or neck of uterus is a ring of muscle which, until birth, keeps the fetus in the uterus. The cervix relaxes to allow birth of the baby.
Vagina or birth canal receives the penis during intercourse, and is the 'way out' for the baby at birth.
Opening of vagina
Female gametes (ova):
▪ are much larger than sperm
▪ are produced in small numbers (one per month)
▪ do not move, but have a large food store.
Jelly coat: changes at fertilisation to allow entry of male nucleus
Haploid nucleus
Cytoplasm: contains food stores for early development of zygote

Q

1 Explain the importance of the following in the human life cycle:
- **a** fertilisation
- **b** puberty
- **c** gamete formation.

2 Describe the pathway followed by:
- **a** spermatozoa at ejaculation
- **b** an ovum at ovulation.

3 Where does fertilisation occur?

4 Why is reproduction necessary?

5 This figure shows the female reproductive system.

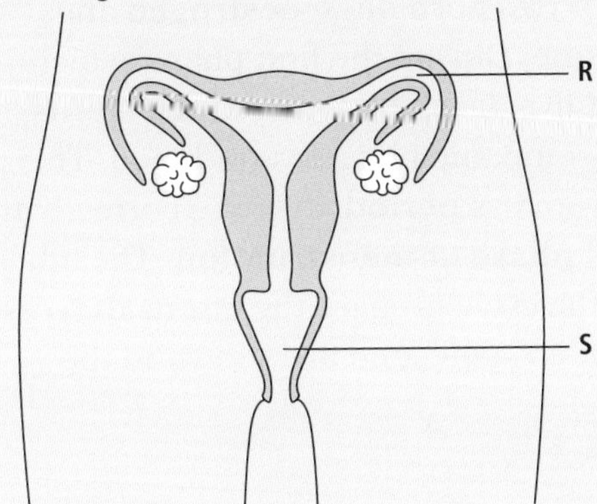

- **a** Name structures **R** and **S**.
- **b** **i** Copy the diagram and label, with a line and a letter **F**, where fertilisation occurs.
  - **ii** Copy the diagram and label, with a line and a letter **I**, where implantation occurs.
- **c** During puberty, the secondary sexual characteristics develop.
  - **i** Name the hormone that controls these developments in a female and state which organ produces it.
  - **ii** State two secondary sexual characteristics that develop in females, in parts of the body other than in the reproductive organs.

Cambridge IGCSE Biology 0610
Paper 2 Q6 May 2008

6 The figure below shows the reproductive organs of the human male after an operation called a 'vasectomy' has been performed. Following a vasectomy, the man can still ejaculate fluid produced by the prostate and Cowper's glands.

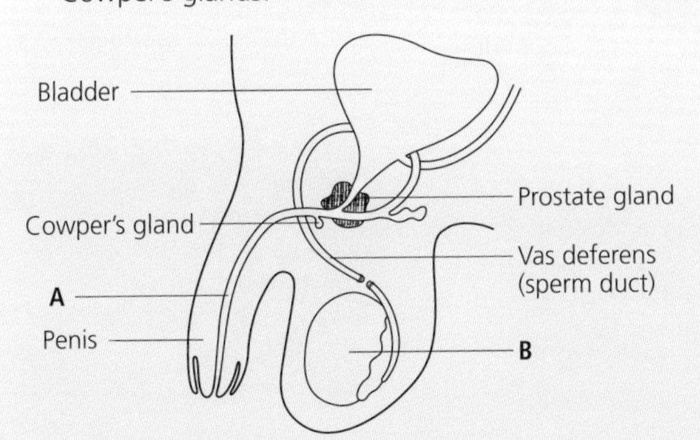

- **a** Name parts **A** and **B**.
- **b** Put an 'X' on the diagram to show where sperms are made.
- **c** **i** In what way are the reproductive organs of the male with the vasectomy different from those of a normal, untreated male?
  - **ii** Explain how this will act as a method of contraception.
- **d** A man infected with the human immuno-deficiency virus (HIV) may transmit AIDS (acquired immuno-deficiency syndrome) to another person. The virus is transmitted in body fluids. Following a vasectomy, is it still possible for an infected man to pass AIDS to another person? Explain your answer.

7 This figure shows a human egg cell and a human sperm cell.

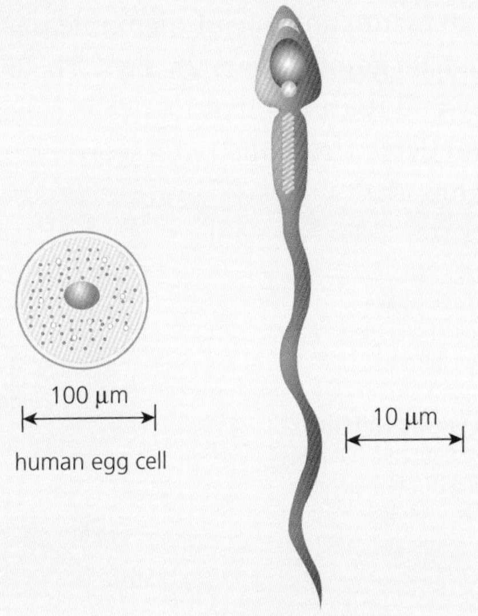

- **a** **i** What is the name given to the release of eggs from the ovary?
  - **ii** Sperm cells and egg cells are haploid. State the meaning of the term *haploid*.
- **b** Complete the table to compare egg cells with sperm cells.

feature	egg cells	sperm cells
site of production		
relative size		
numbers produced		
mobility		

3.7 The menstrual cycle

OBJECTIVES

- To understand the female monthly cycle of ovulation
- To know the role of hormones in the menstrual cycle
- To describe the sequence of events in follicle development and ovulation

The testes produce sperm continually at a rate of about 100 000 000 per day from puberty to old age. Women produce only one ovum per month during their reproductive life, from puberty to middle age. The two ovaries take it in turns to produce an ovum, and one of them releases a mature female gamete every 28 days. The cycle of producing and releasing mature ova is called the **menstrual cycle** (from the Latin word *menstrua* meaning month).

The menstrual cycle is a long-term process controlled by a number of hormones, which:

- prepare the uterus to receive any fertilised ova
- control the development of mature ova.

Hormones affect the wall of the uterus

During the menstrual cycle the wall of the uterus goes through four phases, under the influence of two hormones, **oestrogen** and **progesterone**. During the first phase, which lasts about five days, the lining of the uterus is shed, accompanied by a loss of blood. This time is a woman's **period**, or more correctly the **menstrual phase** or **menstruation**. The other phases of the cycle prepare the uterus to receive and protect a zygote, and are shown in the diagram below.

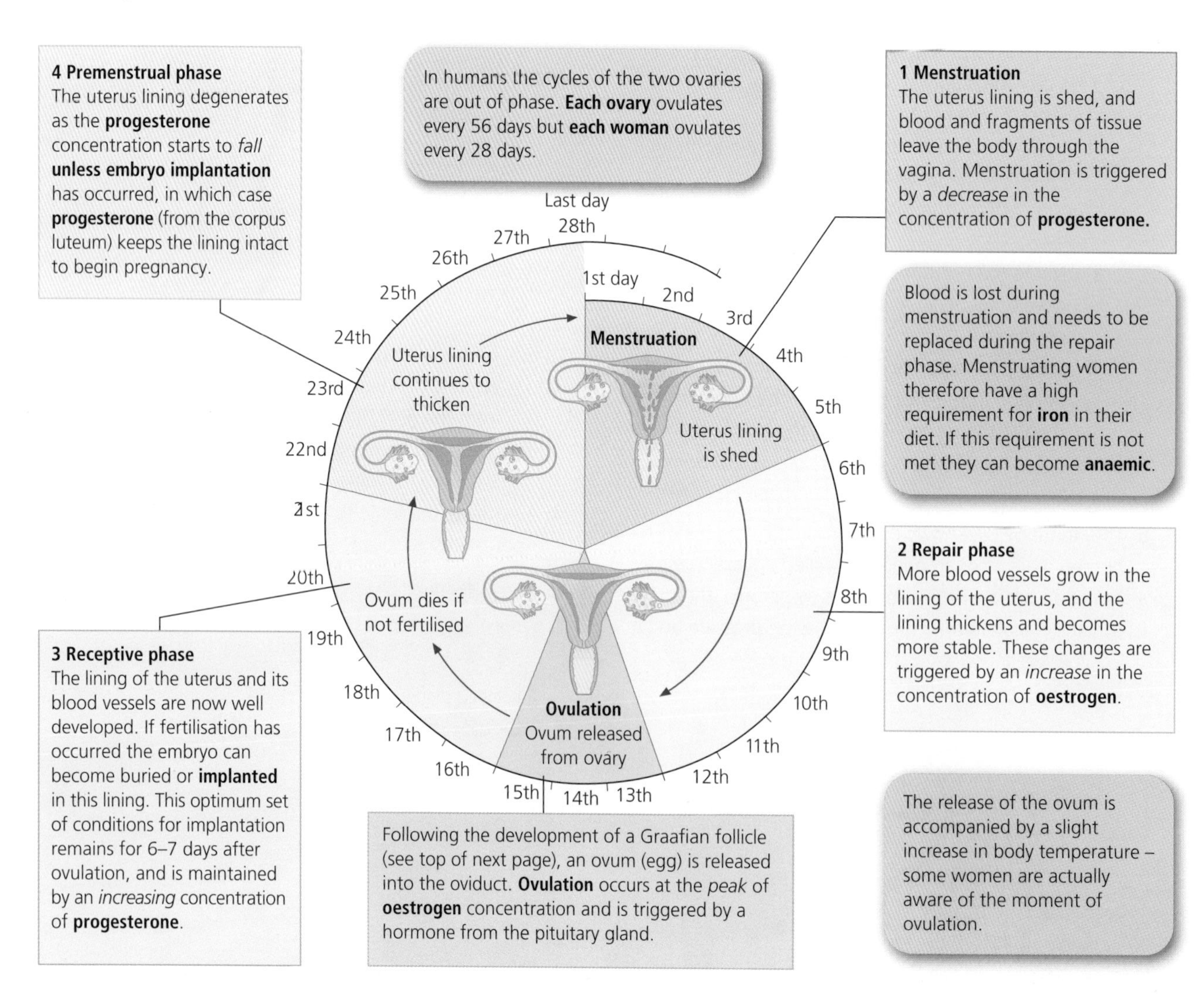

4 Premenstrual phase
The uterus lining degenerates as the progesterone concentration starts to fall unless embryo implantation has occurred, in which case progesterone (from the corpus luteum) keeps the lining intact to begin pregnancy.
In humans the cycles of the two ovaries are out of phase. Each ovary ovulates every 56 days but each woman ovulates every 28 days.
1 Menstruation
The uterus lining is shed, and blood and fragments of tissue leave the body through the vagina. Menstruation is triggered by a decrease in the concentration of progesterone.
Blood is lost during menstruation and needs to be replaced during the repair phase. Menstruating women therefore have a high requirement for iron in their diet. If this requirement is not met they can become anaemic.
Last day
28th
27th
26th
25th
24th
23rd
22nd
21st
20th
19th
18th
17th
16th
15th
14th
13th
12th
11th
10th
9th
8th
7th
6th
5th
4th
3rd
2nd
1st day
Menstruation
Uterus lining is shed
Uterus lining continues to thicken
Ovum dies if not fertilised
Ovulation
Ovum released from ovary
2 Repair phase
More blood vessels grow in the lining of the uterus, and the lining thickens and becomes more stable. These changes are triggered by an increase in the concentration of oestrogen.
3 Receptive phase
The lining of the uterus and its blood vessels are now well developed. If fertilisation has occurred the embryo can become buried or implanted in this lining. This optimum set of conditions for implantation remains for 6–7 days after ovulation, and is maintained by an increasing concentration of progesterone.
Following the development of a Graafian follicle (see top of next page), an ovum (egg) is released into the oviduct. Ovulation occurs at the peak of oestrogen concentration and is triggered by a hormone from the pituitary gland.
The release of the ovum is accompanied by a slight increase in body temperature – some women are actually aware of the moment of ovulation.

Hormones control the development of ova

The ova develop from cells lining the ovary. This is triggered by follicle stimulating hormone (**FSH**) released from the pituitary gland. **FSH** causes a special cell in the ovary to produce a sac around itself. The fluid-filled sac and the developing ovum inside it are together called a **Graafian follicle**. Once the follicle is mature, and there is a high concentration of oestrogen, it moves to the surface of the ovary and bursts, releasing the ovum into the funnel of the oviduct. This process is called **ovulation**. The remaining cells of the Graafian follicle become a structure known as the **corpus luteum**, which produces the hormone progesterone. This hormone keeps the wall of the uterus in good condition for the development of a zygote if implantation has occurred. It also prevents FSH secretion which prevents the release of any more mature ova by feedback inhibition. This ensures that only one fertilised ovum develops in the uterus at any one time. The processes taking place in the uterus and the ovary and their control by hormones are summarised in the diagram below.

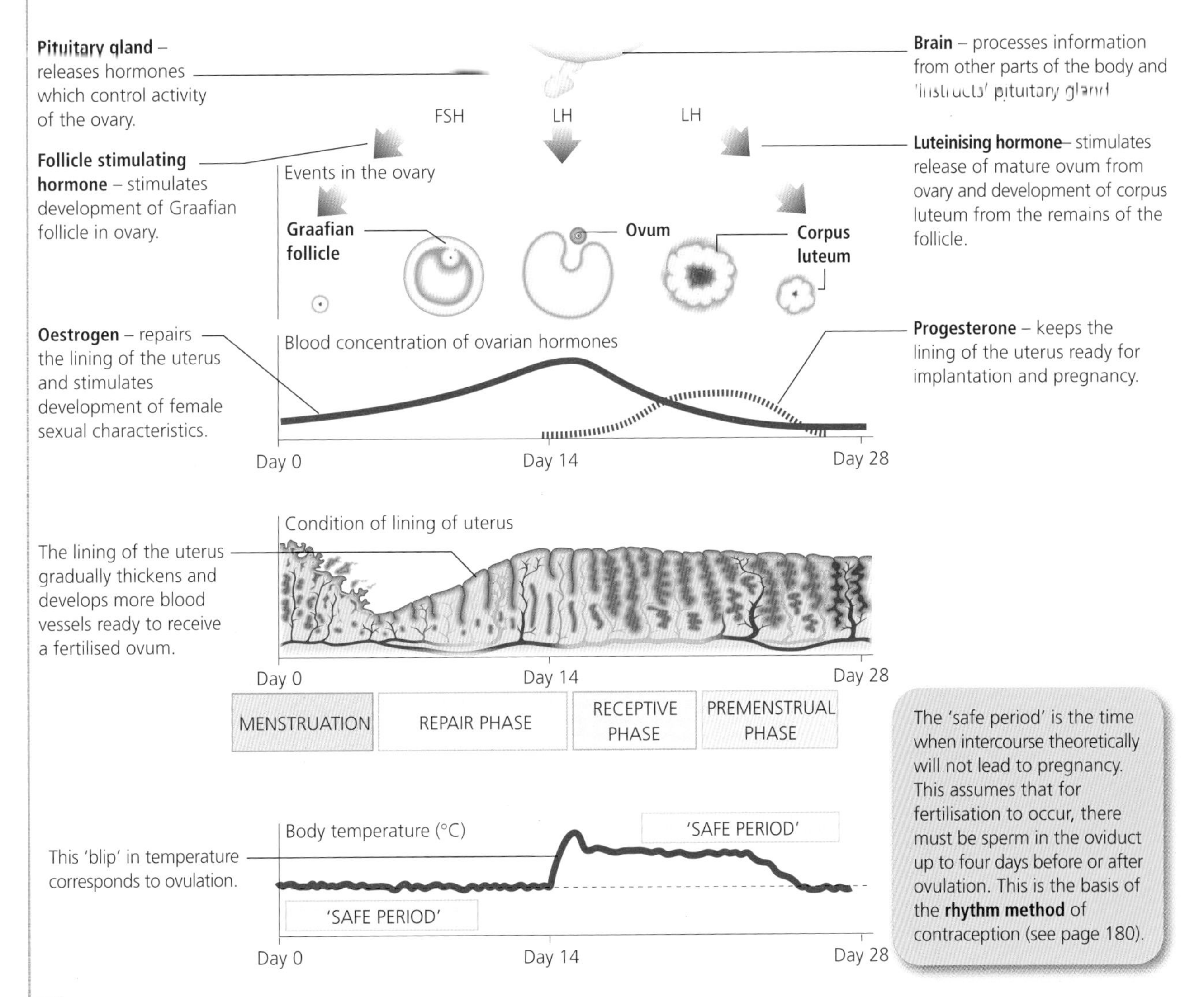

The 'safe period' is the time when intercourse theoretically will not lead to pregnancy. This assumes that for fertilisation to occur, there must be sperm in the oviduct up to four days before or after ovulation. This is the basis of the **rhythm method** of contraception (see page 180).

1. Define the terms **menstruation** and **ovulation**. What is the link between these processes?
2. Describe the role of the hormones FSH, LH, oestrogen and progesterone in the control of the menstrual cycle.
3. Use the term **feedback inhibition** to explain why contraceptive pills contain the hormone progesterone.
4. **a** List the phases of the menstrual cycle.
   b How long does the cycle last?
   c At what time in the cycle does ovulation occur?

3.8 Copulation and conception

OBJECTIVES

- To understand the difference between copulation and conception
- To describe the events of fertilisation
- To describe the processes of *in vitro* fertilisation and artificial insemination by donor
- To understand some of the moral and ethical questions posed by human intervention in reproductive processes

Ovulation provides a female gamete

During ovulation each month, an ovum is released from one of the ovaries. The ovum moves slowly along the oviduct towards the uterus. This movement is brought about by:

- **peristalsis** – rhythmic contractions of muscles in the wall of the oviduct
- **cilia** – fine hair-like structures on the lining of the oviduct which sweep the ovum along.

It takes 4–7 days for the ovum to reach the uterus, and during this time fertilisation may take place in the oviduct (see below).

Copulation delivers male gametes

Before intercourse, sexual stimulation causes blood to flow into the man's penis. The penis becomes erect and hard enough to enter the woman's vagina (helped by lubricating fluids released by the walls of the vagina). This is called **copulation** or **sexual intercourse**. The rubbing of the tip of the penis (the **glans**) against the wall of the vagina sets off a reflex action that releases stored sperm from the testes, and squeezes them by peristalsis along the sperm ducts and the urethra. As the sperm pass along these tubes, seminal fluid is added to them and the complete **semen** is ejaculated in spurts from the tip of the penis. About 3 or 4 cm^3 of semen is ejaculated, and this contains about 300 000 000 sperm. The diagram below illustrates how the male and female gametes arrive at the same place.

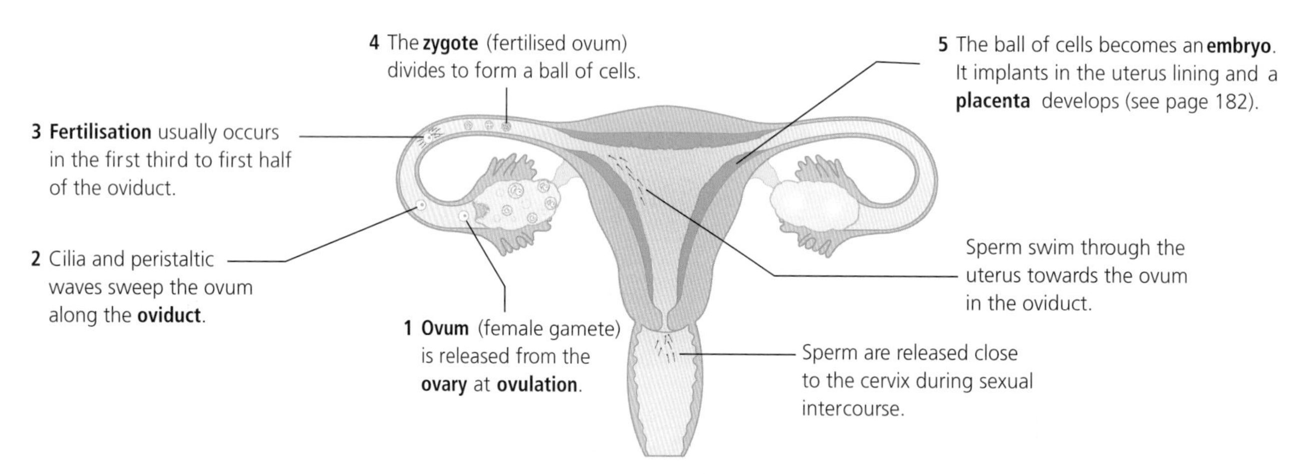
4 The **zygote** (fertilised ovum) divides to form a ball of cells.
5 The ball of cells becomes an **embryo**. It implants in the uterus lining and a **placenta** develops (see page 182).
3 **Fertilisation** usually occurs in the first third to first half of the oviduct.
2 Cilia and peristaltic waves sweep the ovum along the **oviduct**.
1 **Ovum** (female gamete) is released from the **ovary** at **ovulation**.
Sperm swim through the uterus towards the ovum in the oviduct.
Sperm are released close to the cervix during sexual intercourse.

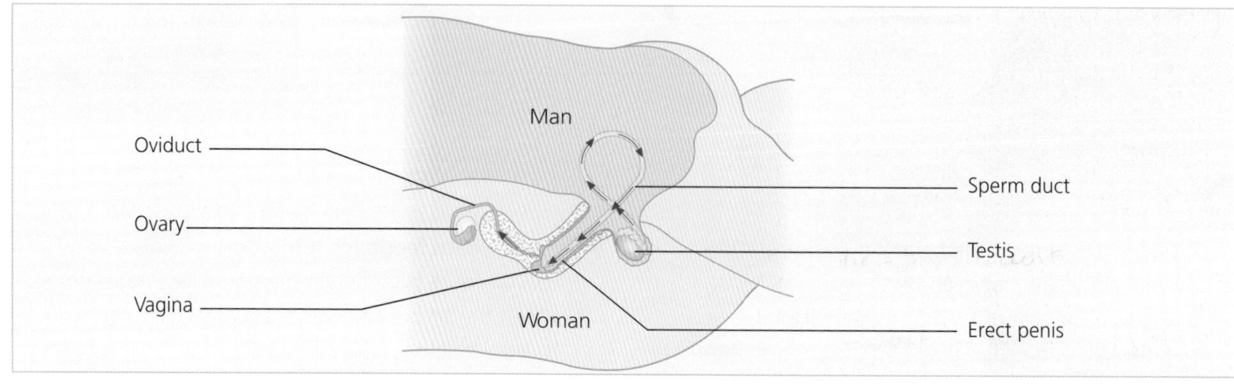
Man
Oviduct
Ovary
Vagina
Woman
Sperm duct
Testis
Erect penis

▲ Sexual intercourse or copulation delivers male gametes to the female reproductive system

1 Sperm swim towards ovum.

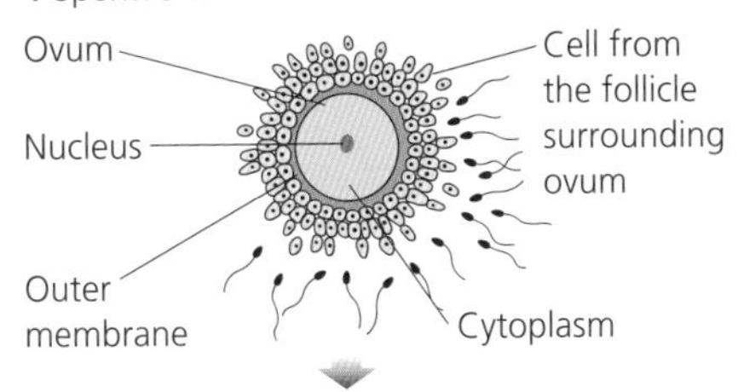

2 Remaining follicle cells are scattered by sperm.

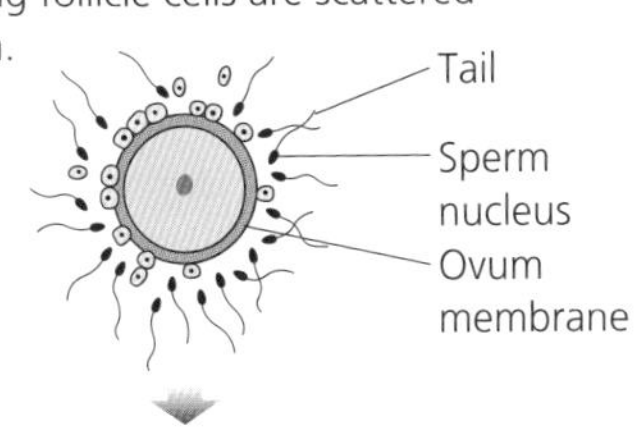

3 One sperm passes through the outer membrane. A barrier forms to prevent entry of more than one sperm.

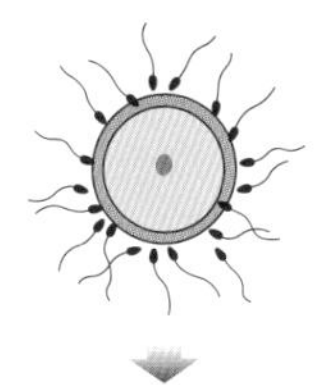

4 Head of sperm crosses ovum cells membrane.

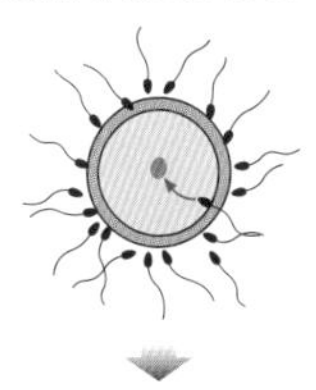

5 Sperm nucleus and ovum nucleus fuse at fertilisation. A zygote has been formed.

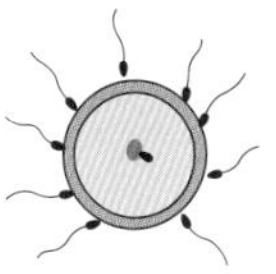

▲ Fertilisation is the fusion of ovum and sperm

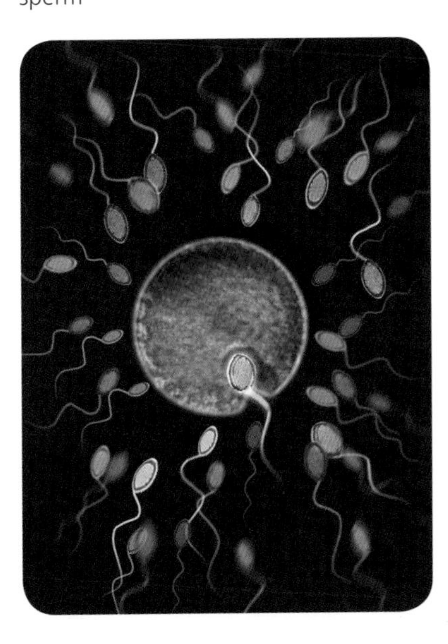

Fertilisation is the fusion of ovum and sperm

Fertilisation is the joining together (fusion) of an ovum and a sperm. The new cell contains a set of genetic material from the mother and a set from the father. Fertilisation takes place in the oviduct, and although several hundred sperm may reach the ovum, only one of them will penetrate the membrane that surrounds it. Once this has happened a series of changes takes place:

- The ovum membrane alters to form a barrier to the entry of other sperm.
- The head of the sperm (the male nucleus) moves towards the nucleus of the ovum and the two fuse (join together).
- The fertilised ovum or **zygote** now starts to divide, first into two cells, then into four, and so on. It continues to move towards the uterus.

The events of fertilisation are summarised in the diagram opposite.

Conception is the implantation of the ball of cells

About six days after fertilisation the ball of cells, now called an **embryo**, becomes embedded in the thickened lining of the uterus. **Conception**, the beginning of the development of a new individual, takes place when this embedding or **implantation** is complete. Once the embryo is attached to the lining of the uterus, some of its outer cells combine with some of the mother's cells and a **placenta** begins to develop.

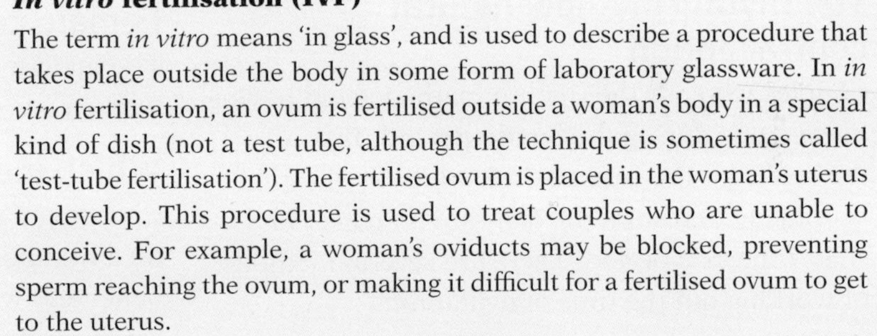

Infertility treatment

In vitro fertilisation (IVF)

The term *in vitro* means 'in glass', and is used to describe a procedure that takes place outside the body in some form of laboratory glassware. In *in vitro* fertilisation, an ovum is fertilised outside a woman's body in a special kind of dish (not a test tube, although the technique is sometimes called 'test-tube fertilisation'). The fertilised ovum is placed in the woman's uterus to develop. This procedure is used to treat couples who are unable to conceive. For example, a woman's oviducts may be blocked, preventing sperm reaching the ovum, or making it difficult for a fertilised ovum to get to the uterus.

Artificial insemination by donor (AID)

If a couple is unable to conceive naturally due to a problem with the man's sperm, they may try AID. Sperm from a donor is obtained from a sperm bank (where it is stored) and is inserted into the woman's uterus close to her time of ovulation.

Q

1 Define the terms conception, copulation and fertilisation. In what order do these events occur?

2 It is possible for humans to intervene in the process of reproduction. Suggest how IVF and AID raise ethical problems for the medical profession.

3.9 Contraception

OBJECTIVES

- To define contraception
- To know about different methods of contraception
- To evaluate the effectiveness of different methods of contraception

Preventing pregnancy

Contraception (which literally means 'against conceiving') is defined as 'deliberately preventing pregnancy'. It is quite natural for two people who love each other to want to have intercourse (to 'make love' or 'have sex'). However, they may not want to have a baby. Using contraception allows a couple to choose when to have children and is an essential part of **family planning**.

Contraception may depend on:

- an understanding of the body's natural cycles (e.g. the **rhythm method**)
- a physical barrier between ovum and sperm (e.g. the **condom**)
- a chemical (e.g. a **pill**, or a **spermicide**)
- a surgical procedure (e.g. **vasectomy**).

The rhythm method: natural contraception

Couples who use the rhythm method avoid having intercourse during the **fertile period**. Instead they wait for the **safe period**, when fertilisation is unlikely to result. The diagram on page 177 shows the safe period. There are several methods of working out the time of ovulation:

- **Calendar method** – the time of ovulation is calculated after keeping a record of when the last six or more periods started.
- **Temperature method** – the woman measures her body temperature and looks for changes at ovulation.
- **Mucus method** – the woman notes changes that take place in the mucus lining of the vagina and cervix at ovulation.

The table opposite shows that the rhythm method is not very reliable. However, it is acceptable to religious groups opposed to artificial methods of contraception.

Male contraceptive methods

There are three methods of contraception for men:

- **Withdrawal (coitus interruptus)** – the penis is withdrawn from the vagina before ejaculation. This method is unreliable as small amounts of semen can leak out before the ejaculation.
- **Vasectomy** – the sperm ducts are cut or tied in a surgical operation (see page 174).
- **Condom or sheath** – this is a thin rubber covering that is fitted over the erect penis before intercourse. It prevents sperm being released into the vagina, and also protects against sexually transmitted infections (see page 189).

◀ The condom or sheath is unrolled over the erect penis, and semen is collected in the end.

Female contraceptive methods

The main methods of contraception used by women are shown in the diagram opposite.

Which method is best?

Choosing a method of contraception depends on a number of factors, including age, religious belief and whether a long- or short-term method is required. The table compares the failure rates of the different methods in the form of **HWY**, the number of pregnancies likely if a **H**undred **W**omen used the same method for one **Y**ear.

Method of contraception	Failure rate / HWY
No method used	54
Rhythm method	15
Condom	8–10
Diaphragm with spermicide	2–3
IUD Causes baby to die = Abortifacient	1
Contraceptive pill	1
Female sterilisation Permanent	0
Vasectomy Permanent	0

Female sterilisation is difficult to reverse and should only be considered by couples who are sure that they do not want any more children.

Female sterilisation
The oviducts are tied and cut during an operation. Released ova cannot reach the part of the oviduct where sperm are present.

- A permanent method which is 100% reliable.

IUD (intrauterine device) or coil
This is a plastic-coated copper coil which may be left in the uterus for months or even for years. Strings attached to the lower end allow the coil to be removed via the vagina.

- Quite reliable, particularly for women who have already had children.
- Irritates the lining of the uterus so that implantation of the zygote does not occur.

A baby is conceived, then aborted. = Abortion

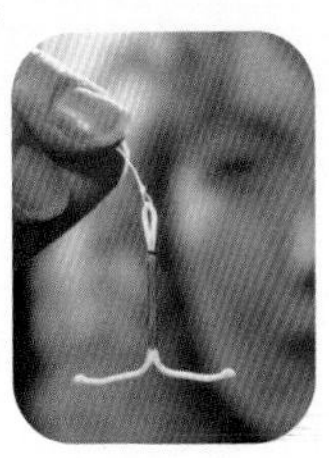

A capsule inserted just beneath the skin can release progestogen over a 3 or 4 month period. This is useful for women who might forget to take the contraceptive pill.

The contraceptive pill
There are two kinds of contraceptive pill. The **mini-pill** contains **progestogen** (synthetic progesterone) which causes changes in the uterus lining so that implantation of a zygote is difficult. The **combined pill** contains **oestrogen** and **progestogen**, and prevents ovulation.

- Almost 100% reliable if used according to instructions.
- Diarrhoea or vomiting can remove the pill from the gut before it has been absorbed fully, reducing its effectiveness.

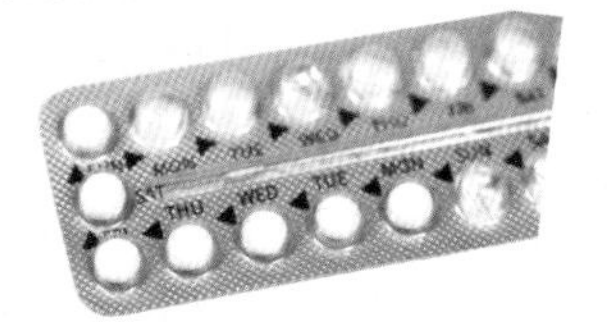

Spermicide
A chemical, applied as a cream, gel or foam, which kills sperm.

- Very unreliable on its own but makes barrier methods of contraception more effective.

Diaphragm or cap
A thin rubber barrier with a springy outer ring to ensure a close fit. Prevents sperm from entering the uterus.

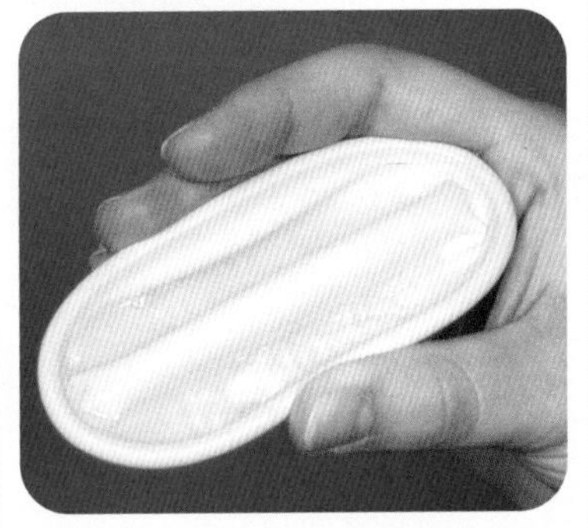

- Very reliable if properly fitted and used with a spermicide.
- Correct size must be used – requires a trial fitting at a family planning clinic.
- Must be fitted before intercourse and remain in place for 6 hours afterwards.

Female condom or 'femidom'
A thin sheath which lines the vagina, and prevents entry of sperm. The 'closed' end has a ring to make fitting easier, and the ring at the open end lies flat against the labia.

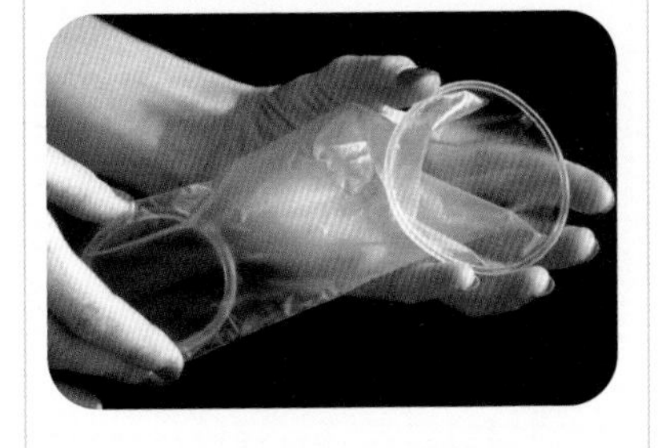

Morning-after contraception
These are pills used after intercourse has taken place. They are not for regular use, but a doctor might prescribe them if there is a risk of unwanted pregnancy which might lead to an abortion at a later date. The morning-after pill contains hormones which cause the uterus lining to be shed. It is taken 48–72 hours after intercourse. A **coil** fitted within 72–96 hours of intercourse usually prevents pregnancy as well.

▲ Methods of contraception for women include physical, chemical and surgical methods

1. List the methods of contraception mentioned on these two pages. Mark each one P, C or S for physical, chemical or surgical.
2. Which contraceptive method(s) might be suitable for a short-term relationship?
3. Why is sterilisation unsuitable for a short-term relationship?
4. Why do you think the rhythm method has such a high rate of failure?

3.10 Pregnancy: the role of the placenta

OBJECTIVES

- To know the sequence of events in the development of a baby
- To understand the role of the placenta

Growth and development

From the time of conception it takes about nine months, or 40 weeks, for a fertilised ovum to become a fully formed baby. This progress involves two closely linked processes:

- **growth** – the repeated division of the zygote to provide the many cells that make up the baby
- **development** – the organisation of the cells into tissues and organs.

During growth, the zygote divides into many identical cells – one zygote at conception becomes 30 million million cells at birth! This type of cell division is called **mitosis** (see page 198). Each cell also takes up its correct position in the embryo. The cells become organised into tissues (see page 19) and start to take on special functions such as nerve cells and skin cells.

A controlled environment

The time taken for the development of a baby from an implanted zygote is called the **gestation period**. The developing fetus needs a stable environment which is provided by the **placenta**, a structure that is only found in mammals. The placenta forms early in pregnancy, partly from the lining of the uterus, and partly from the outside cells of the developing embryo. The fetus is attached to the placenta by the **umbilical cord** as shown below. It is surrounded by the **amniotic sac** which is filled with **amniotic fluid**; this protects the fetus from knocks and bumps.

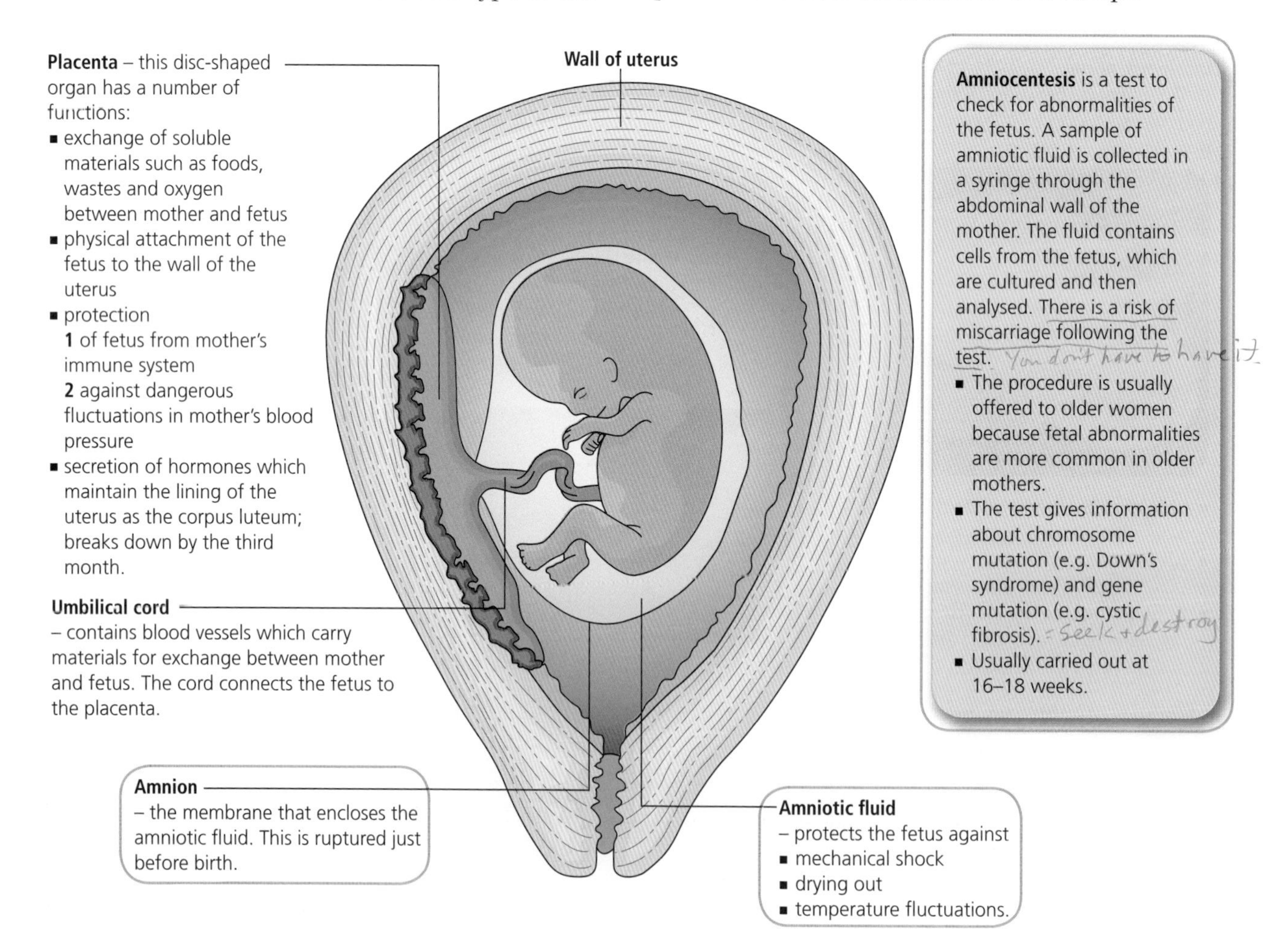

Placenta – this disc-shaped organ has a number of functions:
▪ exchange of soluble materials such as foods, wastes and oxygen between mother and fetus
▪ physical attachment of the fetus to the wall of the uterus
▪ protection
1 of fetus from mother's immune system
2 against dangerous fluctuations in mother's blood pressure
▪ secretion of hormones which maintain the lining of the uterus as the corpus luteum; breaks down by the third month.
Wall of uterus
Umbilical cord – contains blood vessels which carry materials for exchange between mother and fetus. The cord connects the fetus to the placenta.
Amnion – the membrane that encloses the amniotic fluid. This is ruptured just before birth.
Amniotic fluid – protects the fetus against
▪ mechanical shock
▪ drying out
▪ temperature fluctuations.
Amniocentesis is a test to check for abnormalities of the fetus. A sample of amniotic fluid is collected in a syringe through the abdominal wall of the mother. The fluid contains cells from the fetus, which are cultured and then analysed. There is a risk of miscarriage following the test.
▪ The procedure is usually offered to older women because fetal abnormalities are more common in older mothers.
▪ The test gives information about chromosome mutation (e.g. Down's syndrome) and gene mutation (e.g. cystic fibrosis).
▪ Usually carried out at 16–18 weeks.

▲ The placenta protects and nourishes the developing fetus

The placenta begins to develop at implantation and after about 12 weeks it is a thick, disc-like structure with finger-like projections called **villi** that extend deep into the wall of the uterus. The placenta continues to grow to keep pace with the developing fetus and is about 15 cm across, weighing about 500 g, at the time of birth. After the baby has been born the placenta, amniotic sac and umbilical cord are expelled from the uterus as the **afterbirth**. The structure of the placenta, and some of its functions, are illustrated in the diagram below.

Exchange of materials across the placenta

At the placenta, materials are exchanged quickly and selectively between the mother's blood and that of the fetus to keep a constant internal environment inside the fetus. The placenta has adaptations that make this process efficient, as outlined in the diagram below. Towards the end of pregnancy, protective antibodies also cross the placenta so that the baby has some immunity to certain diseases.

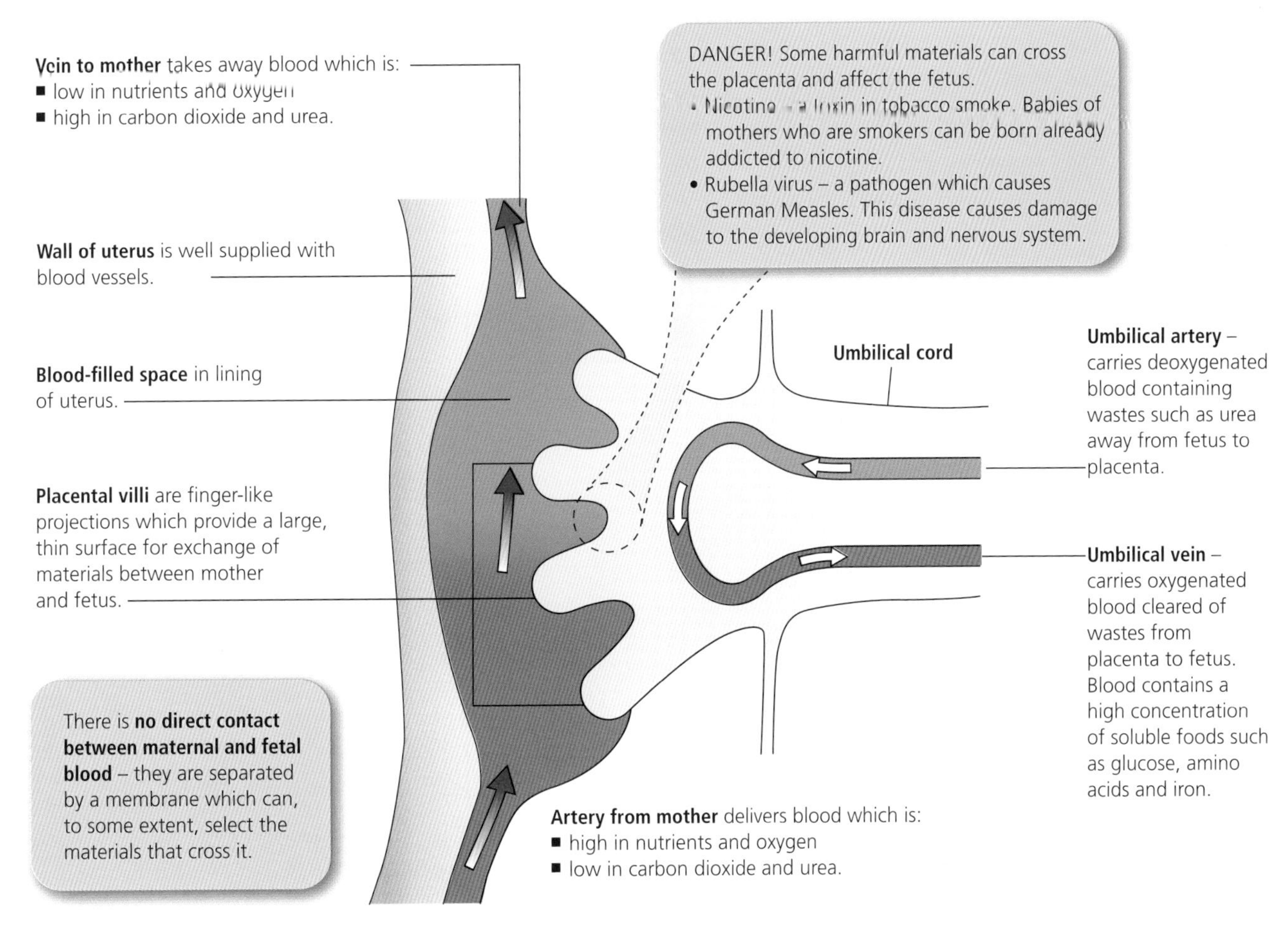

▲ The placenta is the site of exchange – useful substances such as glucose and oxygen pass from mother to fetus and wastes such as urea and carbon dioxide move in the opposite direction

Q

1. What name is given to the complete period from fertilisation to birth?
2. The growth of the fetus is due to an increase in the number of cells of which it is made. A newborn baby has about 30 million million cells. Remember that each cell divides into two, and each of those two into two more (a total of four), and so on. Calculate approximately how many divisions were necessary to produce the baby from the zygote.
3. Cell division is only part of the overall process of production of the baby. Which other process runs alongside cell division? Define this process.

3.11 Pregnancy: development and antenatal care

OBJECTIVE

- To know about prenatal care

Antenatal care

Antenatal care includes
- advice on diet
- guidance on motherhood
- checks on fetus and mother

Signs of pregnancy
- first sign – a missed period
- second sign – another missed period, perhaps with feelings of nausea, tender breasts and more frequent urination.

Testing for pregnancy
Measures the amount of HCG hormone in the urine, using monoclonal antibodies.

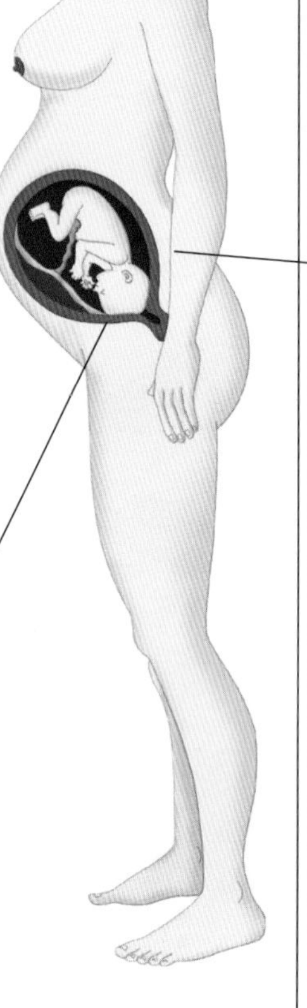

Checks on the fetus

Size and **position** can be felt by gentle pressure on the uterus. Towards the end of pregnancy the best position for the baby is head downwards and facing the mother's back.

Heartbeat can be measured, using a stethoscope, during the second half of pregnancy. The fetal heart rate is usually 120–160 bpm, about twice as high as its mother!

Ultrasound scanning is used to produce a picture of the fetus in the uterus. It provides information on
- baby's age, size, sex and position
- position of the placenta
- whether there are twins!

Amniocentesis (see page 182) provides valuable early information about the developing fetus

Checks on the mother

Weight check: from about the 3rd month of pregnancy a woman gains about 1 lb (450 g) per week. During the whole pregnancy she will gain about 30 lb (12 kg). This includes the weight of the baby, the placenta and the amniotic fluid, as well as some fat under the skin. If she gains too much weight she will be advised to diet.

Blood tests include
- Haemoglobin – to check that the mother is not anaemic (see page 83)
- Blood group – in case an emergency blood transfusion is needed, and to check for Rhesus compatibility
- German measles antibodies – to check that she is immune and there is no risk of the Rubella virus crossing the placenta (see page 183)

Urine tests
- Blood sugar (glucose) – to check for diabetes (see page 133)
- Protein (albumin) – to check for any damage to the kidneys (page 124)

Vaginal examination – on the first visit to the ante-natal clinic the vagina is checked for
- presence of infection e.g. thrush
- to obtain a cervical smear to identify cancer cells.

Late in pregnancy the vagina is checked to make sure that it will be big enough for the baby's head to pass through.

Blood pressure: checked at every visit as high b.p. may indicate **toxaemia of pregnancy** which can be very serious for both mother and baby.

Hormones: the level of **oestrogen** in the blood indicates how well the placenta is functioning to supply the fetus with food and oxygen.

From conception to birth: development of the human fetus

1 month actual length 7 mm

nose
eye
heart
umbilical cord
forelimb
hindlimb

By this time the basic pattern of the body has been laid down – the nerve tube is forming, the heart begins to beat, the limb buds appear and the three regions of the brain are formed. The developing organism is an **embryo**.

2 months actual length 13 mm

Drugs administered during this period may severely affect development e.g. thalidomide leads to loss of or abnormal limbs.

All adult organs are present, though tiny and immature. Muscles begin to differentiate and embryo is capable of movement. Sex organ distinguishable as ovary or testis. Bones begin to harden. From now on the developing organism is know as the fetus.

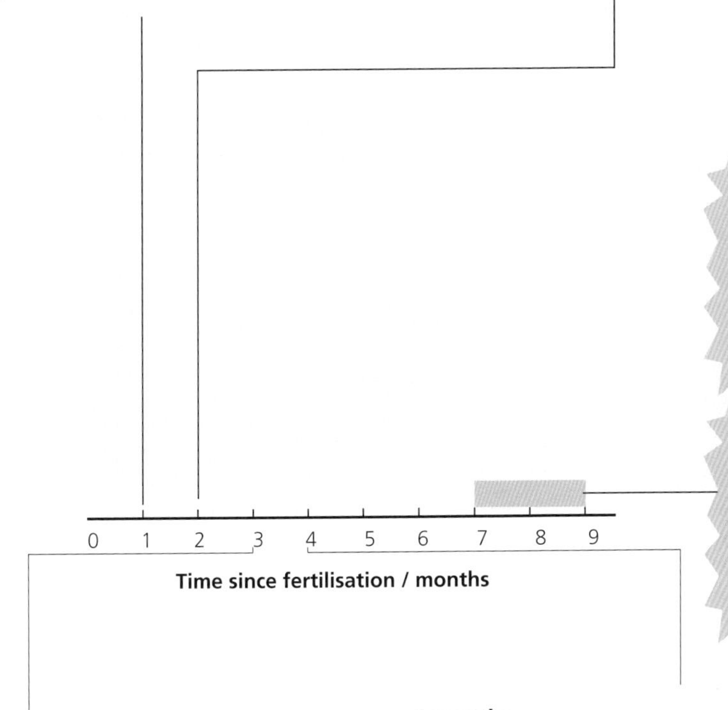

Birth is a compromise – the fetus has outgrown the placenta's supply systems but must adjust to a much more variable environment.

At this stage, **preterm** babies have a good chance of survival with appropriate respiratory and thermoregulatory assistance.

3 months actual length 70 mm

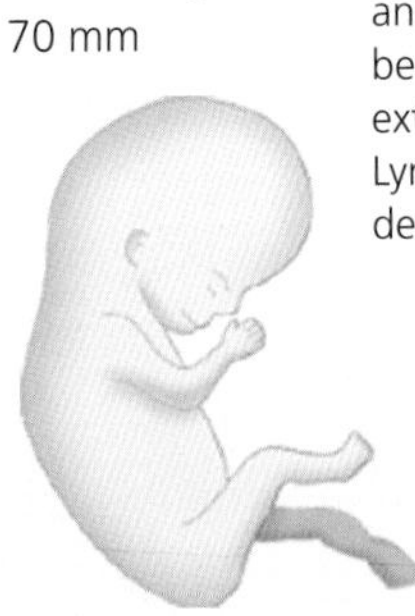

Fetus may kick, curl toes, squint and frown. Sex can be determined by external inspection. Lymph nodes develop.

4 months actual length 95 mm

Face begins to look 'human'. Lobes of the forebrain develop. Eye, ear and nose begin to look 'normal'. Movements may begin to be vigorous, and may be detected by mother (the fetus 'kicks').

3.12 Birth and the newborn baby

OBJECTIVES

- To describe the events leading to birth
- To describe the early care of a young baby
- To appreciate the benefits of breast feeding

Hormones are responsible for birth

The sequence of events that leads to the birth of a baby is called **labour**. Labour begins with the **contractions** of the uterus muscle. These contractions are:

- prevented by **progesterone** – the level of this hormone *falls* as birth approaches
- stimulated by **oxytocin**, a hormone released from the pituitary gland of the mother
- helped by **oestrogen** (which makes the uterus more sensitive to oxytocin) – the level of this hormone *rises* as birth approaches.

Labour

By the end of pregnancy the baby normally lies with its head against the cervix. At first the contractions come every 20 minutes or so, but as birth approaches they become more frequent and more powerful. The contractions cause the amniotic membrane to break and release the amniotic fluid – this is known as breaking the waters – and the cervix to dilate (get wider). The first stage of labour is complete when the cervix is wide enough for the baby's head to pass through. Labour continues as the baby's head is pushed past the cervix into the vagina, which is now acting as a birth canal. From now on the process is quite rapid and needs only gentle contractions by the mother, helped by the midwife or the obstetrician (doctor who specialises in birth). The birth process is quite traumatic for the baby, and it may become short of oxygen as the umbilical cord is compressed by the walls of the birth canal. The baby's heartbeat is monitored during birth, and the blood soon reoxygenates once the baby begins to breathe. When the baby is breathing properly the umbilical cord is clamped (to prevent bleeding) and cut. The placenta comes away from the wall of the uterus and leaves the vagina as the **afterbirth**.

Feeding the newborn baby

When a newborn baby is placed close to its mother's breast it sucks at the nipple. This is known as the **suckling reflex**. Suckling stimulates the mother's brain to release more oxytocin. The hormone causes tiny muscles in the **mammary glands** to squeeze out milk. This process is **lactation.**

The mother's milk is an ideal food for the baby – it contains all the nutrients the baby needs in the correct proportions. It also contains some antibodies from the mother which help to protect the baby during its early months. Milk made in the first few days is called **colostrum.** It contains mainly antibodies and very little food. The mother continues to produce milk as long as the baby suckles. A newborn baby cannot eat solid food because it has no teeth and its digestive system is not developed enough to deal with solids. At around four to six months when the first teeth are starting to appear, the baby can begin to eat some solid food. The gradual changeover from milk to a solid food diet is called **weaning**.

Bottle feeding

The artificial or formula milks intended for use in bottle feeding are based on cows' milk. Various sugars and other substances are added to the dried powder to make it more like human milk. The differences between cow and human milk are shown opposite.

The main advantages of bottle feeding are that the exact quantity of the baby's food intake can be measured, and that people other than the baby's mother can help with feeding. The main disadvantages of bottle feeding are that formula milk is expensive and it is not as easily digested as breast milk. Also, unless the bottles are carefully cleaned and sterilised and unless the milk is made with boiled cooled water, microbes can be passed to the baby.

In some parts of the world, bottle feeding is a leading cause of gastrointestinal upset and deficiency disease in babies. There is an ongoing campaign to encourage breast feeding.

Human milk is low in bacteria and contains antimicrobial factors so that breast-fed infants suffer fewer infections, particularly the dangerously dehydrating gastroenteritis. Bottle-fed infants in the developing countries may have a 10–15 times greater mortality than breast-fed babies.

Human milk is low cost, is delivered at body temperature, requires no preparation and breast feeding encourages a social bond between mother and infant. Suckling may have a contraceptive effect, although this is not certain.

Artificial milk ('formula') is based on cow's milk, but it must be modified since the gastrointestinal tract and kidneys of the human infant are immature and incapable of dealing with the 'richness' of cow's milk.

Cow's milk compared with human milk

Fat	High in long-chain, saturated fatty acids	Difficult to digest, inhibits calcium absorption
Protein	Three times higher	Infant kidneys cannot cope, excess amino acids may cause brain damage
	High casein : whey ratio	Causes hard curd in stomach – not easily digested
Minerals	Very high (particularly sodium)	Infant kidney cannot cope – severe (often fatal) dehydration
Lactose (milk sugar)	Very high	Lactose intolerance – may lead to severe disease, the symptoms of which may include brain damage

Postnatal care of mother and baby*

Mother Soon after the baby is born, the mother should start doing regular exercises to strengthen the muscles which have stretched out of shape during birth.

Where the placenta was attached inside the uterus, a wound remains which takes four to six weeks to heal. During this time a substance called lochia is discharged from the wound and passes out of the vulva. This area needs to be kept very clean because, until the uterus heals, the reproductive and urinary passages can easily become infected with microbes. Regular bathing is essential, preferably with a little salt added to the water.

Baby It takes a few weeks before the baby's temperature-regulating mechanism begins to work efficiently and before the baby builds up a layer of fat under its skin. At first, the baby's room needs to be heated to about 20 °C. If a heater is used, the air may become very dry, which can damage the baby's respiratory passages. So it is beneficial to take a baby out of doors in warm clothing for an hour or two of fresh air each day.

A baby should be bathed once a day, ideally before a meal. The mother should wash her hands before and after she feeds her baby. If bottles are used they should be cleaned in hot water and detergent with a bottle brush and placed in sterilising solution. Teats should be washed and boiled after use.

Q

1. Copy and complete the following paragraph about the birth of a human baby.
An expectant mother knows when she is about to give birth because her ______ begins to experience waves of contraction. These contractions are caused by an increased release of the hormone ______ from the pituitary gland, and become more and more powerful as the concentration of the hormone ______ falls. Eventually the contractions are so powerful that the ______ dilates, the ______ bursts and the 'waters' are released. Further powerful contractions push the baby through the ______ or birth canal (usually head first, but occasionally feet or bottom first in what is called a breech birth). Once the baby has been delivered it is important that it takes deep breaths because it may have been deprived of ______ as the ______ cord is compressed during delivery. This cord is clamped and cut, and relatively mild contractions of the uterus cause the ______ to come away from the wall of the uterus and pass out of the vagina as the ______.

Twins

Humans usually give birth to a single baby. Occasionally two embryos develop together, each with its own placenta and umbilical cord, resulting in **twins**. There are two kinds of twins, and they arise in different ways, as shown in the diagram below.

Very occasionally three or more ova are released and fertilised at the same time, resulting in a multiple birth. This is quite common in women who have been treated with a **fertility drug.** The uterus cannot expand enough to contain several fetuses growing and developing at the same time, and the mother often gives birth early, usually around the seventh month of the pregnancy. Increasingly, medical care (including the use of incubators) means that the babies may survive.

The development of twins.

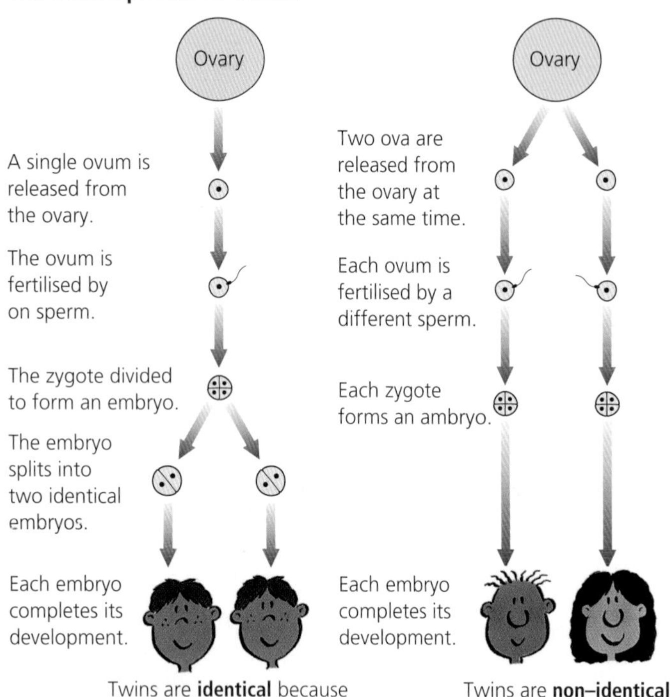

A VERY RARE EVENT!

Ovary

One ovum is released.

but is fertilised late in the menstrual cycle.

A second ovum is released before the ovary is 'switched off' by feedback inhibition (see page 177).

The ovum could be fertilised a month later, **maybe by a sperm from a different man.**

Two children are born as a result of **conjugative conception.** These children may be very different, as they may receive genes from different fathers.

▲ Identical twins have the same genes

▲ Non-identical twins are no more alike than any other brothers and sisters

▲ Multiple births can happen naturally, or following treatment with fertility drugs

Q

2 Describe the role of hormones in the birth and early growth of a human baby.

3 Consider this list of statements about identical twins, and say whether each is true or false.

a They each have the same genes.

b They are formed from two separate ova.

c They each have their own placenta and umbilical cord.

d They may be of the same sex or different sexes.

e They are also known as fraternal twins.

f They are formed from a single fertilised egg that splits in two.

3.13 Sexually transmitted infections

OBJECTIVES

- To understand that disease-causing organisms can be transmitted between partners during sexual activity
- To name some organisms responsible for sexually transmitted infections

The control of **sexually transmitted infections** (STIs) requires the interaction of individuals and communities. These diseases, some of which are listed below, are most likely to be transmitted by body fluids through sexual contact (the responsibility of the individual) but can only be controlled by a concerted effort both locally and worldwide.

Individuals must take care with sexual habits. An STI can be transmitted quickly through a population, so individuals should:

- know the sexual history of their partners
- use a condom for barrier protection
- have a medical check if any symptoms occur.

Communities must offer testing and treatment, e.g. by family doctors.

- Individuals at greatest risk, (e.g. drug users), can be offered testing for HIV.
- Sexual contacts can be traced to identify sources of infection.

Worldwide involvement can include, for example,

- education programmes to prevent infection
- provision of antibiotics
- development of vaccines and antiviral drugs.

STIs can be caused by bacteria, viruses or fungi.

Gonorrhea is a common STI caused by a bacterium.

Bacterial STIs are treated with antibiotics (e.g. penicillin) but resistant strains are developing.

Viral STIs are increasing in frequency. These include AIDS (acquired immune deficiency syndrome).

The causes, effects and treatment for gonorrhea and AIDS are summarised in the table below.

Name of STI and infective organism	Signs and symptoms	Transmission	Treatment
Gonorrhea (caused by bacterium)	1 Pain or burning when passing urine 2 A creamy discharge from the penis or vagina 3 Inflammation of the testicles	Usually by penetrative sex – i.e. when the penis enters the vagina, mouth or anus	Once diagnosed (an easy test in a clinic), treatment is straightforward, involving a course of **antibiotics**
AIDS (caused by virus, HIV–human immunodeficiency virus)	A flu-like illness in the early stages. Many AIDS-related conditions may follow as the immune system begins to fail – e.g. fungal infection of the lungs. The virus reduces the number of lymphocytes and decreases the ability to produce antibodies.	1 Unprotected sex with an infected person 2 Contact with an infected person's blood 3 From mother to child, during pregnancy or childbirth 4 Sharing syringes while injecting drugs	**There is no cure.** Antiviral treatment slows down the progression from HIV+ status to full-blown AIDS. Modern treatments inhibit the enzymes which the virus uses to copy itself.

1. Suggest two steps an individual can take to reduce the risk of named sexually transmitted infections.
2. What are the responsibilities of a community health service?
3. Name one viral and one bacterial sexually transmitted infection (STI).
4. For any one named STI suggest how individuals, local communities and scientists worldwide might be involved in its control.

Questions on human reproduction

1 The following table provides information about the body temperature of a young woman on each day throughout a single menstrual cycle.

a Plot this information in the form of a graph.

b i What is the range of the body temperature during the cycle?

day of menstrual cycle		body temperature / °C
1	Menstruation	36.5
2		36.2
3		36.2
4		36.2
5		36.2
6		36.2
7		36.3
8		36.2
9		36.4
10		36.3
11		36.2
12		36.1
13		36.2
14	Ovulation	36.2
15		36.5
16		36.7
17		36.8
18		36.7
19		36.7
20		36.8
21		36.7
22		36.6
23		36.6
24		36.7
25		36.6
26		36.6
27		36.7
28		36.6

ii These measurements were made at the same time each day, in the same room and while the woman was wearing the same clothes. Why is this important?

c i The times of menstruation and ovulation are shown in the table. What changes in temperature coincide with these events?

ii How could this information be used:

a as a method of contraception

b to increase the chance of conception?

d Which other methods of contraception are available to a woman? Why are these methods more successful than the one outlined in part **c**?

2 A newspaper headline incorrectly stated, 'The use of condoms can result in erectile dysfunction'. Erectile dysfunction is a medical problem which results in problems with sexual intercourse. Scientists are concerned that this incorrect statement could lead to an increase in HIV.

a Describe the process of sexual intercourse in humans.

b Condoms are used as one form of birth control.

i What name is used to describe this method of birth control?

ii Explain how a condom acts as a method of birth control.

c Some readers of the newspaper may believe the newspaper and stop using condoms during sexual intercourse.

i Explain how a decrease in the use of condoms may lead to an increase in the incidence of HIV.

ii State two ways by which a person who does not have sexual intercourse might still become infected with HIV.

iii Explain why the immune system is less effective in a person with HIV.

d Another sexually transmitted disease is gonorrhoea. For this disease, state

i one sign or symptom,

ii one effect on the body,

iii the treatment.

Cambridge IGCSE Biology 0610
Paper 3 Variant 1 Q4 November 2008

3 The temperature of the human fetus whilst in the uterus is about 0.5 °C above that of its mother. At birth it emerges into a relatively cool, dry atmosphere and immediately encounters a problem of temperature control.

a Suggest why the temperature of the fetus is above that of its mother.

b Explain how the following help the newborn baby to control its temperature:

i from about the fifth month of pregnancy onwards a layer of subcutaneous fat is developed by the fetus

ii at birth the blood vessels to the baby's skin constrict very quickly.

c A baby born prematurely is less able to control its body temperature and must be kept in an incubator (see photograph).

i A constant temperature is maintained within the incubator, using a thermostat and an electric heater. Use this example to explain the meaning of the term negative feedback.

ii Suggest two functions of the hood that covers the incubator.

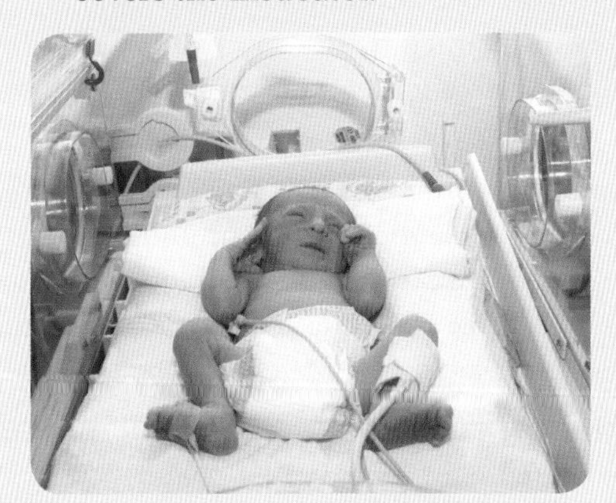

▲ An incubator provides a constant warm environment for a premature baby

d The premature baby must be fed through a plastic tube going through its nostril into its stomach.

i The food supplied to the baby contains substantial quantities of carbohydrate and protein. Suggest one function of each of these substances in the body of the newborn baby.

ii Name one substance normally supplied across the placenta in the late stages of pregnancy which the artificial diet is unlikely to supply. What problem might this cause for the newborn baby?

4 The diagram shows the structure of the placenta and parts of the fetal and maternal circulatory systems.

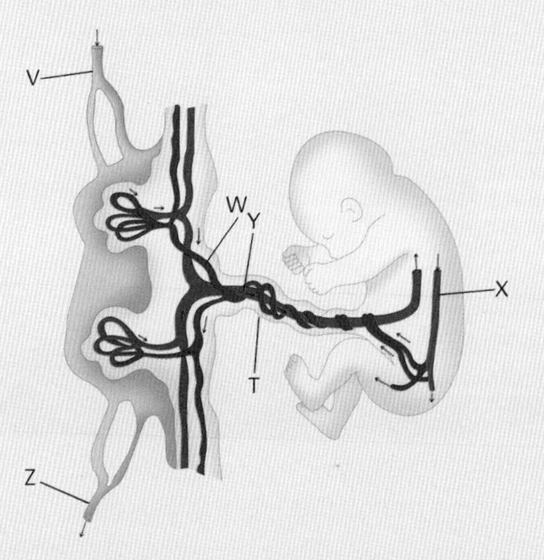

a i Complete Table 1 by listing the blood vessels that carry oxygenated blood. Use the letters in the diagram to identify the blood vessels.

circulatory system	blood vessels that carry oxygenated blood
maternal	
fetal	

▲ Table 1

ii Name structure **T** and describe what happens to it after birth.

iii The placenta is adapted for the exchange of substances between the maternal blood and the fetal blood.

Describe the exchanges that occur across the placenta to keep the fetus alive and well.

b The placenta secretes the hormones oestrogen and progesterone.

Describe the roles of these hormones during pregnancy.

3.14 Variation and inheritance

OBJECTIVES

- To know that there is variation between individuals of the same species
- To appreciate that some differences are inherited whereas others are acquired during the life of the individual
- To understand that genetics has a major impact on the lives of humans
- To understand that chromosomes, located in the nucleus, carry information about inherited characteristics

Inherited and acquired characteristics

Living organisms vary in many ways. For example, humans all have the same general shape and the same set of body organs, but some features differ from one person to the next, such as height, weight, eye and hair colour, shape of nose, language, knowledge and skills. These are examples of **individual variation**.

Some of these features that vary from person to person may be inherited from the parents. **Inheritance** is the transmission of genetic information from generation to generation. Examples of these **inherited** (or **hereditary**) **characteristics** include the tendency to develop some diseases, such as cystic fibrosis, and the permanent colour of the skin. Some, such as a temporary suntan or a scar, cannot be inherited – these are called **acquired characteristics**. Many acquired characteristics can be changed (for example, body mass can be changed by an adjustment to the diet). Inherited characteristics cannot usually be altered (except temporarily).

Genetics

The study of inherited characteristics, and the way they are passed on from one generation to another, is called **genetics**. Our knowledge of the subject of genetics is expanding extremely rapidly, and this knowledge depends upon our understanding of the molecule DNA (see page 194).

When we study inheritance we are looking for answers to several important questions:

- What is a characteristic? Why does a cell or organism develop a certain characteristic?
- How can characteristics be passed on accurately from one cell to another?
- How are the characteristics of two different organisms combined at fertilisation?
- How do characteristics vary from one organism to another and from one generation to another?

Answers to questions such as these help us in many ways, for example to increase our understanding about genetic diseases, and to develop techniques to 'add' desirable characteristics to our domestic animals and plants.

Sexual reproduction and inheritance

The production of offspring by sexual reproduction always involves the **production of gametes** and **fertilisation** (see page 172). In sexual reproduction:

- The only part of the male gamete (sperm in mammals) that goes to form the zygote is the **nucleus**.
- The gametes are formed by **cell division**.
- The young organism develops from a single fertilised egg by **cell division**.

So, if we are to understand how characteristics are passed on during reproduction, we should look carefully at the structure of the nucleus, and how the nucleus behaves during cell division.

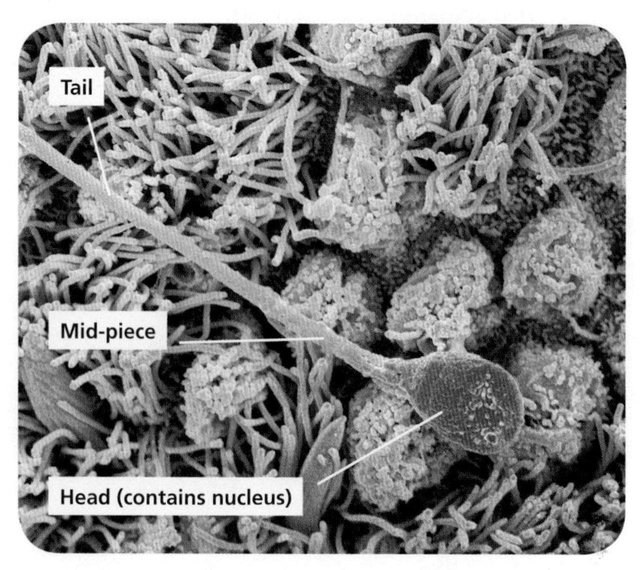

▲ Human sperm ×1000. Only the 'head' (containing the nucleus) enters the egg at the time of fertilisation.

The contents of the nucleus

Special stains can be used to show up the contents of the nucleus. If the cell is not actually dividing, these contents are rather unclear but as the cell begins to divide the contents show up as a series of thread-like structures. As the threads shorten they take up the stain. For this reason they were called **chromosomes** (literally 'coloured bodies'). The structure of a chromosome is outlined below.

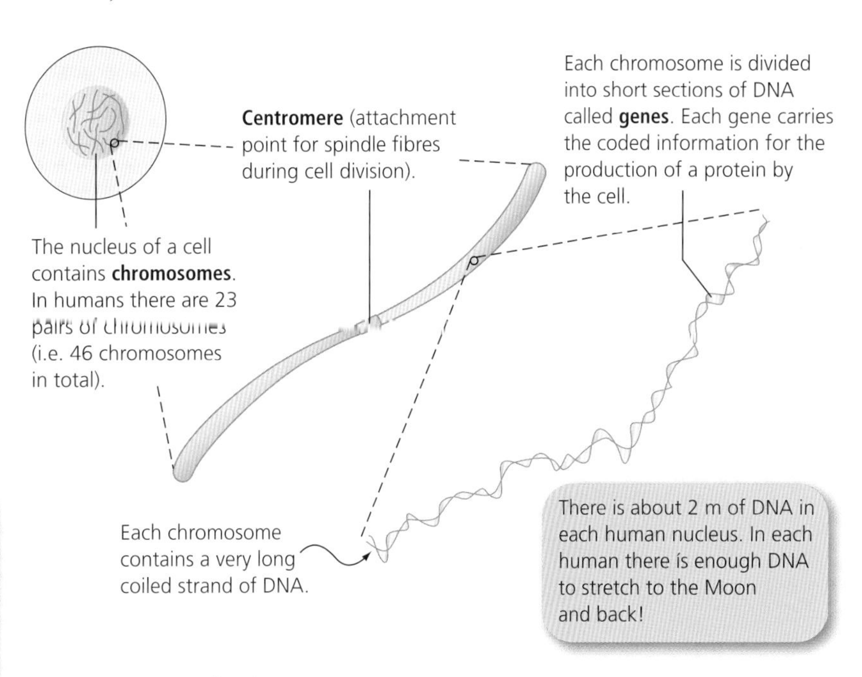

▲ The structure of a chromosome

Q

1. Look around your class. This group of humans shows many variations between individuals. Suggest one variation that is inherited and one that is acquired.
2. Find three newspaper articles that include the words 'gene', 'genetic' or 'inheritance' in their headings. Summarise one of the articles in three or four sentences.
3. a What is a chromosome? What evidence is there that chromosomes carry genetic information?
   b When can chromosomes be observed? Explain why this is possible.

A great deal of evidence suggests that the chromosomes carry **genetic information** – the information that gives the particular characteristics to a cell:

- If sections of chromosome are transferred from one cell to another, the characteristics of the recipient cell change.
- If chromosomes are deliberately damaged, the characteristics of the cell change.
- In some cells, the chromosomes are seen to swell when proteins are being manufactured in the cell.
- The only difference between the nuclei of male and female cells is the presence of one particular chromosome (see page 206). (Males and females certainly have different characteristics!)

Remember these definitions!
Chromosome: a thread-like structure of DNA, carrying genetic information in the form of genes.
Gene: a length of DNA that codes for a specific protein.

3.15 DNA, proteins and the characteristics of organisms

OBJECTIVES

- To know that cell characteristics depend on proteins
- To understand the principle of the genetic code
- To be able to describe the replication of DNA

Characteristics depend on proteins

It has been discovered that the characteristics that a cell or organism possesses depend on the **proteins** that the cell can manufacture. For examples of this, look at the picture at the top of the opposite page.

For cells to specialise in the many different ways that they do, they must make different proteins. The instructions as to which proteins should be manufactured at any one time in a cell are carried as **genes** on the **chromosomes**. Chemical tests have shown that chromosomes are largely composed of the enormous molecule called **deoxyribonucleic acid** or **DNA** for short. In other words:

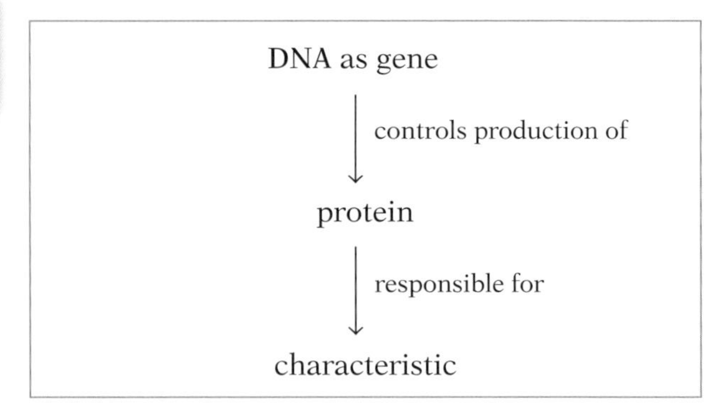

DNA carries its instructions as coded messages using just four different chemical compounds called **nucleotide bases** or organic bases (see page 29). The names of the bases are shown in the diagram below, but you only need to remember their initial letters (A, T, G and C) to understand how the code works.

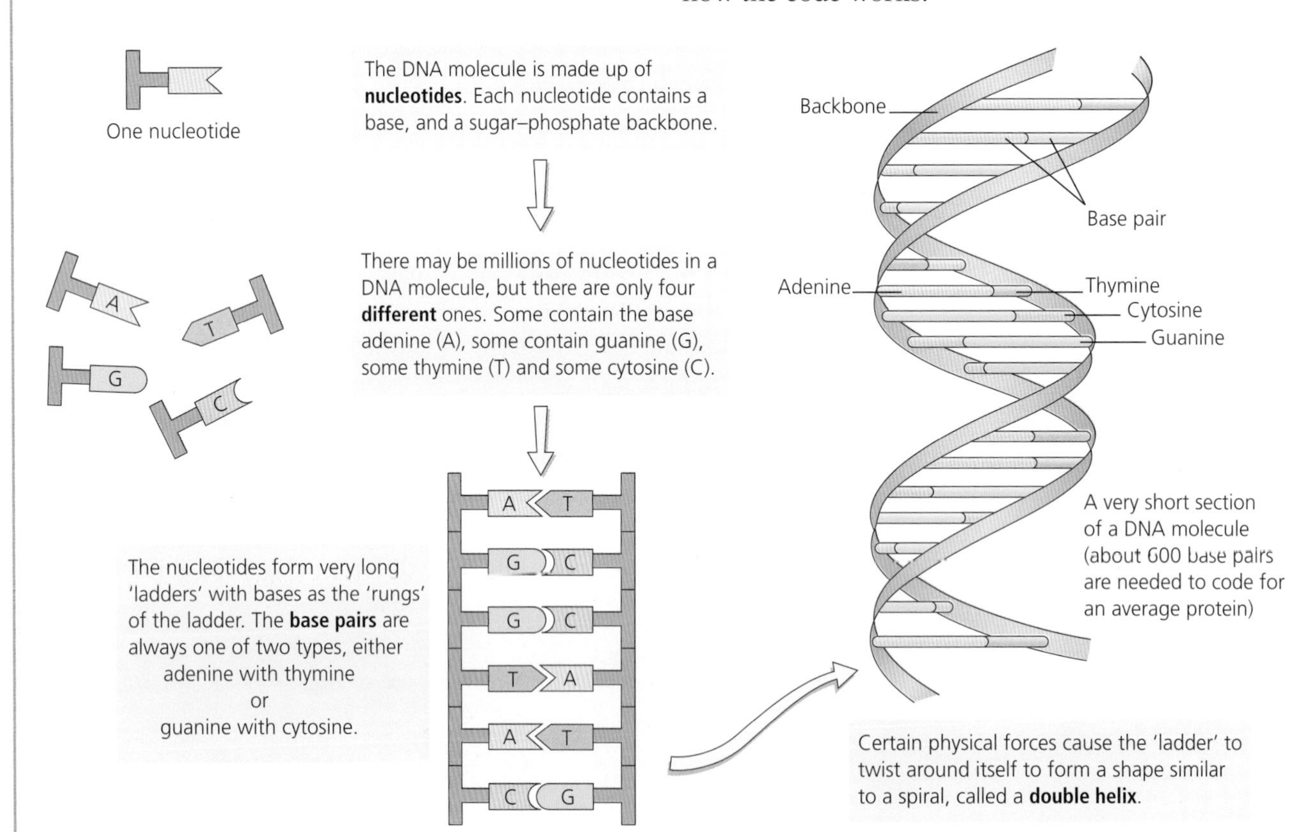

▲ The structure of DNA. The exact length of the DNA molecule is not known even for humans. It is very important to remember that: **adenine** always pairs with **thymine**, and **guanine** always pairs with **cytosine**.

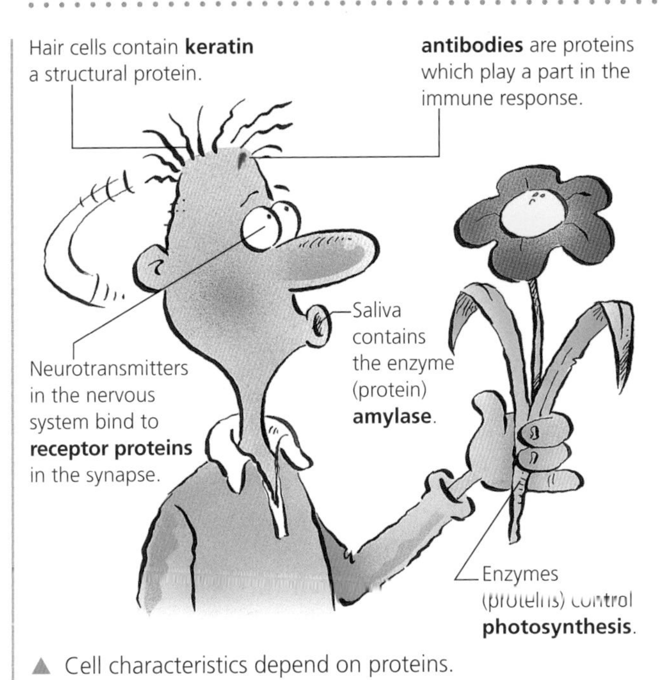

▲ Cell characteristics depend on proteins.

Base pairing can explain how DNA is replicated*

For one organism to pass on characteristics to its offspring, it must be able to copy the coded instructions for these characteristics and hand them on. In other words, DNA in the chromosomes must be copied or **replicated**. This replication must be carried out with great accuracy, since a change in characteristics might be harmful to the organism. The base pairing rule means that the coded sequence on one chain of the double helix automatically determines the coded sequence on the other chain, ensuring accurate replication. The principles of DNA replication are outlined below.

The replication of DNA is a vital part of cell division. It is particularly obvious in copying division (mitosis), as we shall see on page 198.

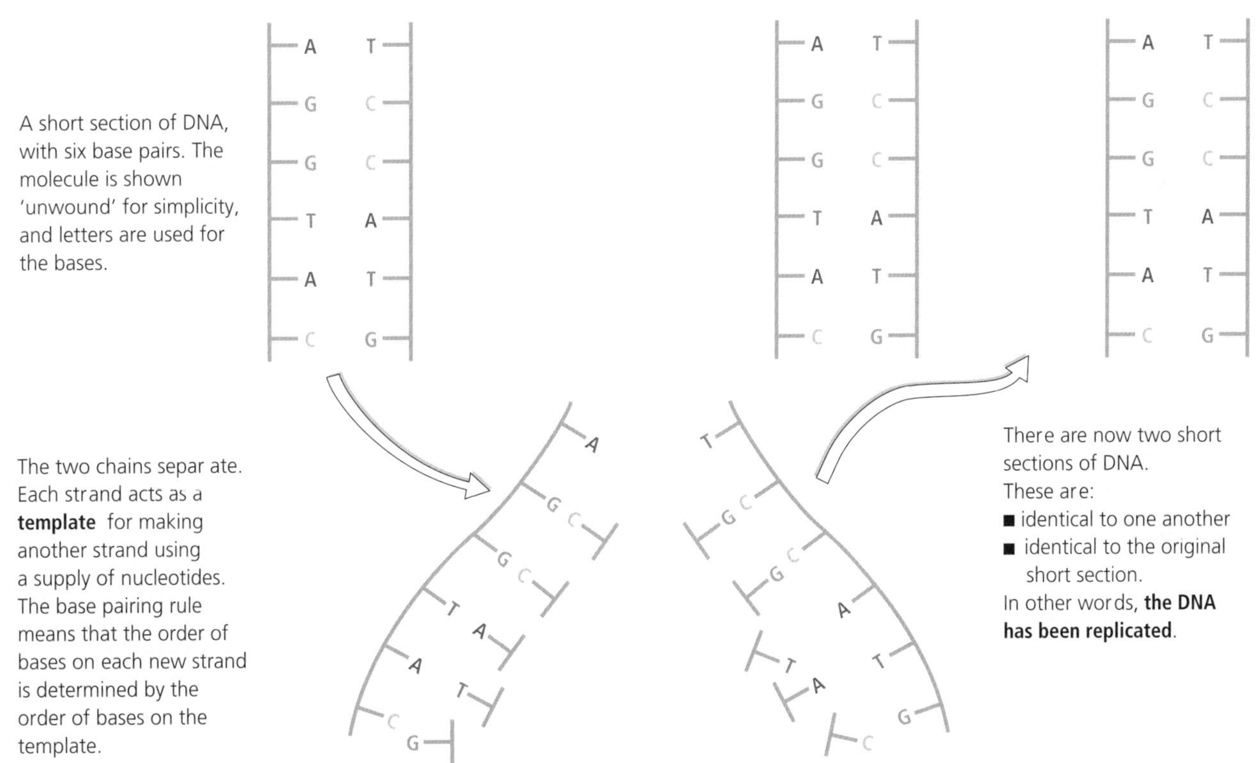

▲ Replication is essential for characteristics to be passed from one generation of cells to the next.

The discovery of DNA

The discovery of DNA structure depended on the work of many scientists. These included

- **Rosalind Franklin:** she made many measurements on DNA, using patterns obtained by directing a beam of X-rays onto crystals of this molecule.
- **James Watson and Francis Crick:** used the measurements made by Franklin, and results of the chemical analysis of DNA which showed that the number of (adenine + guanine) bases was always equal to the number of (thymine + cytosine) bases, to produce a working model of the DNA molecule. This model was the double helix with which we are familiar.

Watson and Crick were awarded the Nobel prize (the highest scientific award) for their work of DNA. Sadly, Rosalind Franklin died before her part in the discovery of DNA structure was properly recognised.

3.16 How the code is carried

OBJECTIVES

- To understand that the genetic code is carried as a sequence of bases on the DNA molecule
- To understand the need for a messenger molecule in protein synthesis
- To define the terms transcription and translation

DNA 'code words' for amino acids

How does the DNA in the genes instruct the cell to make particular proteins? The following points are important in understanding this link:

- Each gene carries a series of coded instructions ('code words') for the synthesis of proteins.
- Each 'code word' on the DNA is made up of three bases (three 'letters') in a certain sequence.
- Each 'code word' – called a triplet – corresponds to a **single amino acid in a protein**.

The sequence of bases in DNA is therefore a series of coded instructions for the building up of amino acids into proteins. The proteins then give the cell or organism a particular characteristic. This relationship between DNA bases and amino acids is called the genetic code, and is outlined below.

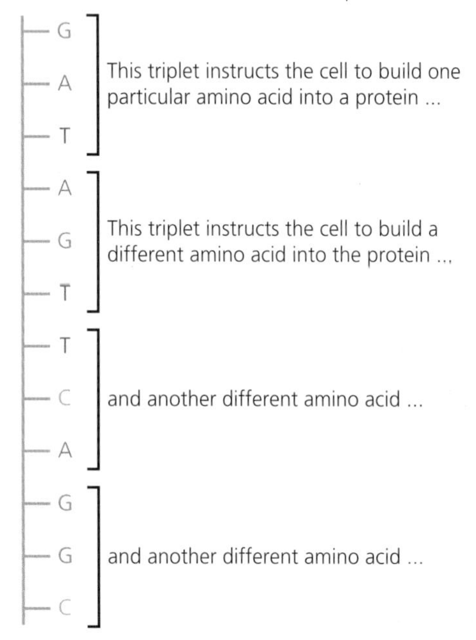

▲ The four different bases in DNA can be arranged in enough different triplets to code for all 20 amino acids normally found in a cell.

Passing the messages to the ribosomes

This coded information in the genes is located on the chromosomes, which are in the nucleus. You may remember that the protein-manufacturing stations, called **ribosomes**, are found outside the nucleus, in the cytoplasm. How does the code pass from the nucleus to the ribosomes in the cytoplasm? It is carried by another type of nucleic acid, called **messenger RNA** (mRNA).

Transcription and translation

The mRNA is made by a process called **transcription**, which literally means 'cross writing'. The base sequence in the DNA is **transcribed** into another base sequence in the mRNA, using very similar base pairing rules to those used in the replication of DNA. There is one important difference – RNA never contains the base thymine (T). Thymine is replaced by a fifth base called **uracil** (U) so instead of the base pair A–T used in DNA replication, in transcription we have the base pair A–U.

Once it has been made, the mRNA leaves the nucleus and travels to the ribosomes. The sequence of bases in the mRNA is used to build up a sequence of amino acids into a protein in the ribosome. This process is called **translation** – it involves rewriting the language of bases into a language of amino acids. The processes of transcription and translation are outlined in the diagram on the next page.

Summary of replication, transcription and translation

It is quite easy to confuse the various processes involving nucleic acids. Remember:

- **Replication** makes a DNA copy, using DNA.
- **Transcription** makes mRNA, using DNA.
- **Translation** makes protein, using mRNA.

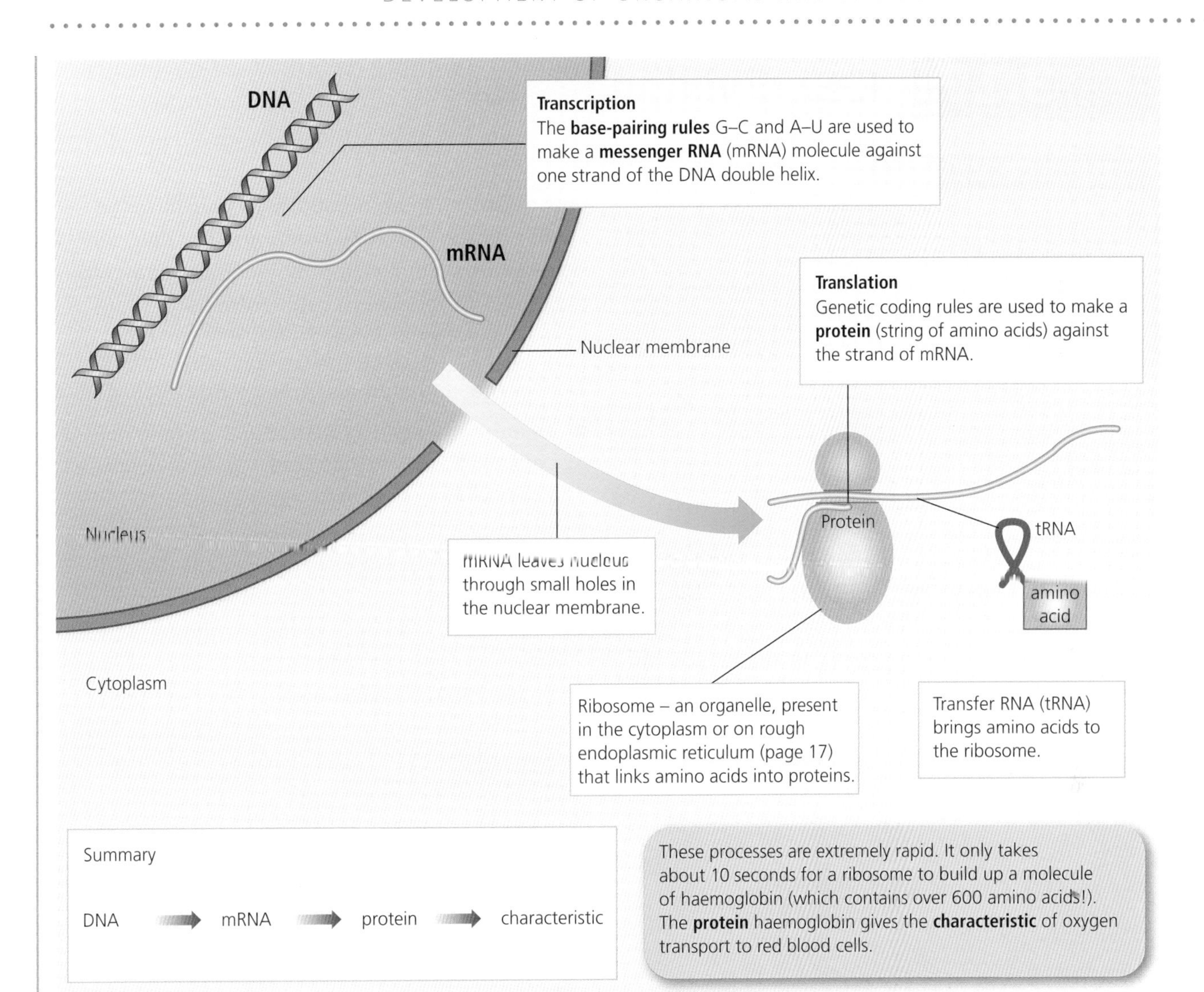

DNA
Transcription
The **base-pairing rules** G–C and A–U are used to make a **messenger RNA** (mRNA) molecule against one strand of the DNA double helix.
mRNA
Nuclear membrane
Translation
Genetic coding rules are used to make a **protein** (string of amino acids) against the strand of mRNA.
Nucleus
mRNA leaves nucleus through small holes in the nuclear membrane.
Protein
tRNA
amino acid
Cytoplasm
Ribosome – an organelle, present in the cytoplasm or on rough endoplasmic reticulum (page 17) that links amino acids into proteins.
Transfer RNA (tRNA) brings amino acids to the ribosome.
Summary
DNA → mRNA → protein → characteristic
These processes are extremely rapid. It only takes about 10 seconds for a ribosome to build up a molecule of haemoglobin (which contains over 600 amino acids!). The **protein** haemoglobin gives the **characteristic** of oxygen transport to red blood cells.

▲ How information in DNA codes for characteristics in cells.

Cell specialisation depends on proteins.

- All body cells in an organism contain the same genes but many genes in a particular cell are not expressed.
- The cell only makes the specific proteins it needs to carry out its specialised function.

Q

1. What are the subunits of a nucleic acid called?
2. Name the four bases in DNA.
3. DNA exists as a double helix. Name the base pairs that hold the double helix together. Why is base pairing important?
4. Define the term DNA replication.
5. Name four proteins that give particular characteristics to named cells.
6. Name one process of which DNA replication is a vital part.
7. The diagram on the right represents the behaviour of DNA strands during the early part of cell division. Use the information in the diagram to help you answer the following questions.
   a. Identify the organic bases X and Y.
   b. Name the process shown in the diagram.
   c. What is the importance of the SEQUENCE of organic bases along a DNA strand?

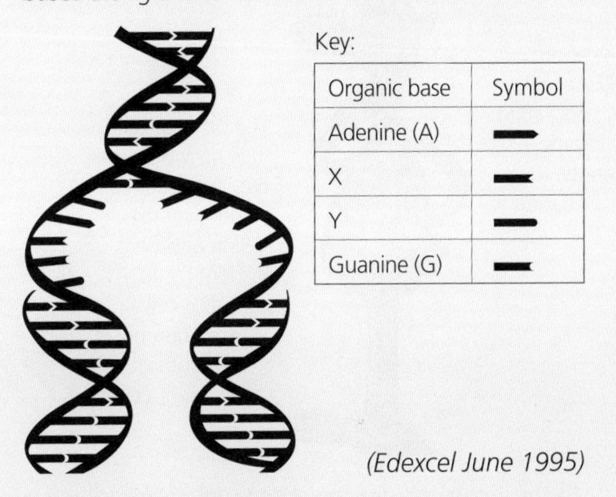

Key:
Organic base | Symbol
Adenine (A)
X
Y
Guanine (G)

(Edexcel June 1995)

3.17 Cell division

OBJECTIVES

- To understand why it is necessary to copy genetic material accurately
- To know that copying division is called mitosis, and results in cells with an identical number and type of chromosomes as their parent cells
- To know how chromosomes behave during mitosis
- To know where mitosis takes place in the bodies of mammals and flowering plants
- To understand the need for a special cell division in the formation of haploid gametes

Mitosis is copying division

Characteristics are transmitted from one generation of cells to the next. For this to happen, the chromosomes must be accurately copied and passed on when cells divide. Each chromosome has a partner, forming **homologous pairs**. Both chromosomes in an homologous pair have the same genes in the same positions. The diagram below shows cell division in which the new cells are copies of the parent cell – **mitosis**.

Mitosis is for growth

Both plants and animals grow by mitosis.

- In animals each tissue provides its own new cells when they are needed, e.g. liver produces liver cells.
- **Stem** cells are unspecialised cells that divide by mitosis to produce daughter cells that can become specialised for specific functions.
- In plants, cell division in the **cambium** increases the plant girth (the plant gets thicker), and cell division in the **meristems** at the tips of the roots and shoots leads to an increase in length (see page 9).

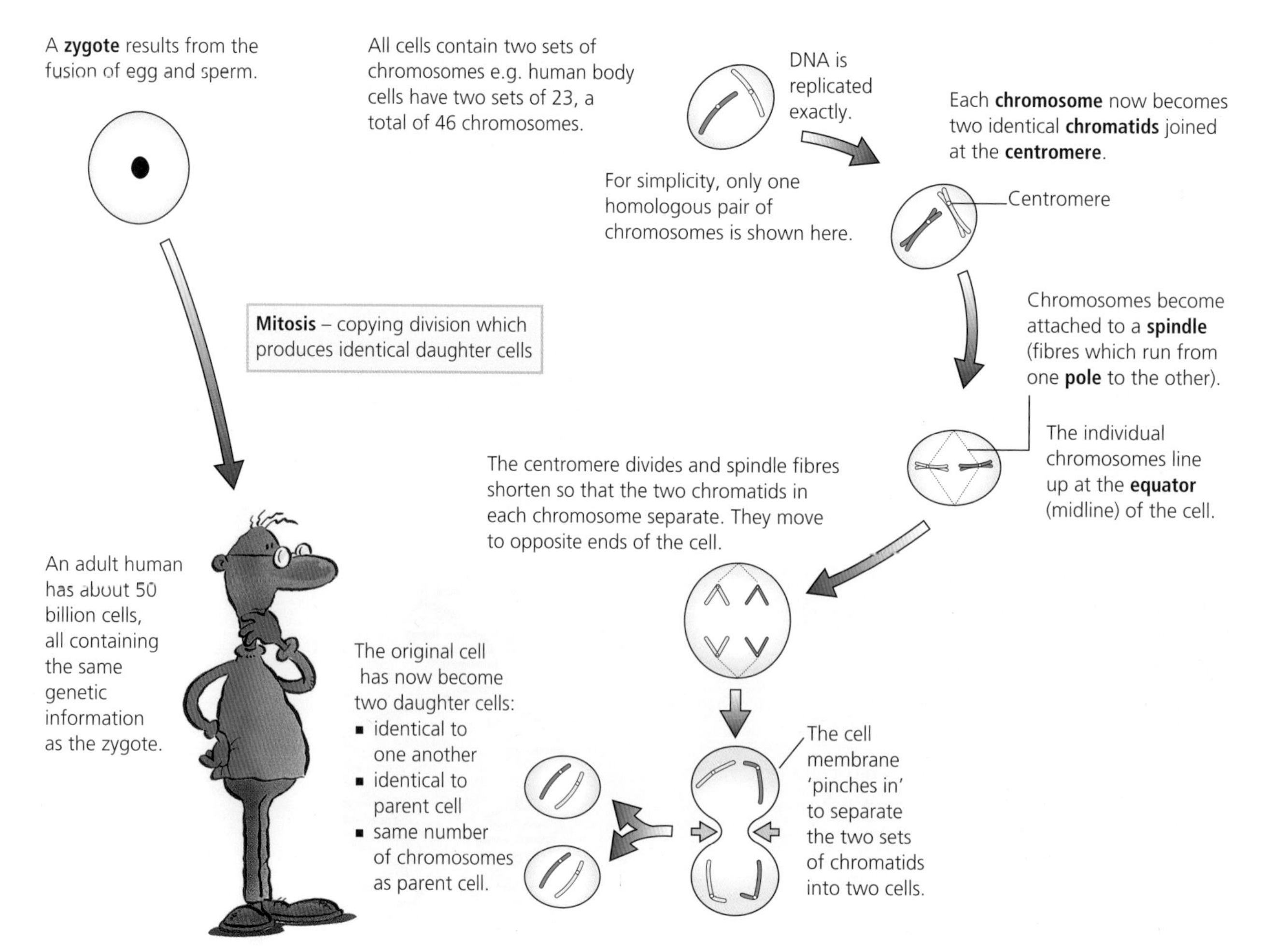

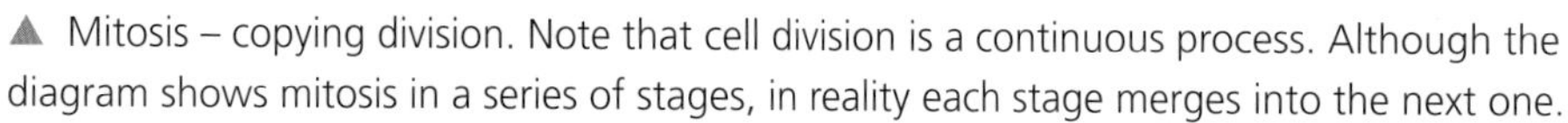

▲ Mitosis – copying division. Note that cell division is a continuous process. Although the diagram shows mitosis in a series of stages, in reality each stage merges into the next one.

Meiosis is reduction division

During sexual reproduction two gametes fuse to form the zygote. The gametes must contain only one set of chromosomes, otherwise the zygote would have twice as many chromosomes as it needed! This principle is outlined in the diagram on the right.

Each cell of an organism has a fixed number of chromosomes within the nucleus. The number of chromosomes in a normal body cell is the **diploid number** (or $2n$); the number of chromosomes in a gamete is the **haploid number** (or n). **Fertilisation** is the **fusion of haploid gametes to restore the diploid number in the zygote.** Gametes are formed by a type of cell division called **meiosis** or **reduction division**, shown in the diagram below.

Meiosis only happens in the gamete-producing organs: the testes and the ovaries in animals, and the pollen sacs of the stamens and the ovules in the ovary in plants.

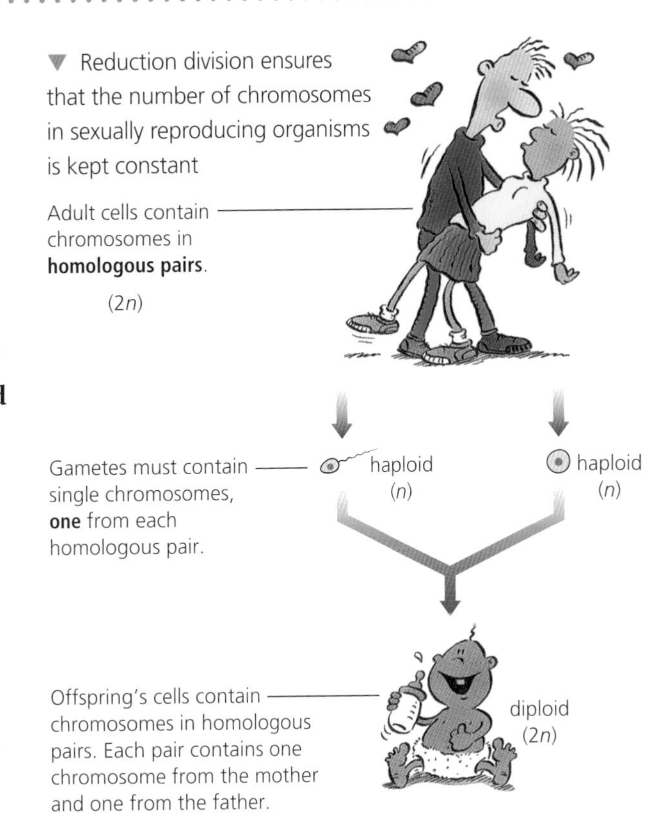

▼ Reduction division ensures that the number of chromosomes in sexually reproducing organisms is kept constant

Meiosis is called reduction division because it halves the number of chromosomes in cells.

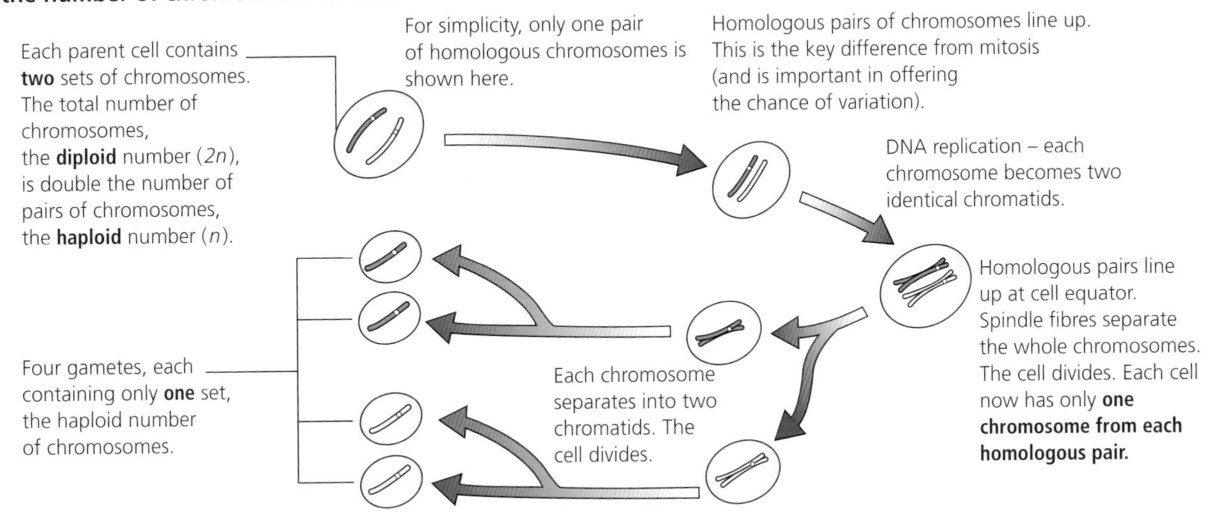

Each parent cell contains **two** sets of chromosomes. The total number of chromosomes, the **diploid** number (2n), is double the number of pairs of chromosomes, the **haploid** number (n).
For simplicity, only one pair of homologous chromosomes is shown here.
Homologous pairs of chromosomes line up. This is the key difference from mitosis (and is important in offering the chance of variation).
DNA replication – each chromosome becomes two identical chromatids.
Homologous pairs line up at cell equator. Spindle fibres separate the whole chromosomes. The cell divides. Each cell now has only **one chromosome from each homologous pair.**
Each chromosome separates into two chromatids. The cell divides.
Four gametes, each containing only **one** set, the haploid number of chromosomes.

Q

1 Cells in bone marrow undergo mitosis. Some of these cells become red blood cells, to replace those lost as they wear out. A typical red blood cell lasts for 120 days before it is removed from the circulation.

a Suggest why red blood cells last for a short time.

b Which organ removes red blood cells from blood?

c This organ stores the main metal ion that forms part of the red blood cell. Which ion is this, and what molecule is it a part of inside the red blood cell?

d The human circulation contains about 5 dm^3 of blood. Each mm^3 of blood contains 5 000 000 red blood cells. (1 dm^3 = 1 000 000 mm^3.)

i Calculate how many red blood cells there are in the human circulation.

ii The total number of red blood cells is replaced every 120 days. Calculate how many cells are replaced each day. How many are replaced each second?

3.18 Inheritance

OBJECTIVES

- To recall the features of sexual and asexual reproduction
- To be able to define the terms gene and allele, homozygous and heterozygous, dominant and recessive

Reproduction: a reminder

Living organisms can pass on their characteristics to the next generation (**reproduce**) in two ways (page 160).

Remember, for sexual reproduction:

- two organisms of the same species, one male and one female are required
- each individual produces sex cells (**gametes**)
- sexual reproduction always involves **fertilisation** – the fusion of the gametes
- offspring receives some genes from each parent, so shows a mixture of parental characteristics.

In sexual reproduction, a mixture of genes is passed from parents to offspring. This handing down of genes is not random, and there are certain rules that govern how genes will be passed on and which ones will show up in the offspring.

The inheritance of characteristics

Chromosomes carry genetic information as a series of **genes**, such as the gene for eye colour, the gene for earlobe shape and the gene for hair texture. Each chromosome in the nucleus of a diploid organism has a partner that carries the same genes. Such a pair of chromosomes is called an **homologous pair**.

Each chromosome in a pair may carry alternative forms of the same gene. These alternative forms are called **alleles**. For example, the gene for eye colour has alleles that code for blue or brown. If both alternative alleles are present in a particular cell nucleus, then the cell is **heterozygous** for that characteristic. On the other hand, if the nucleus carries the same allele on both members of the homologous pair, then the cell is **homozygous**. The meaning of these genetic terms is outlined in the diagram at the top of the next page.

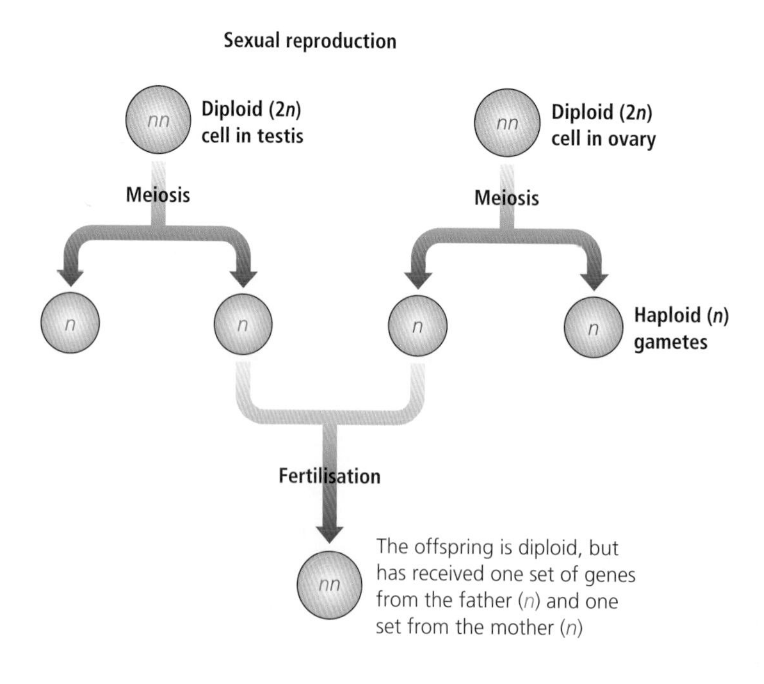

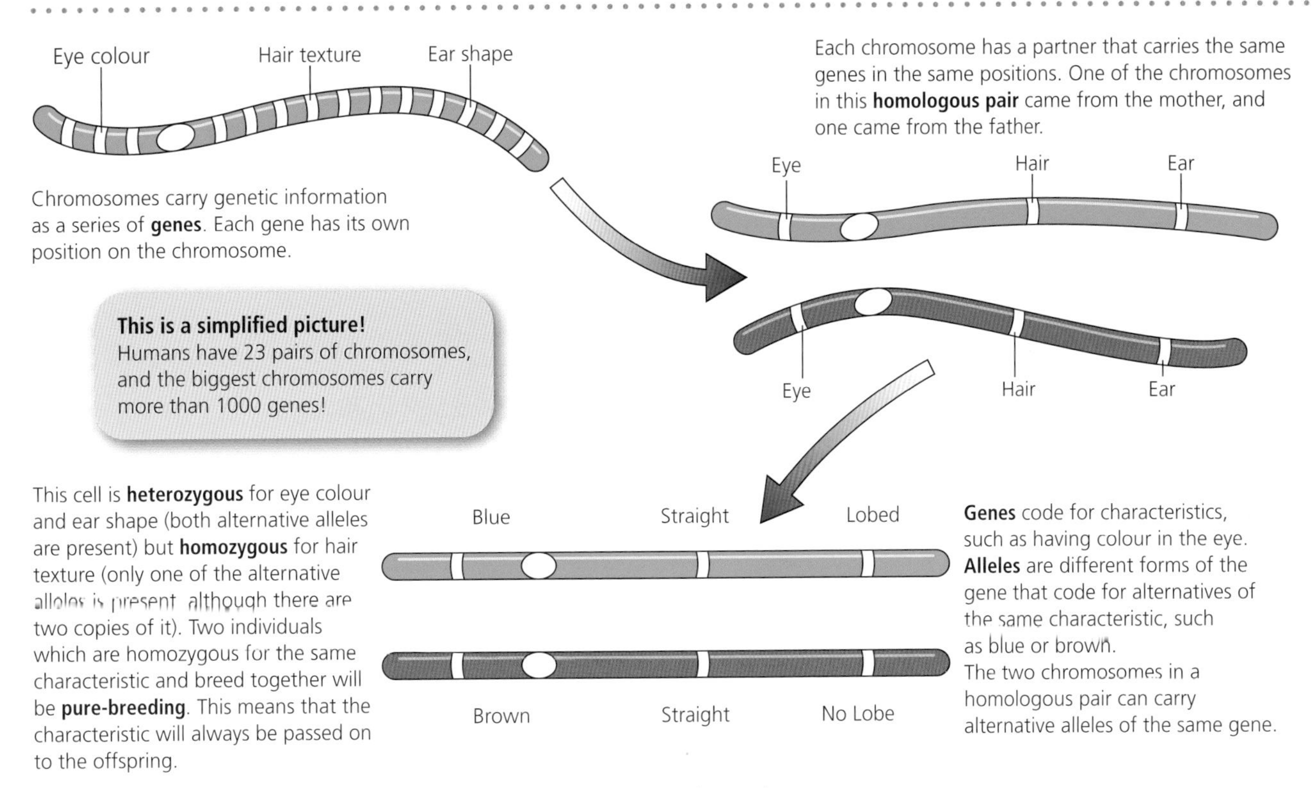

Eye colour
Hair texture
Ear shape
Chromosomes carry genetic information as a series of **genes**. Each gene has its own position on the chromosome.
Each chromosome has a partner that carries the same genes in the same positions. One of the chromosomes in this **homologous pair** came from the mother, and one came from the father.
Eye
Hair
Ear
Eye
Hair
Ear
This is a simplified picture!
Humans have 23 pairs of chromosomes, and the biggest chromosomes carry more than 1000 genes!
This cell is **heterozygous** for eye colour and ear shape (both alternative alleles are present) but **homozygous** for hair texture (only one of the alternative alleles is present, although there are two copies of it). Two individuals which are homozygous for the same characteristic and breed together will be **pure-breeding**. This means that the characteristic will always be passed on to the offspring.
Blue
Straight
Lobed
Brown
Straight
No Lobe
Genes code for characteristics, such as having colour in the eye. **Alleles** are different forms of the gene that code for alternatives of the same characteristic, such as blue or brown.
The two chromosomes in a homologous pair can carry alternative alleles of the same gene.

▲ Diploid organisms have homologous pairs of chromosomes in the nucleus.

In a heterozygous cell, when the members of the homologous pair separate during meiosis the gametes will contain different alleles. This means that when gametes fuse at fertilisation, the resulting zygote may have a number of different possible allele combinations. The production of gametes and the formation of different zygotes in this way are explained in the diagram below.

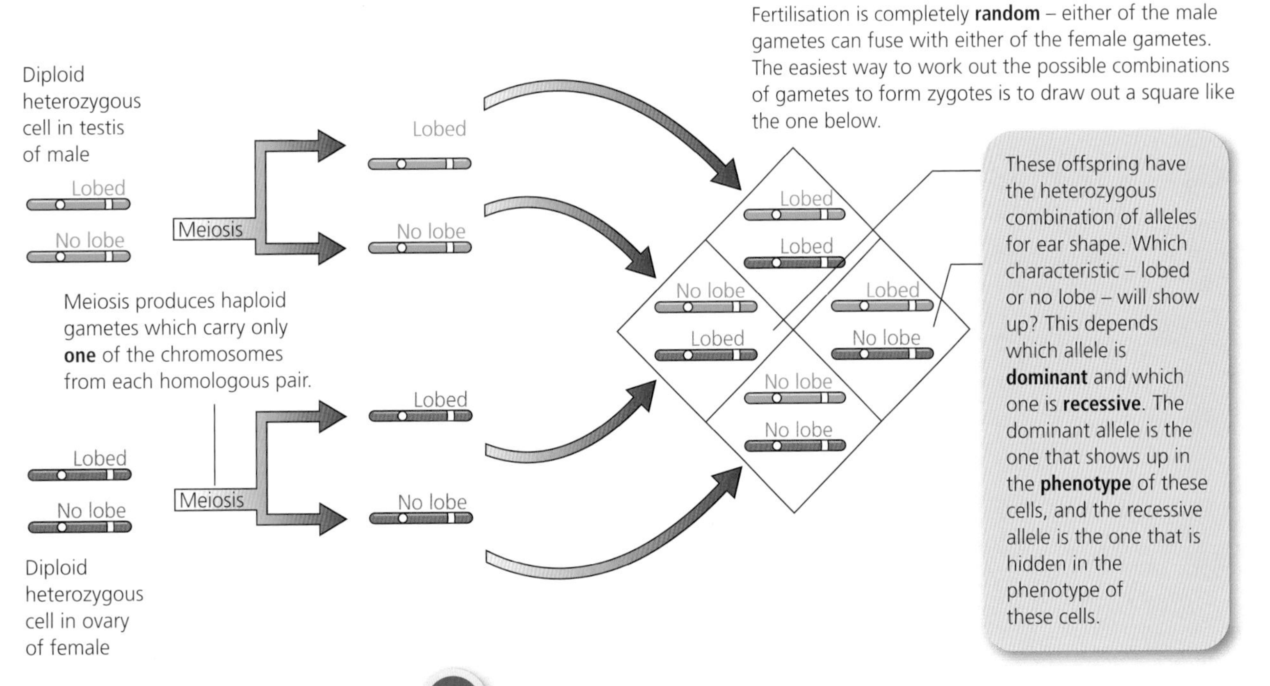

▲ The recombination of alleles during sexual reproduction gives rise to a great variation among offspring. This leads to the evolution of new strains, races and, eventually, new species.

Q

1. Define the terms homologous pair, heterozygous and homozygous.
2. What is the difference between a gene and an allele?
3. What is a meant by a dominant allele?

3.19 Studying patterns of inheritance

OBJECTIVES

- To understand the method for describing genetic crosses
- To know the result of crosses involving two heterozygous parents
- To understand the principle of the test cross

Scientists called **geneticists** study the inheritance of characteristics by carrying out breeding experiments. There is a conventional pattern for describing the results of such experiments – a sort of genetic shorthand, shown in the example below.

At fertilisation, any male gamete can fuse with any female gamete

Drawing out chromosomes carrying alleles of genes is very time-consuming. Geneticists write out the stages of crosses using symbols to replace the chromosomes and genes. These symbols should always be identified at the start of a cross.

Let **B** = brown and **b** = blue

The **capital** letter is used for the **dominant** allele.

Brown (**B**) and blue (**b**) are **alleles** of the **gene** for eye colour.

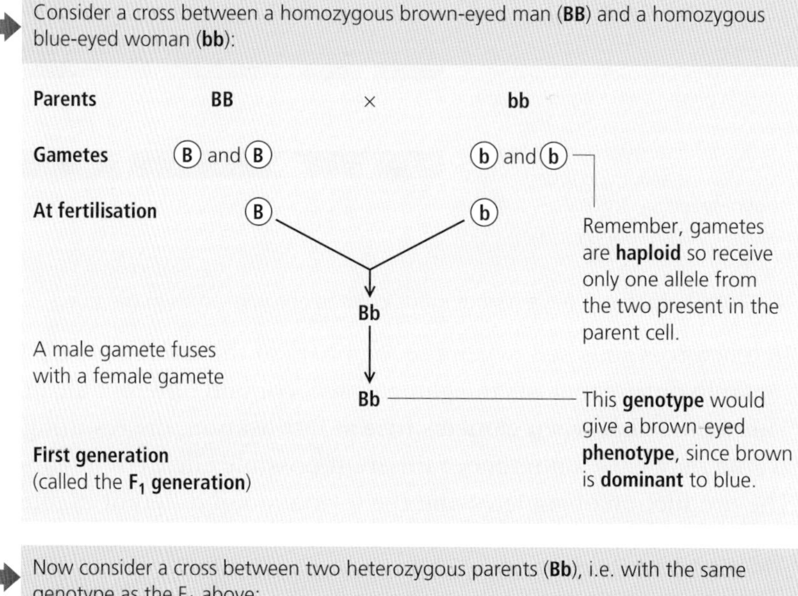

Now consider a cross between two heterozygous parents (**Bb**), i.e. with the same genotype as the F_1 above:

Parents Bb × Bb

Gametes (B) and (b) (B) and (b)

At fertilisation, any male gamete can fuse with any female gamete.

At fertilisation

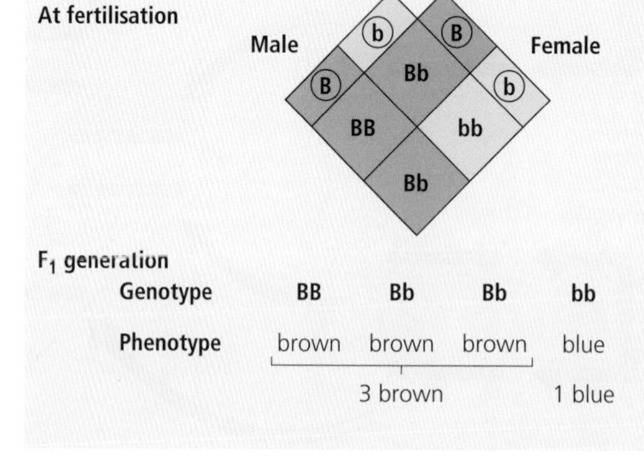

In theory a cross between two heterozygous parents should produce offspring in the ratio of 3 showing dominant to 1 showing recessive. This can be restated as: 'The probability of any offspring showing dominant is 3/4 or 75%; the probability of it showing recessive is 1/4 or 25%'.

The inheritance of eye colour. There are two important points about this cross.

1. These results are **probabilities** (chances). The offspring *should* be in the 3:1 ratio shown, but each fertilisation is random so they may not be. This ratio is more likely to be seen in very large numbers of offspring. For human families a 3:1 ratio is unlikely since very few mothers give birth to four children at one time.
2. Every cross between the same two parents is a different event. If two heterozygous parents produce a child with blue eyes (a 1/4 probability) there is still a 1/4 probability that their next child will have blue eyes.

Test cross

In the eye colour example, both of the genotypes **BB** and **Bb** give the same phenotype – brown eyes. It might be important to know whether a particular organism is homozygous or heterozygous, particularly in the breeding of domestic animals. To do this, geneticists use a **test cross** (often called a **back cross to the recessive**). The principle is outlined on the right.

Reminder!	
Ratio: dominant to recessive	**Phenotypes of parents**
3:1	Both heterozygous
1:1	One heterozygous, one homozygous recessive

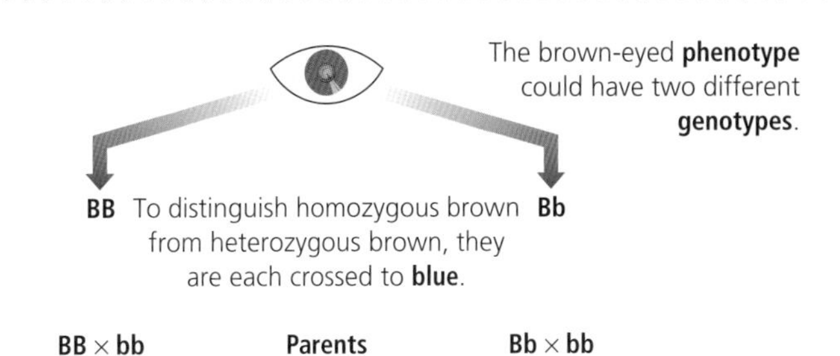
The brown-eyed **phenotype** could have two different **genotypes**.
BB
Bb
To distinguish homozygous brown from heterozygous brown, they are each crossed to **blue**.

BB × bb	**Parents**	**Bb × bb**
Ⓑ or Ⓑ ⓑ or ⓑ	**Gametes**	Ⓑ or ⓑ* ⓑ or ⓑ
All **Bb**	**At fertilisation**	**Bb** or **bb**
All brown eyed	**F_1 generation**	**1 brown eyed: 1 blue eyed**

*The heterozygous brown can supply the **b** allele for a gamete even though it is 'hidden' in the phenotype of the diploid parent.

If **any** offspring showing the recessive characteristic result from a test cross, the parent **must** have been heterozygous.

▲ A test cross can distinguish different genotypes with the same phenotype

The results of genetic crosses are sometimes shown as a **pedigree**.

Rules for showing a pedigree ('family tree')

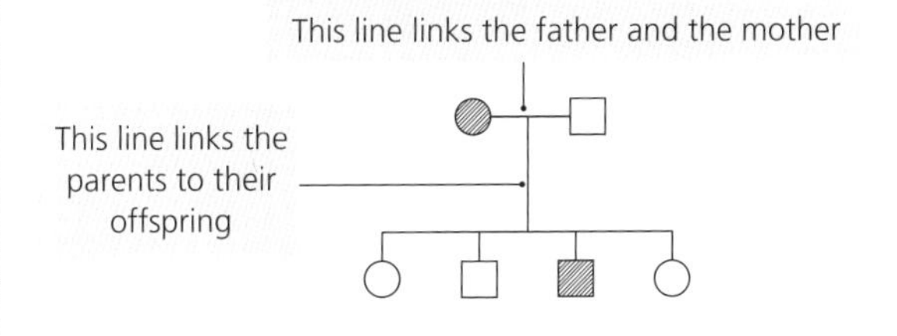
This line links the father and the mother
This line links the parents to their offspring
This shows the offspring from the same mother and father.

A key is used to identify different phenotypes:

○ = 'normal' female
● = 'affected' female
□ = 'normal' male
■ = 'affected' male

SOLVING PEDIGREES – AN EXAMPLE

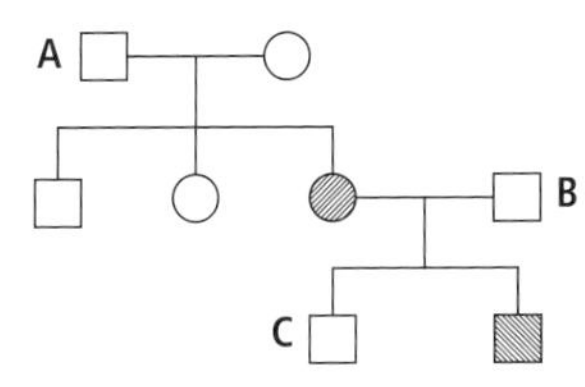
A
B
C

◀ This family tree ('pedigree') refers to the inheritance of fur colour in rabbits. Brown fur is dominant to white fur. What can you deduce about the genotype of individuals A, B and C?

- Let B = brown and b = white
- ● and ■ must be bb, this is the starting point since the only certainly is that animals with the 'recessive' phenotype must have the homozygous recessive genotype.
- A must be Bb because he has brown fur but can pass on the b allele to one of his daughters (his partner must be Bb too!).
- B must be Bb for the same reason as A is.
- C must be Bb because he has brown fur but must have received the b allele from his mother.

Q

1 Draw a diagram to explain how two brown-eyed parents can have a blue-eyed child.

2 Gregor Mendel suggested that a cross between two heterozygous individuals produces offspring in a ratio of 3 showing the dominant characteristic to 1 showing the recessive characteristic. Explain why such crosses rarely give an exact 3:1 ratio.

3 Use a suitable example to explain the value of a test cross.

3.20 Inherited medical conditions and codominance

OBJECTIVES

- To know some examples of inherited conditions
- To understand inheritance when neither allele is dominant

Inherited medical conditions

There are several important medical examples of monohybrid inheritance, including albinism and sickle cell anaemia.

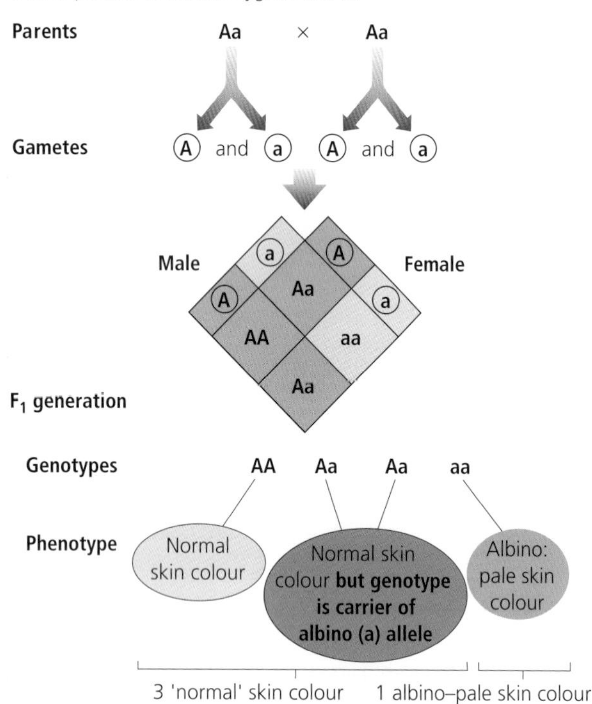

- **Albinism** is caused by a recessive allele. Heterozygous individuals are not affected by the condition, but they are **carriers** of the mutant, recessive allele.
  People with albinism do not produce the pigment **melanin**. This pigment normally makes skin, the iris of the eye and hair a dark colour. These people have pale skin, light hair and pink eyes (the pink is actually the blood in the retina showing up). They are very sensitive to bright light and the skin burns very easily in sunlight.
- **Sickle cell anaemia** is a condition in which a homozygote has the disease, but a heterozygous individual may gain some benefit in certain environments (see also page 205).

Sickle cell anaemia – carriers are anaemic but are resistant to malaria

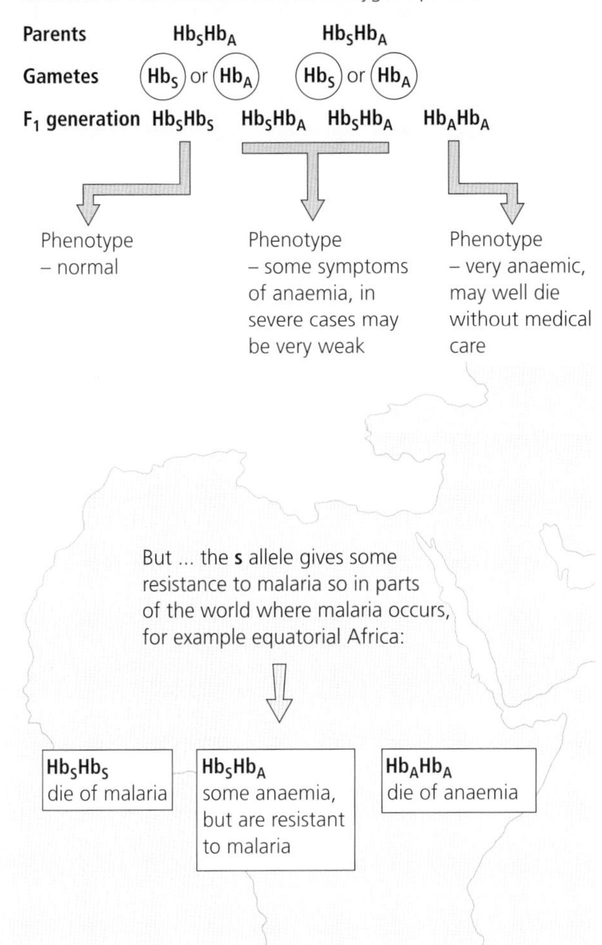

The inheritance of these two conditions is described on these two pages.

Codominance

Some genes have more than two alleles. For example, the gene controlling the human ABO blood groups has three alleles, given the symbols $\mathbf{I^A}$, $\mathbf{I^B}$ and $\mathbf{I^O}$. Neither of the $\mathbf{I^A}$ and $\mathbf{I^B}$ alleles is dominant to the other, although they are both dominant to $\mathbf{I^o}$. This is called **codominance**. It results in an extra phenotype when both alleles are present together. The genotypes and phenotypes are shown in the table.

Genotype	Phenotype
I^AI^A or I^AI^O	Blood group A
I^BI^B or I^BI^O	Blood group B
I^AI^B	Blood group AB
I^OI^O	Blood group O

The human blood groups are easily detected by a simple test on a blood sample.

Q

1 Cystic fibrosis is a common inherited disease caused by a recessive allele. A blood test can detect this allele. A man and his wife were both found to be carriers of the cystic fibrosis allele.

a What is meant by a carrier of the allele?

b Draw a genetic diagram to show the inheritance of cystic fibrosis in any children of this couple. What is the chance that a child will have cystic fibrosis?

2 Huntington's disease (HD) is caused by a dominant allele (**H**) – the other allele (**h**) results in normal working of the central nervous system. This diagram shows how one family is affected by HD.

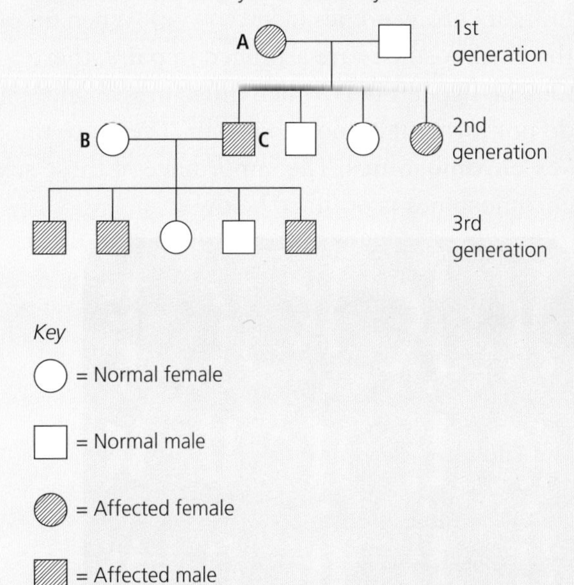

a What was the genotype of the grandmother, **A**? Explain your reasoning.

b Draw a genetic diagram to show how parents **B** and **C** passed on HD to their children.

c If parents **B** and **C** have a sixth child, what are the chances that it will have HD?

d Most serious genetic diseases are caused by recessive alleles. Explain why a dominant allele that causes a serious disease may quickly disappear from a population.

3 **a** Draw a genetic diagram to explain the inheritance of blood group in the Wilson family. Mr Wilson has the genotype I^AI^B and Mrs Wilson has the genotype I^OI^O.

b What is the probability of the Wilsons' first child being female?

c What is the probability of this child being female and having blood group A?

d A person with alleles I^A and I^B shows the effect of both alleles in the phenotype. What term is used to describe this?

3.21 Sex is determined by X and Y chromosomes

OBJECTIVES

- To understand how sex is determined and inherited in humans
- To know that some genes are sex linked
- To understand the pattern of inheritance of sex-linked genes

The photograph below shows a complete set of human chromosomes (a karyotype). When all of the chromosomes are arranged in pairs, there may be two left over which differ in size and do not form an homologous pair. These are the **sex chromosomes**. The importance of these sex chromosomes is outlined below.

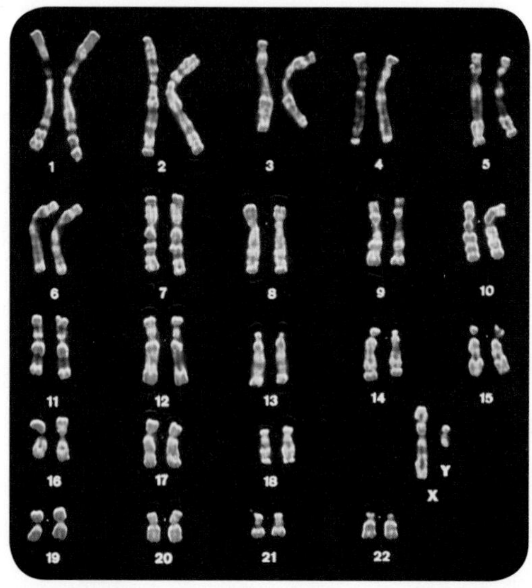

▲ A **karyotype** – a photograph of human chromosomes arranged in pairs. The 23rd pair is an X and a Y chromosome – this cell comes from a male.

Note that:

- The man's sperm determines the sex of his children since the woman can only produce gametes with an X chromosome.
- The sex chromosomes carry genes concerned with sexual development, such as development of the sex organs and position of fat stores.
- The sex chromosomes also carry a few genes that code for characteristics that are not concerned with sex. Since these genes are carried on the sex chromosomes, they may show their characteristics only in one sex.

For example the gene for **colour blindness** is carried on the X chromosome. We say that colour blindness is **sex linked**.

▼ **The inheritance of sex**

Each human body cell has 46 chromosomes. There are 22 **pairs** of chromosomes plus another two chromosomes which may not look alike. These are the **sex chromosomes**. Female cells have two sex chromosomes that are alike (called XX) and male cells have two sex chromosomes that are not alike (XY).

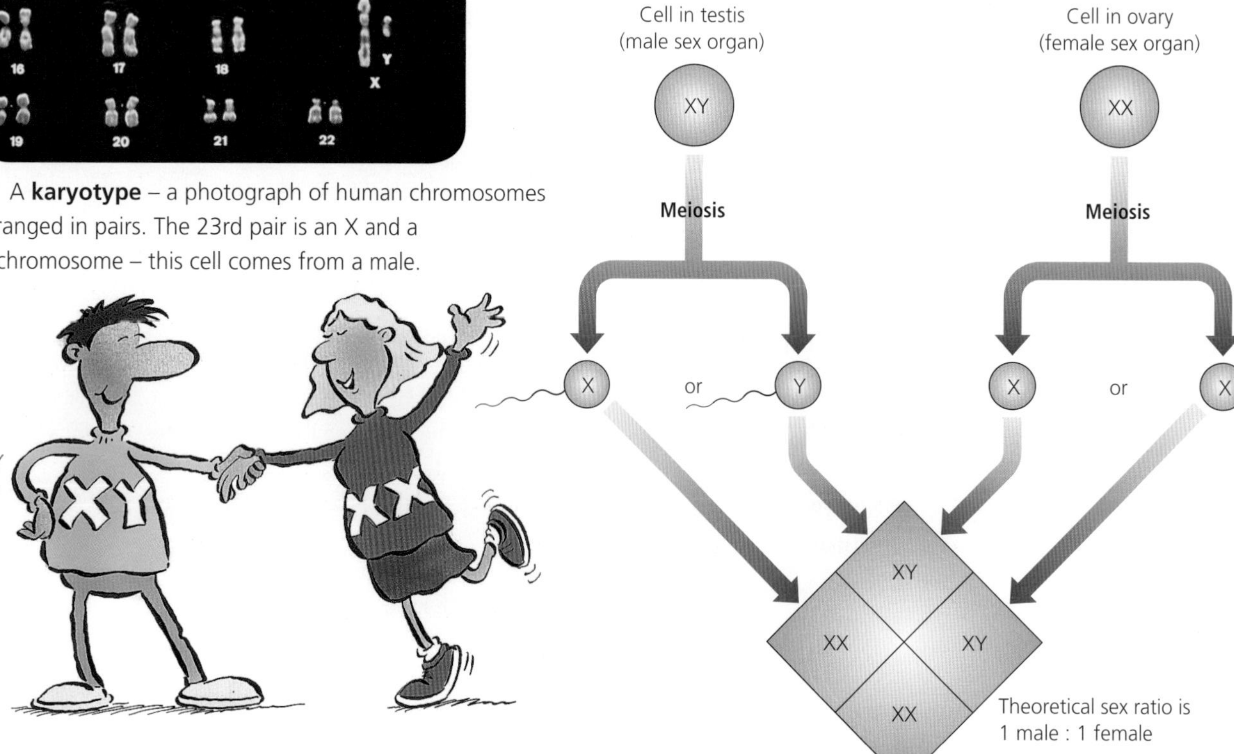

The inheritance of sex-linked characteristics

One well-known sex-linked characteristic is red-green colour blindness. This is a disease in which the person cannot tell the difference between red and green. It is an X-linked condition. The inheritance of this disease is shown below. Haemophilia (a failure in blood clotting) is inherited in a similar way – there are many more males than females who cannot clot their blood efficiently.

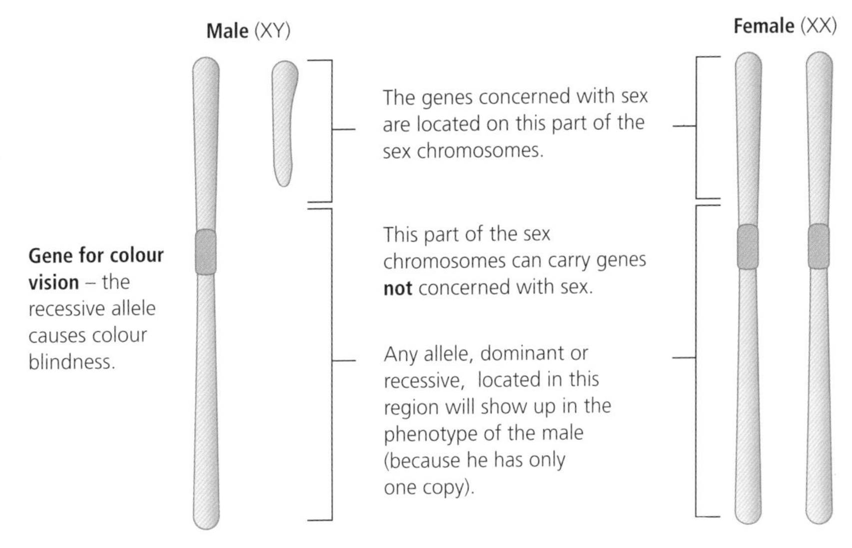

Male (XY)
Female (XX)
The genes concerned with sex are located on this part of the sex chromosomes.
This part of the sex chromosomes can carry genes **not** concerned with sex.
Any allele, dominant or recessive, located in this region will show up in the phenotype of the male (because he has only one copy).
Gene for colour vision – the recessive allele causes colour blindness.

▲ Any gene carried on the sex chromosomes is sex-linked

The inheritance of colour blindness

Let **C** = normal allele and
c = mutant allele.

Because these alleles are carried on the X chromosome, it is necessary to show the sex chromosomes in the pattern of inheritance. For example:

Female

X^CX^c

Female, so two X chromosomes — Heterozygous for colour blindness gene

Male

X^CY — Remember – no 'colour blindness' gene on the Y chromosome

Male, so only **one** X chromosome

Airline and military pilots must be tested for colour blindness, red and green signals mean very different things!

Consider a cross between a carrier woman and a normal man:

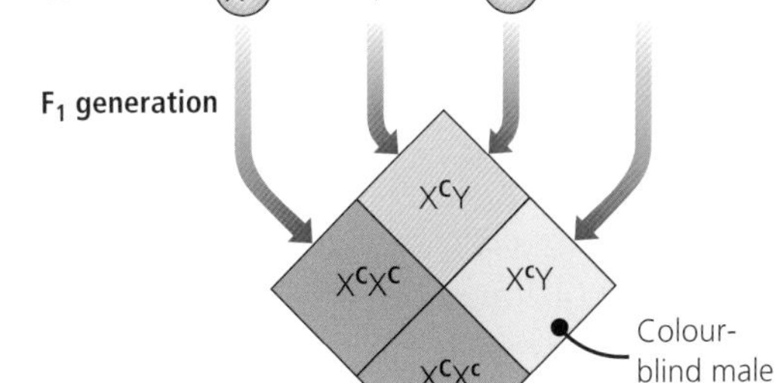

Parents X^CY × X^CX^c
Male: not colour blind
Female: carrier
Gametes X^C Y X^C X^c
F_1 generation
X^CY
X^CX^C
X^cY
X^CX^c
Colour-blind male
Carrier Female

For haemophilia

Parents X^HX^h × X^HY

Female – normal phenotype but a carrier of the mutant allele

Male – normal phenotype

Gametes X^H X^h X^H Y

F_1 generation

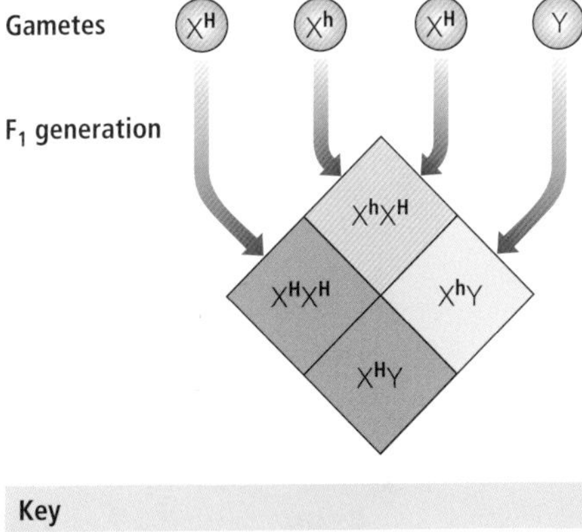

X^hX^H
X^HX^H
X^hY
X^HY

Key

Carrier female
Normal male/female
Male with haemophilia

1 Use a simple genetic diagram to explain why there are approximately equal numbers of male and female babies.

2 Why are males more likely to have red–green colour blindness than females? How could you explain, genetically, a colour-blind female?

3 Haemophilia is a sex-linked characteristic. The diagram above shows how the allele for haemophilia is inherited.

a Explain why the mother is described as a carrier of this condition.

b If she has one child, what is the probability of her having a haemophiliac son?

Questions on inheritance

1 The figure below shows family tree for a condition known as nail-patella syndrome (NPS).

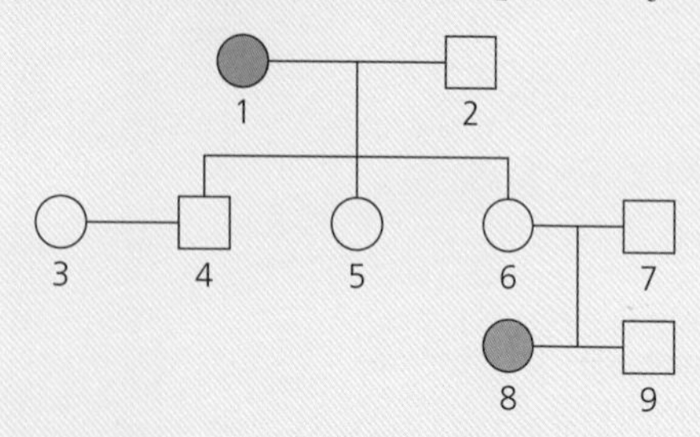

Key

- female without NPS
- female with NPS
- male without NPS
- female with NPS

a **i** State whether NPS is controlled by a dominant or a recessive allele.

ii Explain which evidence from the family tree confirms your answer to **(i)**.

b Explain what the chances are for a third child of parents 6 and 7 having NPS. You may use a genetic diagram to help your explanation.

Cambridge IGCSE Biology 0610
Paper 2 Q7 May 2008

2 Choose words from the list to complete each of the spaces in the paragraph. Each word may be used once only and some words are not used all.

allele	**diploid**	**dominant**
gene	**haploid**	**heterozygous**
homozygous	**meiosis**	**mitosis**
recessive		

In humans there is a condition known as cystic fibrosis.

This is controlled by a single ________________ which has two forms. One form causes cystic fibrosis while the other does not.

Gametes are formed by ________________. When two humans reproduce, their gametes fuse at fertilisation to form a ________________ zygote.

Neither of the two humans has cystic fibrosis but one of their three children does have the condition. This means that cystic fibrosis is controlled by a ________________ allele and that each of the parents is ________________.

Cambridge IGCSE Biology 0610
Paper 2 Q4 November 2008

3 **a** Describe the effect sickle cell anaemia has on red blood cells.

b **i** The allele for normal haemoglobin production is $\mathbf{I^N}$. The allele for sickle cell haemoglobin production is $\mathbf{I^S}$. Two parents who are heterozygous have a child. With the help of a genetic diagram, predict the probability that this child would be heterozygous.

ii Explain why, under some circumstances, people who are heterozygous for this condition have a greater chance of survival than homozygous people.

Cambridge IGCSE Biology 0610
Paper 3 Q7 November 2005

4 **a** Select the correct term from the list below and write it in the box next to its description.

allele	**dominant**	**gene**
genotype	**heterozygous**	**homozygous**
phenotype	**recessive**	

description	term
a form of a gene that always has its effect when it is present	
a form of a gene that codes for one of a pair of contrasting features	
an organism having two different forms of a gene for a particular feature	
the alleles that an organism has in its chromosomes	

b Two red flowered plants were crossed. The seeds produced were germinated and grew into 62 white flowered plants and 188 red flowered plants.

i Which flower colour is controlled by the recessive form of the gene?

ii Using the symbols **R** and **r**, construct a genetic diagram to explain the results of this cross.

iii One of the white flowered offspring was crossed with a red flowered offspring. Predict the two possible ratios of red and white flowered plants that their seeds would produce.

Cambridge IGCSE Biology 0610
Paper 2 Q5 May 2005

5 The figure shows a method used by a student to understand how characteristics are inherited when two plants of species **X** are crossed. Both cubes had three of their faces marked with the letter **T** and three with the letter **t**.

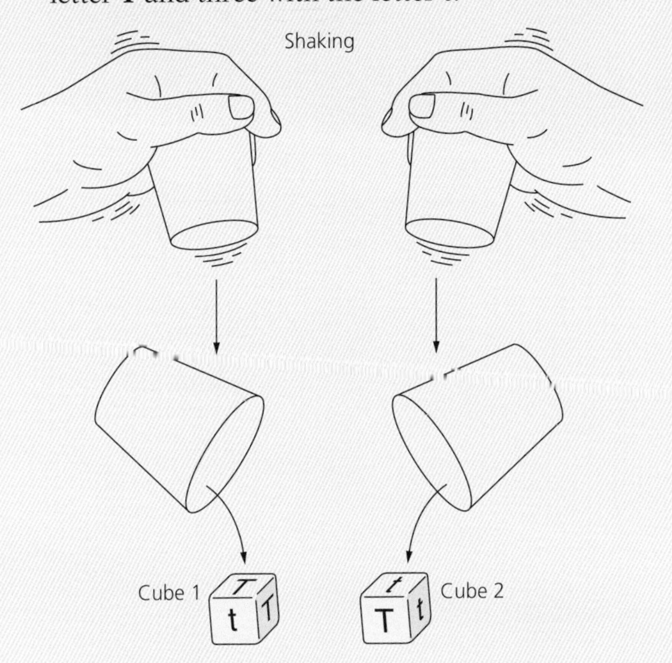

In this example, the letters appearing on the upper faces are **Tt**.

The student shook each container and then tipped both cubes out at the same time and recorded the letters appearing on the upper faces of the cubes.

The student tipped both cubes out a total of 405 times.

a **i** Copy and complete the table to show the results obtained.

letters appearing on the upper faces of the cubes	**tt**	**TT**	**Tt**
number of times each pair of letters appeared	98	....	202

ii State what the letters on the faces of the cubes represent.

iii State the reason for shaking the containers.

b In living plants of species **X**, items **T** and **t** control flower colour. The genetic diagrams below show a sequence of crosses (**V** and **W**) between plants of species **X**, some with red flowers and some with yellow flowers.

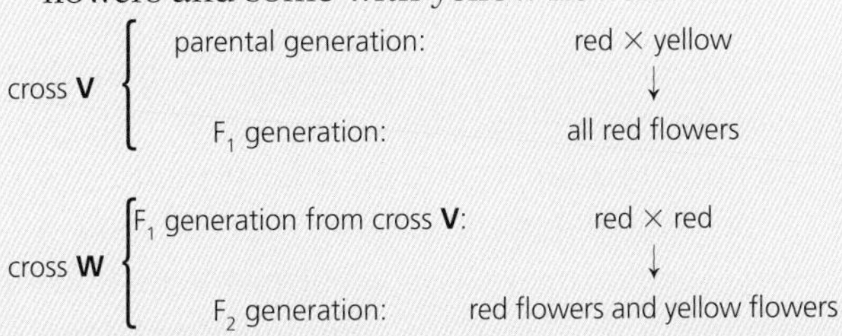

Using the information in the table above and crosses **V** and **W**,

i state which colour is dominant and which of cross **V** and cross **W** could produce the results seen in the table.

ii Explain why you chose this cross.

c **i** Using the letters **T** and **t**, draw a full genetic diagram for another cross, **Y**, between plants of species **X** shown below.

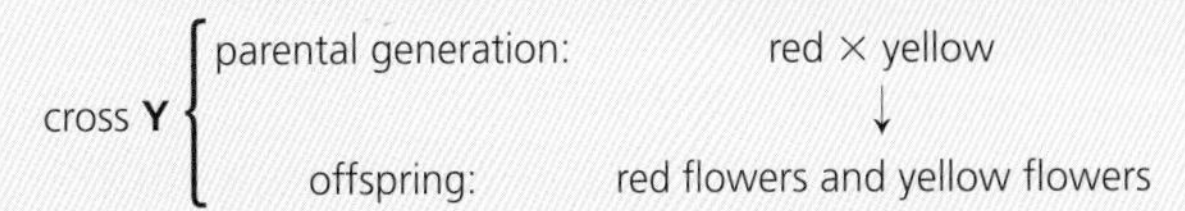

ii In the method shown in the figure above, the student put **T** on three faces and **t** on the other three faces of each cube, so that cube 1 had **T**, **T**, **T**, **t**, **t**, **t** and cube 2 **T**, **T**, **T**, **t**, **t**, **t**.

State what letters the student should write on the faces of each cube to represent the cross in **(c) (i)**.

Cambridge O Level Biology 5090 Paper 2 Q5 May 2006

6 The gene for the ABO blood group has three alleles, I^A, I^B and I^o.

a A person with blood group O has parents who have blood groups A and B. Complete the genetic diagram to show how this is possible.

Use the symbols, I^A, I^B and I^o, for the blood group alleles.

parental phenotypes	blood group A	×	blood group B
parental genotypes	....	×	....
gametes	....	+	....
offspring genotype		....	
offspring phenotype		blood group O	

b Use your answer to **a** to give examples of the following. The first one has been completed for you.

term	example
a dominant allele	I^A
Heterozygous genotype	....
codominant alleles	....
phenotype	....

3.22 Variation

OBJECTIVES

- To recall that living organisms differ from one another
- To distinguish between continuous and discontinuous variation

Living organisms differ from one another. Even members of the same species have slightly different sets of characteristics. Some of these differences are inherited from their parents, and others are the result of the environment. The differences between individuals of the same species are called **variations**. Scientists who study variation are interested in questions such as:

- Are all of the variations of the same type?
- How do the variations come about?
- How are the variations in characteristics passed on from parents to offspring?
- What is the importance of these variations?

There are two types of variation – **discontinuous variation** and **continuous variation**.

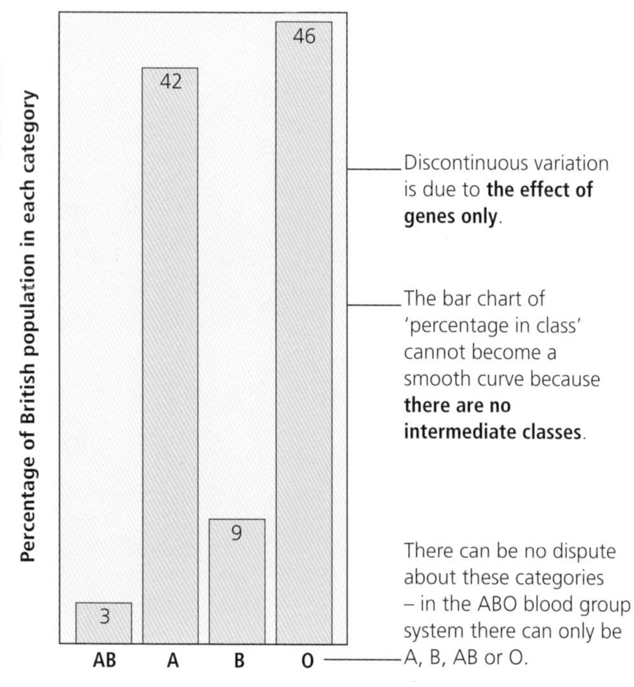
Percentage of British population in each category
3
42
9
46
AB
A
B
O
Discontinuous variation is due to the effect of genes only.
The bar chart of 'percentage in class' cannot become a smooth curve because there are no intermediate classes.
There can be no dispute about these categories – in the ABO blood group system there can only be A, B, AB or O.

▲ Human blood groups are an example of discontinuous variation. Another example is gender – you are either male or female.

Discontinuous variation

Characteristics that show discontinuous variation have several features:

- An organism either has the characteristic or it doesn't have it. There is no range of these characteristics between extremes. An organism can easily be placed into definite categories, and there is no disagreement about the categories.
- These characteristics are usually **qualitative** – they cannot be measured.
- They are the result of genes only – they are not affected by the environment.

An example of discontinuous variation is shown in the diagram above right.

Continuous variation

Characteristics that show continuous variation have different features:

- Every organism within one species shows the characteristic, but to a different extent. The characteristic can have any value within a range. Different scientists might well disagree about which category any single organism falls into.
- These characteristics are usually **quantitative** – they can be measured.
- They result from several genes acting together, or from both genes and the environment.

An example of continuous variation is shown in the diagram at the top of the opposite page.

Characteristics can be both discontinuous and continuous

Some characteristics are difficult to classify as either discontinuous or continuous variation. Human hair colour, for example, appears in a range from black to blond with many intermediate colours (it shows continuous variation as a result of the involvement of many genes). However, the gene for red hair is masked by every other hair colour gene (which gives a discontinuous situation – hair is either red or not red). Eye colour can be identified as brown or not brown, which would classify it as a discontinuous variation, or it can be put into a range of many intermediate classes, which would make it a continuous variation. A simple guideline is: 'if it can be measured and given a numerical value, it is a continuous variation'.

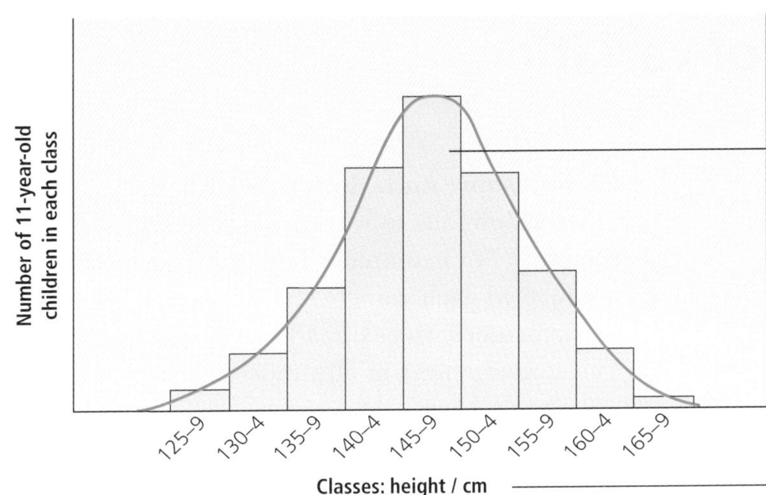

Continuous variation is due to the combined effects of **genes** and the **environment**.

The bar chart of 'number in class' can be redrawn as a smooth curve because there are **many possible intermediate classes between the two extremes** (the curve becomes smoother if the classes become smaller, e.g. 1 cm rather than 5 cm).

There could be a dispute about the boundary of these classes. One observer might use, for example, 127–31, 132–6, etc. rather than the boundaries shown.

▲ Human height is an example of continuous variation. Other examples include quantitative characteristics such as chest circumference, body mass and hand span.

Francis Galton was a cousin of Charles Darwin. He investigated intelligence as an example of phenotype, and believed that 'nature' (genotype) was much more important than 'nurture' (environment). He suggested that:

a children from families where the parents were uneducated would always have low intelligence

b females were less intelligent than males – he suggested the 'ideal' child would be born to 'a man of genius and a woman of wealth'.

His ideas have rightly been updated, but he does deserve some credit – his studies showed that identical twins were genetically the same.

Phenotype and genotype

Variations in characteristics allow us to recognise different organisms, and to place them in different categories. The overall appearance of an organism is a result of the characteristics that it has inherited from its parents and the characteristics that result from the effects of the environment. The following equation summarises this:

phenotype	=	**genotype**	+	**effects of the environment**
the observable characteristics of an organism		the full set of genes it possesses		

Some characteristics result from both genes and environment. For example, a bean seedling has the genes to develop chlorophyll (and turn green) but it won't do so unless it receives enough light. A young mammal has the genes to develop a rigid bony skeleton, but it won't do so unless it receives calcium in its diet.

Q

1 Copy and complete the following paragraph.
Variation occurs in two forms, ____, which shows clear-cut separation between groups showing this variation, and ____, in which there are many intermediate forms between the extremes of the characteristic. The first of these is the result of ____ alone, whilst the second is also affected by ____ factors. The sum of the genes that an organism contains is called its ____ and the total of all its observable characteristics is called the ____.
The two are related in a simple equation: ____ equals ____ plus ____.

2 Which of the following is an example of discontinuous variation?
Body mass, chest circumference, blood group, hairstyle, height
Explain:
a why you chose one of these characteristics
b why you rejected the others.

3 Two students in the first year of secondary school were carrying out a mathematical investigation. They decided to measure the heights of all of the other students in their class. The results are shown in the table below.
a Plot these results as a bar chart.
b Does this illustrate continuous or discontinuous variation? Explain your answer.
c Suggest one characteristic that the students could have recorded to illustrate the other kind of variation.

Height category / cm	Number in category
121–125	2
126–130	4
131–135	9
136–140	6
141–145	4
146–150	1

3.23 Causes of variation

OBJECTIVES

- To identify mutation and sexual reproduction as sources of variation
- To understand that mutations may involve whole chromosomes or genes within them
- To recognise that environmental factors may increase the likelihood of mutation

Permanent changes to the phenotype

Permanent characteristics that can be inherited are due to the genetic make-up of an organism. This may be altered, thereby increasing variation, as a result of **mutation** or of **sexual reproduction**.

Mutation

A **mutation** is a change in a gene or chromosome and can arise because of:

- mistakes in the copying of DNA as cells get ready to divide – pairing with the 'incorrect' base
- damage to the DNA – some environmental factor might alter the bases present in the DNA
- uneven distribution of chromosomes during the division of cells.

Chromosome mutations occur when cell division fails to work with complete accuracy. For example, when human gametes are formed each gamete should receive 23 chromosomes. Occasionally an error occurs. The consequences of chromosome mutations are often serious.

Gene mutations occur when part of the base sequence of the DNA on a single chromosome is changed. As a result a defective protein may be produced, or no protein at all. This can lead to a considerable change in a characteristic. There are many examples, including **sickle cell anaemia**, shown below.

Beneficial mutations

Not all mutations are harmful. Many of them give benefits to the organisms that have them, and aid adaptation to the environment (see page 214). Some may cause harm in one environment but be a benefit in another! Sickle cell anaemia is an example of this (see page 205) people who are heterozygous for sickle cell anaemia are resistant to malaria.

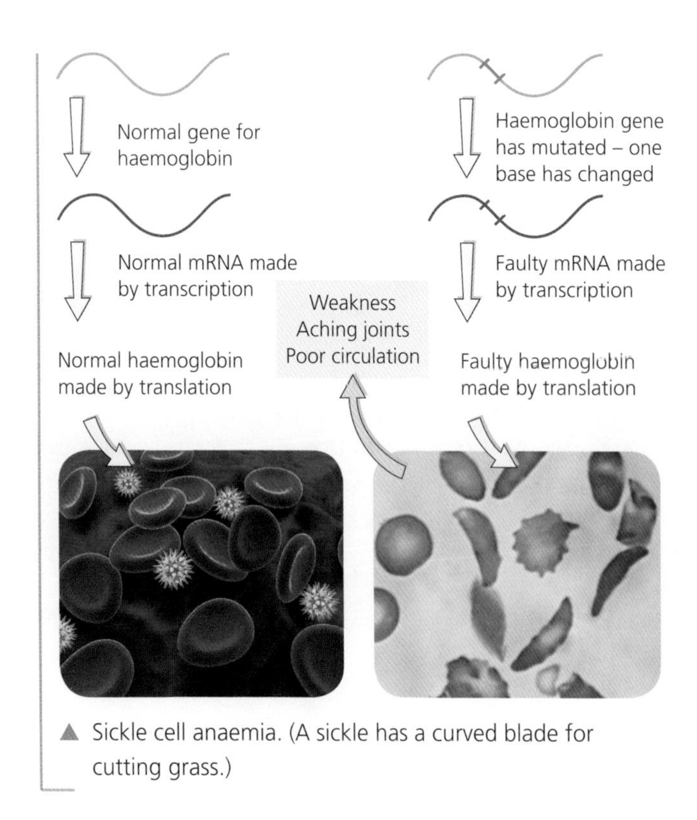

▲ Sickle cell anaemia. (A sickle has a curved blade for cutting grass.)

Radiation can increase mutation rates

Mutations occur spontaneously (for no apparent reason), though they are very rare events. However, a number of factors (called **mutagens**) can increase the rate of mutation. Important mutagens are:

- **radiation** – gamma, ultraviolet and X-radiation can all damage DNA and so cause mutations
- **chemicals** – tars in tobacco smoke, high concentrations of some preservatives and some plant control hormones can cause mutation.

Mutations may be linked with cancer. A mutagen that causes uncontrolled cell division is called a **carcinogen** ('cancer maker').

Sexual reproduction leads to variation

Sexual reproduction mixes up genetic material in three ways, as shown below, producing new genotypes and so variations in phenotype.

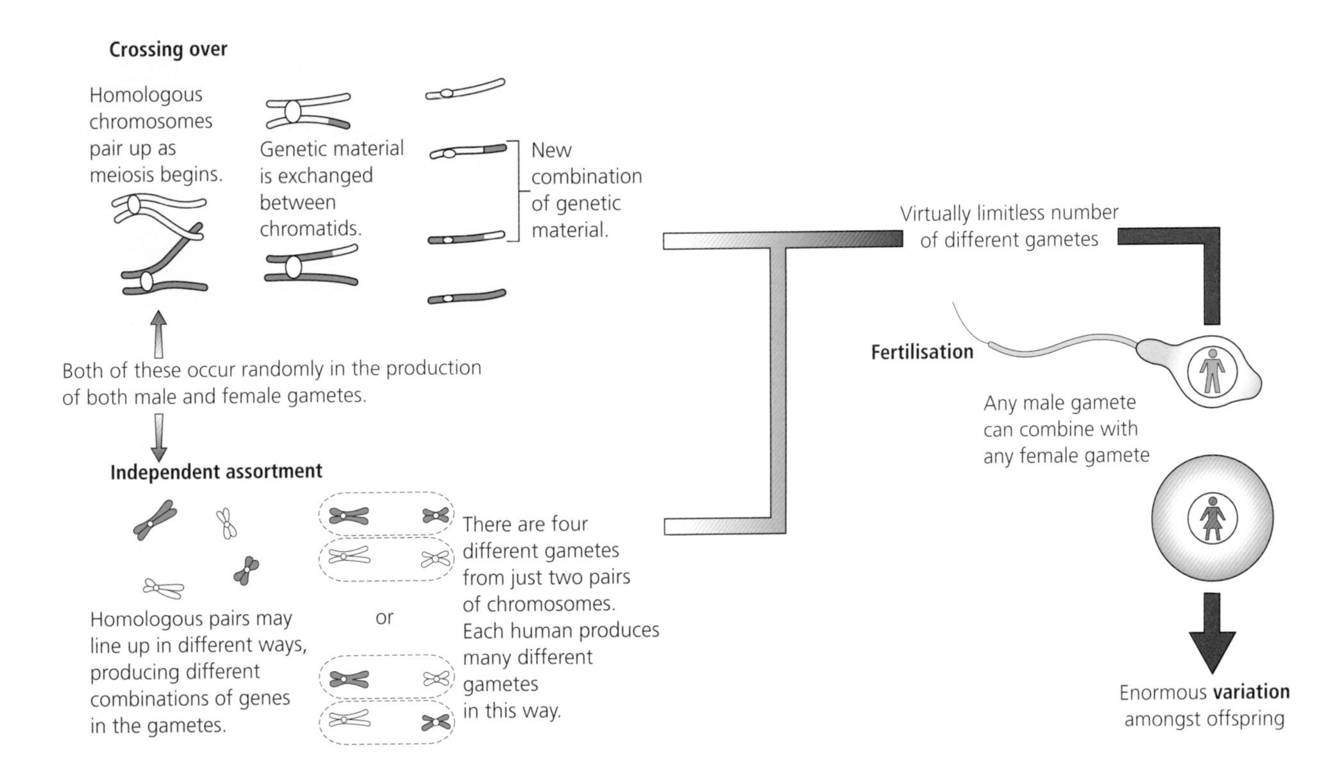

▲ **Crossing over**, **independent assortment** and **fertilisation** all lead to variation. There is very little chance that any two gametes from one individual will be identical. The combined effects of mutation and sexual reproduction lead to enormous variation between individuals.

Q

1 What is a gene mutation? Give an example of:

a a harmful

b a beneficial

gene mutation.

2 What is a carcinogen? Give two examples of carcinogens.

3 How does sexual reproduction lead to variation amongst members of a population?

3.24 Variation and natural selection: the evolution of species

OBJECTIVES

- To understand the meaning of adaptation, and to provide examples of this
- To realise that Darwin's theory benefited from the ideas of other scientists

Adaptation

As we have seen, living organisms differ from one another. Some of these variations make an organism well suited to its environment, some make no difference, and others make the organism *less* well suited to its environment. An organism that is well suited to make the most of the limited resources within its environment is said to show **adaptation** to its environment, as shown opposite. The cactus (a xerophyte) and the water lily (a hydrophyte) (see page 81) are both well adapted to their environments.

▲ A lion has adaptations that enable it to capture prey efficiently. These are:

- structural (e.g. teeth and claws)
- biochemical (e.g. extra protein-digesting enzymes)
- behavioural (e.g. hunting in groups).

The work of Charles Darwin*

Charles Darwin (1809–82) was a British naturalist who took part in a world voyage on a ship called HMS *Beagle*. The voyage, which began in 1831 and lasted for five years, allowed Darwin to see many examples of adaptations. His most famous observations were made on the Galapagos Islands off the west coast of South America. Some species seemed to have adaptations to life on particular islands, but had similarities, and Darwin suspected that they all originated from a single species.

Organisms that are well adapted show high **fitness** - the probability that an organism will survive and reproduce in the environment in which it is found.

An **adaptive feature** (adaptation) is an inherited functional feature of an organism that increases its fitness.

Other species similar to the large ground finch had **adapted** to take advantage of the feeding opportunities on the different islands. Darwin suspected that they had **evolved** from the large ground finch.

Evolution by natural selection

At the time that Darwin lived, most people believed that each species was fixed, and had been put on Earth in its current form by a creator – God. Darwin thought that species were not fixed, and that they changed through time to produce new species. He called this changing through time **evolution**. At first he was unable to convince other scientists because he could not suggest a **mechanism for evolution**. However, he eventually published his ideas in a famous book, *The Origin of Species by Means of Natural Selection*. An outline of the events leading to this publication is shown on the opposite page.

As a result of many observations made during his voyage on HMS *Beagle*, Charles Darwin began to think that living organisms could change in structure. Darwin suggested that species became adapted to meet the challenges of their environment.

The Rev. Thomas Malthus was a mathematician who showed that populations of living organisms would increase indefinitely unless they were kept in check by limited resources.

Charles Lyell was a geologist who showed that the Earth's rocks were very old. He pointed out that rocks of different ages contained the fossilised remains of different animals and plants.

Darwin proposed a mechanism of evolution by natural selection

Alfred Russell Wallace was a professional animal collector working in Malaysia. He wrote to Darwin, and Darwin realised that Wallace had reached the same conclusions about natural selection.

Darwin was stung into action, and published his ideas in *The Origin of Species*.

▲ Darwin and *The Origin of Species*

Update! We can now define natural selection as the greater chance of passing on genes by the best adapted organisms.

Q

1. What is meant by the term adaptation?
2. Name one animal and one plant with which you are familiar, and describe how each is adapted to its environment.

3.25 Natural selection

OBJECTIVES

- To understand how adaptation leads to natural selection

Darwin's ideas about natural selection can be summarised under a number of headings, as shown in the diagram below.

Natural selection may lead to new species

Some organisms are better suited to their environment than others. These organisms will be more likely to survive and breed than some of their competitors. Because of this, the characteristics they possess will become more common in the species over successive generations. If we consider one such characteristic, for example neck length in antelopes, we should be able to draw a graph showing how this characteristic is distributed in the population (see page 211). If the environment applies a **selection pressure** such as a limited availability of leaves for food, one part of the population may be favoured. In a few generations' time the graph may look rather different, as shown on the opposite page.

If two populations of this antelope were separated from one another, natural selection might favour different adaptations in the two environments. Eventually the two different populations of antelope could have so many different adaptations that they can no longer interbreed – they are said to be different **species**.

Alternative theories of evolution

Early scientists and religious leaders believed that a creator had placed all living organisms on the Earth in their present-day forms. Even once the evidence for evolution had been accepted, not everyone agreed with Darwin's idea of natural selection. Jean Baptiste Lamarck was a Frenchman who lived about 70 years before Darwin. He believed that organisms adapted to their environment by 'use and disuse', that is, a giraffe might gain a long neck by stretching up for food in a tall tree. However, Lamarck was never able to show how these acquired characteristics could be passed on from one generation to another. The major difference between the theories of Darwin and Lamarck was that:

- **Lamarck** suggested that the environment *caused* the variations
- **Darwin** suggested that the environment *selected* the variations, which had arisen purely by chance.

Over-production – all organisms produce more offspring than can possibly survive, and yet populations remain relatively stable.
e.g. a female peppered moth may lay 500 eggs, but the moth population does not increase by the same proportion!

Struggle for existence – organisms experience environmental resistance i.e. they compete for the limited resources within the environment.
e.g. several moths may try to feed on the same nectar-producing flower.

Variation – within the population there may be some characteristics that make the organisms that have them more suited for this severe competition.
e.g. some moths might be stronger fliers, have better feeding mouthparts, be better camouflaged while resting or be less affected by rain.

Survival of the fittest – individuals that are most successful in the struggle for existence (i.e. that are the best suited/adapted to their environment) are more likely to survive than those without these advantages.
e.g. peppered moths: dark-coloured moths resting on soot-covered tree trunks will be less likely to be captured by predators than light-coloured moths.

Advantageous characteristics are passed on to offspring – the well-adapted individuals are more likely to breed than those that are less well-adapted – they pass on their genes to the next generation. This process is called **natural selection.**
e.g. dark-coloured moth parents will produce dark-coloured offspring.

▲ Evolution by means of natural selection

Height is a characteristic that shows **continuous variation** in antelopes.

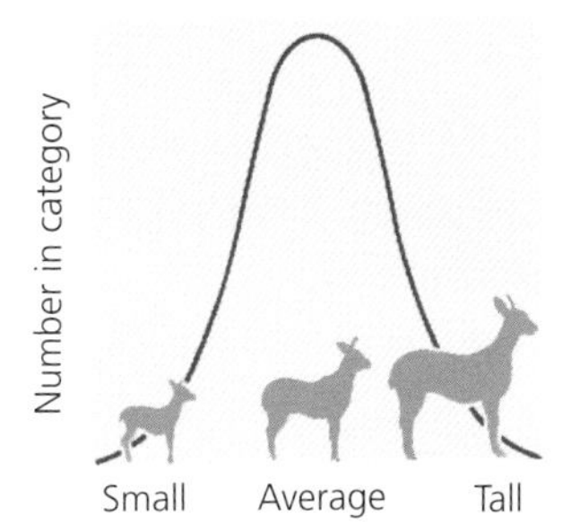

When food is only available from high branches, **natural selection** picks out the taller antelopes.

Many generations later

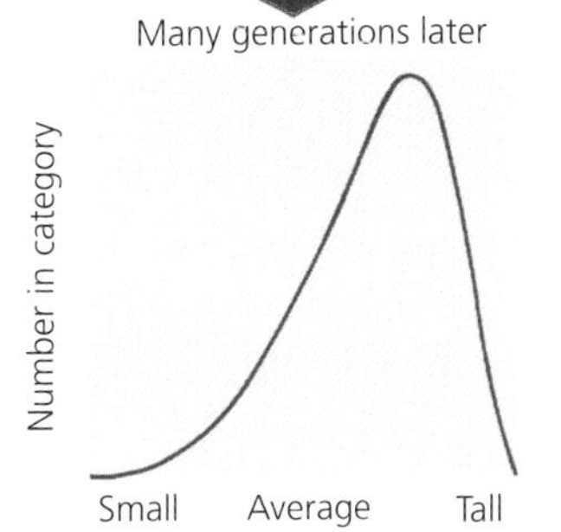

Forming new species
Two populations of the same species could be separated, e.g. by a mountain range. Different **selection pressures** might exist on opposite sides.

Natural selection by **predators**, who can see tall antelopes more easily, favours the small antelopes.

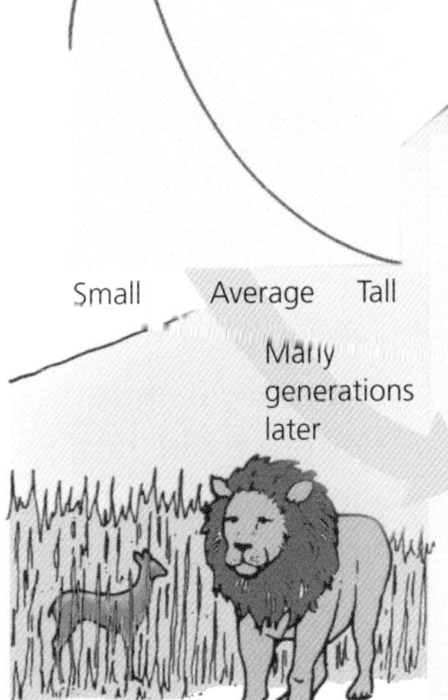

Natural selection by **food availability** on high trees only, favours the tall antelopes.

The two populations now have so many different adaptations that they cannot interbreed. They have now become two species, e.g. the dik-dik and giraffe in east Africa.

Rubbish
Totally untrue.

Dik-diks are antelopes,
Giraffes are NOT antelopes & never have been.

The formation of new species

▲ The development of antibiotic-resistant strains of bacteria (see page 248) is a well-known example of natural selection

Yes the fittest survive, but they DO NOT change into a new species.

Q

1 Using only the following information, answer questions **a** to **d**.

Cepaea is a type of snail which shows considerable variation in its shell colour. The basic colour can be yellow, brown, fawn, pink, orange or red. Over the top of this basic colour up to five bands of a darker colour may occur, around the shell. Colour of shell provides camouflage for the snail because some colours are more difficult to see than others against the background.

Cepaea is an important part of the diet of thrushes. These birds collect snails and break open their shells by banging them on a stone. Thrushes tend to use the same stone, called an 'anvil', whenever feeding in a particular area.

It is possible to collect the remains of the shells and count the number of each shell type. It is also possible to collect the live snails in the same area and count the numbers of each shell type.

Collections of both live snails and broken shells were made in an area where the ground layer plants gave a fairly evenly coloured background. The results are shown in the table.

	Number of snails			
	Banded	**Unbanded**	**Total**	**% Banded**
Live snails	264	296	560	47.0
Shell remains from 'anvils' in the area	486	377	863	56.0

a How many more live, unbanded *Cepaea* were collected than banded?

b Suggest an explanation for thrushes taking more banded snails even where there appear to be more unbanded snails in the live populations.

c Which type of shell, banded or unbanded, would you expect to occur most frequently in a live snail population

i amongst dead leaves in a wood?

ii amongst grasses growing on a sand dune?

d The main points of the theory of evolution by natural selection are listed below.

A The number of offspring is far greater than the number surviving to adult stage.

B Variation exists among the offspring.

C Some variations are useful and help the organisms to survive.

D Competition occurs between the offspring.

E Only those surviving can breed.

Natural selection can change the proportions of the different colours in a snail population. Use the five points A to E above to describe how this change might come about.

3.26 Artificial selection

OBJECTIVES

- To understand the process of artificial selection

Making organisms useful to humans

Variation occurs naturally and randomly in all living organisms, but the natural environment is not the only agent of selection. Ever since early humans began to domesticate animals and plants, they have been trying to improve them. This improvement is brought about by selecting those individuals that have the most useful characteristics and allowing only these individuals to breed. This process is called **artificial selection**. Humans have replaced the environment as the agents of selection. There are many important examples of selective breeding:

- **Jersey cattle** have been bred to produce milk with a very high cream content.
- All domestic **dogs** are the same species, but some have been bred for appearance (e.g. Pekinese), some for hunting (Springer Spaniels) and some as aggressive guards (Rottweilers).
- **Wheat** has been bred so that all the stems are the same height (making harvesting easier) and the ears separate easily from the stalk (making collection of the grain easier).

Some examples of artificial selection are shown below and opposite. The same species can be bred in different ways for different purposes. Early horses have become specialised as carthorses or for racing, for example.

Maintaining variation

What appears to humans as a valuable characteristic might not always be valuable in a natural situation. For example, a Chihuahua dog would probably not survive in the wild because its hunting instincts have been bred out to make it more suitable as a pet. It is very important that humans preserve animal and plant genes for characteristics that do not offer any advantage to us at the moment. A cow with a limited milk yield may carry a gene that makes it resistant to a disease which is not yet a problem in domestic herds, for example. This resistance gene might be extremely valuable if ever such a disease did become established. For this reason many varieties of animals and plants are kept in small numbers in rare-breed centres up and down the country. Plant genes may be conserved as seeds, which are easy to store, and some animal genes may be kept as frozen eggs, sperm or embryos.

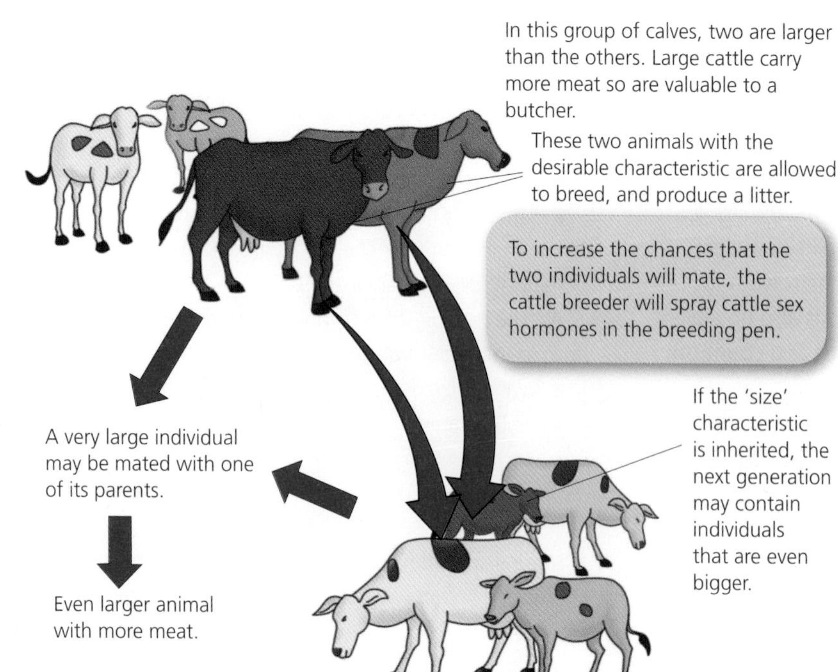

In this group of calves, two are larger than the others. Large cattle carry more meat so are valuable to a butcher.
These two animals with the desirable characteristic are allowed to breed, and produce a litter.
To increase the chances that the two individuals will mate, the cattle breeder will spray cattle sex hormones in the breeding pen.
If the 'size' characteristic is inherited, the next generation may contain individuals that are even bigger.
A very large individual may be mated with one of its parents.
Even larger animal with more meat.

Artificial insemination

Male animals, no matter how many useful characteristics they have, **cannot give birth to young animals!** So:

- Most male offspring will be fattened up for selling as meat.
- 'Desirable' males may be electrically stimulated to ejaculate – the sperm is collected and frozen. One male can easily produce several hundred samples.
- The sperm can be taken to a cattle breeding farm and used to inseminate many females.

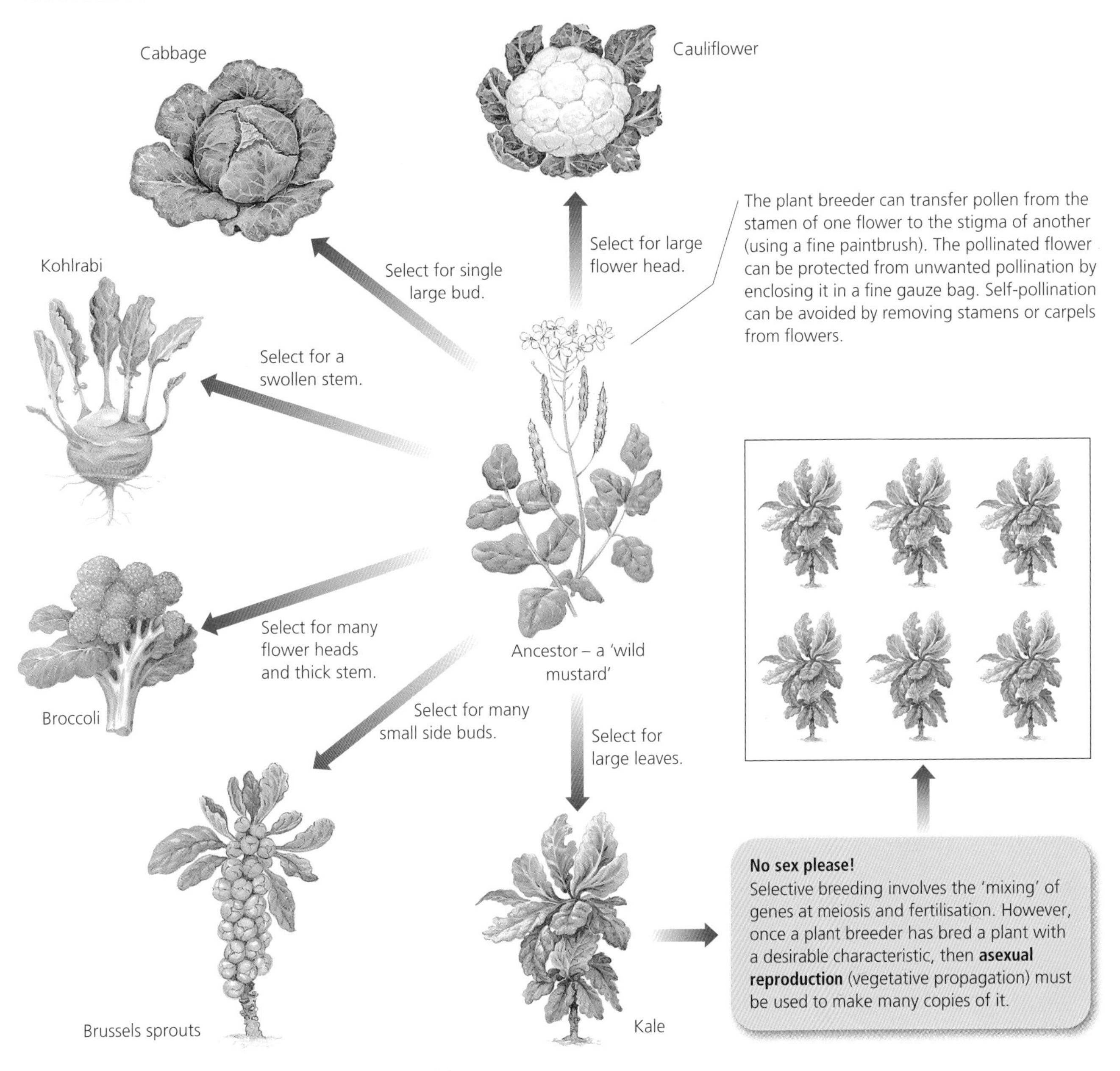

▲ Artificial selection has produced many vegetables from one ancestor species

Three major differences between artificial and natural selection

- In artificial selection, humans are the agents of selection, while natural selection depends upon the natural environment.
- Artificial selection is much quicker than natural selection.
- Artificial selection offers no advantage to the animal or plant in its natural environment.

It is likely that selective breeding will be replaced by genetic engineering in the future. This technique, explained on page 252, is much more predictable than selective breeding and produces useful results more quickly.

Q

1 A potato grower wanted to produce a new variety of potato which grows quickly and makes good chips. She has one variety which grows quickly but makes poor chips, and another which grows more slowly but makes good chips.
When plants of these two varieties produce flowers she crosses the two varieties. Later she collects the seeds and plants them.

a To cross the two varieties, the grower pollinates a flower of one variety with pollen from the other. Describe how the grower should do this.

b From the seeds she collects, she finds that one of the new plants grows very quickly and produces potatoes which make good chips. How would she produce a crop of potatoes which are exactly the same?

Questions on variation

1 Three species of zebra are shown below.

Equus burchelli

Equus grevyi

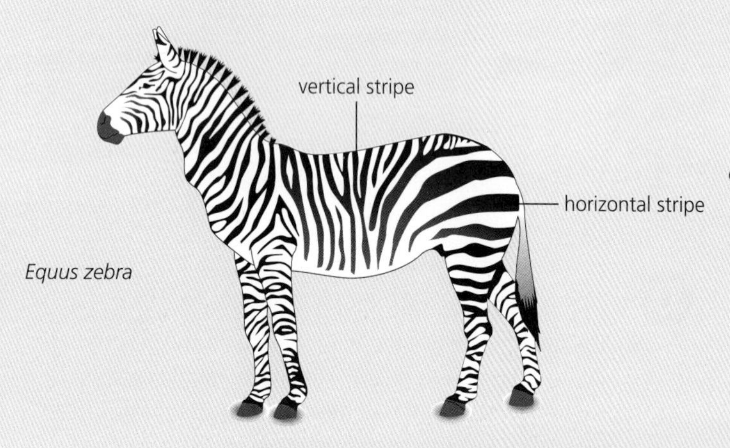

Equus zebra

a Describe **one** method a scientist could use to show that the zebras shown are different species.

b Studies have shown that the hotter the environment, the more stripes zebras have.

i State the type of variation which would result in different numbers of stripes.

ii Suggest which species of zebra lives in the hottest environment.

c Occasionally, zebras are born that are almost completely black. The change in appearance is the result of mutation.

i State the term that is used to describe the appearance of an organism.

ii Define the term *mutation*.

d Tsetse flies attack animals with short fur, sucking their blood and spreading diseases. A tsetse fly is shown below. This fly is an insect, belonging to the arthropod group.

i State **one** feature, **visible in this diagram**, which is common to all arthropods.

ii State two features, **visible in this diagram**, which distinguish insects from other arthropod groups.

e Scientists have discovered that zebras with more horizontal stripes attract fewer tsetse flies.

i Suggest why the stripes on the head and neck of the zebra would be an advantage when it feeds on grass on the ground.

ii Describe how a species of zebra could gradually develop more horizontal stripes.

Cambridge IGCSE Biology 0610 Paper 3 Variant 1 Q4 May 2008

2 After an accident at a nuclear power plant in 1986, particles containing radioactive strontium were carried like dust in the atmosphere. These landed on grassland in many European countries. When sheep fed on the grass they absorbed the strontium and used it in a similar way to calcium.

a Explain where in the sheep you might expect the radioactive strontium to become concentrated.

b Suggest the possible effects of the radiation, given off by the strontium, on cells in the body of the sheep.

Cambridge IGCSE Biology 0610
Paper 2 Q3 November 2008

3 a The graph below shows the variation in the height of human adults in an African population.

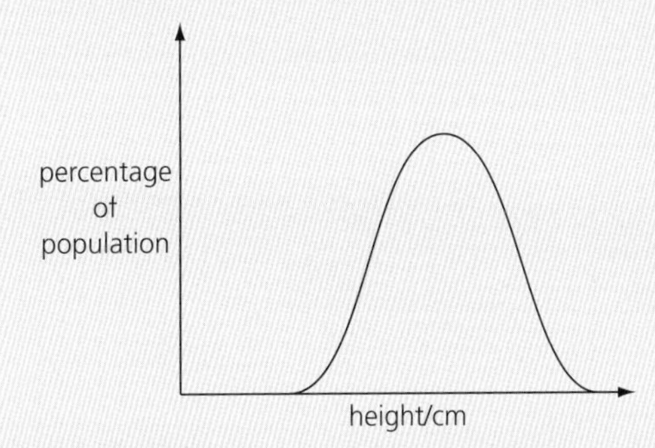

State the type of variation shown by this data.

b In Britain 42% of the population have blood group A. The frequency of the other blood groups is: B (9%), AB (3%) and O (46%).

i Plot the data as a bar chart.

ii Complete the following sentence.

Height is controlled by environment and by genes but human blood groups are controlled only by ____.

c Sometimes human characteristics are altered by mutations.

i Define the term *mutation*.

ii Suggest two factors that could increase the rate at which mutations occur.

Cambridge IGCSE Biology 0610
Paper 2 Q3 May 2007

4 A survey of berries from a number of bushes of one species in a school grounds showed variation in their mass. Berries were collected at random and 50 had their mass determined. This table shows the results of their investigation.

mass of berry / g				
1.3	0.6	1.6	1.3	1.2
1.0	1.3	1.2	0.4	1.1
1.3	0.9	0.4	1.4	1.2
1.0	1.0	0.6	1.5	1.2
1.1	0.5	1.1	1.3	1.1
0.3	1.3	0.5	1.2	0.5
1.1	1.3	1.0	0.6	1.4
1.4	1.2	1.4	1.2	1.3
0.6	1.3	1.2	0.7	1.2
0.5	0.6	1.3	1.3	1.4

a i Complete the table below for the number of berries of mass 1.2 g and 1.3 g.

mass of berry / g	number of individuals
0.3	1
0.4	2
0.5	4
0.6	5
0.7	1
0.8	0
0.9	1
1.0	4
1.1	5
1.2	
1.3	
1.4	5
1.5	1
1.6	1

ii Plot the data in the second table as a histogram.

b State, with a reason, the type of variation illustrated by the berries with masses between 0.3 g and 0.7 g.

Cambridge IGCSE Biology 0610
Paper 2 Q4 November 2006

4.1 Ecology and ecosystems

OBJECTIVES

- To understand that living organisms require certain conditions for their survival
- To understand that living organisms interact with one another, and with their non-living environment
- To define population, community and ecosystem
- To realise that available resources change through the year

Environmental survival kit

All living organisms depend upon their environment for three 'survival essentials'. These are a **supply of food**, **shelter** from undesirable physical conditions and a **breeding site**. The living organism **interacts** with its environment – for example, a living plant:

- removes carbon dioxide, water and light energy from its habitat
- may be eaten by an animal or a parasite
- depends upon soil for support.

Factors in the environment affect the growth of the plant. Some of these factors are **biotic** – other living organisms – and some are **abiotic** – the non-living components of the habitat. **Ecology** is the study of living organisms in relation to their environment. The interactions between the organism and its environment are summarised below.

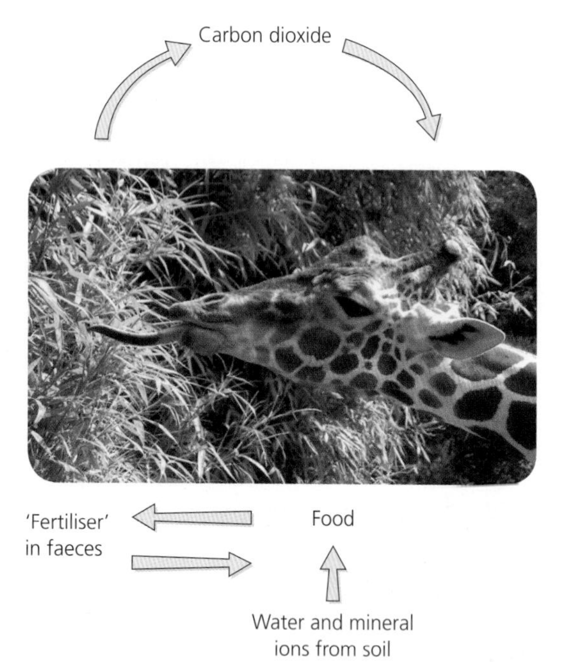

A giraffe feeds on a thorn tree. The tree requires water, mineral ions, carbon dioxide and light to grow. The giraffe may provide carbon dioxide from respiration, and ions from decomposition of its faeces.

Changing with the seasons

The ability of the habitat to supply living organisms with their requirements may vary at different times of year. The ecosystem in the photograph opposite will only exist for a certain period of time – as food or water becomes exhausted some animals may leave. These will then be followed by the predators which feed on them. The great animal **migrations** seen in East Africa result from the changing conditions in the animals' environment, for example:

- poor rain means little growth of grass
- herbivores leave for areas of fresh growth
- carnivores follow herbivores
- (then scavengers follow carnivores!).

Living together

Living organisms normally exist in groups. The names given to these groups, and the way they interact with the abiotic environment, are explained opposite.

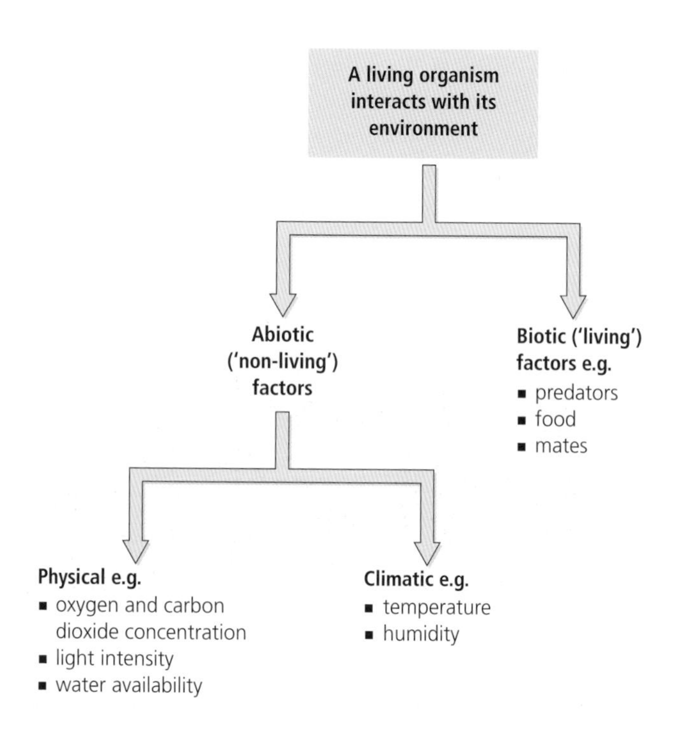

A **population** is all of the members of the same species (e.g. wildebeest) in a particular area.

A **community** is all of the populations of living organisms in one area (e.g. acacia trees, zebra, wildebeest and grass). The community is the **biotic environment**.

Air, water and soil make up the **abiotic environment**.

An **ecosystem** is all the living organisms and the non-living factors interacting together in a particular part of the environment

A **habitat** is a part of the environment that can provide food, shelter and a breeding site for a living organism (e.g. a patch of grassland).

▲ Organisms exist in groups within an ecosystem

1 Define the terms population, community and ecosystem.

2 Name two abiotic factors that might determine whether or not a habitat is suitable for a living organism.

3 Suggest two ways in which a plant and an animal in the same habitat may interact.

4 What must a habitat provide?

5 How are the following observations related?

- Very few flying insects are found in Britain during the winter.
- Swallows migrate to Africa when it is winter in the UK.
- Hobbies (small bird-eating falcons) leave Britain in late autumn.

6 What is meant by the term ecology?

7 a A group of students were studying a forest. They noticed that the plants grew in two main layers. They called these the tree layer and the ground layer.

tree layer

ground layer

The students measured the amount of sunlight reaching each layer at different times in the year. Their results are shown on the graph.

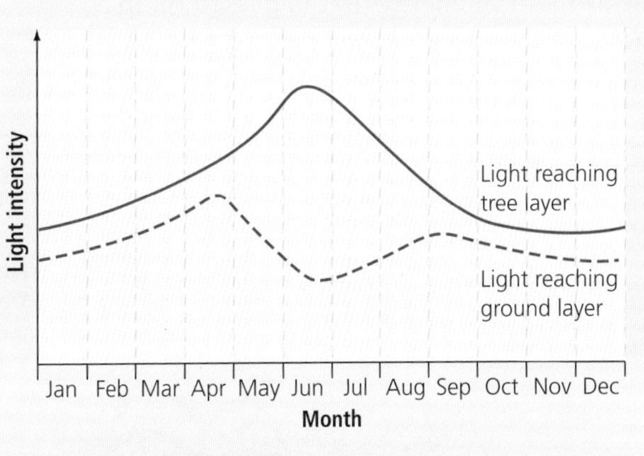

i During which month did most light reach the tree layer?

ii During which month did most light reach the ground layer?

iii Suggest why the amount of sunlight reaching the ground layer is lower in mid-summer than in the spring.

b The pupils found bluebells growing in the ground layer. Bluebells grow rapidly from bulbs. They flower in April and by June their leaves have died.

i Suggest why bluebells grow rapidly in April.

ii Suggest why the bluebell leaves have died by June.

4.2 Flow of energy: food chains and food webs

OBJECTIVES

- To know that the feeding relationships in an ecosystem can be expressed as food chains
- To understand why energy transfer through an ecosystem is inefficient
- To understand why complex food webs are the most stable

Food chains

The most obvious interaction between different organisms in an ecosystem is feeding. During feeding, one organism is obtaining food – energy and raw materials – from another one. Usually one organism eats another, but then may itself be food for a third species. The flow of energy between different organisms in the ecosystem can be shown in a **food chain**, as in the diagram below.

Energy transfer is inefficient

The amount of energy that is passed on in a food chain is reduced at every step. Since energy can be neither created nor destroyed, it is not lost but is converted into some other form. During respiration, some energy is transferred to the environment as heat. The flow of energy through a food chain, and the heat losses to the environment, are illustrated in the diagram opposite.

Food webs

Since so little energy is transferred from the base to the top of a food chain, a top carnivore must eat many herbivores. These herbivores are probably not all of the same species. In turn, each herbivore is likely to feed on many different plant species. All these interconnected food chains in one part of an ecosystem can be shown in a **food web**.

The more complicated a food web, the more stable the community is. For example, in the forest food web shown opposite, if the number of squirrels fell, the owls could eat more worms, mice and rats. The mice and rats would have less competition for food from squirrels, and so might reproduce more successfully.

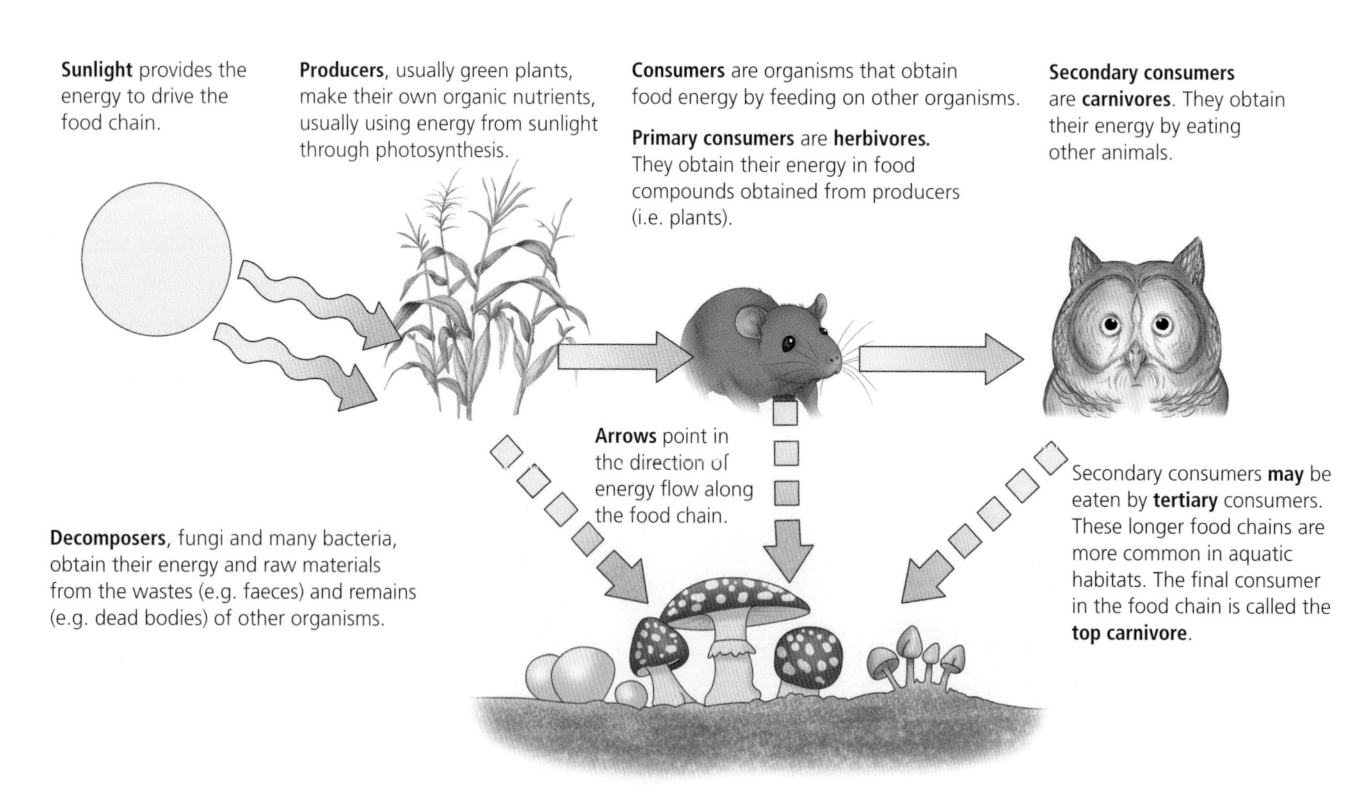

▲ Food chains show energy flow through an ecosystem. The position of each organism in the food chain represents a different **trophic** (feeding) **level**.

Energy transfer

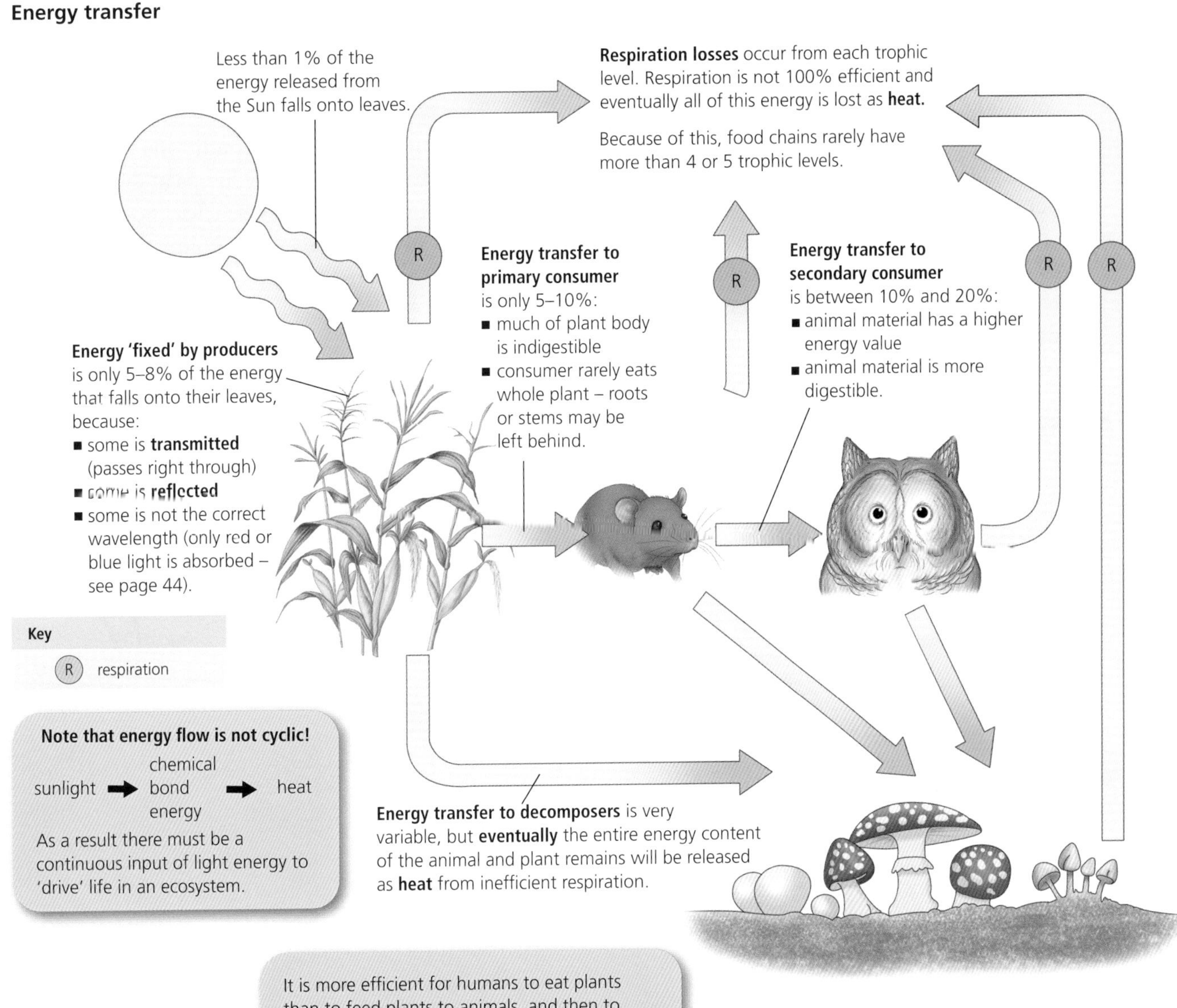

Note that energy flow is not cyclic!

sunlight ➔ chemical bond energy ➔ heat

As a result there must be a continuous input of light energy to 'drive' life in an ecosystem.

It is more efficient for humans to eat plants than to feed plants to animals, and then to eat the animals. Every step in the chain loses 80–90% of the available energy as heat, so THE FEWER 'STEPS' THE BETTER!

A simple forest food web

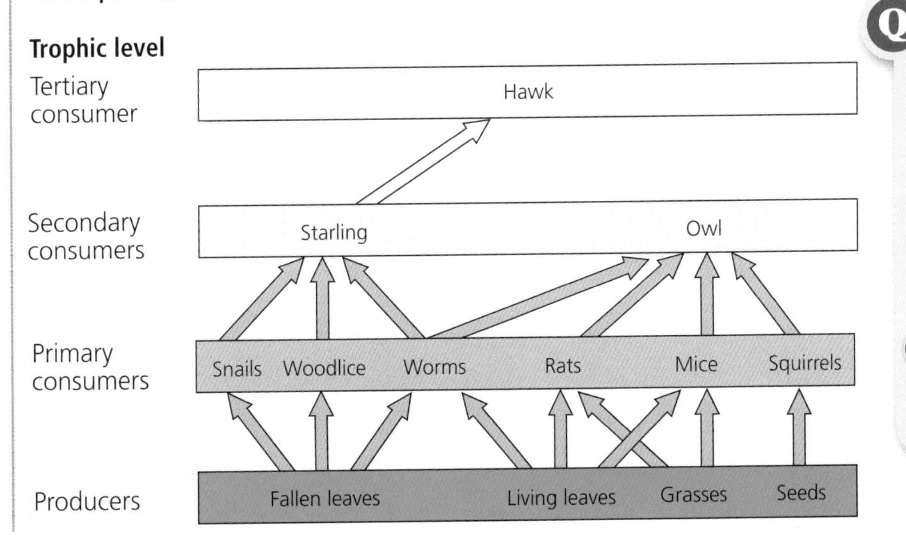

Q

1. Define the terms producer, consumer and decomposer. Which of these could be omitted from an ecosystem? Explain your answer.
2. Write out a food chain from a named ecosystem which you have studied.
3. Why are food chains usually restricted to three or four trophic levels?

Humans and food webs

Humans can have negative effects on food chains and food webs

- by over-harvesting food species such as cod (so species eaten by cod increase in numbers, and species which eat cod may switch to eating other foods).
- by introducing foreign species to a habitat. Rabbits, for example, were introduced to Australia for hunting but bred very quickly. This reduced grass available for native Australian species.

More examples of feeding relationships

Food chains and food webs in aquatic (watery) environments can be longer than those on the land. This is because this type of environment has space and ideal growth conditions for many producers. Even with energy losses at every stage there is enough 'trapped' energy for more steps in the chain. Many of these food chains begin with phytoplankton (tiny green plants) or algae.

A freshwater food chain

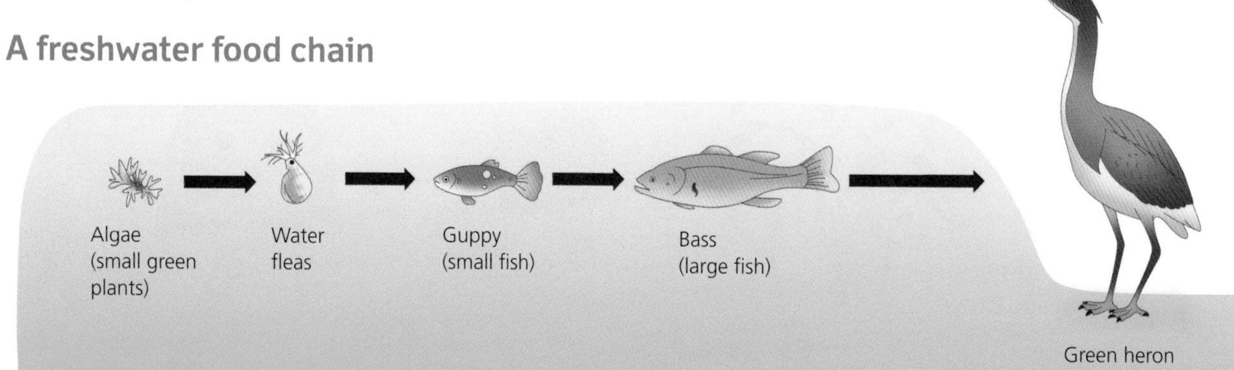

A seashore food chain

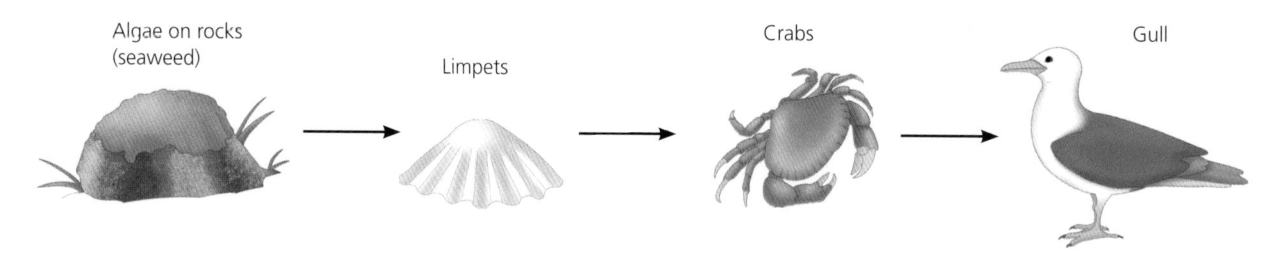

The seashore is an excellent environment for animals, at least as far as food is concerned, because fresh supplies are delivered with every tide!

Some of the top predators on seashores need so much food that they need to travel between different parts of the habitat. A gull, for example, might have to fly to several different parts of the same shore.

Q

1 Look at the three aquatic feeding relationships shown on this page and the next. Make a table like this one:

Producers	Herbivores	Carnivores	Top carnivores

2 Shark fishing is a popular sport. Explain what might happen if all of the sharks living around a section of reef were captured by fishermen.

Coral reef food web

The most complex food webs are found in the ocean.

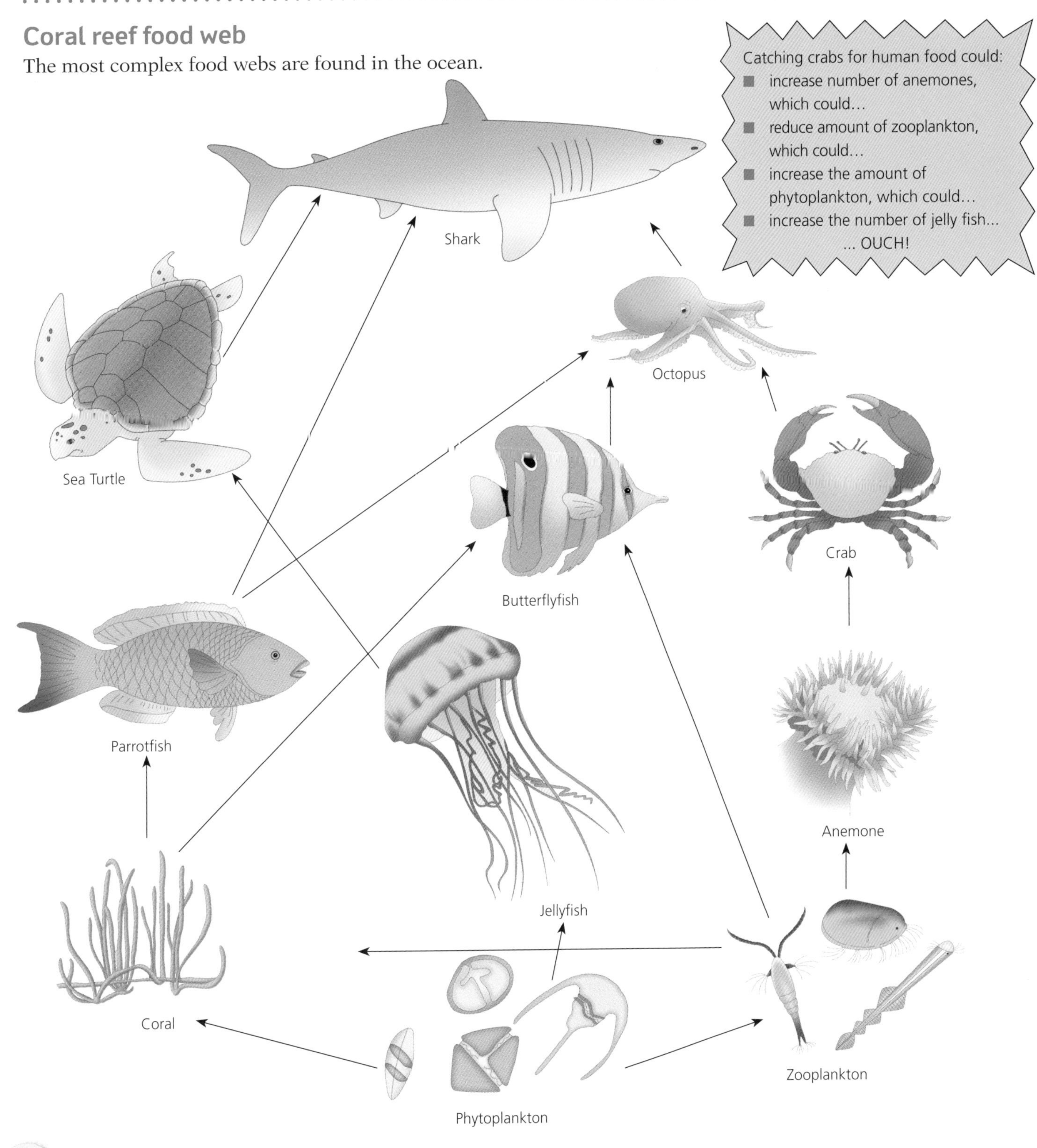
Catching crabs for human food could:
increase number of anemones, which could…
reduce amount of zooplankton, which could…
increase the amount of phytoplankton, which could…
increase the number of jelly fish…
... OUCH!
Shark
Sea Turtle
Octopus
Butterflyfish
Crab
Parrotfish
Anemone
Jellyfish
Coral
Zooplankton
Phytoplankton

Q

3 Use words from the following list to complete the paragraph about ecosystems. You may use each word once, more than once or not at all.

decomposition, producer, chemical, carnivore, consumer, photosynthesis, energy, light, elements, decomposers, herbivore.

In each ecosystem there are many feeding relationships. A food chain represents a flow of _____ through an ecosystem, and always begins with an organism called a _____ which is able to trap _____ energy and convert it to _____ energy. An organism of this type is eaten by a _____, which is a kind of _____ that feeds only on plant material. This type of organism is, in turn, eaten by a _____ (an organism that consumes other animals).

4.3 Feeding relationships: pyramids of numbers, biomass and energy

OBJECTIVES

- To be able to describe pyramids of numbers, biomass and energy
- To understand how data can be gathered to make ecological pyramids

Pyramids of numbers

Look at the food chain on page 224. Two things should be clear:

- The organisms tend to get bigger moving along the food chain. Predators, such as the owl, need to be large enough to overcome their prey, such as the mouse.
- Energy is 'lost' as heat on moving from one trophic level to the next, so an animal to the right of a food chain needs to eat several organisms 'below' it in order to obtain enough energy. For example, a rabbit eats many blades of grass.

Food chains and food webs provide **qualitative** information about an ecosystem – they show which organism feeds on which other organism. How do we show **quantitative** information, for example how many predators can be supported by a certain number of plants at the start of the chain? We can use a **pyramid of numbers** or a **pyramid of biomass**, as shown in the diagram below.

Pyramids of energy*

A pyramid of biomass describes how much biomass is present in a habitat *at the time the sample is taken*. This can be misleading, because different feeding levels may contain organisms that reproduce, and so replace themselves, at different rates. For example, grass in a field would replace itself more quickly than cattle feeding on the grass, so when the pyramid of biomass is constructed there would be more 'cattle biomass' than 'grass

Pyramid of numbers – a diagrammatic representation of the number of different organisms at each trophic level in an ecosystem **at any one time**

Note

1. The number of organisms at any trophic level is represented by the length (or the area) of a rectangle.
2. Moving up the pyramid, the **number** of organisms generally **decreases**, but the **size** of each individual **increases**.

Problems

a. The range of numbers may be enormous – 500 000 grass plants may only support a single top carnivore – so that drawing the pyramid to scale may be very difficult.
b. Pyramids may be **inverted**, particularly if the **producer is very large** (e.g. an oak tree) or **parasites feed on the consumers** (e.g. bird lice on an owl).

Pyramid of biomass – which represents the **biomass** (number of individuals × mass of each individual) at each trophic level **at any one time**. This should solve the scale and inversion problems of the pyramid of numbers.

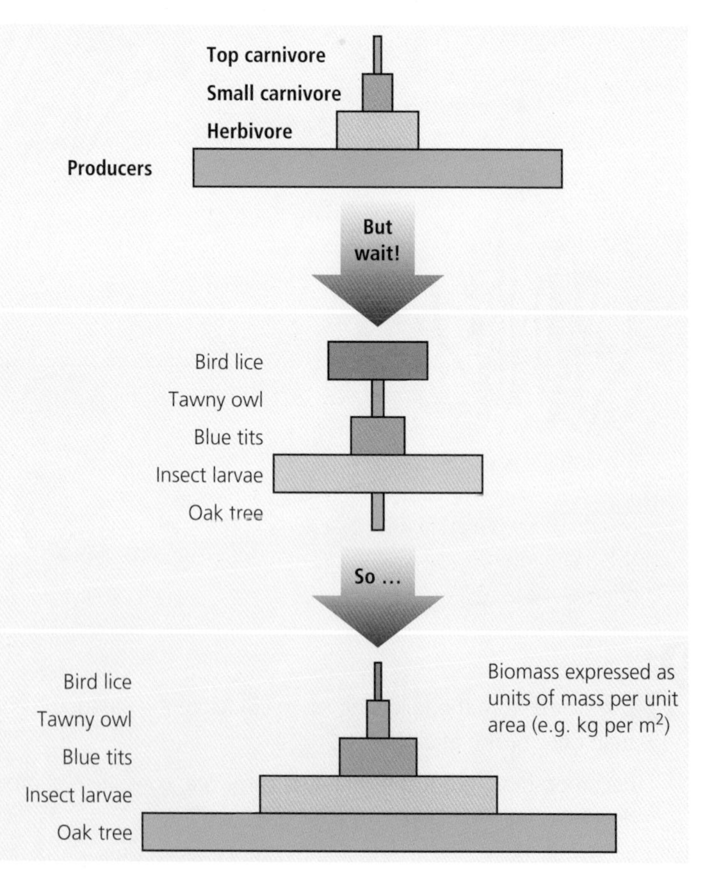

▲ Ecological pyramids represent numerical relationships between successive trophic levels. The pyramid of biomass is useful because the biomass gives a good idea of how much energy is passed on to the next trophic level.

biomass' and the pyramid would be inverted. To overcome this difficulty a **pyramid of energy** can be constructed. This measures the amount of energy flowing through an ecosystem *over a period of time*. The time period is usually a year, since this takes into account the changing rates of growth and reproduction in different seasons. It is even possible to add an extra base layer to the pyramid of energy representing the solar energy entering that particular ecosystem.

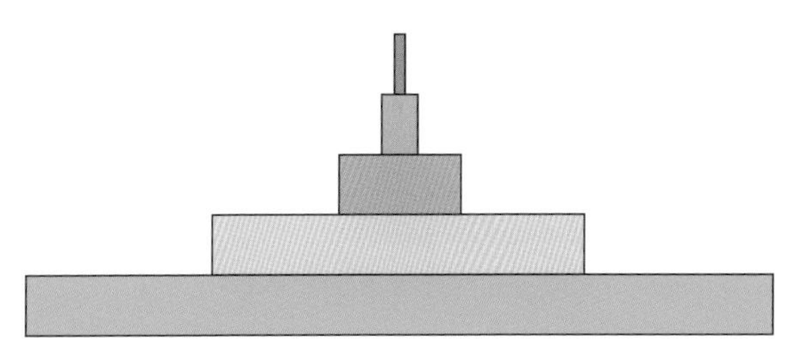

▲ **Pyramid of energy:** energy values are expressed as units of energy per unit area per unit time (e.g. kJ per m^2 per year)

GATHERING DATA FOR ECOLOGICAL PYRAMIDS

To construct a pyramid of numbers or of biomass, organisms must be captured, counted and (perhaps) weighed. This is done on a **sample** (a small number) of the organisms in an ecosystem. Counting every individual organism in a habitat would be extremely time-consuming and could considerably damage the environment.

The sample should give an accurate estimate of the total population size. To do this:

- The sampling must be **random** to avoid any bias. For example, it is tempting to collect a large number of organisms, by looking for the areas where they are most common. To avoid this, the possible sampling sites can each be given a number and then chosen using random number generators on a computer.
- The sample must be the **right size** so that any 'rogue' results can be eliminated. For example, a single sample might be taken from a bare patch of earth, whereas all other sites are covered with vegetation. The single sample from the bare patch should not be ignored, but its effects on the results will be lessened if another nine samples are taken. A **mean value** can then be used.

Sampling plants and sessile animals

Once the organisms in a sample have been identified and counted, the population size can be estimated. For example, if 10 quadrats gave a mean of 8 plants per quadrat, and each quadrat is one-hundredth of the area of the total site, then the total plant population in that area is $8 \times 100 = 800$.

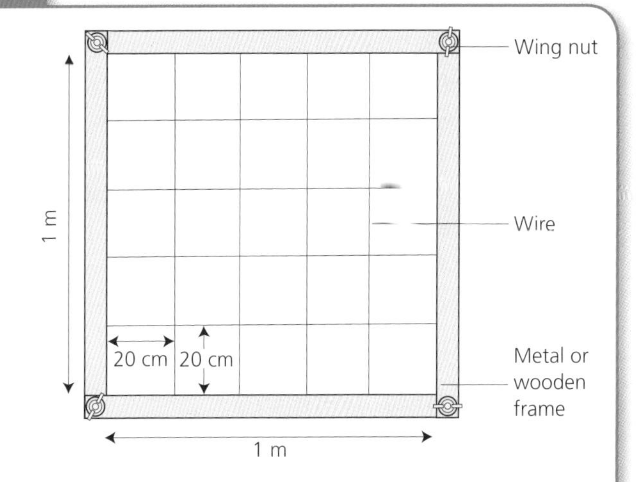

▲ **A quadrat** is a square frame made of wood or metal. It is simply laid on the ground and the number of organisms inside it is counted.

▼ A quadrat is used most commonly for estimating the size of plant populations, but may also be valuable for the study of populations of sessile or slow-moving animals (e.g. limpets).

4.4 Decay is a natural process

OBJECTIVES

- To understand that nutrients in dead organisms are recycled
- To know that the process of decay often begins with the activities of scavengers
- To know how saprotrophic nutrition is responsible for decomposition

Recycling nutrients

Humans have an unusual skill – they can modify their environment to suit themselves. For example, we cut down forests and plant crops, and we build houses. Many building materials are natural, such as wood and straw, and the environment treats these materials as the dead remains of once-living organisms – the environment reclaims the nutrients and returns them to the ecosystem.

Starting with scavengers

When an organism dies, the nutrients in its body are returned to the environment to be reused. The nutrients are recycled by a series of processes carried out by other living organisms. The first ones to appear are usually the **scavengers** which break up the dead bodies into more manageable pieces. Scavengers eat some of the dead body, but leave behind blood or small pieces of tissue.

▲ Scavengers such as the vulture feed on dead bodies

Decomposition by microorganisms

The remains that are left are **decomposed** by the feeding activities of microorganisms. These fungi and bacteria feed by secreting enzymes onto the remains and absorbing the digested products. This form of nutrition is called saprotrophic feeding.

The diagram on the opposite page illustrates some of the features of the decomposition process. The decay process provides energy and raw materials for the decomposers. It also releases nutrients from the bodies of dead animals and plants, which can then be reused by other organisms, for example:

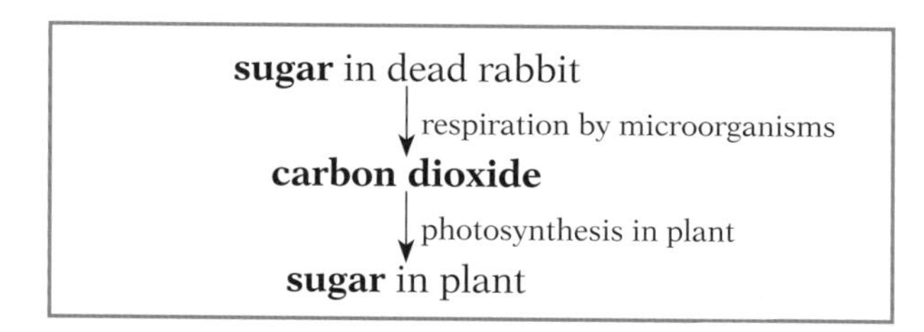

In this way substances pass through **nutrient cycles** as microbes convert them from large, complex molecules in animal and plant remains to simpler compounds in the soil and the atmosphere. The next sections describe the recycling of the elements carbon and nitrogen.

Importance of decomposition processes to humans

- Organic waste in sewage is decomposed and made 'safe' in water treatment plants (see page 275).
- Organic pollutants such as spilled oil may be removed from the environment by decomposing bacteria (see page 233).
- Food is spoiled due to decomposition by fungi and bacteria. Many food treatments alter physical conditions to inhibit enzyme activity.
- Wounds may become infected by saprotrophs, leading to tissue loss or even to death. Many medical treatments inhibit the multiplication or metabolism of saprotrophs.

Saprotrophs cause decay.

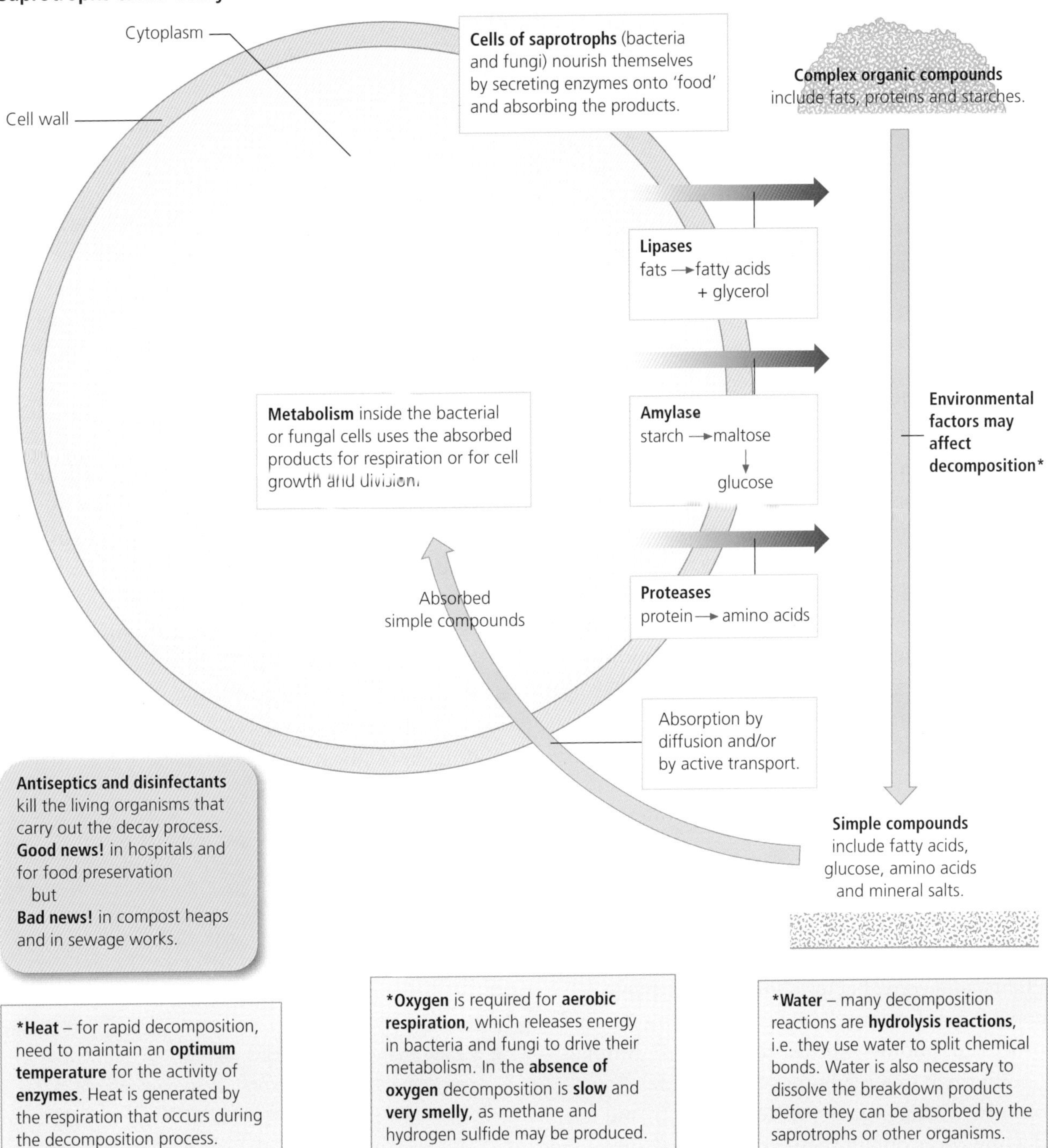

Antiseptics and disinfectants kill the living organisms that carry out the decay process.
Good news! in hospitals and for food preservation
but
Bad news! in compost heaps and in sewage works.

***Heat** – for rapid decomposition, need to maintain an **optimum temperature** for the activity of **enzymes**. Heat is generated by the respiration that occurs during the decomposition process.

***Oxygen** is required for **aerobic respiration**, which releases energy in bacteria and fungi to drive their metabolism. In the **absence of oxygen** decomposition is **slow** and **very smelly**, as methane and hydrogen sulfide may be produced.

***Water** – many decomposition reactions are **hydrolysis reactions**, i.e. they use water to split chemical bonds. Water is also necessary to dissolve the breakdown products before they can be absorbed by the saprotrophs or other organisms.

Q

1 Copy and complete the following paragraph.
During the process of decay, ____ and ____ convert complex chemicals into ____ ones. For example, proteins are converted to ____ , and ____ to fatty acids and glycerol. These decay processes involve the biological catalysts called ____ , and so the processes are affected by changes in ____ and ____ . Humans exploit decay, for example in the treatment of ____ to provide drinking water, and may deliberately limit decay, for example in the preservation of ____.

2 Gardeners often place vegetable waste on a compost heap. Over the course of time the waste will be decomposed.
- **a** What do gardeners gain from the decomposed waste?
- **b** Why do gardeners sometimes spray water over the heap in warm summer weather?
- **c** Why do gardeners often build compost heaps on a pile of loose-fitting sticks or bricks?

4.5 The carbon cycle

OBJECTIVES

- To recall why living organisms need carbon-containing compounds
- To appreciate that carbon is cycled between complex and simple forms by the biochemical processes of photosynthesis and respiration
- To understand that formation and combustion of fossil fuels may distort the pattern of the carbon cycle

Carbon-containing nutrients – a reminder

The Sun keeps supplying *energy* to food chains. However, the supply of *chemical elements* to living organisms is limited, and these elements must be recycled. The nutrient elements are cycled between simple forms in the non-living (abiotic) environment and more complex forms in the bodies of living organisms (the biotic component of an ecosystem). Living organisms require carbon-containing compounds as:

- **a source of energy**, released when carbon-containing compounds are oxidised during respiration (particularly carbohydrates and fats)
- **raw materials** for the growth of cells (particularly fats and proteins).

Recycling carbon compounds

Plants, and some bacteria, manufacture these compounds from carbon dioxide during the process of photosynthesis (see page 38). Animals obtain them in a ready-made form by feeding on other living organisms (see page 52), and decomposers obtain them as they break down the dead bodies or wastes of other living organisms. These processes of feeding, respiration, photosynthesis and decomposition **recycle** the carbon over and over again. Theoretically, the amount of carbon dioxide fixed by photosynthesis should equal the amount released by respiration. As a result the most accessible form of carbon in the non-living environment, that is **carbon dioxide**, remains at about the same concentration year after year after year (about 0.03% of the atmosphere). Other processes may affect this regular cycling of carbon.

- Sometimes conditions are not suitable for respiration by decomposers, and carbon dioxide remains 'locked up' in complex carbon compounds in the bodies of organisms. For example, anaerobic, low pH or extreme temperature conditions will inhibit decomposition – this is how fossil fuels have been laid down in environments where decomposition is not favoured.
- Over millions of years the formation of fossil fuels has removed carbon dioxide from the environment. Humans have exploited fossil fuels as a source of energy over a relatively short time, and the **combustion** of oil, gas, coal and peat has returned enormous volumes of carbon dioxide to the atmosphere. As a result carbon dioxide concentrations are increasing (see page 265).
- The burning of biomass fuels such as wood and alcohol uses up oxygen also returns carbon dioxide to the atmosphere, and can have a very severe local effect although worldwide it is less significant than the combustion of fossil fuels.

The way in which these different processes contribute to the cycling of carbon is illustrated opposite.

1. Refer to the carbon cycle opposite.
   - **a** Name the simple carbon compound present in the abiotic part of the ecosystem.
   - **b** Name two compounds present in the biotic part of the ecosystem.
   - **c** Which processes raise the concentration of carbon dioxide in the atmosphere?
   - **d** Which process reduces carbon dioxide concentration in the atmosphere?
   - **e** Name the process that distributes carbon dioxide throughout the atmosphere from places where it is released.
   - **f** Suggest a reason why some fossil fuels were formed as sediments at the bottom of ancient seas.

The processes of photosynthesis, feeding, death, excretion and respiration lead to the cycling of carbon between living organisms and their environment. Fossil fuel formation and combustion affect the concentration of carbon dioxide in the atmosphere.

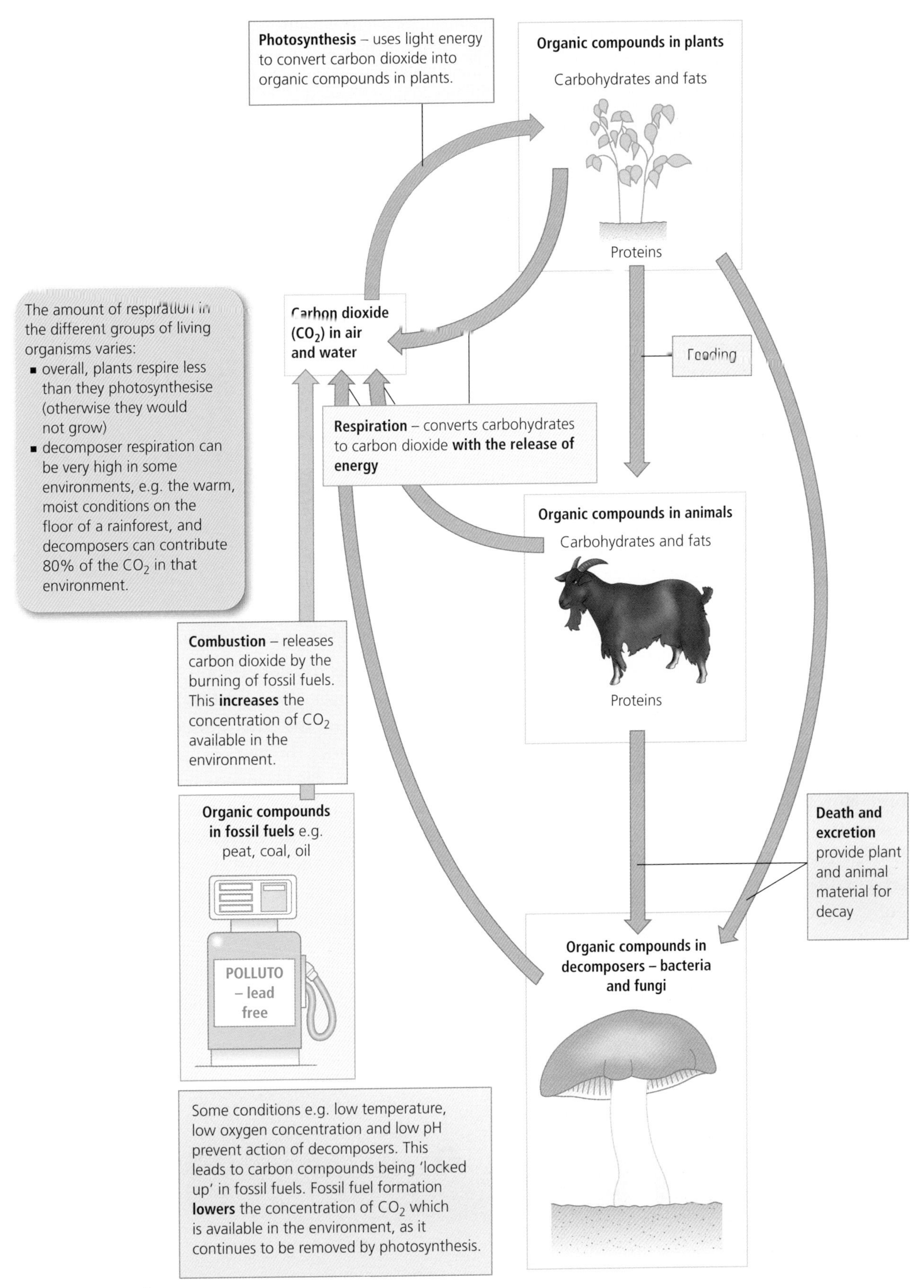

▲ The carbon cycle

4.6 The nitrogen cycle

OBJECTIVES

- To recall why nitrate is an essential mineral for plant growth
- To know how nitrate is made available in the soil
- To understand that a series of biochemical processes results in the cycling of nitrogen between living organisms and the environment
- To appreciate the part played by microorganisms in the cycling of nitrogen

Plants need nitrate

Plants need nitrogen for the synthesis of proteins and other compounds, including DNA and vitamins. Nitrogen gas makes up about 80% of the Earth's atmosphere, but plants do not have the enzymes necessary to use the nitrogen directly – instead they must absorb it as **nitrate**. Nitrate is formed by two sets of processes carried out by microorganisms – **nitrogen fixation** and **nitrification**.

Nitrogen fixation

In **nitrogen fixation**, nitrogen and hydrogen are combined to form ammonium ions and then nitrate. The process depends upon enzymes that are only possessed by certain bacteria called **nitrogen-fixing bacteria**. Some of these bacteria live free in the soil, but one important species lives in swellings called **nodules** on the roots of leguminous plants such as peas, beans and clover. Nitrogen fixation only happens if oxygen is present. It also occurs naturally in the atmosphere when the energy from lightning combines nitrogen directly with oxygen. Farmers can plant legumes in a crop rotation scheme to avoid having to use so much nitrogen-containing fertiliser. This saves money, and also limits pollution of water (see page 268).

Nitrification

In **nitrification**, ammonium ions produced by the decomposition of amino acids and proteins are oxidised, first to **nitrite** and then to nitrate. The process is carried out by **nitrifying bacteria** which live in the soil. Nitrification only happens if oxygen is present. In the absence of oxygen the process is reversed, and **denitrifying bacteria** obtain their energy by converting nitrate to nitrogen gas. This is why waterlogged soils, for example, tend to lose nitrate as nitrogen gas.

Recycling nitrogen

Once nitrate has been formed by either nitrogen fixation or nitrification, it can be absorbed by plants through their roots. Eventually the plant dies, and its body is added to the animal wastes and remains in the soil. Decomposers break down the nitrogen compounds in these wastes and remains and the formation of nitrate can begin again.

In a typical ecosystem the processes shown opposite recycle nitrogen between living organisms and the environment. However, some processes cause the loss of nitrate from the environment. This happens naturally as a result of **denitrification** (see above), and less naturally when crops are **harvested** and removed from the site where they have grown. These losses of nitrate can be made up either by nitrogen fixation or by adding nitrate in the form of fertilisers.

Q

1 Use your knowledge of the nitrogen cycle to explain how the following farming practices might improve soil fertility.

- **a** ploughing in stubble rather than burning it
- **b** draining waterlogged fields
- **c** planting peas or beans every third year
- **d** adding NPK fertiliser
- **e** adding well-rotted compost

2 Explain why farmers drain waterlogged fields.

The nitrogen cycle

The processes of nitrification, absorption, feeding, death, excretion and decay lead to the cycling of nitrogen between living organisms and their environment. In a natural ecosystem nitrogen fixation can 'top up' the cycle and make up for losses by denitrification.

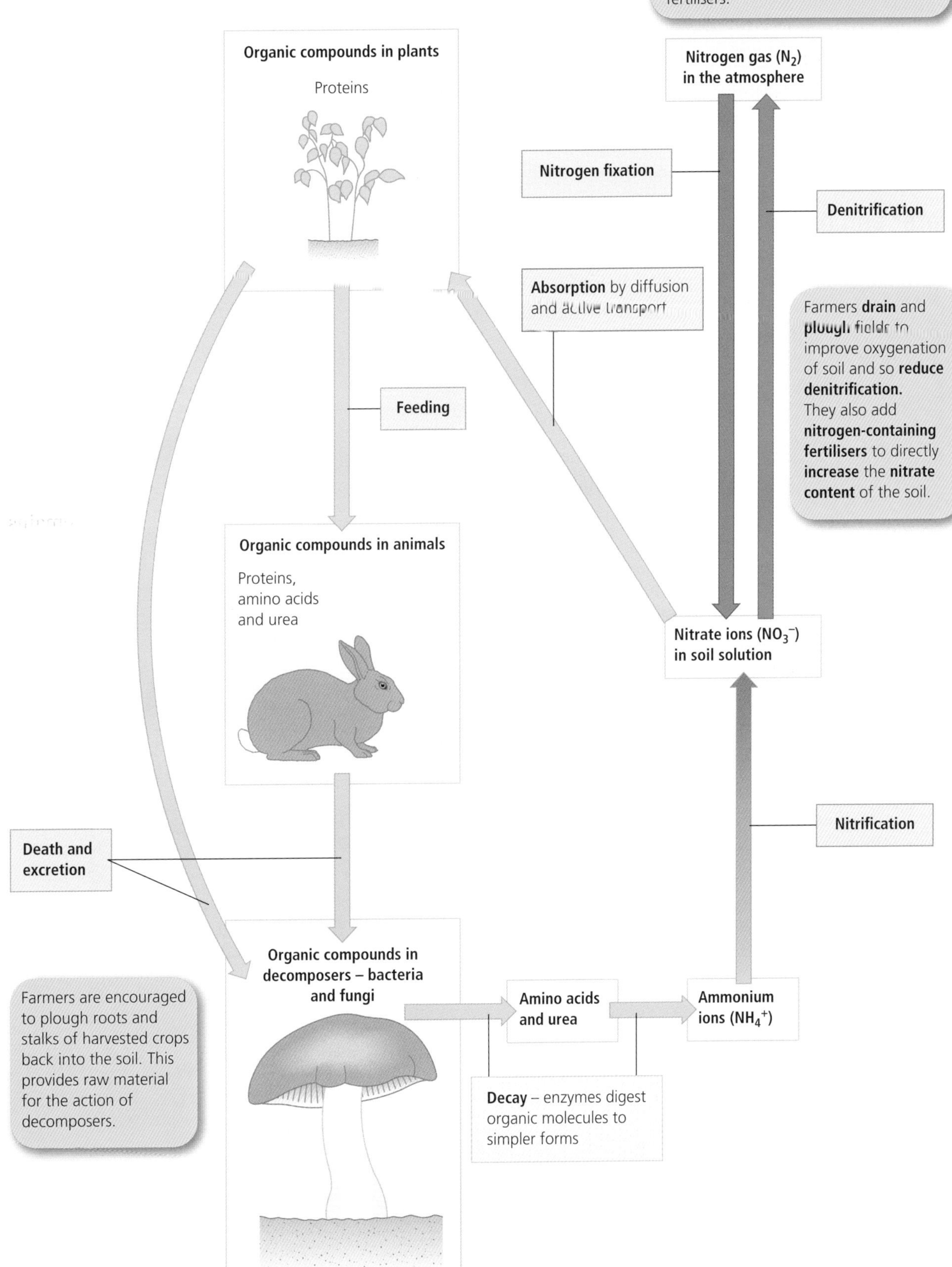
Some plants, called **legumes** (beans and peanuts are examples), have swellings on their roots. These **root nodules** contain bacteria, which can convert nitrogen gas to nitrate ions. These plants reduce the need for artificial fertilisers.
Organic compounds in plants
Proteins
Nitrogen gas (N_2) in the atmosphere
Nitrogen fixation
Denitrification
Absorption by diffusion and active transport
Farmers **drain** and **plough** fields to improve oxygenation of soil and so **reduce denitrification.** They also add **nitrogen-containing fertilisers** to directly **increase** the **nitrate content** of the soil.
Feeding
Organic compounds in animals
Proteins, amino acids and urea
Nitrate ions (NO_3^-) in soil solution
Nitrification
Death and excretion
Organic compounds in decomposers – bacteria and fungi
Amino acids and urea
Ammonium ions (NH_4^+)
Farmers are encouraged to plough roots and stalks of harvested crops back into the soil. This provides raw material for the action of decomposers.
Decay – enzymes digest organic molecules to simpler forms

4.7 Water is recycled too!

OBJECTIVES

- To know that all living organisms are largely water, and that biological reactions always take place in an aqueous (watery) environment
- To understand that the biological properties of water result from the structure of the water molecule
- To list some of the biological functions of water

Water and life

Life first evolved in water for a number of reasons:

- The molecules that were used by living organisms, and that made up their structure, were dissolved in the first seas.
- In the muddy estuaries and shallow seas of the primitive Earth, the molecules could become concentrated enough to react together.
- Water acted as a protective shield for the first living organisms against the damaging ultraviolet rays from the Sun.

Recycling water

Life continues on this planet because water has special properties. In particular, all three states of water – solid ice, liquid water and gaseous water vapour – exist at the temperatures found on the Earth's surface. The temperature varies at different times and at different places on the planet, but the average temperature over the Earth's surface is about 16.5 °C. This means that ice, liquid water and water vapour are all present and are continually interchanging. Water is recycled between different parts of the environment, as shown in the water cycle opposite.

The water cycle

All of the elements that make up living organisms, not just carbon and nitrogen, are recycled. The water cycle is different to the cycles of carbon and nitrogen because:

- only a tiny proportion of the water which is recycled passes through living organisms
- the most important factor in water recycling is heat energy from the Sun. This evaporates water, and also creates the temperature gradients which lead to winds.

The steps involved in the water cycle are shown in the diagram opposite.

The special properties of water

The picture of the kangaroo shows the importance of the properties of water to living things.

The **high specific heat capacity** of water means that cells or bodies with a high water content tend to resist heating up or cooling down, even when the temperature of their environment changes.
Evaporation of water from a surface allows loss of heat. Water has a **high latent heat of vaporisation**.
Because water is **incompressible**, it provides excellent support. Water helps support a whole organism (e.g. a fish), or part of an organism (e.g. the eyeball,
Water is an excellent **lubricant**, for example in saliva or in the synovial fluid of movable joints.
Water can be a **biological reagent**, for example in the processes of photosynthesis and digestion.
Water is an excellent **transport medium** for many biological molecules, such as oxygen, glucose, amino acids, sodium ions and urea.

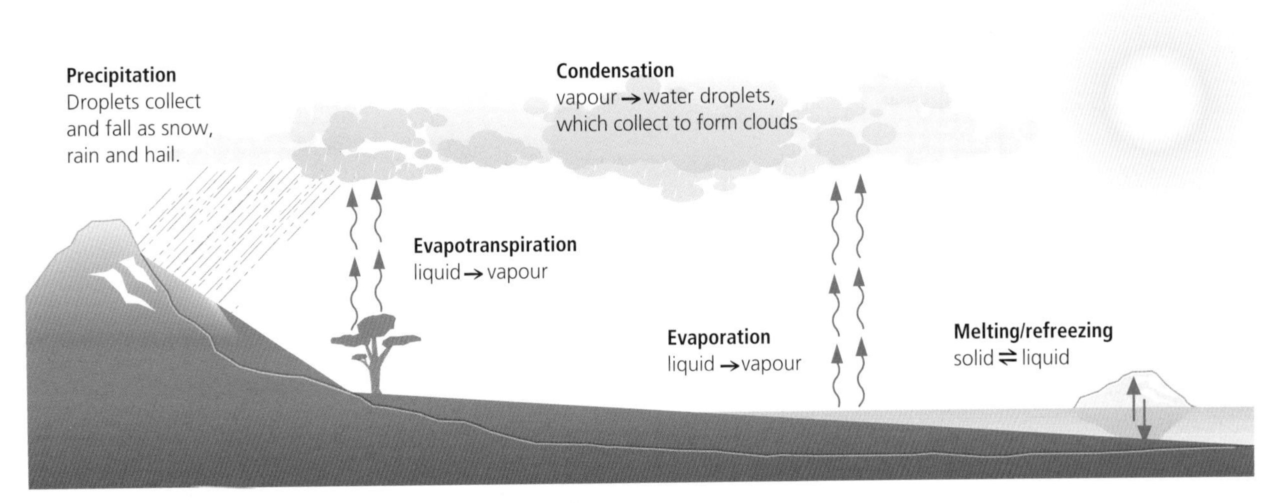

▲ The water cycle is maintained by heat energy from the Sun

1 **a** Explain how nitrogen in the muscle protein of a herbivore may be recycled to form protein in another herbivore some years later.

b Explain how the activities of some bacteria form a part of both the carbon and nitrogen cycles.

Cambridge IGCSE Biology 0610 Paper 2 Q8 June 2004

2 Use words from the following list to complete the paragraphs about ecosystems. You may use each word once, more than once or not at all.

respiration, decomposition, producer, chemical, carnivore, consumer, photosynthesis, energy, light, elements, decomposers, herbivore

In each ecosystem there are many feeding relationships. A food chain represents a flow of ____ through an ecosystem, and always begins with an organism called a ____ which is able to trap ____ energy and convert it to ____ energy. An organism of this type is eaten by a ____, which is a kind of ____ that feeds only on plant material. This type of organism is, in turn, eaten by a ____ (an organism that consumes other animals).

The process in which light energy is transferred into a chemical form is called ____ – eventually the energy is released from its chemical form during the process of ____ This process provides energy for all living organisms, including ____ which are microbes that feed on the remains of animals and plants.

3 In Africa, mammals called jackals are quite common. They feed on small herbivores such as young springboks and dik-diks, hunting in packs to catch their prey. They will also eat larger herbivores such as kudu that have been killed by larger predators such as lions.

A farmer in South Africa found that a number of his sheep, while feeding on grassland, were being killed by jackals. He noted that jackals always kill sheep by attacking their necks. He designed a plastic collar for the sheep that covered their necks. None of his sheep have been killed since fitting these collars. Other farmers are now buying the collars to protect their sheep from jackal attack.

a The prey species of the jackal are usually primary consumers. State the type of food that all primary consumers eat.

b Name the two carnivores identified in the text.

c Construct a food chain for the jackal to show its relationship with sheep.

d Suggest a reason why jackals survive better when they hunt in packs.

e When the farmer started to use collars on his sheep, although none of his sheep were being killed, the population of jackals did not decrease. Suggest why the number of jackals did not decrease.

f Name two structures, found in the neck of a sheep, that could be damaged when jackals attack it.

g Some of the protected sheep die of old age and their remains are eaten by other animals. Suggest and explain why the collars of the dead sheep could create an environmental problem.

Cambridge IGCSE Biology 0610 Paper 3 Q1 June 2004

Questions on ecosystems, decay and cycles

1 Over a period of several months, a student recorded some activities of the wildlife in a particular habitat. The following observations appeared in her notebook.

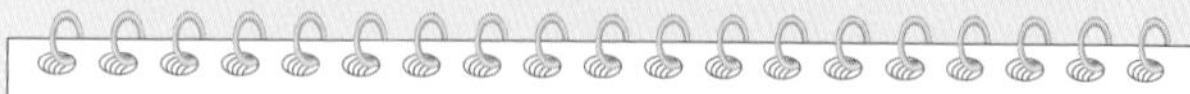

1. *Young shoots of a crop of bean plants covered with greenflies (aphids) sucking food from the stems.*
2. *Saw a large bird (hawk), which usually catches mice, swoop to take a small bird visiting the bean field to eat some of the aphids or butterflies.*
3. *Flowers of beans being visited by many different species of butterfly.*
4. *Mice seen nibbling at some dispersed bean seeds.*
5. *Spider's web constructed between two bean plants with 5 large black flies caught in it. Rotting body of a mouse nearby attracting similar flies.*

a Copy and complete the figure by filling in the names of the organisms to show the feeding relationships in this community.

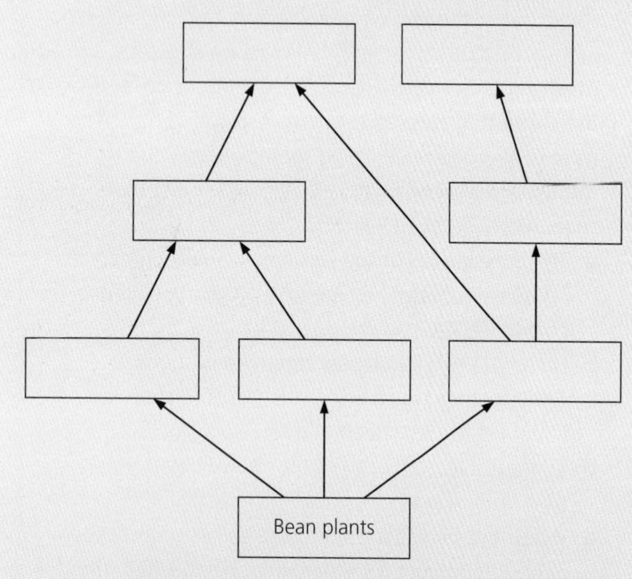

b **i** What name is given to a chart of feeding relationships as shown in the figure?

ii Name two top carnivores observed by the student.

c **i** Draw and label a pyramid of biomass for the hawks, mice and bean plants in this habitat.

ii Draw and label a pyramid of numbers for a bean plant, small birds and aphids.

Cambridge O Level Biology 5090
Paper 2 Q3 May 2008

2 **a** Cape buffalo graze on grass. While the buffalo are grazing, two or three oxpecker birds are often seen standing on the backs of each buffalo. These birds eat ticks that are parasites on the buffalo's skin.

i Draw a pyramid of numbers to represent these feeding relationships. Label the pyramid with the names of the organisms.

ii Draw a pyramid of biomass to represent the same feeding relationships. Lable the trophic levels on this pyramid.

b Explain how the nutrition of consumers differs from that of producers.

Cambridge IGCSE Biology 0610
Paper 2 Q6 November 2008

3 The figure shows parts of some natural cycles in the environment.

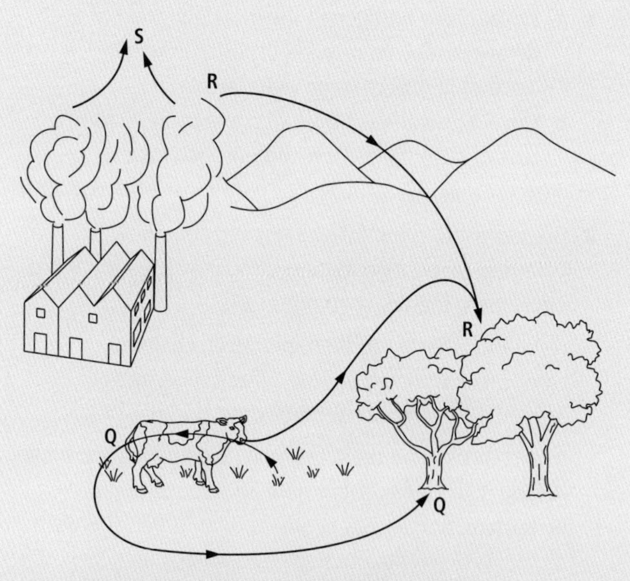

a With reference to the carbon and nitrogen cycles, explain what is happening at **Q** and **R**.

b Identify two gases that may be released at **S** and describe the possible harmful effects they may have on the environment.

Cambridge O Level Biology 5090
Paper 2 Q6 May 2007

4 Caribbean farmers sometimes:

a Plant peas and corn together to ensure that the corn plants could grow well. Use your knowledge of the nitrogen cycle to explain this practice.

b Use manure from farm animals as fertiliser for their crops, which are sold in the organic produce section at the greengrocer's store. Comment on this practice.

5 a The figure below shows the carbon cycle.

i Name the processes that cause the changes shown by the arrows labelled **A–D**.

ii Name **one** type of organism that brings about decomposition.

b Over the last few decades, the carbon dioxide concentration in the atmosphere has been rising.
Suggest how this has happened.

Cambridge IGCSE Biology 0610
Paper 2 Q7 November 2008

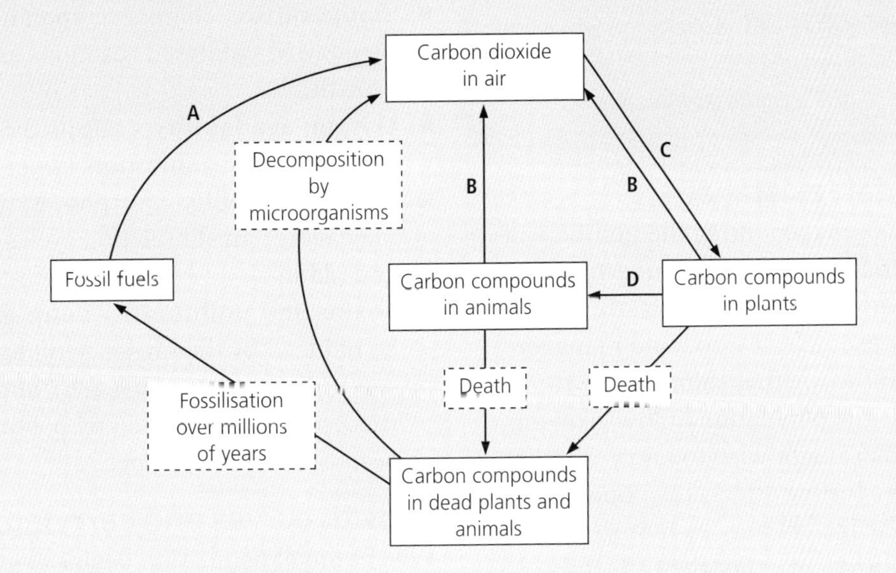

6 The figure below shows the water cycle.

a i The arrows labelled **P** represent evaporation. Which type of energy is needed for this process?

ii State what causes the formation of clouds at **Q**.

b i What process is represented by the arrows labelled **R**?

iii Name three factors that could alter the rate at which process **R** happens.

c A logging company wants to cut down the forest area.

i Suggest what effects this deforestation might have on the climate further inland. Explain your answer.

ii State two other effects deforestation could have on the environment.

Cambridge IGCSE Biology 0610
Paper 2 Q4 May 2009

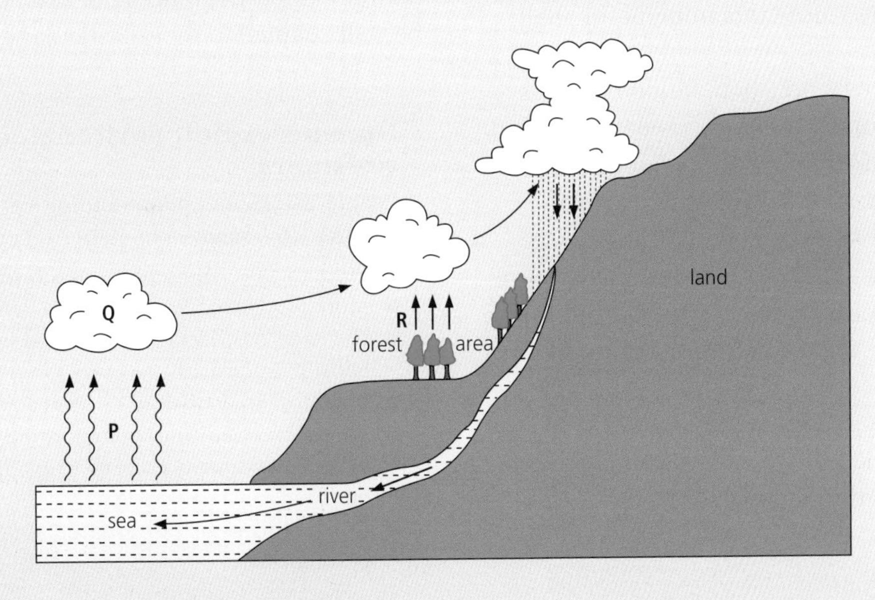

4.8 Factors affecting population size

OBJECTIVES

- To understand what is meant by environmental resistance
- To list examples of biotic and abiotic factors which limit population growth
- To explain the form of a typical growth curve

Environmental resistance

By taking samples and counting the numbers of organisms in a particular habitat, an ecologist can study how any factor affects the size of a population. For example, a woodland manager might wish to know whether thinning the trees in the wood affects the population of low-growing plants, or a farmer might wish to know whether the time of grass-cutting affects the population of bank voles. These factors may be **living** (**biotic**) or **non-living** (**abiotic**). Together they affect the rate at which a population grows, and also its final size. All the factors that affect population growth and size together make up the **environmental resistance**.

Some significant **biotic** factors that affect population growth include:

- **Food** – both the quantity and the quality of food are important. Snails, for example, cannot reproduce successfully in an environment low in calcium, no matter how much food there is, because they need this mineral for shell growth.
- **Predators** – as a prey population becomes larger, it becomes easier for predators to find the prey. If the number of predators suddenly falls, the prey species might increase in number extremely quickly.
- **Competitors** – other organisms may require the same resources, such as food, from the environment, and so reduce the growth of a population.
- **Disease** – often caused by parasites, and may slow down the growth and reproductive rate of organisms within a population.

Important **abiotic** factors affecting population growth include:

- **Temperature** – higher temperatures speed up enzyme-catalysed reactions and increase growth.
- **Oxygen availability** – affects the rate of energy production by respiration.
- **Light availability** – for photosynthesis. Light may also control breeding cycles in animals and plants.
- **Toxins and pollutants** – tissue growth can be reduced by the presence of, for example, sulfur dioxide (see page 266), and reproductive success may be affected by pollutants such as oestrogen-like substances.

Growth curves and carrying capacity

When a small population begins to grow in a particular environment, the environmental resistance is almost non-existent – there may be plenty of food and no accumulation of poisonous wastes. The diagram opposite shows how environmental resistance eventually limits population growth, and the environment reaches its **carrying capacity**. Unless the environmental resistance is changed, perhaps by a new disease organism, the size of the population will only fluctuate slightly. Organisms that are able to maintain their population, or even increase it, must be well adapted to their particular environment.

Humans exploit environmental resistance

People use their understanding of environmental resistance to manage populations. For example:

- Predators are eliminated from farm situations.
- More food is made available to domestic animals.
- Nitrogen fertilisers and artificial light are used to boost plant growth.
- Predators may be used to control pests.
- Anaerobic conditions or low temperatures are used to prevent populations of microbes from consuming our food.
- Competitors are eliminated from crops using pesticides.

An example can be found on page 248.

Factors affecting population growth

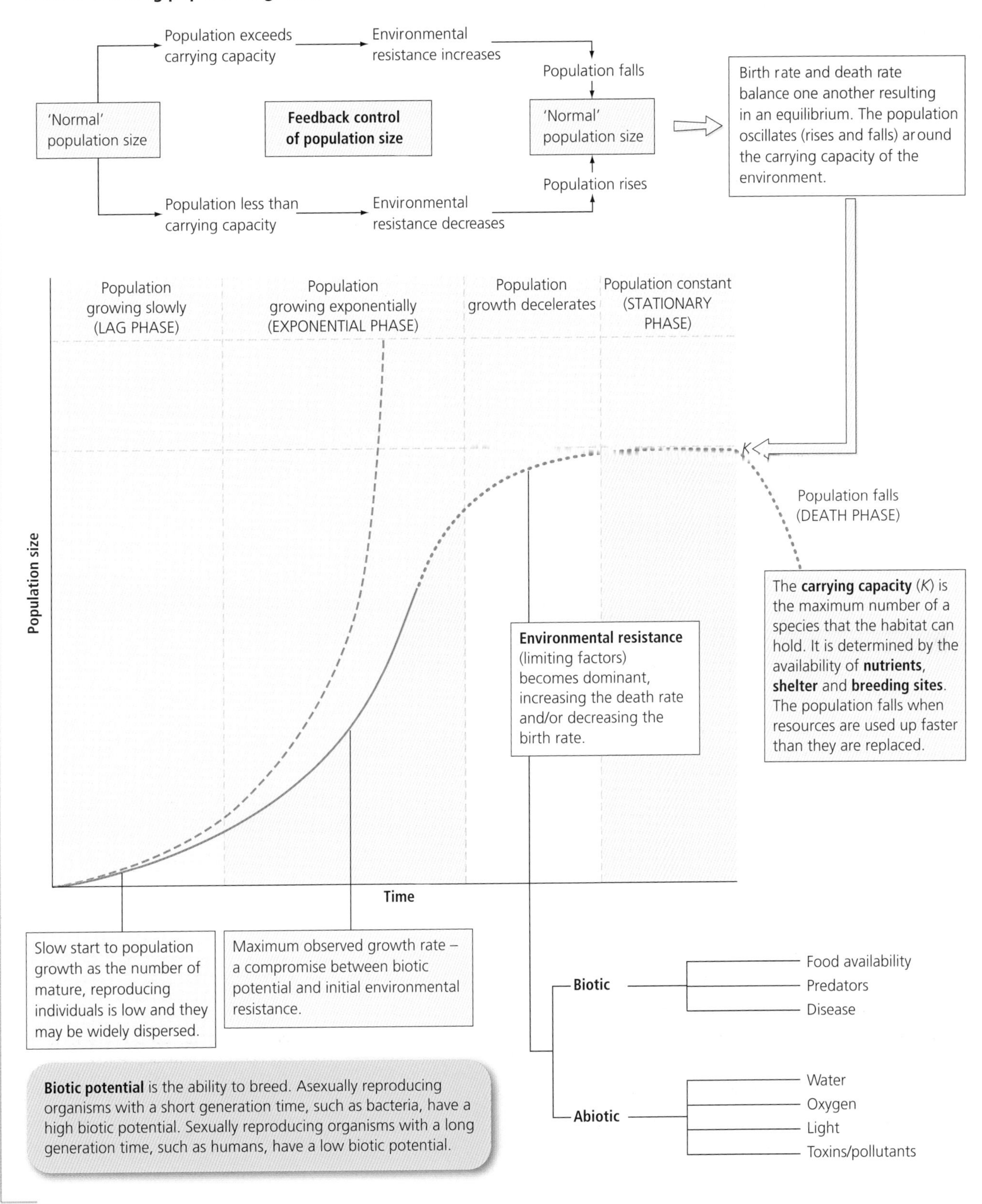

Biotic potential is the ability to breed. Asexually reproducing organisms with a short generation time, such as bacteria, have a high biotic potential. Sexually reproducing organisms with a long generation time, such as humans, have a low biotic potential.

1. What is meant by the term environmental resistance? Give an example.
2. Define the terms biotic and abiotic factors, and give examples of each.
3. Give examples of the ways in which humans exploit their knowledge of the factors affecting population growth.

4.9 Human population growth

OBJECTIVES

- To know that the evolution of humans from hunter-gatherers to permanent settlers caused changes in the environment
- To understand the form of a human population growth curve

Humans, like other organisms, must find **food**, **shelter** and a **place to breed**. The first humans were hunter-gatherers who moved from place to place, taking what they needed to satisfy these requirements (but allowing the environment to recover once they moved on). After the most recent Ice Age this method of living became more difficult and humans began to settle in the most suitable areas. This meant that the environment did not have time to recover.

Growth of the human population

As with other organisms, the growth of the human population can be presented as a population growth curve. The number of humans increases, by reproduction, until the carrying capacity of the environment has been reached. Humans have the ability to alter their environment to raise the carrying capacity. Three major changes in human activities led to significant surges in the world population over the past 300 years, as shown below.

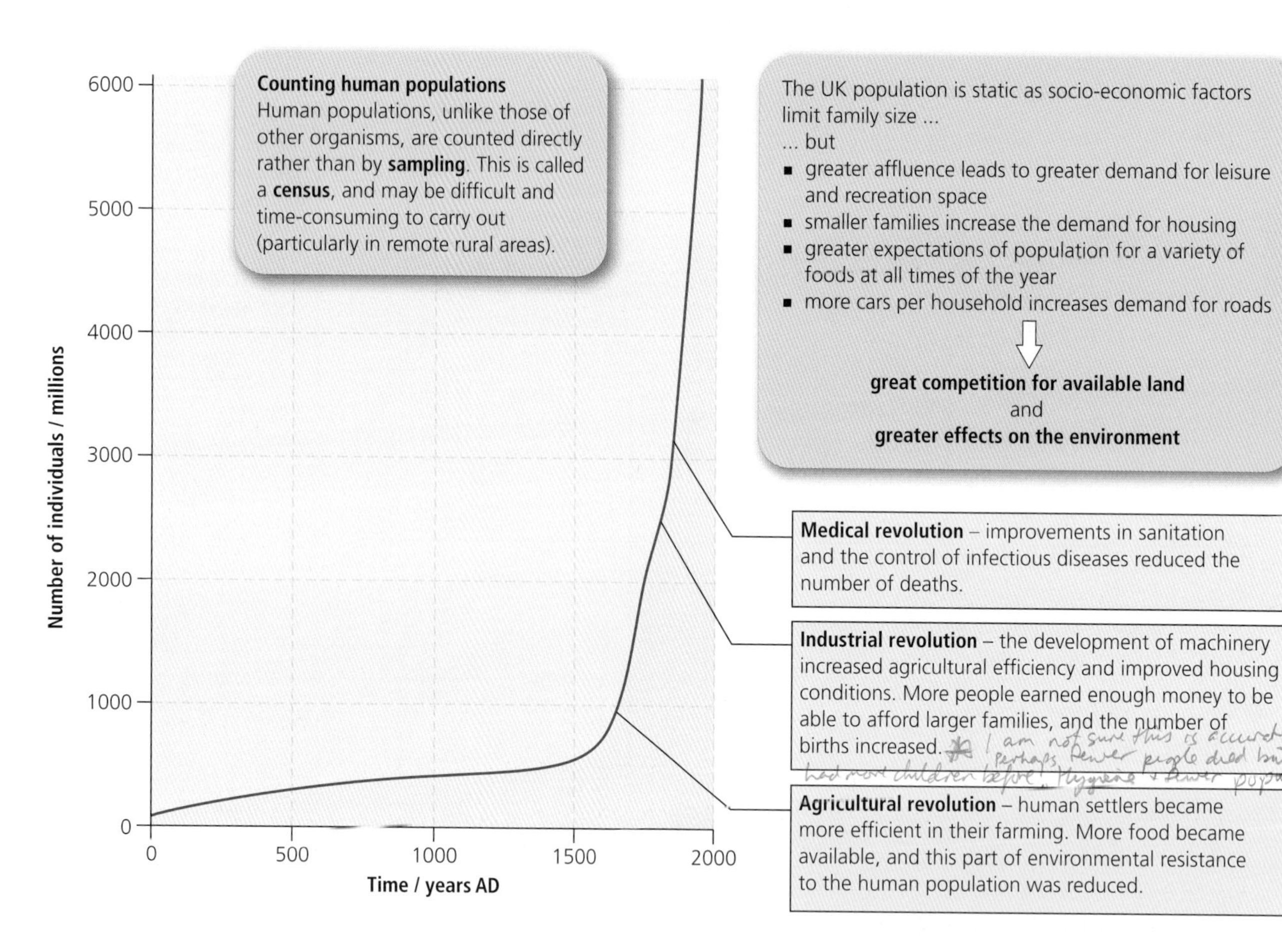

▲ Human population of the world

The shapes of populations

A population grows if the number of births plus the number of immigrants is greater than the number of deaths plus the number of emigrants. This relationship is shown in the figure below.

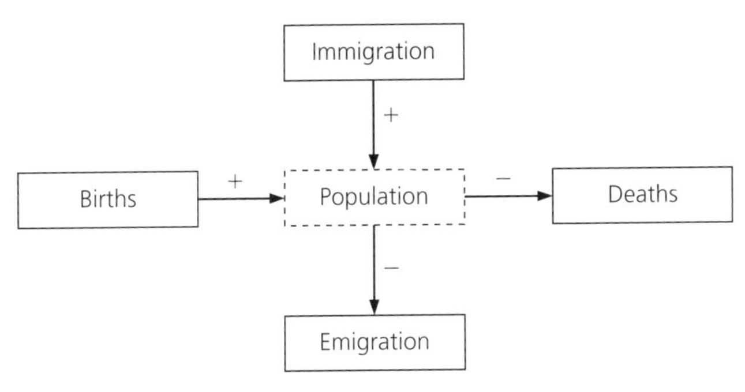

▲ Factors affecting population size

If immigration and emigration remain stable, the population size will depend on the relative numbers of births and deaths. If the population size is measured, and then the data is broken down into different age groups, a **population pyramid** can be drawn. A population pyramid shows the **age structure** of a population – the percentage of males and females in each age group in a country.

This age structure is affected by:

- **Food availability:** this is particularly significant in those countries where a balanced diet might be difficult to obtain. Deficiencies in protein and vitamins can lead to a very high death rate amongst infants.
- **Hygiene:** food poisoning and water-borne infections can cause an increased death rate from diarrhoea and dehydration. Provision of a 'safe' water supply has a huge impact on the stability of a population.
- **Medical provision:** vaccination programmes can prevent many deaths, and care during childbirth can dramatically reduce deaths among young women. The availability of combination therapy for HIV infections has significantly reduced death rates in the 25–40 age groups in sub-Saharan Africa.
- **Working conditions:** unsafe work, for example mining, can increase death rates amongst young people (usually men), particularly if medical care is poor.
- **Political instability:** this can sometimes lead to civil war, with serious effects on death rates.

Less economically developed nations (LEDNs) typically have a high birth rate but a high mortality rate amongst young people. The population pyramid tends to have a wide base (lots of births) but sloping sides (high death rate in younger age groups). The population pyramids for more economically developed nations (MEDNs) have a narrower base but almost vertical sides, reflecting a lower birth rate and few deaths amongst young people.

All population pyramids show a higher proportion of females in the older age groups (the two X chromosomes offer genetic benefits against sex-linked conditions).

Population pyramids for two contrasting nations are shown here.

Age (population) pyramids show the age structure of a population.

They can be compared with respect to:

- width of base (proportion of young people in population)
- height (number of individuals surviving to old age)
- angle of sides (death rate affects slope).

LEDN

- wide base
- low height
- sloping sides

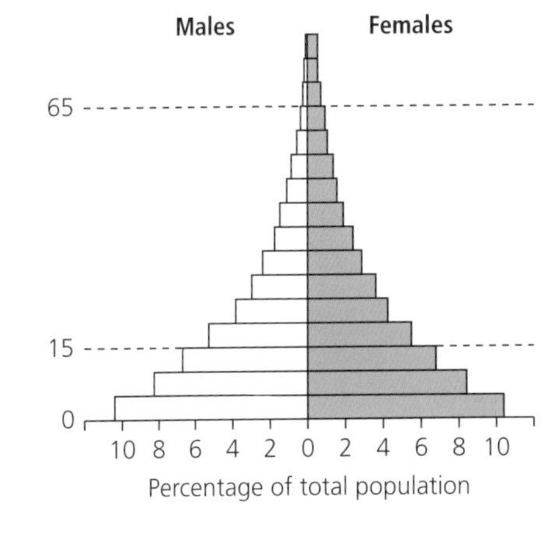

MEDN

- narrow base
- considerable height
- vertical sides

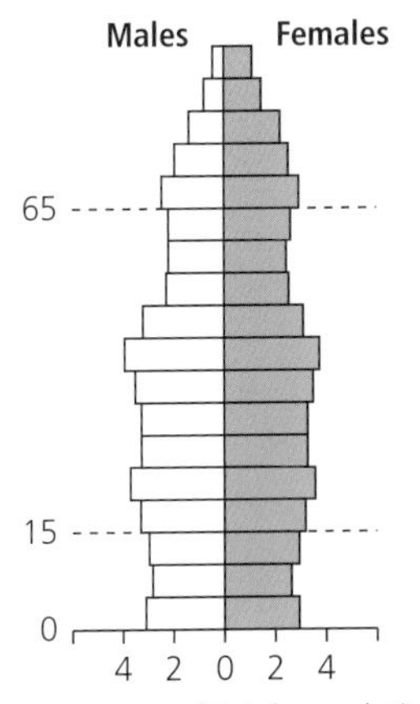

4.10 Bacteria are useful in biotechnology and genetic engineering

OBJECTIVES

- To know the structure of a typical bacterial cell
- To know the requirements for bacterial growth
- To know how bacteria reproduce
- To understand some of the ways in which bacteria can be used in human activities

Requirements of bacteria

Bacteria have certain requirements that their environment must provide. An understanding of these requirements has been important in biotechnology and in the control of disease. If the environment supplies these needs, the bacteria can multiply rapidly by **binary fission** (see opposite). In this process each bacterium divides into two, then each of the two divides again and so on, until very large populations are built up. A bacterial colony can quickly dominate its environment, making great demands on food and oxygen and perhaps producing large quantities of 'waste' materials. Some of these waste materials can be useful to humans. For example, bacterial enzymes are used in a number of industrial processes.

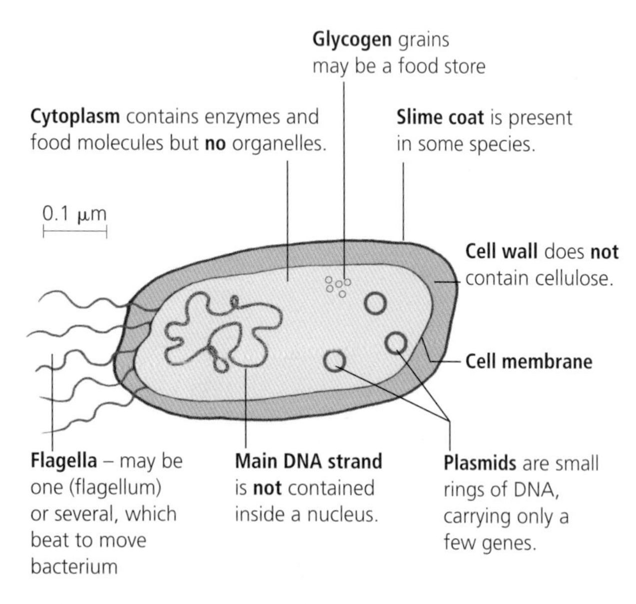

▲ Bacteria have a cell wall but do not have a nucleus or organelles

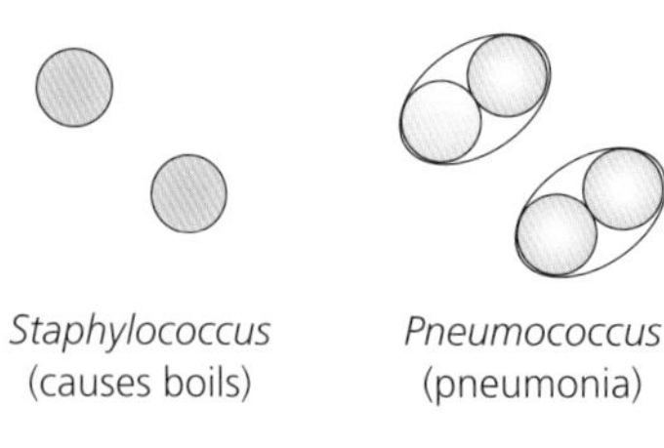

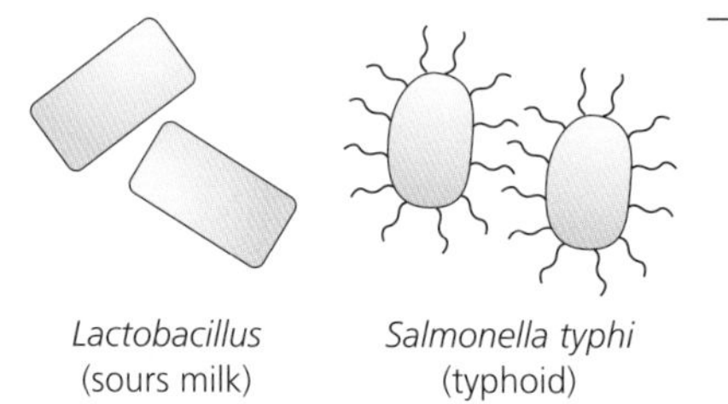

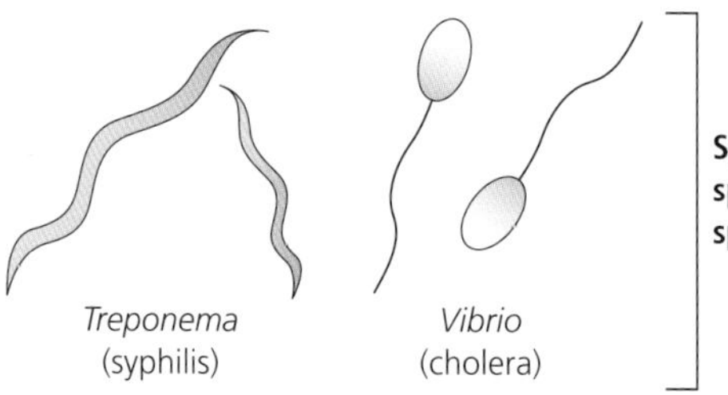

▲ Bacterial shapes

Bacterial structure

Bacteria (singular: bacterium) are single-celled organisms that **have no true nucleus**. Bacterial cells do not contain organelles like those found in typical animal and plant cells (see page 17), but are able to carry out all of their life processes without them. A few can photosynthesise, but most feed off other organisms. They may be **parasites**, feeding off living organisms, or **saprotrophs**, feeding off dead organisms.

Bacteria are very small, usually about 1–2 μm in length, and so are only visible using a high-powered microscope. The structure of a typical bacterium is shown in the diagram below.

Bacteria exist in a number of different shapes, some of which are shown opposite above. Shape can be used to classify bacteria.

The **generation time** (time taken for each cell to divide into two) can be as little as 20 minutes under ideal conditions. One *E. coli* cell in the human gut could **theoretically** become 2^{72} cells in 24 hours – this number of cells weighs about 8000 kg!

The number of cells in a population, starting from just one cell, is 2^n (where n = number of generations)

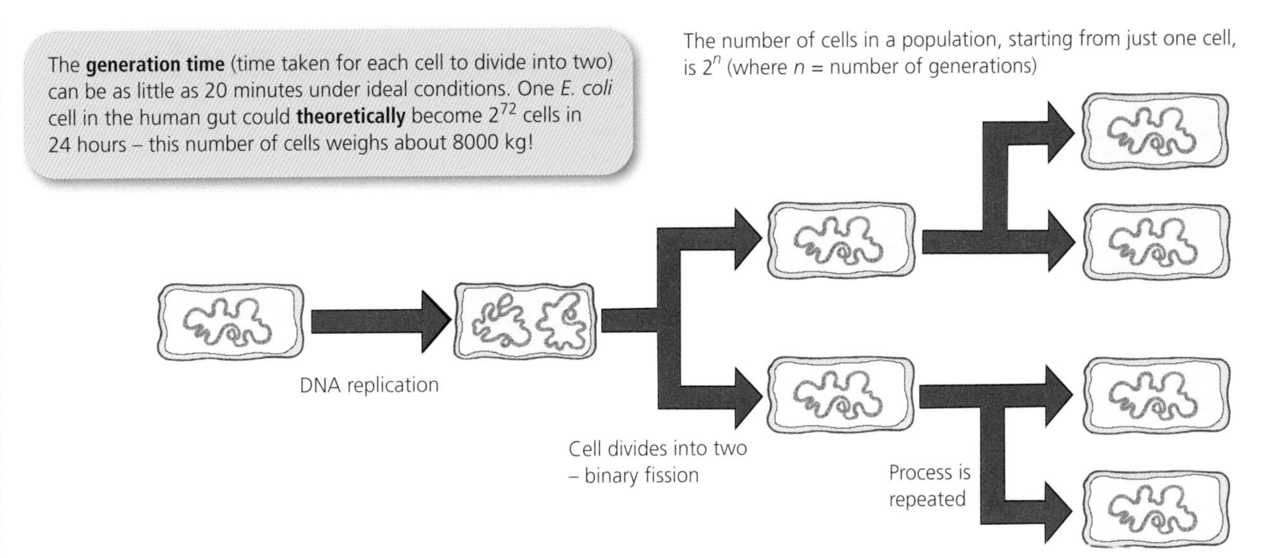

This can be so rapid that a bacterial population can grow very quickly

When conditions prevent growth, some bacterial cells survive by producing **hard-coated spores**.
This can make some species very difficult to kill.

- Sterilisation kills all bacteria and their spores

Number of bacteria

Time

Stationary phase – population remains constant as the number that die is balanced by the number produced by binary fission. The rate of binary fission is limited by:

- availability of nutrients
- availability of oxygen, in aerobic species
- temperature
- availability of water
- pH.

The number of deaths may be increased by:

- production of waste which increases in concentration as population becomes more dense.

Death phase – population falls as number of deaths exceeds number produced by binary fission. Caused by shortage of resources affecting stationary phase.

Exponential phase – population grows rapidly, doubling perhaps as often as every 20 minutes.
The bacteria require:

- amino acids, for the synthesis of proteins
- carbohydrate, e.g. glucose, for the release of energy by respiration
- water, as a solvent and a reagent in biochemical reactions
- a suitable temperature, for optimum enzyme activity.

Bacteria are particularly useful in biotechnology and genetic engineering because:

- they reproduce very rapidly so scientists can build up large populations very quickly
- they can produce complex molecules, such as enzymes and the hormone insulin
- their genetic code is the same as that in more complex organisms (even humans!)
- they have extra pieces of DNA – plasmids – which scientists can use to carry genes from one cell to another
- people are less worried about experiments on bacteria than on more familiar, larger, organisms such as mice, rabbits and dogs.

Understanding what bacteria need to reproduce can be useful in:

- preventing food spoilage – see page 99
- controlling infections
- providing the best conditions for growth of useful bacteria.

4.11 Humans use enzymes from bacteria

OBJECTIVES

- To understand that enzymes have many roles which benefit humans
- To know examples of a range of uses of enzymes
- To understand the benefits of enzyme immobilisation

Enzymes in industry

Enzymes are biological catalysts that operate at the temperatures, pressures and moderate pH values found in living organisms (see page 33). Using enzymes for industrial processes therefore does not require extreme (and expensive) conditions. For example, the Haber process for producing ammonia requires a temperature of 750 °C and a pressure of 30 times atmospheric pressure whereas nitrogen-fixing bacteria can perform this process at 25 °C and at atmospheric pressure.

The majority of enzymes used commercially are obtained from microbial sources, usually fungi or bacteria.

Microorganisms that produce those valuable enzymes are grown under carefully controlled conditions in a fermenter (bioreactor), like the one shown on the opposite page.

Food production

- A **protease** from bacteria softens gluten, making the rolling of biscuits easier.
- **Pectinase** breaks down clumps of plant cells: this changes fruit juices from cloudy to clear.

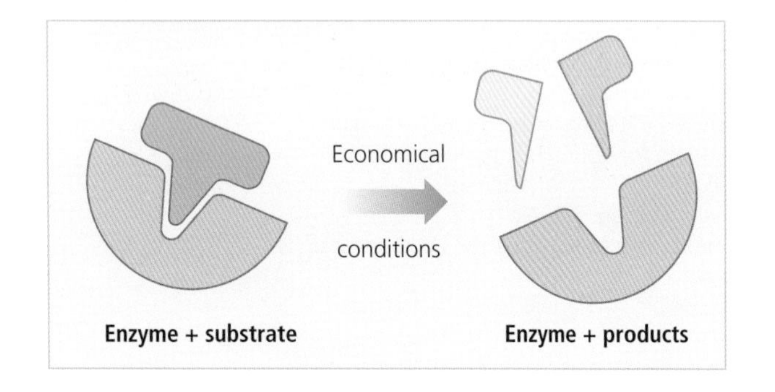

Biological washing powders

The most difficult stains to remove from clothes often include **lipids** (e.g. from fatty foods or greasy fingerprints) or **proteins** (e.g. from blood). These can be washed out using biological washing powders that contain **lipase** and **protease**. These powders have the added advantage of working at lower temperatures (less water heating needed, and clothes don't shrink).

Commercial applications of enzymes

Some enzymes are **immobilised** on a carrier of fibres. They can be re-used many times eg. lactase for lactose-free milk

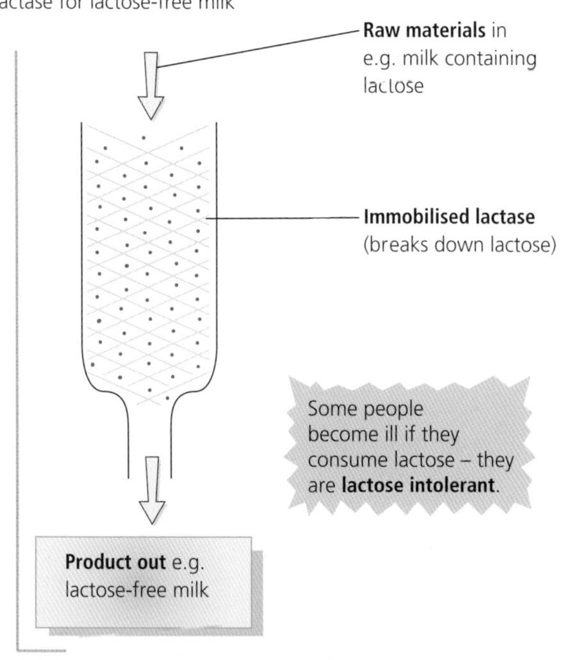

Some people become ill if they consume lactose – they are **lactose intolerant**.

FUNGI CAN BE USEFUL TOO!
A **lipase** from fungi helps to make chocolate flow to cover biscuits and sweets!

A fermenter (or bioreactor)

Paddle stirrers continuously mix the contents of the bioreactor:
- ensures microorganisms are always in contact with nutrients
- ensures an even temperature throughout the fermentation mixture
- for aerobic (oxygen-requiring) fermentations the mixing may be carried out by an **airstream**.

Fermenters are also used in fuel production (see page 278) and the manufacture of the antibiotic penicillin (see page 248).

Microbe input – the organisms that will carry out the fermentation process are cultured separately until they are growing well.

Nutrient input – the microorganisms require:
- an energy source – usually carbohydrate
- growth materials – amino acids (or ammonium salts which can be converted to amino acids) for protein synthesis.

Gas outlet – gas may be evolved during fermentation. This must be released to avoid pressure build-up, and may be a valuable by-product, e.g. carbon dioxide is collected and sold for use in fizzy drinks.

Constant temperature water jacket – the temperature is controlled so that it is high enough to promote enzyme activity but not so high that enzymes and other proteins in the microbes are denatured.

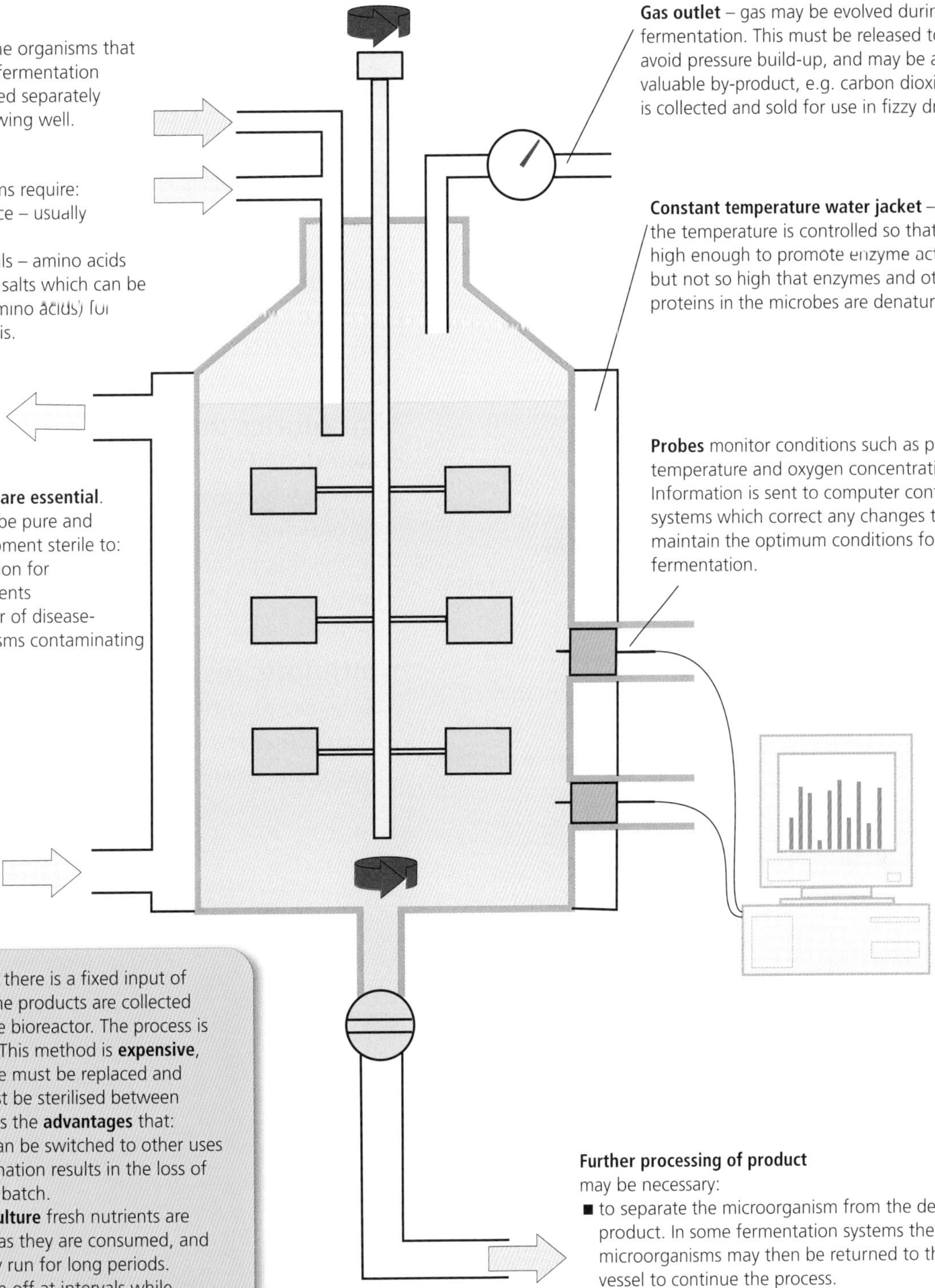

Heating/cooling water out

Sterile conditions are essential. The culture must be pure and all nutrients/equipment sterile to:
- avoid competition for expensive nutrients
- limit the danger of disease-causing organisms contaminating the product.

Probes monitor conditions such as pH, temperature and oxygen concentration. Information is sent to computer control systems which correct any changes to maintain the optimum conditions for fermentation.

Heating/cooling water in

In **batch culture** there is a fixed input of nutrients and the products are collected by emptying the bioreactor. The process is then repeated. This method is **expensive**, since the culture must be replaced and the reactor must be sterilised between batches, but has the **advantages** that:
- the vessels can be switched to other uses
- any contamination results in the loss of only a single batch.

In **continuous culture** fresh nutrients are added as soon as they are consumed, and the reactor may run for long periods. Products are run off at intervals while the process continues. This is **more economical** but has the **disadvantages** that:
- contamination causes greater losses
- the culture may block inlet or outlet pipes.

Further processing of product may be necessary:
- to separate the microorganism from the desired product. In some fermentation systems these microorganisms may then be returned to the vessel to continue the process.
- to prepare the product for sale or distribution – this often involves **drying** or **crystallisation**.

4.12 Using fungi to produce antibiotics: drugs to control bacterial disease

OBJECTIVES

- To recall the mode of nutrition of fungi
- To understand that fungi may produce antibiotics as part of their normal metabolic processes
- To define the term antibiotic
- To understand how penicillin works as an antibiotic
- To describe the large-scale production of antibiotics
- To understand the problems of antibiotic resistance

Natural antibiotics: penicillin

A fungus such as *Penicillium* absorbs food molecules from its environment, and then uses these molecules for its own metabolism. Sometimes a *Penicillium* mould will make substances that it secretes into its environment to kill off any disease-causing or competitive microorganisms. A product made by one microorganism to kill off another microorganism is called an **antibiotic**.

The first antibiotic to be discovered was called **penicillin** after the organism *Penicillium* that produced it. Antibiotics used in medicine work in various ways to inhibit the development of bacterial infections, without harming human cells. Penicillin prevents the bacterial cell walls forming (human cells do not have cell walls). Antibiotics only affect bacterial cells – there is no benefit in taking antibiotics to treat viral or fungal diseases.

Production of penicillin

The large-scale production of penicillin takes place in industrial fermenters. Penicillin is a **secondary metabolic product** – it is made when growth of the producer organism is slowing down rather than when it is at its maximum, as shown in the diagram below.

Antibiotics do not always kill bacteria

Bactericidal antibiotics kill the pathogen directly, whereas **bacteriostatic** ones prevent it reproducing, leaving the host's defences to kill the existing pathogens. The diagram opposite illustrates the different effects of a bactericidal agent and a bacteriostatic agent.

Antibiotic resistance

As the use of antibiotics increases, strains of bacteria that are **resistant** to the antibodies are developing, as shown opposite.

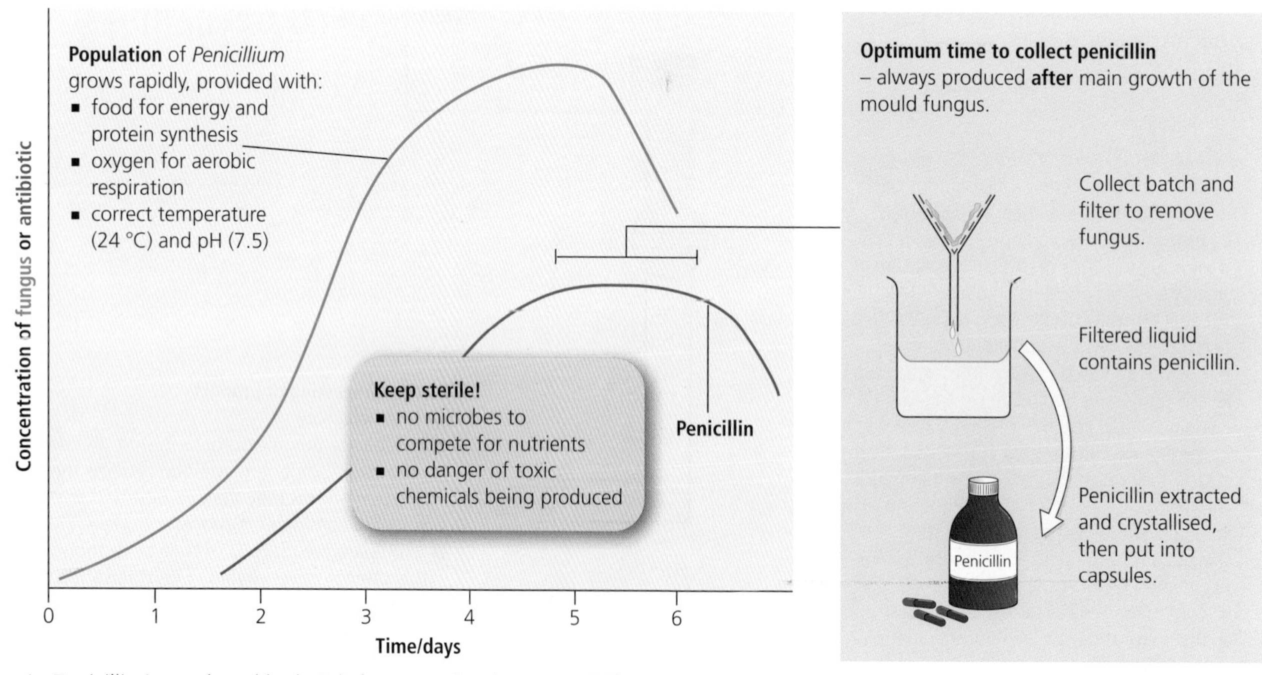

▲ Penicillin is produced by batch fermentation (see page 35)

Antiseptics are not the same as antibiotics

An **antibiotic** is used to **treat** a bacterial infection. An **antiseptic** can be used to **prevent** infection ('anti' = prevent, 'sepsis' = infection). Antiseptics are chemicals applied to the outside of the body, such as the skin, to kill microbes.

Disinfectants are also chemicals that kill microbes, but they are more powerful and are generally used on worksurfaces or lavatory bowls, where potentially harmful bacteria may be growing.

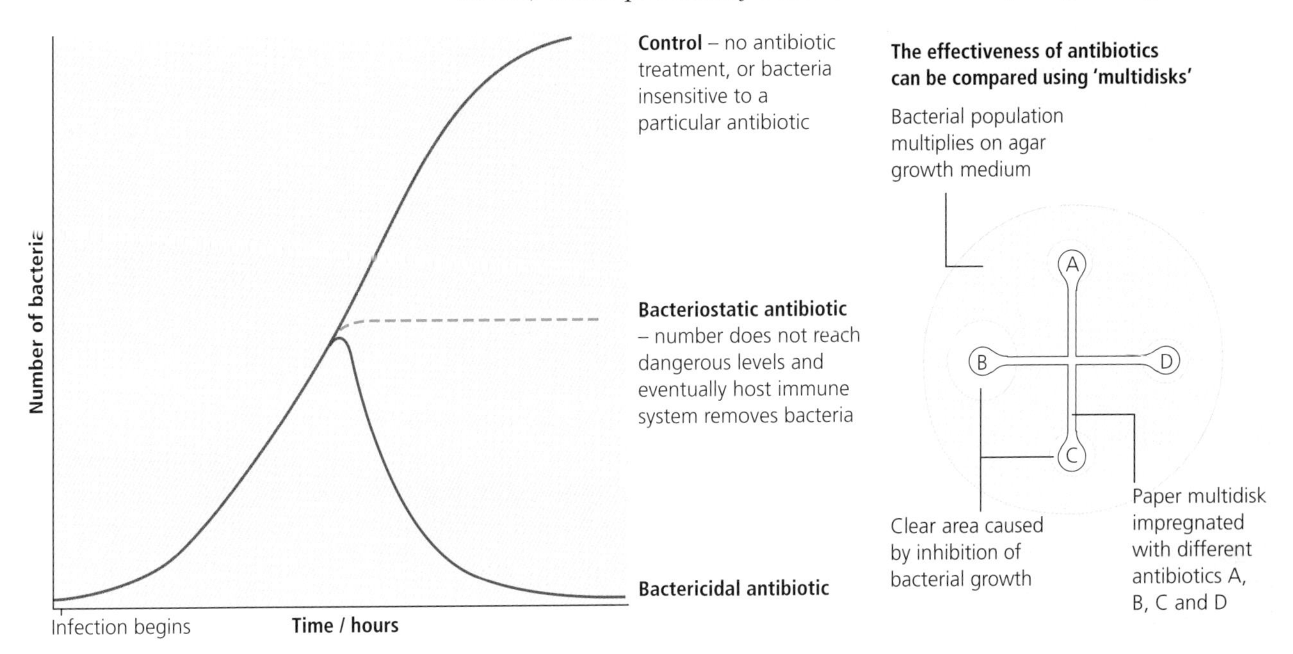

▲ Penicillin is bactericidal at high concentrations, but these may have some side-effects in humans (and may lead to **resistance** – see below) so penicillin is usually given at bacteriostatic doses.

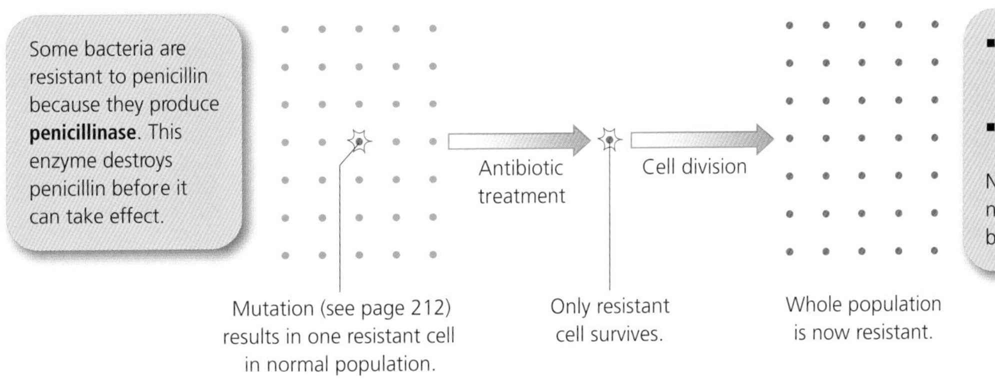

- The antibiotic is the **selection pressure** (see page 216).
- Resistance is the **survival advantage**.

NB The antibiotic does not **create** the resistance, but **selects** it in a population.

▲ Antibiotic-resistant strains of bacteria are formed by artificial selection. This is more likely to happen if antibiotics are used unnecessarily, or if people do not finish a course of prescribed antibiotics. This allows some of the bacterial population to recover, and possibly develop strains that are resistant to the antibiotic.

Q

1. Distinguish clearly between the following pairs of terms:
   - **a** antiseptic and antibiotic
   - **b** antibiotic and antibody
   - **c** resistance and immunity.
2. Purified penicillin is normally taken orally (by mouth). The crystals of the drug are enclosed in a capsule for distribution. Suggest three important properties of the material used to make the capsules.

4.13 Baking and brewing: the economic importance of yeast

OBJECTIVES

- To appreciate that some microorganisms are useful to humans
- To recall an equation for anaerobic respiration
- To understand the industrial production of alcohol and bread

Bread production

Flour, sugar, water and salt are mixed with **yeast**. This process is called **kneading** and produces **dough**.

The dough is rolled into shape, then kept in a warm (about 28 °C), moist environment. The yeast ferments the sugar, and the bubbles of carbon dioxide are trapped by the sticky proteins of the dough. The dough expands or **rises**.

Cooking at 180°C – **baking**:
- kills the yeast and stops fermentation
- causes alcohol to evaporate
- hardens the outer surface to form a crust.

The product is **bread**, which should be:
- spongy in texture, because of trapped carbon dioxide
- fresh smelling (partly alcohol vapour!).

If a loaf is not left to rise, it will be small and dense

If a loaf is left to rise for too long too much carbon dioxide is produced and there are too many large holes

The perfect loaf

'Wholemeal' bread is made with flour which has **not** had the husk removed from the wheat – hence 'whole' meal. It is rich in B vitamins and dietary fibre.

Bread may be fortified – in Britain, bread often has added protein, calcium and vitamins A and D – not to mention preservatives and whitening agents.

Yeast performs alcoholic fermentation

Much of the commercial use of microorganisms is based on the process of **fermentation**. This term covers *any* metabolic process carried out by microorganisms using carbohydrate as a starting material. The best-known fermentation reaction is the anaerobic respiration of glucose by the single-celled fungus **yeast**:

$$\text{glucose} \longrightarrow \text{carbon dioxide} + \text{ethanol} + \text{energy}$$

Since one of the products is ethanol (ethyl alcohol), this process is often known as **alcoholic fermentation**. As well as making alcohol, the carbon dioxide produced is valuable commercially. For the yeast, the useful product is energy – the alcohol and carbon dioxide are by-products.

Alcohol has a high energy content and is toxic

Alcohol contains a great deal of energy, a fact that is exploited when alcohol is used as a fuel source (see page 278). Alcohol is also toxic (poisonous) and the yeast excretes it into the surrounding liquid medium. If the alcohol concentration in the medium gets higher than 8–9%, the yeast is killed and fermentation stops. Alcohol is toxic to humans too (page 152)

Baking and brewing use fermentation

The processes of **baking** and **brewing** have both been around for thousands of years. The process of baking is outlined on the left.

Fine tuning the fermentation process

Because of the enormous commercial significance of these processes, much research has gone into making them as efficient as possible. For example:

- strains of yeast that are tolerant to higher concentrations of alcohol have been developed
- high-yielding strains of wheat have been selectively bred
- genetic engineers have developed yeasts which can convert starch to maltose, and thus remove the need for the 'malting' stages in brewing.

Questions on bacteria

1 Yeast ferments glucose ($C_6H_{12}O_6$) to ethanol (C_2H_5OH) and carbon dioxide. Write out a balanced chemical equation for this process.

2 Bread may be fortified with vitamins A and D, and the mineral calcium. What benefit would these give to the consumer? Suggest why bakers often add: **a** preservative and **b** whitener to their product.

3 Mycoprotein is similar to single cell protein and is sold as an alternative to meat such as beef. The table shows the composition of mycoprotein and beef.

nutrient	dry mass / g per 100 g	
	mycoprotein	uncooked beef
protein	49.0	51.4
fat	9.2	48.6
fibre (roughage)	19.5	0.0
carbohydrate	20.6	0.0

a i State two differences in composition between mycoprotein and beef.

ii Using data from the table, suggest two reasons why eating mycoprotein is better for health than eating beef. Explain your answers.

b i Calculate the dry mass of mycoprotein **not** represented by protein, fat, fibre or carbohydrate. Show your working.

ii Suggest **one** nutrient that this dry mass might contain.

c The antibiotic penicillin is produced by fungi that are grown in a fermenter, as shown in the figure. The process is similar to the manufacture of enzymes.

i Name the two raw materials likely to be present in the feedstock.

ii State the function of **X**.

iii Suggest the name of the main gas present in the waste gases.

d During the fermenting process, the temperature in the container would rise unless steps are taken to maintain a constant temperature.

i Suggest a suitable temperature for the feedstock.

ii Explain why the temperature rises.

iii Explain why a constant temperature has to be maintained.

iv Using the information from the figure below, suggest **how** a constant temperature is maintained.

Cambridge IGCSE Biology 0610
Paper 3 Variant 1 Q3 November 2008

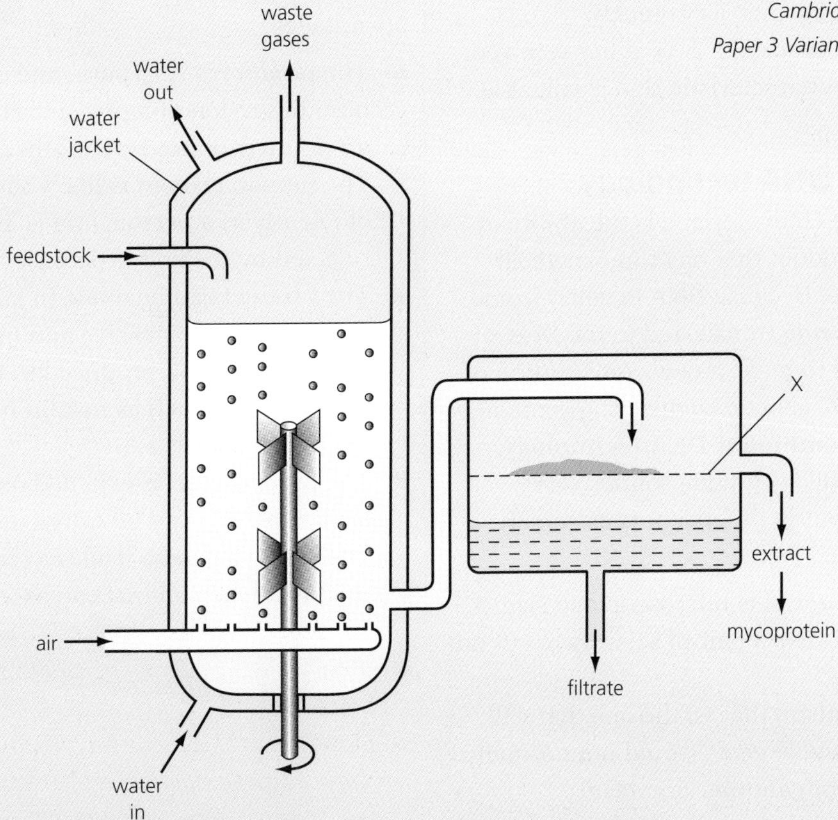

4.14 Genetic engineering

OBJECTIVES

- To understand the term genetic engineering
- To understand the value of enzymes in genetic engineering
- To be able to describe a technique in genetic engineering
- To list some products of genetic engineering that are of value to humans

What is genetic engineering?

The examples of selective breeding described on page 218 are a form of genetic engineering, since humans are interfering with the natural flow of genetic material from one generation to the next when they choose the animals or plants that will be allowed to reproduce (and pass on their genes). However, what we now call genetic engineering is a much more predictable and refined process than selective breeding. In this process:

- Genes that code for characteristics valuable to humans are identified.
- These genes are removed from the animal or plant that normally shows this characteristic.
- The genes are transferred to another organism, usually one that grows very quickly.
- This organism 'reads' the gene it has received, and shows the characteristic that is valuable to humans.

Recombinant DNA technology

In many cases the characteristic is the ability to manufacture a product that has some medical or industrial value. Because DNA (a gene) from one organism is being transferred to the DNA of another organism to make a new combination of DNA, this 'modern' genetic engineering is often referred to as **recombinant DNA technology**. The principle of this technique sounds very straightforward, but in practice it is extremely difficult.

- The genetic material is microscopic in size. A technician can't use a pair of scissors to cut out a gene!
- The **host** organism, that is, the one that will receive the valuable gene, would not normally take in DNA from another organism.
- The host organism might not show the valuable characteristic, for example it may not make a particular protein, even though it now has the DNA that codes for the protein.

Technological advances have overcome these potential problems. Key to success has been the discovery of:

- **enzymes** that can cut DNA from one chromosome and paste it into another, acting like scissors and glue
- **vectors** that can carry the gene from one organism to another
- **culture techniques** that allow large quantities of the valuable product to be produced and collected, even if the host organism would not normally produce it.

Many compounds can be manufactured by genetically engineered bacteria, including insulin and other products which are used directly by humans. These products are otherwise often taken from animals. For example, insulin can be isolated from the pancreas of a pig.

Recombinant DNA technology has several advantages:

- The product is very pure, and can be the human version of a protein rather than a version produced by another animal. The human protein is likely to work more efficiently in a person, and is less likely to be rejected by the body's defences.
- The product can be made in large quantities, making it less expensive and more readily available. Insulin produced in this way costs about 1% as much as insulin produced from pig pancreas.
- The process can be switched on or off easily as the bacteria can be stored until needed again. The product can be made as required rather than just when animal carcasses are available.

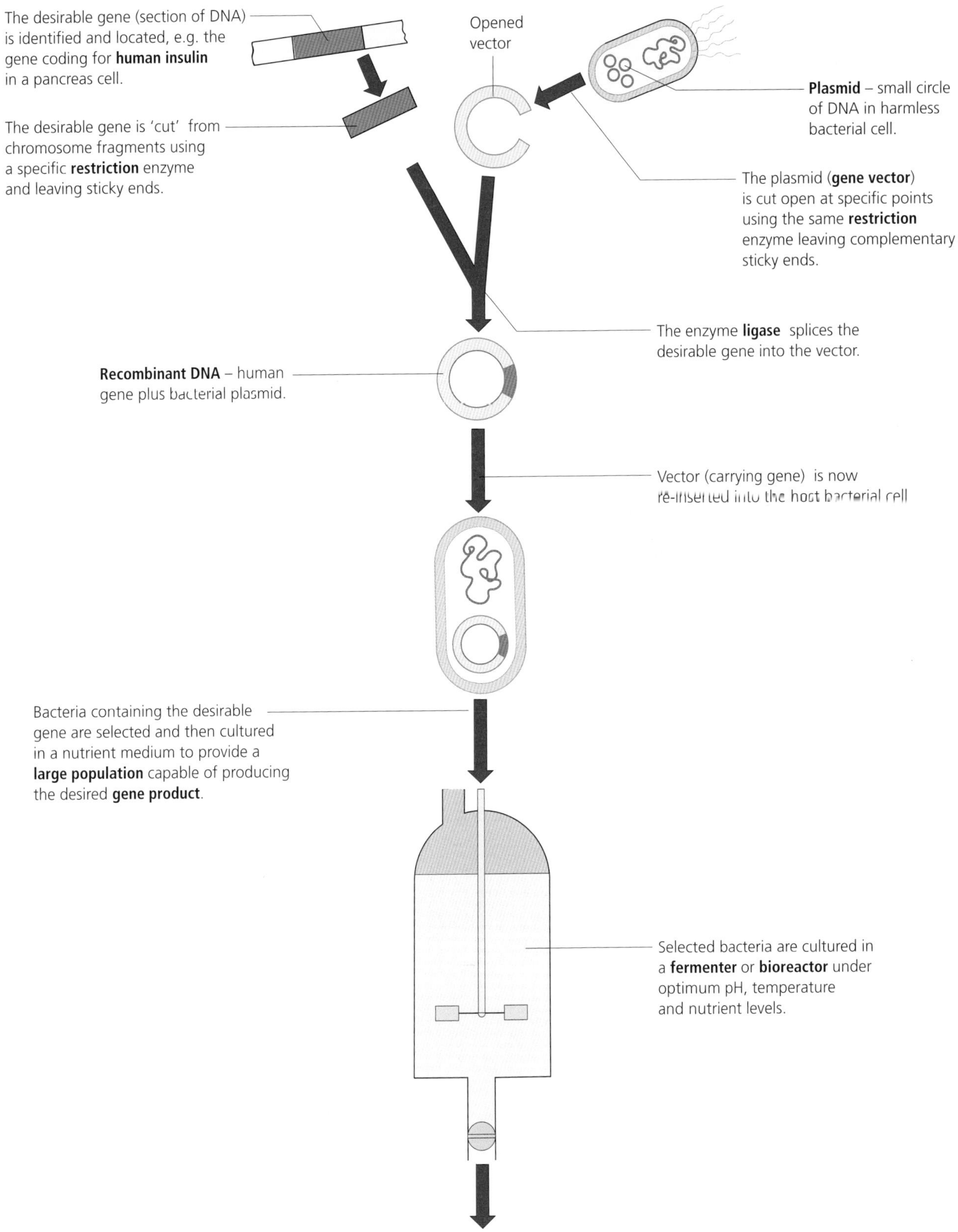

The desirable gene (section of DNA) is identified and located, e.g. the gene coding for **human insulin** in a pancreas cell.
The desirable gene is 'cut' from chromosome fragments using a specific **restriction** enzyme and leaving sticky ends.
Opened vector
Plasmid – small circle of DNA in harmless bacterial cell.
The plasmid (**gene vector**) is cut open at specific points using the same **restriction** enzyme leaving complementary sticky ends.
The enzyme **ligase** splices the desirable gene into the vector.
Recombinant DNA – human gene plus bacterial plasmid.
Vector (carrying gene) is now re-inserted into the host bacterial cell
Bacteria containing the desirable gene are selected and then cultured in a nutrient medium to provide a **large population** capable of producing the desired **gene product**.
Selected bacteria are cultured in a **fermenter** or **bioreactor** under optimum pH, temperature and nutrient levels.

Product

After some processing, for example to remove the bacterial cells for recycling, the product is extremely pure and relatively inexpensive. Important examples of such gene products are:

- **insulin** (required for the treatment of diabetes)
- **human growth hormone**
- **factor VIII** (blood clotting factor for haemophilia)
- **BGH** is an important animal hormone used to speed up the growth of beef cattle (see page 150).

▲ Genetic engineering (recombinant DNA technology) depends on enzymes and the culture of microorganisms

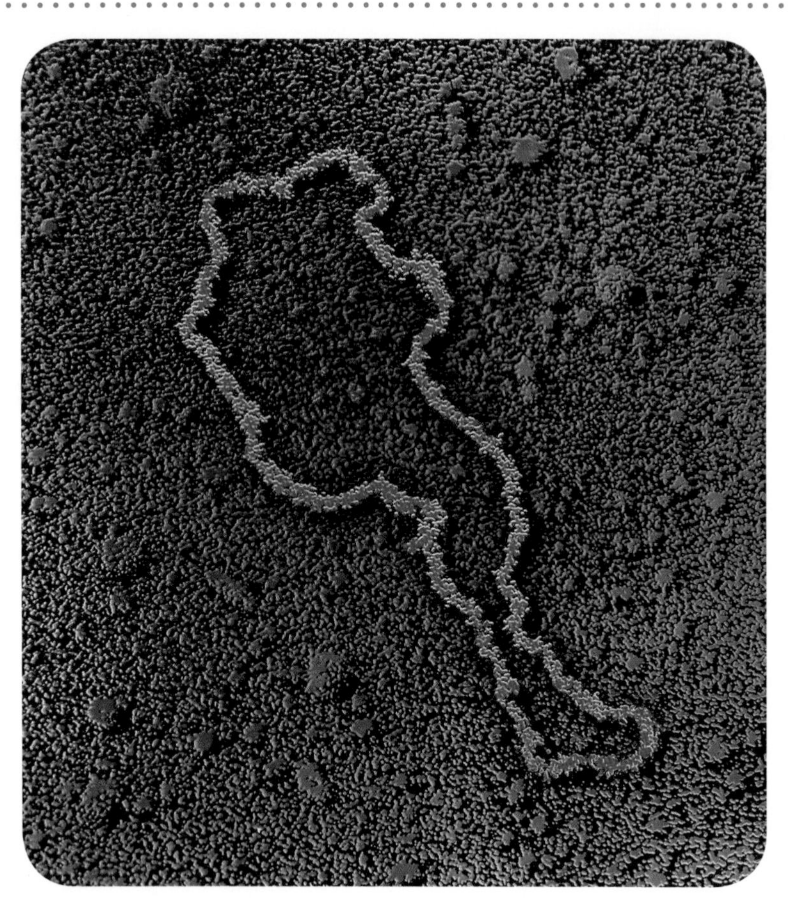

▲ A plasmid of bacterial DNA

1 The following table lists events from the identification of a human gene coding for a hormone "X" to the commercial production of hormone "X". Genes can be transferred into plasmids, tiny circles of DNA which are found in bacteria.
Show the correct sequence of events 1 to 8 by copying the table below and writing the appropriate number in each box provided. The first (number 1) and last (number 8) have been completed for you.

Event	Number
Cutting of a bacterial plasmid using restriction enzyme	
Cutting of human DNA with restriction enzyme	
Identification of the human DNA which codes for hormone "X"	1
Many identical plasmids, complete with human gene, are produced inside the bacterium	
Mixing together human gene and "cut" plasmids to splice the human gene into the plasmid	
Some of the cloned bacteria are put into an industrial fermenter where they breed and secrete the hormone	8
The bacterium is cloned	
Using the plasmid as a vector, inserting it, complete with human gene, into a bacterium	

Moral and environmental concerns about genetic modification

The benefits claimed for genetically modified products, and potential problems, are shown in the table below.

Benefits	Problems
Engineered organisms can offer higher yields from fewer resources. With plants, for example, this might reduce the need for pesticides and fertilisers.	Plants engineered for pesticide resistance could cross-pollinate with wild relatives, creating 'superweeds'
Crops engineered to cope with extreme environmental conditions will open up new areas for cultivation, and reduce the risk of famine	Engineered bacteria may escape from the laboratory or the factory, with unpredictable consequences
Genetic engineering gives much more predictable results than selective breeding	'New' organisms might be patented. A company that has spent a lot of money on developing such an organism might refuse to share its benefits with other consumers, making the company very powerful.
Foods can be engineered to be more convenient, such as potatoes which absorb less fat when crisps are made, or even to contain medicinal products such as vaccines	How far should we allow research into human gene transfer to go? Will we allow the production of 'perfect' children, with characteristics seen to be desirable by parents?

Monsanto

Research.
Glyphosate GMO's
Effects of humans + animals and environmental pollution from GMO's.
Glyphosate (Roundup) → into rives
India Cotton farmers.
GMO Corn USA – effect on cattle that eat it.
GMO Soya → effects on humans. Hormones, etc.

Gene therapy*

Gene therapy is the transfer of healthy human genes into a person's cells that contain mutant alleles which cause disease. The principle of gene therapy is outlined below.

Gene therapy may be able to repair diseased human cells.

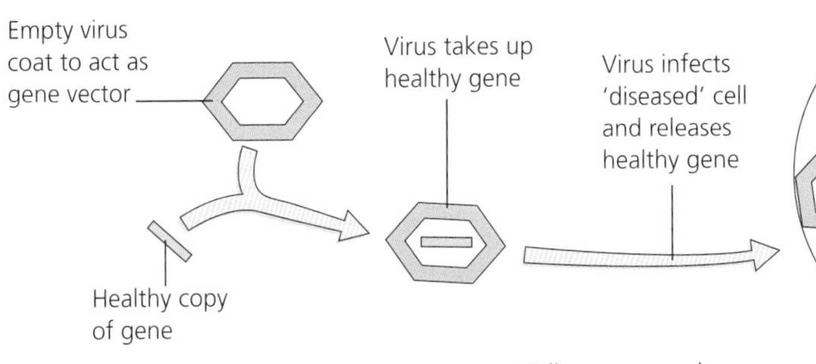

Virus infects 'diseased' cell and releases healthy gene

Cell can now make protein product, and no longer has disease

Healthy gene 'repairs' diseased cell

Cystic fibrosis is a good candidate for gene therapy because:

- the healthy gene has been identified and is easily obtained
- the diseased tissue in the lungs is easy to reach via trachea and bronchi
- the coat of an influenza virus can be used as a gene vector.

Q

1. What might be the benefits of transferring nitrogen-fixing genes from bacteria to other organisms?
2. Describe one example of gene therapy. Explain why this form of treatment is thought likely to be successful in the example which you describe.

4.15 Food supply: humans and agriculture

OBJECTIVES

- To understand why agriculture is a threat to conservation
- To understand how careful management can help to conserve biodiversity

The human population is increasing and may reach 9 billion (9000 000 000) by the end of the century. More people means a demand for more food. This food is mainly provided by agriculture – the management of land for food production.

Around the world, agriculture presents a serious threat to conservation of existing habitats. Although 75% of England is designated as farmland, only 2% of the UK population is involved in farming. In 1850, one farm worker produced enough food for about 4 people; now in advanced agricultural countries the same worker produces enough for 60 people. The enormous rise in productivity of land is due to many factors. These include:

- **increased and more efficient use of machinery** has resulted in increased field sizes and the removal of hedges (see page 258)
- **increased use of inorganic fertilisers** affects plant diversity dramatically and can result in excessive nitrate levels in rivers, leading to eutrophication (page 267)
- **use of pesticides** to eliminate competitors for crop species has caused the reduction in population of many native wildflowers and some non-target insects and birds by direct toxicity or, more often, by removal of their food sources. Pesticides include **insecticides** to improve the quality and yield of crops by killing insects, and **herbicides** which reduce competition with weeds
- the tendency towards **monoculture**
- selective breeding of both crops and livestock.

All of these can have negative effects on the environment.

It all boils down to monoculture!

Farms used to be small and mixed, supplying enough food for just a few families throughout the year, but now tend to be large and specialised. Growing a single crop in this way is called **monoculture** and, as ever, there are pluses and minuses.

Monoculture

+ (Good points)

Specialised harvesting techniques: one type of machine can collect all of the crop.

Highly selected strains: varieties of plants with desirable characteristics (see page 161) can be 'matched' to the conditions.

Mineral/water requirement: scientists can work out exactly what the crop needs, and farmers can make sure they use it.

− (Bad points)

Poor wildlife foods: very little variety of 'weeds' for insects and birds.

Spread of disease: plant pathogens, such as potato blight fungus and tobacco mosaic virus, spread easily: it's not very far to the next ideal food plant!

Loss of genetic variety: may mean that any change in environmental resistance (see page 240) could damage or kill all of the plants.

Damage to soil: the same minerals will be 'drained' away by many copies of the same plant. As the crop is harvested and taken away the minerals will be lost from the soil.

Seedless grapes.

Loss of amenity value!

Monocultures are boring to look at, and few people enjoy walking in such a monotonous environment.

We must aim for sustainable development.
This means that we should not alter our environment so much that we take things away and affect the value of the environment *for future generations*.
We should still provide for the needs of an increasing population.

Some farmers are returning to traditional **crop rotations**, which has several benefits:

Crop rotations

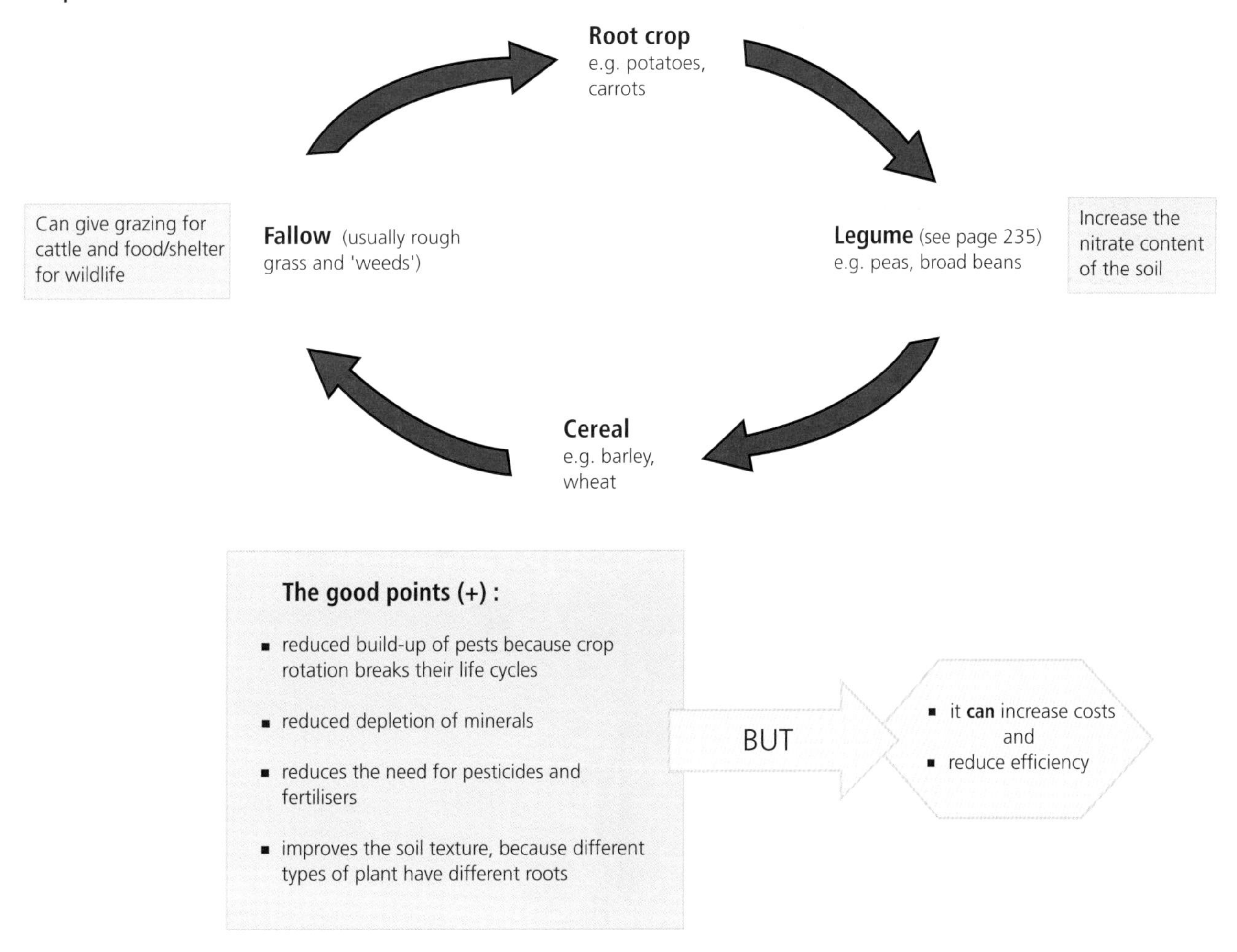
Root crop
e.g. potatoes, carrots
Legume (see page 235)
e.g. peas, broad beans
Increase the nitrate content of the soil
Cereal
e.g. barley, wheat
Fallow (usually rough grass and 'weeds')
Can give grazing for cattle and food/shelter for wildlife
The good points (+) :
reduced build-up of pests because crop rotation breaks their life cycles
reduced depletion of minerals
reduces the need for pesticides and fertilisers
improves the soil texture, because different types of plant have different roots
BUT
it can increase costs
and
reduce efficiency

The problem for the world is how to balance **productivity** (more food for humans) and **biodiversity** (keeping the wide range of wildlife).

4.16 Land use for agriculture

OBJECTIVES

- To understand that chemical discharges may pollute land as well as air and water
- To understand that many human effects on the land involve loss of wildlife habitat
- To know that there are ways to reverse loss of wildlife habitat

Many of the pollutants described on pages 264–8 have an effect on land, as well as on the air or water.

Pollution also includes the loss of wildlife habitat that results from human competition for land.

Removal of hedges

Farmers remove hedges to increase the area where they can grow crops. The benefits of hedgerows as habitats, and the reasons why farmers feel justified in removing them, are outlined below.

Deforestation

The removal of woodland provides firewood, building materials, cleared land for crops or for grazing of cattle. For our first settler ancestors, it also removed the habitat of predators of domestic animals.

Deforestation is also a result of clearing land so that natural resources (e.g. oil in the Amazon basin and copper in Zaire) can be extracted.

The technique used to clear forest today is often called **slash and burn**. The largest trees may be removed for sale as timber for furnishings but the less valuable woods are simply chopped down (slashed) and then burned. The humans using the land gain a short-term benefit, but the damage to

Advantages

Hedges act as **windbreaks** which provide shelter for domestic animals, protect fragile crops, limit soil erosion and reduce water losses by evaporation from soil. A hedge 1 m in height provides these benefits for approximately 2 m to its sheltered side.

Taller hedges offer secure nesting sites for up to 65 **bird species**. These species may be important predators on pest species on local crops.

Fallen leaves and fruits provide **nutrient enrichment** for soil.

Grassy strip provides **shelter** for game birds and overwintering insects, and **nesting sites** for small animals (e.g. voles).

Hedges and associated herbs povide **feeding and breeding** opportunities for **pollinating insects** (including 23 species of butterfly).

Roots improve **soil stability** and limit both wind and water erosion.

Disadvantages

A hedge takes up spaces that could be occupied by crops and reduces economic use of modern agricultural machinery.

Hedge may shade crop species (compete with them for light).

May be a source of insects, and viral and fungal pests (although such species are often specific and are therefore unlikely to be pests of local crop species).

May act as a reservoir of weed species which then invade and compete with crops.

Hedge banks offer burrowing opportunities to rabbits which may then consume crops, especially in the young stages.

Roots may consume water and nutrients which otherwise would be available to crop plants.

Maintenance of hedges is **labour intensive** compared with barbed-wire boundaries.

▲ The removal of hedgerows in Britain has averaged 8000 km per year in the twentieth century. This represents a disaster for wildlife.

wildlife habitats is immediate and humans also suffer in the long term. Newspaper and television headlines emphasise the loss of forest in tropical areas of the world, but it should be remembered that most of the UK was once forested! Some of the penalties of large-scale deforestation are illustrated below.

1. It has been estimated that more than 150 000 km of hedgerow have been removed from around Britain's fields since the Second World War. Why has this been done? What are the possible effects on wildlife?
2. What is deforestation? Give reasons why humans should be anxious about this process.

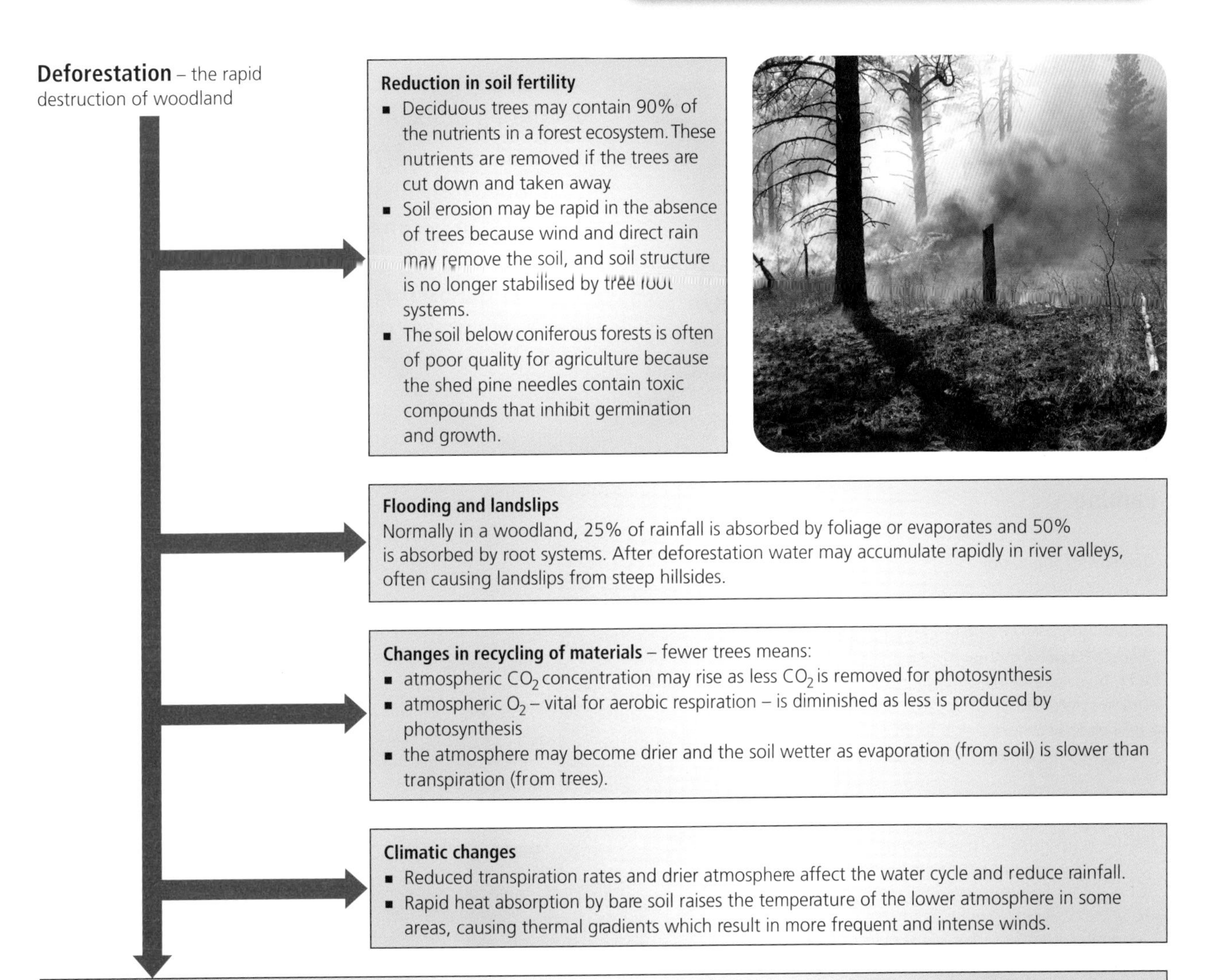

Deforestation – the rapid destruction of woodland

Reduction in soil fertility

- Deciduous trees may contain 90% of the nutrients in a forest ecosystem. These nutrients are removed if the trees are cut down and taken away.
- Soil erosion may be rapid in the absence of trees because wind and direct rain may remove the soil, and soil structure is no longer stabilised by tree root systems.
- The soil below coniferous forests is often of poor quality for agriculture because the shed pine needles contain toxic compounds that inhibit germination and growth.

Flooding and landslips

Normally in a woodland, 25% of rainfall is absorbed by foliage or evaporates and 50% is absorbed by root systems. After deforestation water may accumulate rapidly in river valleys, often causing landslips from steep hillsides.

Changes in recycling of materials – fewer trees means:

- atmospheric CO_2 concentration may rise as less CO_2 is removed for photosynthesis
- atmospheric O_2 – vital for aerobic respiration – is diminished as less is produced by photosynthesis
- the atmosphere may become drier and the soil wetter as evaporation (from soil) is slower than transpiration (from trees).

Climatic changes

- Reduced transpiration rates and drier atmosphere affect the water cycle and reduce rainfall.
- Rapid heat absorption by bare soil raises the temperature of the lower atmosphere in some areas, causing thermal gradients which result in more frequent and intense winds.

Species extinction Many species are dependent on forest conditions:

- orang utans depend on rainforest in Indonesia
- golden lion tamarins depend on coastal rainforest of Brazil
- ospreys depend on mature pine forests in northern Europe.

It is estimated that one plant and one animal species become extinct every 30 minutes due to deforestation.

Many plant species may have medicinal properties, e.g. as tranquillisers, reproductive hormones, anticoagulants, painkillers and antibiotics.
The Madagascan periwinkle, for example, yields a potent drug used to treat leukaemia.

4.17 Damage to ecosystems: malnutrition and famine

OBJECTIVES

- To recall that malnutrition is a condition resulting from a defective diet where certain important food nutrients (such as proteins, vitamins, or carbohydrates) may be absent
- To remember that malnutrition can lead to deficiency diseases
- To understand some of the reasons why famines occur including over-use of land for agriculture.

A high global death rate linked to malnutrition has arisen from famine situations caused by a number of natural problems as well as by socio-political factors, such as alcohol and drug abuse, poverty, and war.

As well as suffering from a deficiency disease, a malnourished person is usually more at risk from other diseases (as the immune system may be very much reduced in efficiency) and a society with many malnourished individuals experiences mounting problems:

- Adults become ill and cannot work, so little money is available to buy food.
- Children become ill, and so adults must attempt to care for them (often going without food themselves).
- Older, weaker people will die and the society is deprived of their experience and knowledge.

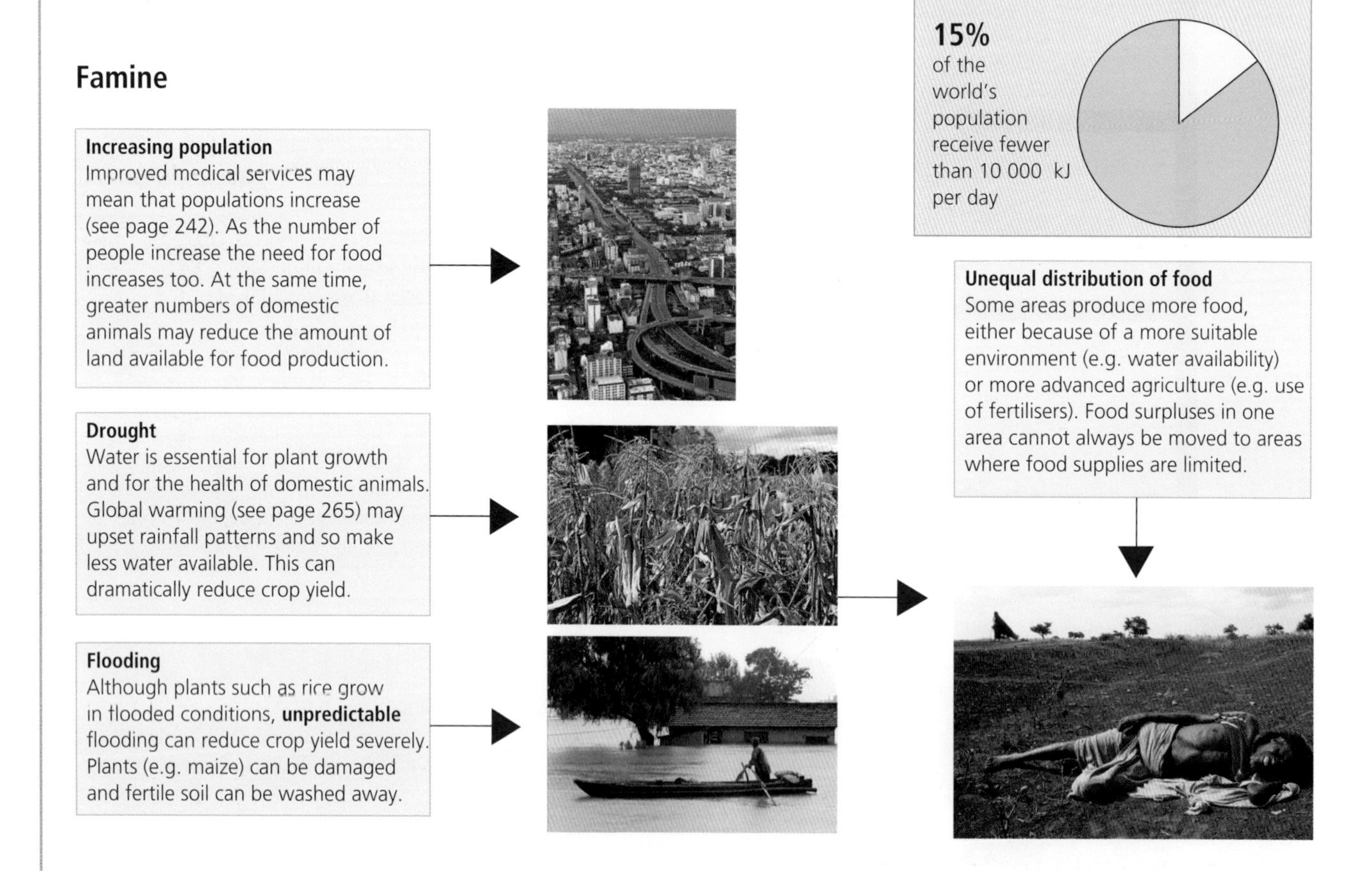

Famine
Increasing population
Improved medical services may mean that populations increase (see page 242). As the number of people increase the need for food increases too. At the same time, greater numbers of domestic animals may reduce the amount of land available for food production.
Drought
Water is essential for plant growth and for the health of domestic animals. Global warming (see page 265) may upset rainfall patterns and so make less water available. This can dramatically reduce crop yield.
Flooding
Although plants such as rice grow in flooded conditions, unpredictable flooding can reduce crop yield severely. Plants (e.g. maize) can be damaged and fertile soil can be washed away.
15%
of the world's population receive fewer than 10 000 kJ per day
Unequal distribution of food
Some areas produce more food, either because of a more suitable environment (e.g. water availability) or more advanced agriculture (e.g. use of fertilisers). Food surpluses in one area cannot always be moved to areas where food supplies are limited.

Cost of fuel and fertiliser
Crop yield may be low in developing countries because poor farmers cannot afford fertiliser, or fuel to run machinery.

Can famine be stopped?

The world probably produces enough food to feed all of its inhabitants. Agriculture in developed parts of the world is extremely efficient (see page 260). Even so, as many as 15% of the world's inhabitants suffer from malnutrition because the available food is not evenly distributed among the world's population. Some people believe that all surplus food from developed countries should be made available to poorer countries, but many others (including EU agricultural ministers) argue that aid to poorer people should be mainly in the form of education and farming equipment. Direct food aid should only be reserved for disaster situations, such as the earthquake in Pakistan and flooding in Central America, where many individuals would quickly die if food were not made available. The problem of unequal food distribution is outlined in the figure.

Transportation
Bulky foods are expensive to transport. Foods with a high water content (e.g. milk) do not give good 'food value per transport cost'. Perishable foods may not remain in good condition during long periods of transport.

Unequal distribution of food

Long distance transport can increase atmospheric pollution and global warming!

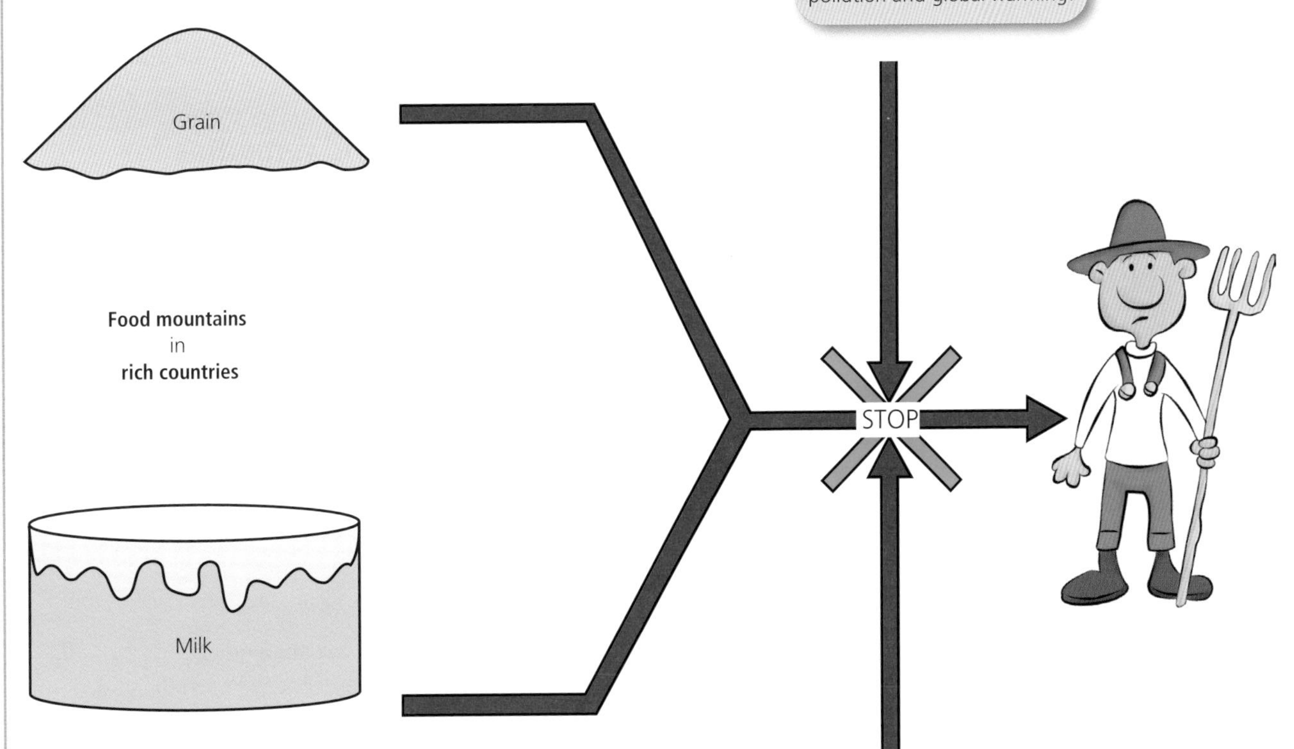

Self-sufficiency
It makes more sense to help poor people to grow food for themselves than to send food surpluses to them. For example

- **Education** about agricultural methods
- **Provision of water supplies** by digging wells

Price controls
Taking cheap food from a rich country to a poor one can artificially lower food prices in the poor region. This can upset the local economy, and reduce the incentive for indigenous people to grow their own food.

4.18 Human impacts on the environment: pollution

OBJECTIVES

- To understand that humans can alter their environment
- To understand how changes in human population have altered our impact on the environment
- To consider human impacts in terms of cause, effect and possible remedies

Humans already are, or could easily become, the most significant biotic factor in every environment. Primitive humans had a *temporary* effect on the environment – hunting, fishing and burning wood – because they were **nomadic** and allowed the environment periods of time to recover. After humans became **cultivators** and **settlers** (thousands of years ago) the following 'effects' on the environment gradually became noticeable:

- The use of tools and the domestication of animals meant a more efficient **agriculture**.
- The greater need for shelter, agricultural land and fuel meant a greater rate of **deforestation** and **desertification**.
- The careless use of pesticides and fertilisers, together with increasing consumption of fossil fuels, produced problems of **pollution**. These have been added to by the development of nuclear energy sources.

Humans may also have positive effects – we are able to use our skills and knowledge in **conservation**.

Our demands on the environment

Human success, measured as an increase in population size, is largely due to our ability to solve complex problems and modify the environment for our benefit. This places great demands on the environment, causing changes in **the atmosphere**, **the aquatic environment** and **the land.**

Humans have also, intentionally or otherwise, seriously upset the balance of populations of other living organisms. (see effects on food webs, page 224).

Causes, effects and remedies for pollution

Pollution is any effect of human activities upon the environment, and a **pollutant** is any product of human activities that has a harmful effect on the environment. When looking at pollution, we shall consider three key points:

- What is the **cause**?
- What are the **effects**?
- What are the **solutions**?

Q

1 The following table contains information on human population changes since the year 1700.

Year	Population / millions
1700	700
1750	800
1800	900
1850	1200
1900	1600
1950	2500
1980	4450
1990	5300

a Plot these data in the form of a graph.

b Calculate the likely world population in the year 2050 if the growth rate shown between 1700 and 1800 had remained constant.

c Calculate the likely world population in the year 2050 if the trend in your graph continues.

d Explain the difference between your answers to **b** and **c**.

Pollution of the atmosphere

Two big problems with the Earth's atmosphere are:

- the **greenhouse effect**
- production of **acid rain**.

Burning of fossil fuels has a major effect on the atmosphere. The combustion process oxidises elements and compounds in the fuel, as shown in the equations on the right.

These oxides affect the atmosphere – carbon dioxide is a greenhouse gas and sulfur dioxide and oxides of nitrogen contribute to acid rain. The other major pollutants of the atmosphere are the chlorofluorocarbons (CFCs).

There are also localised problems with lead compounds, and with smoke. The causes and effects of, and possible solutions to, each of these problems are outlined in the diagrams here and on page 266.

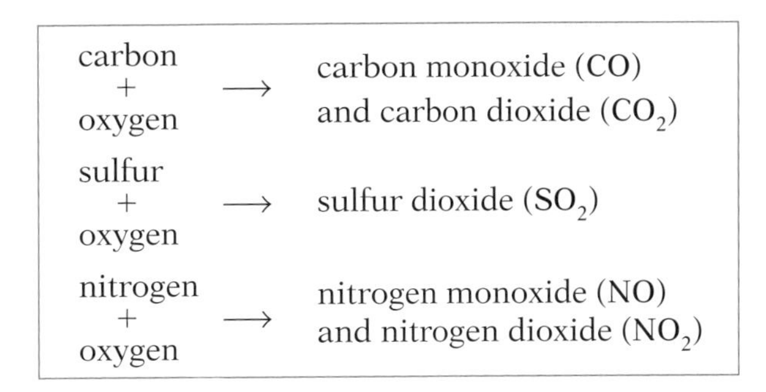

carbon + oxygen $\longrightarrow$ carbon monoxide (CO) and carbon dioxide (CO_2)

sulfur + oxygen $\longrightarrow$ sulfur dioxide (SO_2)

nitrogen + oxygen $\longrightarrow$ nitrogen monoxide (NO) and nitrogen dioxide (NO_2)

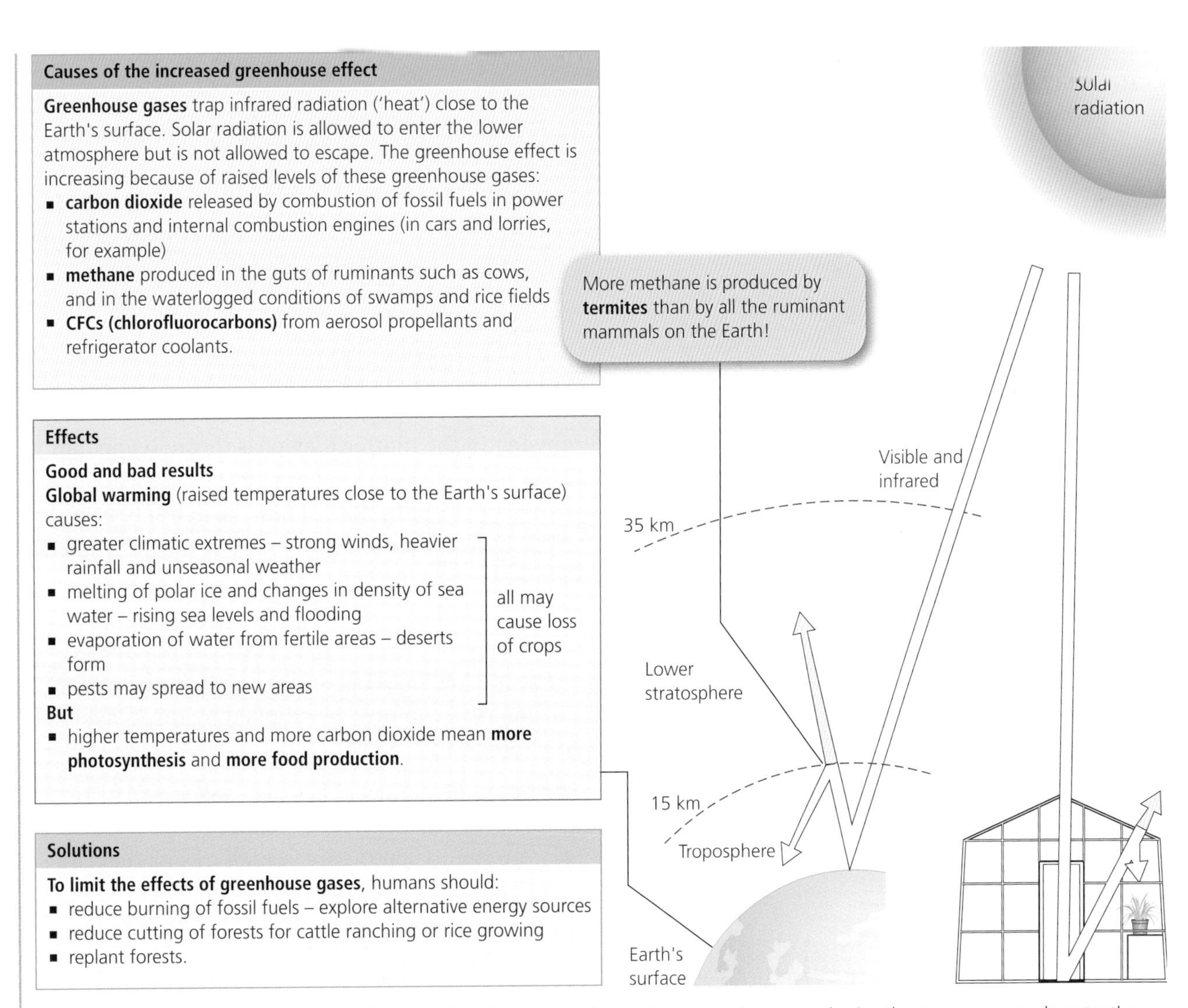

Causes of the increased greenhouse effect
Greenhouse gases trap infrared radiation ('heat') close to the Earth's surface. Solar radiation is allowed to enter the lower atmosphere but is not allowed to escape. The greenhouse effect is increasing because of raised levels of these greenhouse gases:
carbon dioxide released by combustion of fossil fuels in power stations and internal combustion engines (in cars and lorries, for example)
methane produced in the guts of ruminants such as cows, and in the waterlogged conditions of swamps and rice fields
CFCs (chlorofluorocarbons) from aerosol propellants and refrigerator coolants.
More methane is produced by termites than by all the ruminant mammals on the Earth!
radiation
Visible and infrared
35 km
Lower stratosphere
15 km
Troposphere
Earth's surface
Effects
Good and bad results
Global warming (raised temperatures close to the Earth's surface) causes:
greater climatic extremes – strong winds, heavier rainfall and unseasonal weather
melting of polar ice and changes in density of sea water – rising sea levels and flooding
evaporation of water from fertile areas – deserts form
pests may spread to new areas
all may cause loss of crops
But
higher temperatures and more carbon dioxide mean more photosynthesis and more food production.
Solutions
To limit the effects of greenhouse gases, humans should:
reduce burning of fossil fuels – explore alternative energy sources
reduce cutting of forests for cattle ranching or rice growing
replant forests.

▲ Some gases added to the atmosphere act like the panes of glass in a greenhouse and raise the temperature close to the Earth's surface

Acid rain

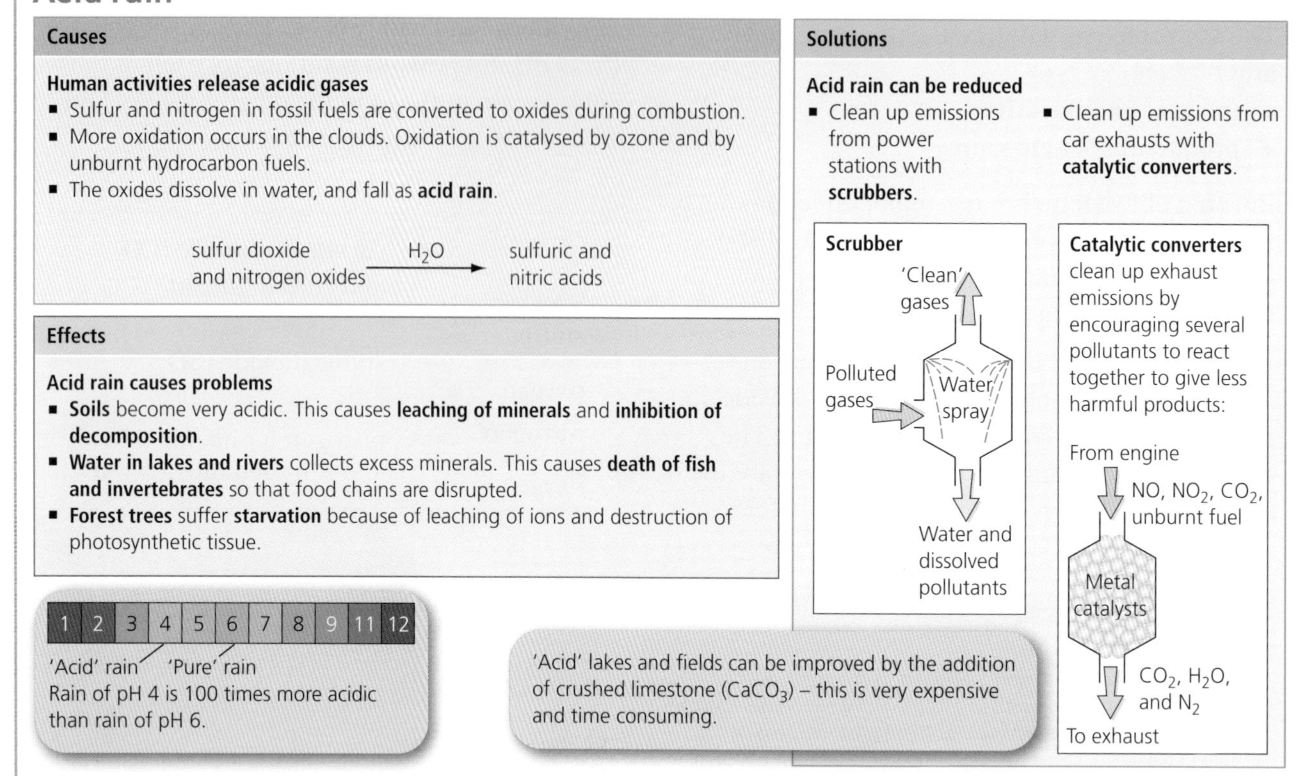

Causes
Human activities release acidic gases
Sulfur and nitrogen in fossil fuels are converted to oxides during combustion.
More oxidation occurs in the clouds. Oxidation is catalysed by ozone and by unburnt hydrocarbon fuels.
The oxides dissolve in water, and fall as acid rain.
sulfur dioxide and nitrogen oxides
H2O
sulfuric and nitric acids
Effects
Acid rain causes problems
Soils become very acidic. This causes leaching of minerals and inhibition of decomposition.
Water in lakes and rivers collects excess minerals. This causes death of fish and invertebrates so that food chains are disrupted.
Forest trees suffer starvation because of leaching of ions and destruction of photosynthetic tissue.
1 2 3 4 5 6 7 8 9 11 12
'Acid' rain
'Pure' rain
Rain of pH 4 is 100 times more acidic than rain of pH 6.
'Acid' lakes and fields can be improved by the addition of crushed limestone (CaCO3) – this is very expensive and time consuming.
Solutions
Acid rain can be reduced
Clean up emissions from power stations with scrubbers.
Clean up emissions from car exhausts with catalytic converters.
Scrubber
'Clean' gases
Polluted gases
Water spray
Water and dissolved pollutants
Catalytic converters
clean up exhaust emissions by encouraging several pollutants to react together to give less harmful products:
From engine
NO, NO2, CO2, unburnt fuel
Metal catalysts
CO2, H2O, and N2
To exhaust

▲ Acid rain damages both living organisms and buildings made by humans. Acid rain also causes pollution of water. Some lakes are almost empty of life as a result of pH levels as low as 4 or 5.

Radioactive pollution

The Earth's environment is naturally radioactive due to cosmic radiation emitted from space, and terrestrial radiation emitted from the Earth's crust. However, this background level of radiation is exceeded in many parts of the world by contamination from human-made sources of radioactivity.

'Radiation' here means ionising radiation (as opposed to other non-ionising radiation, e.g. UV, infra-red, microwave). Radiation is emitted from radioactive substances as they spontaneously decay and takes several forms. The types of most concern are:

- alpha particles (α)
- beta particles (β)
- gamma rays (γ).

Each type of radiation has different properties and different penetrating power; e.g. alpha particles can be stopped by a few centimetres of air or a sheet of paper, however they are intensely ionising in the matter they pass through and can cause more damage to living tissue than particles with a longer path. Like alpha particles, beta particles lose their energy within a short distance and their biological significance is greatest if a beta emitter is taken into the body. Gamma radiation particles are similar to x-rays – they are deeply penetrating and strongly ionising; gamma rays can only be stopped by thick layers of concrete, lead or steel.

The decay of radioactive wastes takes a very long time. Some radioactive substances have a half-life of more than 10000 years, which means they are dangerous for a very long time. (The half-life is the 'period of time required for the disintegration of half of the atoms in a sample of a radioactive substance'.)

There are two major ways in which the atmosphere can become polluted by radiation:

Nuclear power plants

Nuclear power plant accidents can endanger life and the surrounding environment if the radioactive core is exposed and meltdown is occurring and releasing large amounts of radioactivity. This happened at Chernobyl, a nuclear power station in the Ukraine: an accident here destroyed the Chernobyl-4 reactor and killed 30 people, including 28 from radiation exposure. Large areas of Belarus, Ukraine, Russia and beyond were contaminated in varying degrees, and an increased number of birth defects were noted in the years following the disaster.

In 2011 a tsunami severely damaged the nuclear power station at Fukushima, Japan. There are fears that marine life will be contaminated, and a large area near the reactors is still considered uninhabitable.

Nuclear weapons

Nuclear weapon tests that are conducted above ground or under water can release radiation into the atmosphere. A nuclear test is the explosive test of a complete nuclear warhead – some tests in the twentieth century were conducted in areas occupied by soldiers. Most of the troops ordered to take part in the testing programme were not equipped with any specialised protective clothing. They were simply ordered to turn their backs or cover their eyes to avoid being blinded by the flash of the explosion. One former soldier remembers having his hands over his eyes but the flash was so bright that it acted like an x-ray and he could see all the bones in his hands.

Exposure to radioactivity frequently leads to various forms of cancer, including leukaemia. The troops involved have suffered much higher cancer rates than normal. However since cancer may take years to develop, it is difficult to prove that a particular case is linked to a particular cause.

Non-biodegradable plastics

Plastics are widely used as packaging and transporting materials. They are useful because they can be shaped to form different products, they are light and they protect products (especially foods) from fungi and bacteria.

Many plastics are made from large hydrocarbon molecules and cannot be decomposed by normal microbial methods. Because they are so widespread and they are non-biodegradable they can have severe environmental effects. They can pollute both aquatic and terrestrial habitats.

- They can block the passage of water through drainage channels, leading to waterlogging of soils. This reduces oxygenation and so affects soil fertility.
- They can be mistakenly consumed by animals, on both land and in water. They block the animal's digestive system and cause many deaths.
- They do not allow the passage of oxygen, so when they are present in landfill sites they inhibit natural decomposition of other wastes.
- If they are burned in an attempt to remove them they release toxic 'smoky' particles, which can affect breathing and have a long-term effect on health.

These plastics are very light, and so when discarded they easily blow from place to place. They make the environment less attractive (they lower the amenity value of the environment) and often become lined up against natural windbreaks such as hedgerows and stands of trees.

Pesticides

Causes

Over-use of pesticides on agricultural land (e.g. to protect a crop from insects) or directly on water (e.g. to kill an aquatic stage of an insect) can raise pesticide levels in water. The pesticide levels are then **amplified** as they pass through food chains. For example, one stickleback may consume 500 *Daphnia*. The living matter in the *Daphnia* will be used for raw materials or lost as heat but the pesticide remains **concentrated in the tissues** of the stickleback.

DDT concentration in parts per million:

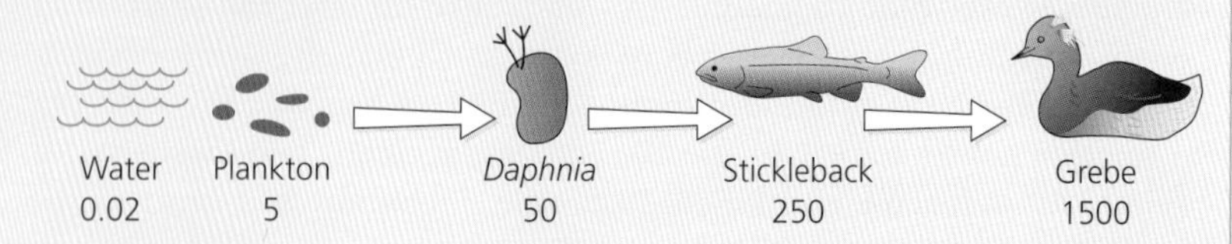

Effects

High concentrations of pesticides may accumulate in the tissues of top carnivores. The pesticide may be toxic (and kill the carnivore) or may affect its metabolism. DDT, used to control mosquitoes in malarial zones, severely reduced breeding success in birds of prey.

Solutions

- Use degradable pesticides – DDT, for example, lasts for a long time and so its use is banned in the UK.
- Explore alternative methods, such as biological pest control.
- Crops that are **genetically modified** to resist attack by insects may reduce the need to use insecticides.

Female contraceptive hormones are washed into water when they excreted in urine. They:

- reduce sperm count in men
- lead to feminisation of aquatic organisms (causing imbalance of gender in fish populations, for example).

1 Sea otters eat fish, and fish eat small crustaceans such as shrimps. The shrimps feed by filtering algae from the water.

a Write out a food chain that links these organisms.

b Pesticides are washed from nearby farmland into rivers and then into the sea. Farmers say that the concentration of the pesticides is too low to directly affect the otters. Explain how the pesticides might still cause the death of the otters.

2 Untreated human sewage should not enter river water, but occasionally an overflow from a water treatment plant occurs. The tables below contain information on the changes that occurred in river water downstream from a sewage overflow.

a Plot these data in the form of a line graph. Choose axes to display the information in the way that best relates the abiotic factor to the biotic factors.

b Suggest why the number of bacteria was high at 0 m.

c Explain the shape of the curve for algae.

d How is it possible for fish numbers to fall to zero and then recover?

e Describe the changes in the concentration of oxygen dissolved in the water downstream from the point of sewage entry.

f Explain what might have caused these changes in oxygen concentration.

Distance downstream / m	Concentration of dissolved oxygen / percentage of maximum
0 (point of sewage entry)	95
100	30
200	20
300	28
400	42
500	58
600	70
700	80
800	89
900	95
1000	100

	Number / arbitrary units of		
Distance downstream / m	**bacteria**	**algae**	**fish**
0 (point of sewage entry)	88	20	20
100	79	8	6
200	74	7	1
300	60	21	0
400	51	40	0
500	48	70	0
600	44	83	0
700	42	90	0
800	39	84	0
900	36	68	4
1000	35	55	20

4.19 Pollution of water: eutrophication

OBJECTIVES

- To recall why water is important to living organisms
- To understand that water supplies oxygen to living organisms
- To know how excess nutrients in water lead to depletion of oxygen levels
- To recall other aspects of water pollution

The causes of oxygen depletion

All living organisms depend on a supply of water, as we saw on page 236. Many organisms actually live in water. Most of these **aquatic** organisms respire aerobically and so require oxygen from their environment. Any change that alters the amount of oxygen in the water can seriously affect the suitability of the water as a habitat. The two pollutants that most often reduce oxygen in water are:

- **Fertilisers** – nitrates and phosphates are added to soil by farmers (see page 258). Some of the fertiliser is washed from the soil by rain into the nearest pond, lake or river. This process is called **leaching**.
- **Sewage** – this contains an excellent source of organic food for bacteria, and also contains phosphates from detergents (see page 74).

How fertilisers and sewage affect the oxygen concentration

Water that contains few nutrients is rich in oxygen and supports a wide variety of living organisms. The oxygen enters the water from the atmosphere by diffusion and from photosynthesising aquatic plants. Simpler forms of life, such as algae and bacteria, are controlled because the low concentration of nutrients such as nitrate is a limiting factor for their growth. If more nutrients are made available, from fertiliser runoff or from sewage, then:

- Algae and other surface plants grow very rapidly, and block out light to plants rooted on the bottom of the river or pond.
- The rooted plants die, and their bodies provide even more nutrients.
- The population of bacteria increases rapidly. As they multiply, the bacteria consume oxygen for aerobic respiration. There is now a biological oxygen demand (or BOD) in the water because of oxygen consumed by these microbes.
- Other living creatures cannot obtain enough oxygen. They must leave the area, if they can, or they will die. Their bodies provide even more food for bacteria, and the situation becomes even worse. This is an example of **positive feedback** – the change from ideal conditions causes an even greater change from ideal conditions.

The lower diagram overleaf shows what happens if a pond or river receives too many nutrients. The process is called **eutrophication**. The pond or river soon becomes depleted of living organisms. Only a few animals, such as *Tubifex* (sewage worms), can respire at the very low oxygen concentrations that are available. The solution to this problem is straightforward – **do not allow excess nutrients into the water**.

Sea water can also be polluted!
Marine environments are affected by human activities:

- spillage of oil which can seriously reduce oxygen levels on the seabed as well as reducing waterproofing properties of seabirds' feathers
- radioactive compounds from cooling of nuclear power stations
- temperature changes caused by global warming.

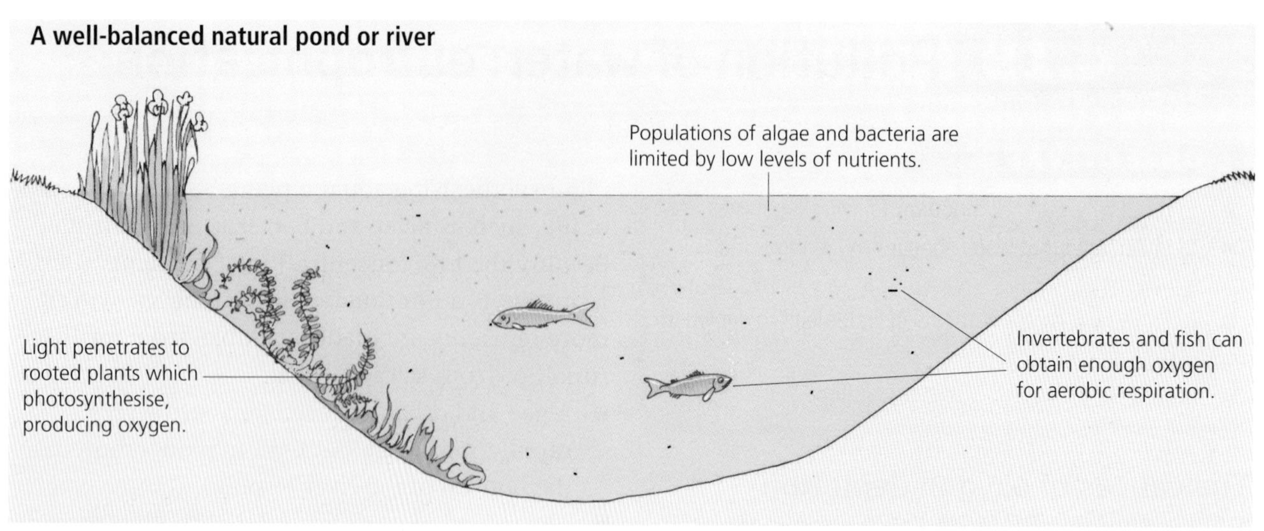
A well-balanced natural pond or river
Populations of algae and bacteria are limited by low levels of nutrients.
Light penetrates to rooted plants which photosynthesise, producing oxygen.
Invertebrates and fish can obtain enough oxygen for aerobic respiration.

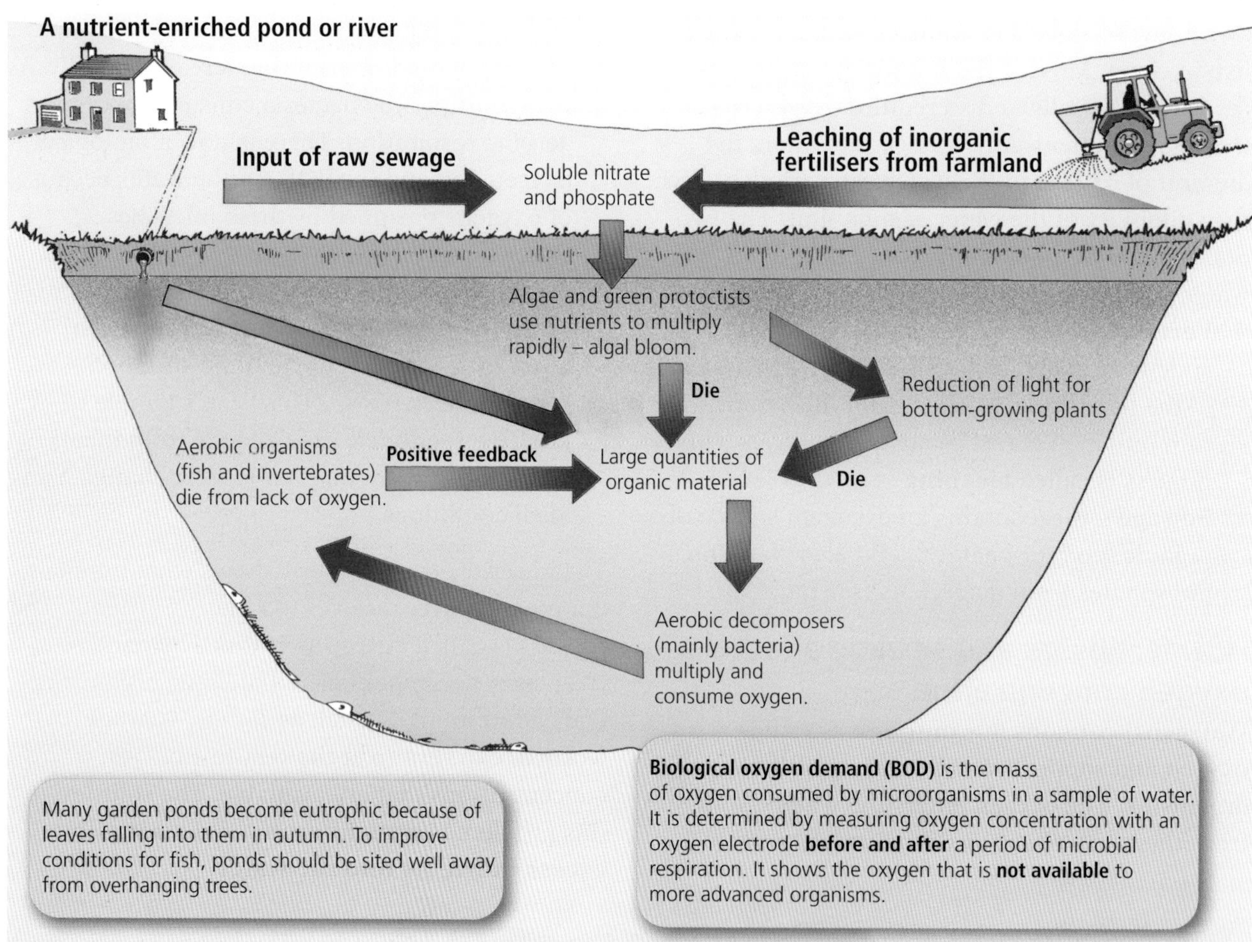
A nutrient-enriched pond or river
Input of raw sewage
Soluble nitrate and phosphate
Leaching of inorganic fertilisers from farmland
Algae and green protoctists use nutrients to multiply rapidly – algal bloom.
Die
Reduction of light for bottom-growing plants
Die
Large quantities of organic material
Positive feedback
Aerobic organisms (fish and invertebrates) die from lack of oxygen.
Aerobic decomposers (mainly bacteria) multiply and consume oxygen.

Many garden ponds become eutrophic because of leaves falling into them in autumn. To improve conditions for fish, ponds should be sited well away from overhanging trees.

Biological oxygen demand (BOD) is the mass of oxygen consumed by microorganisms in a sample of water. It is determined by measuring oxygen concentration with an oxygen electrode **before and after** a period of microbial respiration. It shows the oxygen that is **not available** to more advanced organisms.

Causes of eutrophication

Unnaturally high levels of nutrients:

- from leaching of fertilisers
- from input of raw sewage
- from liquid manure (slurry) washed out of farmyards.

Effects

Depleted oxygen levels in water cause death of fish and most invertebrates. High nitrate levels can be dangerous to human babies.

Solutions

- Treat sewage before it enters rivers (see page 276).
- Prevent farmyard drainage entering rivers and ponds.
- Control use of fertilisers:
  - apply only when crops are growing
  - never apply to bare fields
  - do not apply when rain is forecast
  - do not dispose of waste fertiliser into rivers and ponds.
- Bubble a stream of air through badly polluted ponds.

4.20 Humans may have a positive effect on the environment: conservation of species

OBJECTIVES

- To understand that humans may have a beneficial effect on the environment
- To realise that conservation often involves compromise
- To understand a conservation strategy
- To know some examples of successful conservation

Humans do not always damage the environment – growing numbers of **conservationists** try to balance the human demands on the environment with the need to maintain wildlife habitat. They will try to assess the likely effects of any human activity by producing an **Environmental Impact Statement**.

Forest management

Humans have been responsible for deforestation of much of the Earth's surface. Humans have also set up schemes for the large-scale planting of trees – in areas that have been cleared (**reforestation**) or in a new site (**afforestation**). There are a number of reasons for planting trees:

- as a cash crop, providing timber for building (coniferous plantations in the UK) or for fuel (quick-growing eucalyptus trees are widely planted in central Africa)
- to reverse soil erosion, particularly valuable in areas that have become deserts
- to provide valuable wildlife habitats – for example, Scots pine plantations are important habitats for red squirrels
- as recreational areas, providing leisure activities such as camping and mountain-biking.

A well-managed forest can combine all of these functions. Forestry managers around the world apply biological knowledge to the management of forests, as illustrated in the programme for conservation of red squirrels outlined opposite.

Endangered species

Competition between humans and other living organisms means that many species have disappeared or declined in number. The reasons for this are not always understood, but the following may be to blame:

- **Pest control** – the term pest includes any species that causes inconvenience to humans. Many species have been hunted ruthlessly, such as red deer (which damage trees), and also predators such as cheetahs, and scavengers such as vultures.
- **Commercial exploitation** – species of value to humans have been exploited, such as the tiger which has been hunted and trapped for fur and medicinal compounds.
- **Loss of habitat** – more land is being used for agriculture, including previously unusable land that has been drained. This removes habitats for many species, such as wading birds and amphibians.

Conservationists work to slow down or stop the decline in **biodiversity** (the number of different species), and also to raise public awareness of the need to maintain species and their habitats. The number of different species in a community of living organisms can be described by a formula called the **Species Diversity Index**.

Conservation strategies

Conservation involves management of an area, and may include a number of strategies:

- **Preservation** – keeping some part of the environment unchanged. This might be possible in Antarctica, but is less significant in a densely populated area like Britain than …
- **Reclamation** – the restoration of damaged habitats such as replacing hedgerows or recovering former industrial sites, and …
- **Creation** – producing new habitats, for example by digging a garden pond, or planting a forest.

A **conservation plan** involves several stages:

- **sampling** to assess the number of organisms
- **devising a management plan** – for example, trying to increase a species' population based on knowledge of its breeding requirements
- **carrying out the plan**
- **re-sampling** to assess the number of the 'conserved' species once more, and find out whether the conservation plan has worked.

The red squirrel in the UK

The red squirrel (*Sciurus vulgaris*) used to be widespread in Britain, but in most areas it has now been replaced by the larger grey squirrel (*Sciurus carolinensis*) which was introduced into Britain from North America. There are a number of possible reasons for the decline of the red squirrel:

- **Competition with the grey squirrel** – the red squirrel feeds mainly on seeds from pine cones. The grey squirrel eats a wider range of foods.
- **Disease** – the grey squirrel may carry a virus which harms red squirrels.
- **Habitat loss** – many recent forest plantings have been composed of trees all of the same age, and have often been of sitka spruce, which produces small seeds and sheds them early in the year, leaving little food for red squirrels in the winter.

Conservation plans to support red squirrel numbers are shown in the diagram below. Introducing a species from another country often causes problems for native wildlife.

1. Briefly describe why red squirrels are endangered in Britain. Outline a conservation strategy which might help to guarantee their survival.
2. What is meant by the term conservation? Why is conservation necessary?
3. It is sometimes stated that 'Conservation is a compromise'. Explain whether you think that this statement is a valid one. Use examples to illustrate your answer.

Tree species chosen to provide a food source that favours red squirrels, such as broad-leaved species that produce small seeds and are less attractive to the grey squirrels. Species favoured by the grey squirrels may be removed. Coniferous species might include Norway spruce and Scots pine, both of which shed their seeds late and thus provide a year-round supply of food for the red squirrels.

Competitor grey squirrels can be poisoned using warfarin-baited food in a 'grey-squirrel-selective' hopper. This strategy is restricted by legislation, but many landowners favour it because of damage to trees by grey squirrels.

Immunosterilisation – a vaccine could sterilise both male and female grey squirrels whilst leaving red squirrels unaffected. This could reduce the grey squirrel population in a humane way because it has little effect on the squirrels other than to make them sterile.

Supplementary feeding using hoppers that only allow access to red squirrels. These are placed in clusters of two or three about 20–30 m apart, and filled with a mixture of yellow maize, wheat, peanuts and sunflower seeds. They must be inspected and filled frequently.

Reintroduction of red squirrels – survival and behaviour of the introduced red squirrels must be carefully monitored by radio tracking and field observation.

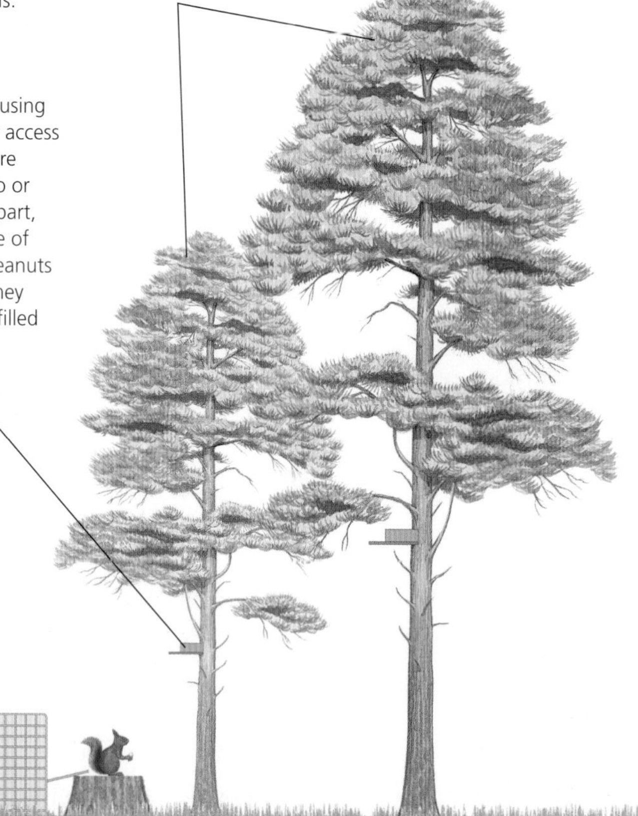

Habitat management – red-squirrel reserves should be surrounded by at least 3 km of conifer forest or open land to act as a buffer against the entry of grey squirrels.

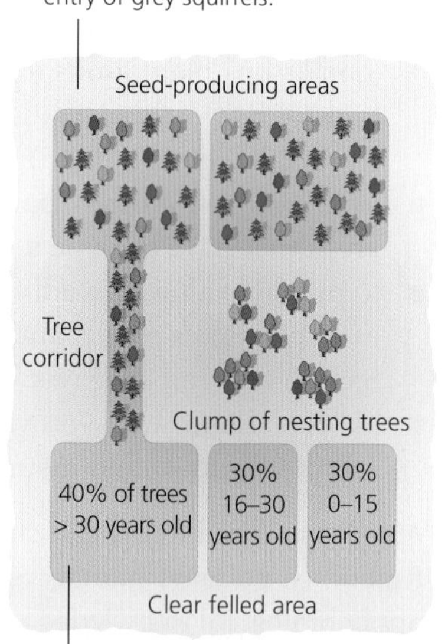

Forest management to provide both food and shelter. When clear felling, some single seed-producing trees should be left, and some small groups as nesting sites. Seed-producing areas should be connected by corridors of trees to prevent isolation and allow movement between one area and the next.

▲ Conservation plans to support red squirrel populations involve habitat management for food provision

Flagship species

Large, attractive and 'cuddly' species attract funding from agencies and donations from the public. These flagship species are often mammals – pandas and dolphins for example – but the funds they raise help other animals and plants which share their habitats, but which may be less attractive to humans.

▲ Giant Pandas are one of the most recognisable of animals. Even though they are solitary and rather bad-tempered, people like them – this means that they attract income from nature lovers.

Captive breeding and the role of zoos

Many animal species are threatened by habitat degradation, fragmentation and destruction. IUCN estimate that only 3% of the planet is designated as a protected reserve and that, on average, one mammal, bird or reptile species has been lost each year for the last hundred years. Certain groups of species are particularly at risk – those with a restricted distribution, those of high economic value, those at the top of food chains and those in climax habitats.

Zoos are areas of confinement keeping samples of species alive under varying degrees of captivity. The role of zoos in captive breeding is becoming more and more important.

Some successes for captive breeding in zoos

Père David's deer (*Elaphurus davidianus*)	First made known to Western science in the 19th century, the only surviving herd was in a preserve belonging to the Chinese emperor. After the remaining population in China was lost the remaining deer in Europe were transported to England: the current population stems from this herd. Two herds of Père David's deer were reintroduced to China in the late 1980s. In spite of the small population size, the animals do not appear to suffer genetic problems.
Przewalski's horse (*Equus przewalskii*)	The wild population in Mongolia died out in the 1960s; a program of exchange between captive populations in zoos throughout the world was started to reduce inbreeding. In 1992, sixteen horses were released into the wild in Mongolia, followed by additional animals later on. These reintroduced horses successfully reproduced, and the status of the animal was changed from 'extinct in the wild' to 'endangered' in 2005. In October 2007 scientists at the Smithsonian Institution's National Zoo realised that a Przewalsk's horse was one of the most genetically valuable Przewalski horses in the breeding programme and successfully reversed a vasectomy on it.
Arabian oryx (*Oryx leucoryx*)	Eliminated in the wild by hunting: restocked in Oman and Jordan from populations in London, Phoenix and San Diego zoos.
Mauritius kestrel (*Falco punctatus*)	Severe decline in the 1950s and 1960s due to indiscriminate DDT use and invasive species like cats, mongooses and crab-eating macaques which killed the kestrels and their eggs. Breeding programme began at Durrell Wildlife Preservation Trust - now up to 800 individuals in the wild.
Golden Lion Tamarin (*Leontopithecus rosalia*)	Population decline due to deforestation in western coastal forests of Brazil. In the 1970s the Brazilian government provided zoos with Golden Lion Tamarins which were successfully bred and disseminated to zoos around the world (especially Jersey Zoo and the Smithsonian Zoo in Washington, US). Over the next 10 years the government succeeded in establishing forest reserves in Brazil that allowed reintroduction of the Tamarins. This captive breeding programme effectively bought the time needed for the original habitat to be conserved. There are now an estimated 1000 wild Tamarins and 600 in captivity.
Lowland Gorilla (*Gorilla gorilla gorilla*)	Critically endangered by war and by bushmeat trade: breeding at Port Lympne Zoo in Kent has released animals into special reserves in the Gabon and the Congo. First 'wild' births to these reintroduced animals in 2008.

4.21 Managing fish stocks: science and the fishing industry

OBJECTIVES

- To understand the value of fish as a food source
- To know how fishing methods increase yield
- To understand how overfishing has reduced fish stocks
- To appreciate that the management of fish stocks depends on scientific research

Fish as a source of food

Fish has been a valuable food source for humans for thousands of years. Fish is an excellent source of protein and, depending on species, of oils. Scientists are becoming more aware of the value of fish oils in preventing some of the 'diseases of affluence', such as coronary heart disease (see page 92). It is very important that humans include certain oily compounds (especially unsaturated fatty acids) in their diet, and these oily compounds are extremely abundant in some fish species.

Science and fishing

Scientists make many contributions to our understanding of the value of fishing. These include

- **Indentification of valuable species:** the 'oily' fish are typically those which swim actively near the surface of the sea (pelagic species). These species have a high lipid content in their bodies because they need to use these stores of energy in their lengthy periods of rapid swimming. Of these species, the herring has been a particularly significant source of food to those nations bordering the North Atlantic Ocean.
- **Study of fish populations:** scientists are able to show how the population of any species changes over the course of time (see page 240). They are able to predict how many fish can be removed from a population without reducing the overall numbers, and they can estimate how long a population will take to recover from overfishing.
- Knowledge of the growth pattern of individual fish: it is possible to study small groups of fish to work out how long they take to reach breeding age. what their maximum size will be and which diseases they might be susceptible to. This allows scientists to advise on net mesh sizes which should be used to catch adult fish, and to suggest fishing methods which will limit the catch of 'trash' fish (fish which are not used for human food). Fishing methods and mesh sizes used in herring fishing are shown below.

Seine ('purse-seine') netting – the net is towed to the school of herring by two boats. The fish are surrounded by the net and the bottom of the 'purse' is sealed by pulling on the rope. The net is slowly tightened so that the fish are concentrated in a small volume of water - the immature fish have time to escape. The captured fish are lifted from the water using hand nets - this limits the damage to the fish and so reduces wastage.

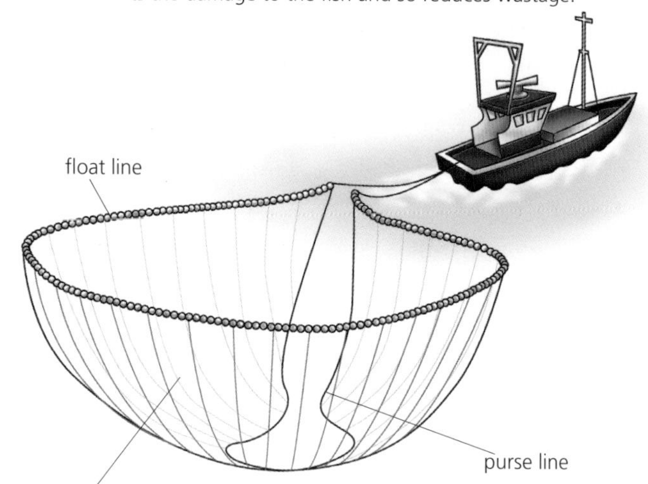

Mesh – square mesh does not close up to trap small fish whereas older diamond mesh did.

120 mm mesh size catches mature herring but allows small cod and haddock to escape.

▲ Mesh and methods used in herring fishing.

- **Understanding of fish migratory patterns:** ecologists have been able to 'tag' fish and release them. These fish are sometimes recaptured at a later date, and the ecologists are able to work out the patterns of migration of the fish. They can then advise the fishing fleet about where and when they are likely to obtain the best catches of mature fish.

Overfishing and the herring population

Herring fishing become extremely efficient in the mid- to late-1960s. As a result the North Atlantic herring population fell dramatically, as shown in the graph below.

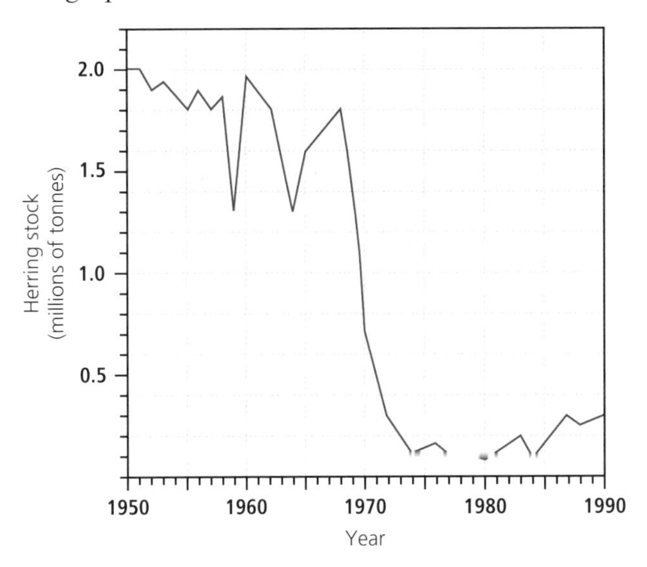

▲ Herring population changes in the North Atlantic

Reasons for this population crash included:

- Shoal location becomes more efficient as radar systems become freely available.
- Larger and safer boats meant that fisheries extended as boats could remain further from port even in bad weather.
- Fishing was often 'mixed' i.e. fishing for one species would often catch other species of similar size. These other species might not be at the same growth stage, and so their population could not be maintained.

The situation become so severe that herring fishing is now very tightly controlled by the European Union. Some of the measures designed to limit overfishing are listed below.

- **Control of mesh:** both the size and the shape of the mesh are designed to allow immature fish to escape.
- **Setting of quotas:** the EU set up a Common Fisheries Policy in 1983 in an attempt to limit the size of catches. Unfortunately scientists now believe that initial quotas were too high, and even the reduced quotas now in force have not allowed herring populations to recover. The fishermen were tempted to catch more than their quota but to throw back the smaller, less marketable fish. Unfortunately these fish were often damaged by capture and did not survive to reproduce.
- **Protection of fisheries:** the EU set a limit of 320 km around their waters in an attempt to control fishing by non-EU countries. These limits have been very difficult to enforce, and have often led to confrontation between fisherman and governments from different countries.
- **Reduction of fishing fleets**: Governments are encouraged to provide subsidies to fishermen whose livelihood has been threatened (or even destroyed) by the measures listed above. The British Government has been reluctant to support these schemes, and many fishing communities have been devastated by the strict control of herring fishing.

As with many situations, there are no clear-cut answers. Fishermen understand that herring stocks must be allowed to recover but are naturally anxious about making a living. The scientists can only advise, the governments must act on this advice!

1 The table shows the chemical composition of some common fish species.

Species	Protein / %	Lipid / %
Cod	17	0.7
Herring	17	19
Mackerel	18	11
Salmon	22	9

a Plot this information in the form of a bar chart.
b Which compound will make up the bulk of the rest of the fish?
c Why are oily fish beneficial to health?
d Fish are also rich in minerals such as iron. How does this help in a healthy diet?

2 a Studies of individual fish species such as herring indicate the age and size at which they reach maturity. Why is it better to catch herring after they reach maturity rather than before?
b What is the advantage in fish conservation of using purse-seine netting?
c Describe how ecologists can provide information that might lead to conservation of fish stocks.
d Suggest three ways in which the British Government can help stocks of herring in the North Atlantic to recover.

4.22 Conservation efforts worldwide

Although the demands on land – for agriculture, mining and building – are increasing in almost every country around the world, many scientists are trying to maintain biodiversity by managing the populations of endangered species. This management is always a compromise, since humans usually place their own requirements above those of wildlife. As a result the conservation managers must be prepared to plan to balance the needs of local people with the scientists' long-term plans to conserve individual species.

CITES	Convention for Trade in Endangered Species: controls the transfer of wildlife and wildlife products between member countries.
IUCN	International Union for the Conservation of Nature and Natural Resources.
Red Data Book	List of all those species at risk: includes categories such as 'vulnerable', 'at risk', 'endangered' and 'critically endangered'. Updated every three years.
WWF	Worldwide Fund for Nature: organisation which attracts public funding for conservation projects.
rare	A species limited to a few areas but in no immediate danger.
vulnerable	A species which is likely to become endangered if the causes of its decline continue.
endangered	A species which is unlikely to survive if the causes of its decline continue.
extinct	A species for which there have been no confirmed sightings for 50 years.

▲ Egrets have suffered for many years as they were shot for their feathers. Conservation measures mean that the Cattle Egret is beginning to breed successfully, even in the UK.

Some examples of species conservation are outlined on the map on the opposite page. In each case note that there is a reason why the species is endangered. There is also a suggested management plan to try to maintain a breeding population of the species under threat. Also see the importance of scientific research into the biology of species, particularly into the breeding requirements of endangered species.

SUSTAINABLE DEVELOPMENT REQUIRES MANAGEMENT OF CONFLICTING DEMANDS

HUMANS **and** **WILDLIFE**

Require sufficient land area to grow crops.

Livestock must be protected from predators.

Reduce risk of extinction by over-hunting

Maintain population size high enough to maintain genetic variation.

Protect suitable habitat, which must provide food, shelter and breeding sites.

PLANNING AND COOPERATION AT DIFFERENT LEVELS

local: farmers, for example
national: government regulations
international: conservation bodies such as WWF

Poison dart frogs in Central America

a Loss of habitat, due to deforestation and drainage of wetlands.

b A series of viral and fungal diseases.

Setting up protected areas, with buffer zones to prevent the spread of contagious diseases. Education programmes which emphasise that amphibians are often a good indicator of the 'health' of our planet. Research into disease, and into the breeding success of these animals.

Tiger in Sumatra, Siberia and India

a Hunting for fur, bones, teeth and blood which are used in Chinese traditional medicines.

b Competition for land with villagers, who cannot afford to lose livestock to predation by the tigers. In some areas, tigers can pose a direct threat to humans as they are such powerful predators.

Setting up protected areas such as fenced sanctuaries, conservation areas and intensive protection zones, especially in National Parks. Trade in tiger products is banned by CITES. Tiger populations in the wild have fallen so dramatically that captive breeding (see page 273) is now responsible for keeping up numbers of most subspecies.

Snow leopard in Pakistan and Nepal

a Hunting for fur, bones, teeth and blood which are used in some traditional medicines.

b Conflict with villagers, who cannot afford to lose livestock to predation by the leopards.

Snow leopard populations are carefully monitored by cameras being set on established trails. Some animal movements can also be tracked using radio collars. Government bodies provide financial compensation to villagers, and set up education programmes which show local people the possible benefits of eco-tourism.

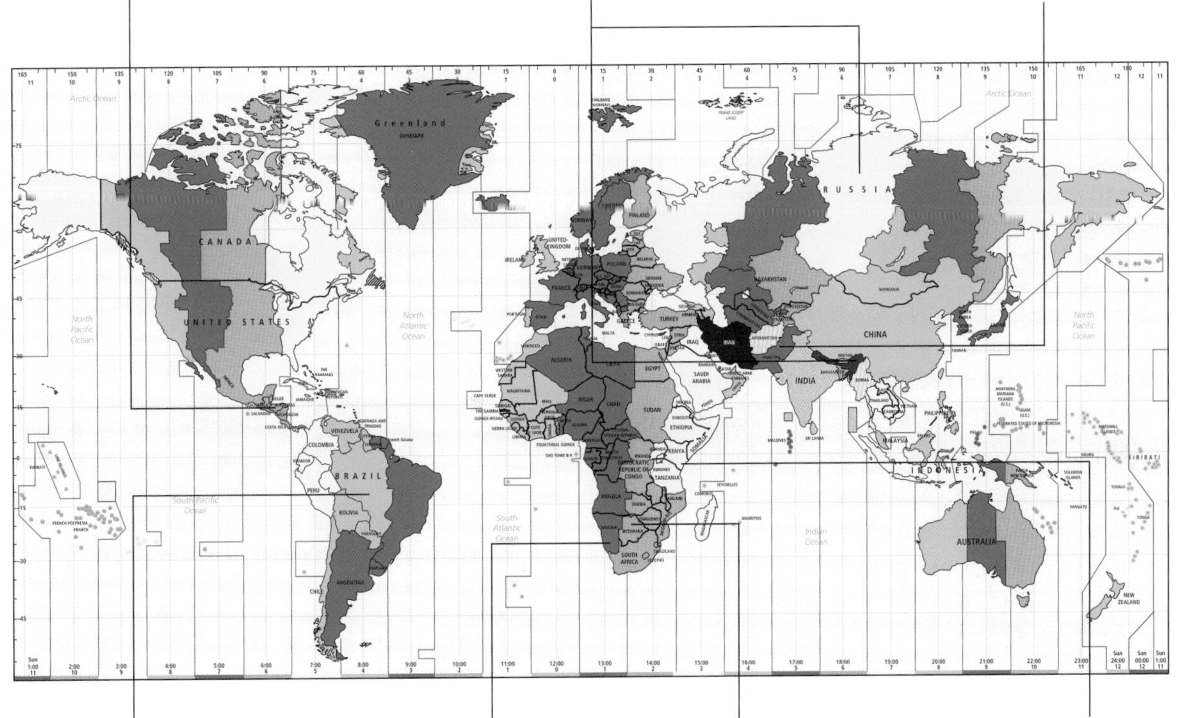

Macaws in Brazil

a The destruction of the rainforest and other Macaw habitats by the logging/timber industries. Without proper nesting sites, these parrots cannot breed.

b The capture of Macaws for the pet trade. As the birds become rarer, their value becomes ever higher. This can be an enormous income to local people.

Protection of habitat is vital to conservation of populations in the wild: the birds can only breed in mature trees and require large areas for foraging for food. Several key areas in Brazil and in Costa Rica have been set aside as suitable habitats.
The trade in Macaws has been banned by CITES, and only licensed owners can keep these birds. Captive breeding has been necessary to try to rebuild some populations (ten years ago there was believed to be only one Spix's Macaw left in the wild).

Cheetahs in Namibia

a Loss of habitat causes competition with local people, who believe that cheetahs kill domestic livestock.

b Hunting for fur.

c The cheetahs' own breeding habits, and low genetic variability.

Trusts have been set up to look after and rehabilitate cheetahs trapped on farmland. CITES has banned trade in fur of spotted cats. Breeding studies have learned more about the unusual breeding requirements of cheetah (several males must 'court' a female for her to become receptive) so captive breeding has become more reliable.

Elephants in Botswana

a Poaching of adults for ivory: apart from death of individuals this upsets family relationships so that younger elephants do not learn survival skills and whole populations are put at risk.

b Animals may be shot if they trample crops or damage fences in villages.

Game wardens can be paid for by the tourist industry. Some people argue that controlled 'trophy' hunting can provide funds to warden the populations more effectively. Trade in ivory can be banned by CITES, or very closely controlled by government organisations.

Rhinoceros in Africa

a Loss of habitat as agricultural development increases.

b Hunting for horn, which is mistakenly believed to have medicinal properties.

Setting up protected areas such as fenced sanctuaries and intensive protection zones especially in National Parks. Surplus animals have been translocated to set up new populations within and outside the species' former range. Controlled sport hunting of surplus males, for example, attracts investment in rhino conservation. Trade in rhino products is banned by CITES.

▲ Worldwide conservation programmes

4.23 Conservation of resources: recycling water by the treatment of sewage

OBJECTIVES

- To recall that pathogens cause disease only when they gain access to body tissues
- To recognise that food and water are possible means of entry to body tissues
- To understand the processes involved in the provision of a safe water supply

Limiting the spread of diseases

Pathogenic microbes can only cause disease if they are able to enter the body and invade the tissues. The body has defences against disease, but there are a number of weaknesses (see page 96). For example, food and water may contain pathogenic microbes. *Salmonella* and *Escherichia coli* (*E. coli*), which cause food poisoning, enter the body in this way, and so do *Cholera vibrio* and *Amoeba* which are waterborne organisms causing cholera and dysentery, respectively. We try to protect ourselves from these diseases by making sure our drinking water, milk and foods are pathogen-free.

Safe water: sanitation and water treatment

Living organisms need water for a number of reasons (see page 236). Water can be lost very rapidly from the human body and we need access to a supply of drinking water. Water supplies are kept **potable** (pathogen-free and drinkable) by both **sanitation** and by **sewage treatment**.

Sanitation is the removal of faeces from waste water so that any pathogens they contain cannot infect drinking water.

Pit latrine

In camps and in isolated houses sewage can be disposed of in a hole. The hole is dug deep enough to accept a large quantity of sewage (ideally several metres deep), and then filled in with soil to keep away flies and rats. This type of disposal system – called a 'long drop' or **pit latrine -** is cheap to produce and does not depend on running water, but has several disadvantages:

- The smell can be extremely unpleasant, especially in tropical areas.
- The sewage can overflow, especially during rainy periods. This allows faeces, perhaps contaminated with human pathogens, to reach water supplies and agricultural land. The sewage (and pathogens) can be washed down through the soil and contaminate nearby water supplies.

Where a good water supply is available, a flush WC is connected to the water carriage system and a flow of water carries the waste away.

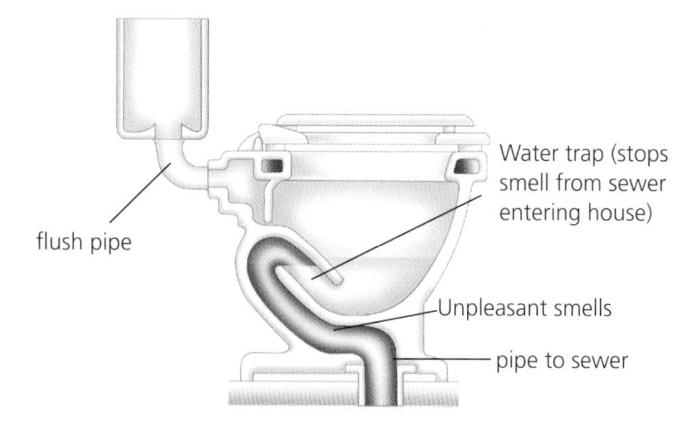

▲ **Diagram of a lavatory pan or water closet (W.C.).** Water is flushed around the rim, down the sides, and through the U-bend. Water retained in the U-bend seals off smells from the sewer.

The waste is then treated at a **sewage treatment plant** so that the valuable water can be recycled.

The page opposite shows one type of sewage treatment plant called the **activated sludge** system. The treatment of the sewage has two functions:

- to destroy or eliminate potential pathogens – either by the high temperature in the anaerobic digestion tank, or by chlorination of the water before it is discharged
- to remove organic compounds (mainly in faeces and urine). These might otherwise contribute to the **biological oxygen demand** (**BOD** – see page 267) of the water into which the treated sewage is discharged. Organic compounds are digested by fungi and bacteria.

In the activated sludge chamber, powerful jets of air keep the sludge aerated so that the processes of decomposition and nitrification (see page 234) can be completed in 8–12 hours. This means that large quantities of sewage can be processed very quickly.

Sewage treatment provides clean water by a combination of physical and biological methods

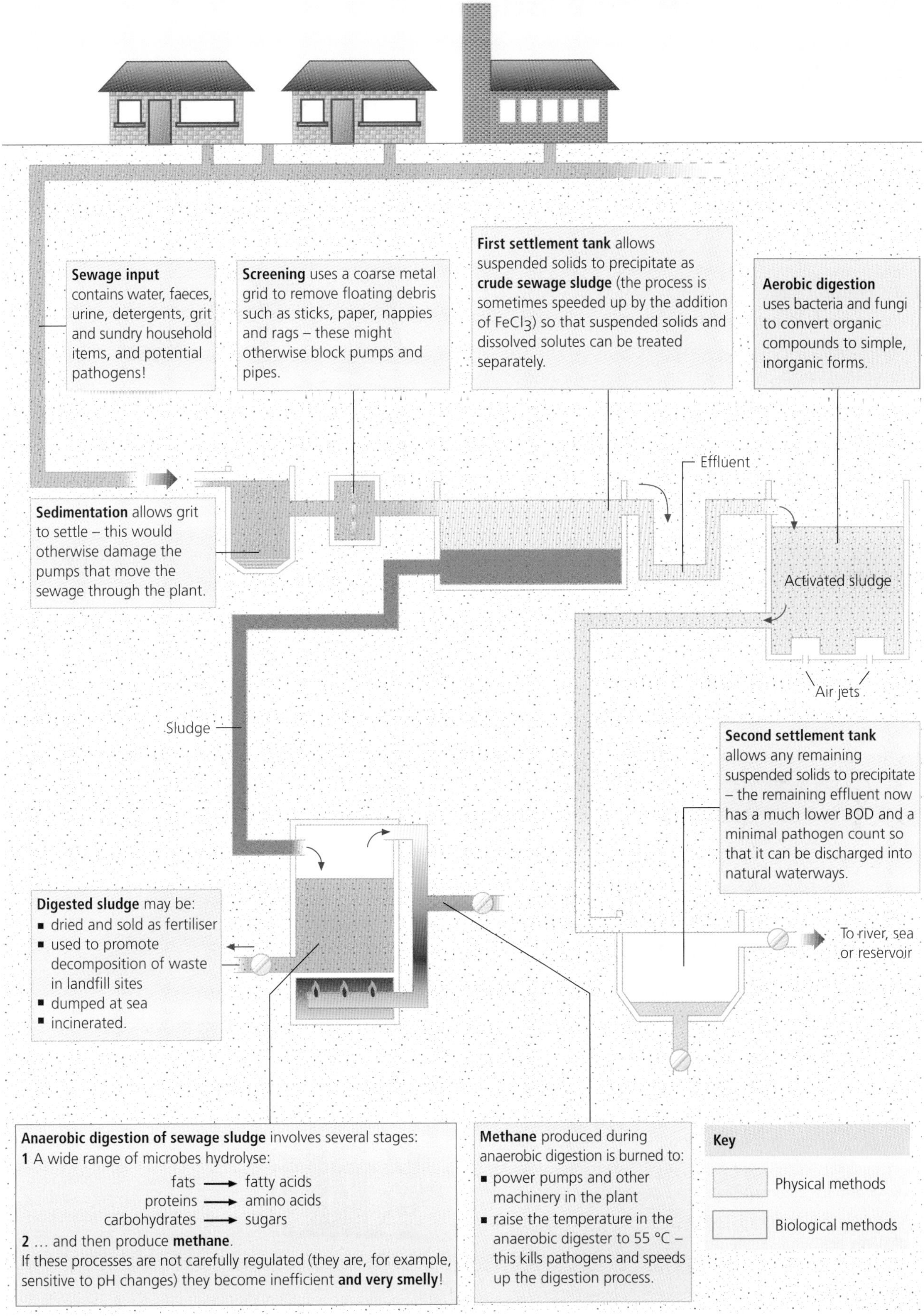

Sewage input contains water, faeces, urine, detergents, grit and sundry household items, and potential pathogens!

Screening uses a coarse metal grid to remove floating debris such as sticks, paper, nappies and rags – these might otherwise block pumps and pipes.

First settlement tank allows suspended solids to precipitate as **crude sewage sludge** (the process is sometimes speeded up by the addition of $FeCl_3$) so that suspended solids and dissolved solutes can be treated separately.

Aerobic digestion uses bacteria and fungi to convert organic compounds to simple, inorganic forms.

Effluent

Sedimentation allows grit to settle – this would otherwise damage the pumps that move the sewage through the plant.

Activated sludge

Air jets

Sludge

Second settlement tank allows any remaining suspended solids to precipitate – the remaining effluent now has a much lower BOD and a minimal pathogen count so that it can be discharged into natural waterways.

Digested sludge may be:

- dried and sold as fertiliser
- used to promote decomposition of waste in landfill sites
- dumped at sea
- incinerated.

To river, sea or reservoir

Anaerobic digestion of sewage sludge involves several stages:

1 A wide range of microbes hydrolyse:

fats ⟶ fatty acids
proteins ⟶ amino acids
carbohydrates ⟶ sugars

2 … and then produce **methane**.

If these processes are not carefully regulated (they are, for example, sensitive to pH changes) they become inefficient **and very smelly**!

Methane produced during anaerobic digestion is burned to:

- power pumps and other machinery in the plant
- raise the temperature in the anaerobic digester to 55 °C – this kills pathogens and speeds up the digestion process.

Key

Physical methods

Biological methods

4.24 Saving fossil fuels: fuel from fermentation

OBJECTIVES

- To recall that microbes carry out fermentation reactions
- To appreciate that fermentation products may be used as fuel rather than as food
- To list examples of fuels generated by microbial fermentation reactions

Fermentations make products which can be used as **fuel**. **Biomass fuels** use raw materials produced by photosynthesis. These materials are from plants and can therefore be regenerated. Biomass fuels include:

- **solid fuels** – wood, charcoal and vegetable waste
- **liquid fuels** – alcohol and vegetable oil
- **gaseous fuel** – biogas (a methane/carbon dioxide mixture).

The production of these biomass fuels is described in the following diagrams: this can help to reduce our use of fossil fuels, which are **non-renewable**.

1. What is a biomass fuel? What advantages might the use of biomass fuels offer?
2. What are the environmental benefits of quick-growing species of tree?
3. Look at the diagram opposite.
   - **a** Suggest three products which could be sold, apart from the alcohol produced.
   - **b** Both amylase and cellulase are involved in the preparation of the glucose feedstock for a gasohol generator. Suggest the exact function of these enzymes.
   - **c** How has genetic engineering helped the gasohol industry?
   - **d** A typical American car travels 10 000 miles a year at 15 m.p.g. This consumes the alcohol generated by the fermentation of 5000 kg of grain. A human on a subsistence diet consumes about 200 kg of grain per year. Comment on these figures.
4. Give two benefits that biogas generators offer to rural communities in poor countries.
5. Why are biogas generators built underground?
6. Why would it be inefficient to add disinfected household waste to the biogas generator?

Solid fuel using natural woodland causes environmental problems but there are alternatives.

Natural woodland – live cutting: see **deforestation** (page 260).

Fallen, dead wood: lesser environmental problem, but still loss of habitat and nutrients.

Cut wood: may have to be carried long distances. Heavy because of water content – much time/energy spent in wood-gathering.

Heating wood without air present produces almost pure carbon, called **charcoal**.

This fuel:
- burns slowly, releasing much heat
- causes very little pollution.

Producing charcoal

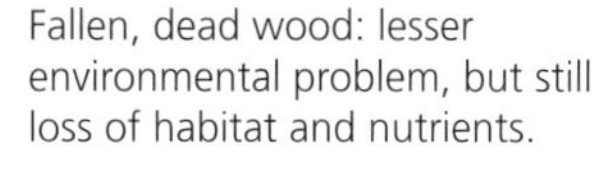

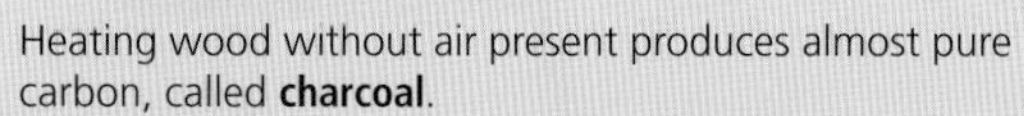

Wood slowly burning

Turf covering wood to keep out air

Quick-growing tree species may provide renewable fuel

e.g. eucalyptus in the Democratic Republic of the Congo …

Erect, regular growth habit means easy cutting and storing.
Rapid growth – 8 m in 3 years.
High resin/oil content means it is clean-burning, giving out much heat.

… and **sweet chestnut** from coppices in Kent may fuel local power stations.

Liquid fuel - ethanol

The raw material is called feedstock

USA uses **Corn**
- **Soak** (then use the liquid as **cattle feed**) → **Seed**
- **Grind** – separates germ, producing **corn oil** → **Starch**
- **Enzymatic hydrolysis** uses amylase → glucose

Brazil uses **Sugar cane**
- Wash and crush → **Cane pulp**
- **Enzymes** amylases and cellulases; **Separation** – waste is **bagasse**. This is burned, giving energy for distillation* → **Syrup**
- **Purification** – impure sucrose solution (molasses) → **cattle feed** → glucose

Fermentation and distillation stages are energy demanding*

Sugar →; Other nutrients → fermenter

Bubble CO_2 to encourage **anaerobic conditions**

→ **Dried yeast** → vitamins 'Marmite®'

→ **Alcohol solution** (15–20%) → DISTILLATION → **Pure ethanol**

$$C_6H_{12}O_6 \xrightarrow{\text{fermentation}} 2C_2H_5OH + 2CO_2$$

(glucose) (alcohol)

Fermentation agent

the yeast *Saccharomyces*.

Useful mutants/engineered strains can:

- cope with high temperatures generated during fermentation
- survive at higher alcohol concentrations (up to 15%, compared with 8–9% for 'normal' strains)
- respire anaerobically even when O_2 is present.

Fuel value of gasohol

- start engine on high petrol/alcohol mixture
- once warm, engine runs well on alcohol alone.

Environmental impact

- alcohol fuels do not require 'anti-knock' additives
- alcohol fuels produce fewer emissions such as CO, SO_2 and NO
- process does not increase CO_2 in atmosphere (photosynthesis of feedstock removes the CO_2 which is generated by combustion)

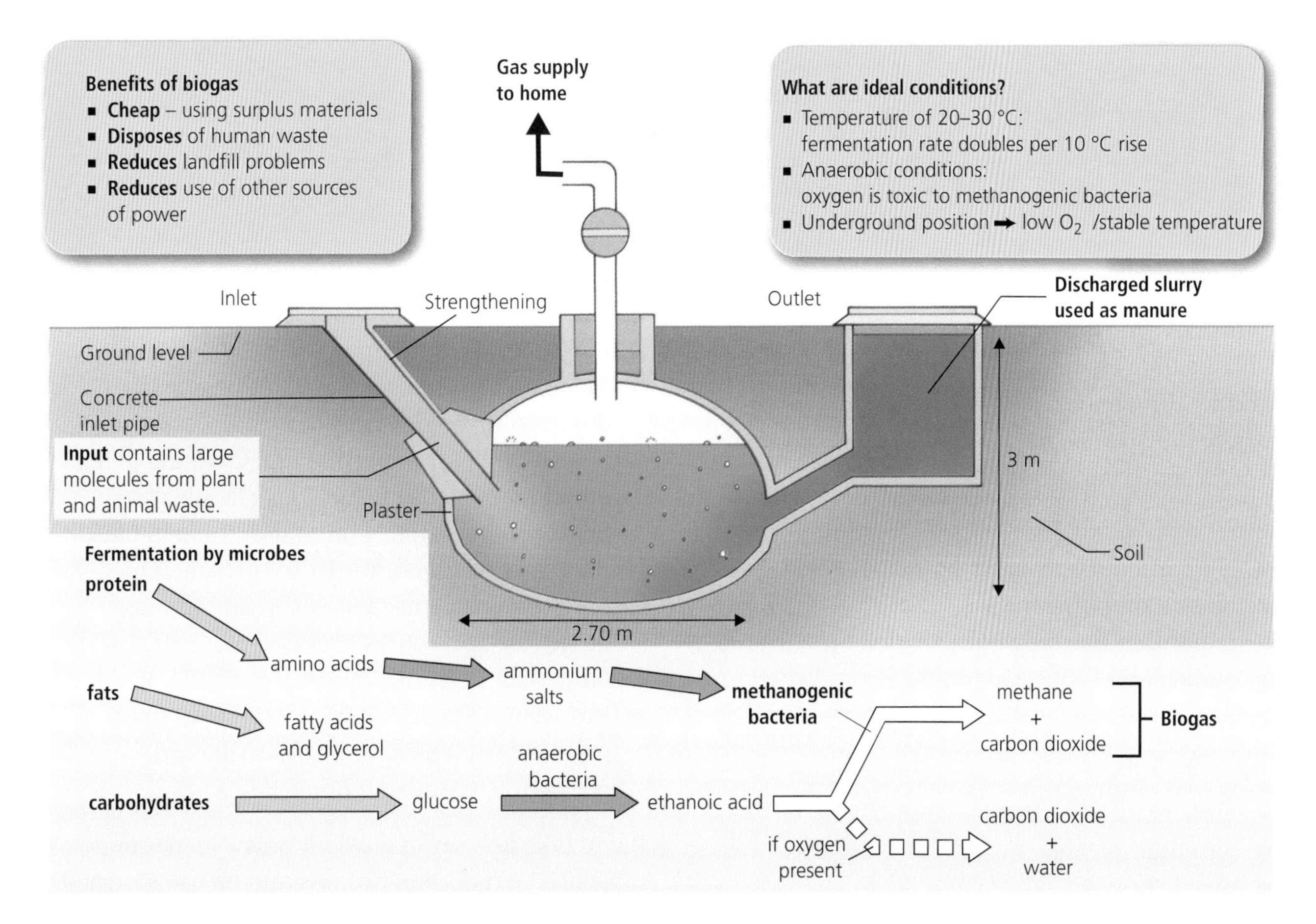

4.25 Recycling: management of solid waste

OBJECTIVES

- To recall that humans produce many waste materials
- To understand that human wastes may become food for disease vectors
- To appreciate that there are different ways of disposal of human refuse
- To understand the benefits of recycling

Human refuse contains such items as metals, plastic bags, glass, remains of food and ash. This refuse must not be allowed to accumulate because:

- It is inconvenient, and gets in the way of people, their vehicles and their animals.
- It is unpleasant to look at.
- It could act as breeding ground for organisms which transmit disease.

To prevent refuse (rubbish) from building up it can be delivered to refuse tips, landfill sites or incinerators. The refuse disposal sites are normally sited away from residential areas, and surrounded by fences to stop rubbish blowing away and prevent children from playing in them. Two alternative ways of disposing of the refuse – **incineration** and **landfill** – are described on page 100. Each of these methods has the advantage that it can provide energy, and so reduce our use of fossil fuels.

Recycling

Recycling involves reusing waste products or materials which would otherwise be thrown away. This includes reusing items in their original form and sending materials away to special centres where they can be melted down or pulped to act as raw materials in industry.

Items that can be reused include:

- glass milk or soft drink bottles
- plastic shopping bags
- paper that has only been written on one side
- clothes

. . . . and materials that can be pulped or melted down include:

- glass bottles
- aluminium cans
- plastic bottles
- paper
- scrap metal.

Plastic bottles can be molten and reused in clothing, for example!

Non-biodegradable plastics cannot be broken down by natural biological processes. They may accumulate and are a danger to fish, birds and mammals. If they can be collected they may be recycled.

Recycling can make scarce resources last for longer, and can reduce the energy requirements of industry. It takes only 5% of the energy to make an aluminium can from recycled aluminium than from aluminium ore. People and companies will only recycle, however, if:

- their products are cheaper than if they don't use recycling
- governments give subsidies for manufacturing products from recycled materials.

Recycling of paper

Paper is one of the easiest materials to recycle. It is collected from our kerbside or recycling banks by local authorities and waste management companies.

After the paper is collected there are several steps in the recycling process.

- First it is sorted, graded and delivered to a paper mill. As it is sorted, contaminants such as plastic, metal, and other rubbish that may have been collected with the paper are removed.
- Once at the paper mill it is added to water and then turned into pulp.
- The paper is then screened, cleaned and de-inked through a number of processes until it is suitable for papermaking. Cleaning and de-inking may use hazardous chemicals: hydrogen peroxide is often used to help bleach the paper when dark inks are present. Once the pulp has been rinsed, it is spread onto large flat racks, and rollers press the water out of the pulp. As it dries, new paper forms.
- It is then ready to be made into new paper products such as newsprint, cardboard, packaging, tissue and office items.

It can take just seven days for a newspaper to go through the recycling process and be transformed into recycled newsprint which is used to make the majority of national daily newspapers.

Questions on human impacts on ecosystems

1 The graph below shows changes in the population of bacteria that take place in a river when untreated sewage is added to it.

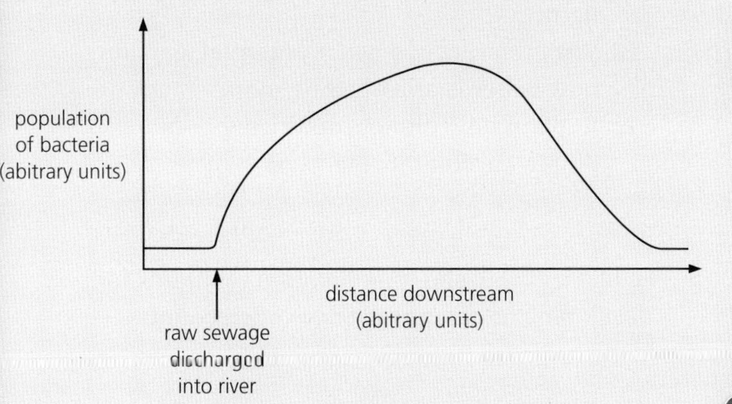

a Describe the changes in the population of bacteria that take place in this river.

b Suggest an explanation for these changes in the population of bacteria.

Cambridge IGCSE Biology 0610
Paper 2 Q8 May 2008

2 The wild dog is one of the smaller African carnivorous mammals. It has disappeared from 25 of the 39 countries where it used to live. Wild dogs hunt in packs, feeding on antelopes, which are grass-eating mammals.
A conservation programme has been started to increase the wild dog population in South Africa. Farmers are worried about numbers getting out of control because wild dogs breed at a very fast rate. However, conservationists are not concerned because the lion is a natural predator of the dogs.

a Wild dogs are carnivorous mammals.
- **i** Define the term *carnivore*.
- **ii** State **one** external feature which distinguishes mammals from other vertebrates.

b **i** Suggest two reasons why numbers of African wild dogs are decreasing.
- **ii** Suggest what could happen to the species if numbers continue to decrease.

c Using the information in the passage above, construct a food chain for a wild dog, including its predator. Label each organism with its trophic level.

d It is important that the wild dog species is conserved.
- **i** Explain the meaning of the term *conservation*.
- **ii** Outline the measures that could be taken to conserve a mammal, such as the wild dog.

e When wild dogs die, nitrogen compounds in their bodies may become available for plants. Outline the processes that occur to make these nitrogen compounds in the bodies of dead animals available for plants to absorb.

Cambridge IGCSE Biology 0610
Paper 3 Variant 1 Q2 May 2008

3 The freshwater mussel, *Margaritifera margaritifera*, is a mollusc which lives in rivers and streams. When the mussel reproduces, gametes are released into the water and fertilisation takes place. The embryos, in the form of larvae, attach themselves to the gills of fish and develop there for a few months. The larvae then release themselves and grow in sand in the river, feeding by filtering food from the water. The number of mussels is falling due to human predation and the species is threatened with extinction.

a The mussel belongs to the group known as the molluscs. State two features you would expect the mussel to have.

b Explain how the species name of the freshwater mussel can be distinguished from its genus.

c State the type of reproduction shown by the mussel. Explain your answer.

d **i** Fish gills have the same function as lungs. Suggest **one** advantage to a mussel larva of attaching itself to fish gills.
- **ii** The mussel develops on the fish gills. Define the term *development*.

e The mussel is threatened with extinction. Name another organism which is also threatened with extinction and outline how it could be conserved.

Cambridge IGCSE Biology 0610
Paper 3 Variant 1 Q1 November 2008

4 The bar graph below shows crop productivity for a range of plants but it is incomplete.

a Complete the bar graph using the data below it.

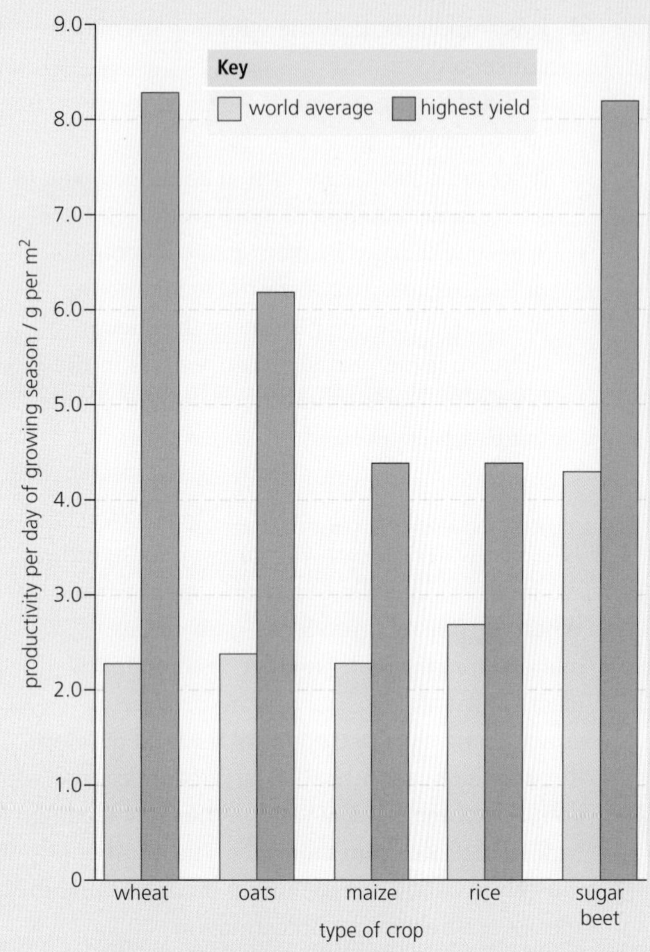

crop	productivity per day of growing season / g per m^2	
	world average	highest yield
potatoes	2.6	5.6

b State which crop has

i the highest average productivity,

ii the greatest difference between the average yield and the highest yield.

c Outline how modern technology could be used to increase the productivity of a crop from the average yield to a high yield.

d When the yield is measured, dry mass is always used rather than fresh mass. Suggest why dry mass is a more reliable measurement than fresh mass.

e Maize is often used to feed cows, which are grown to provide meat for humans. Explain why it is more efficient for humans to eat maize rather than meat from cows that have been fed on maize.

f i Complete the equation for photosynthesis.

$$6CO_2 + 6H_2O \xrightarrow{\text{light energy}} C_6H_{12}O_6 +$$

ii Describe how leaves are adapted to trap light.

iii With reference to water potential, explain how water is absorbed by roots.

iv Explain how photosynthesising cells obtain carbon dioxide.

Cambridge IGCSE Biology 0610 Paper 3 Variant 1 Q2 November 2008

5 The figure below shows population pyramids for a developing country and a developed country.

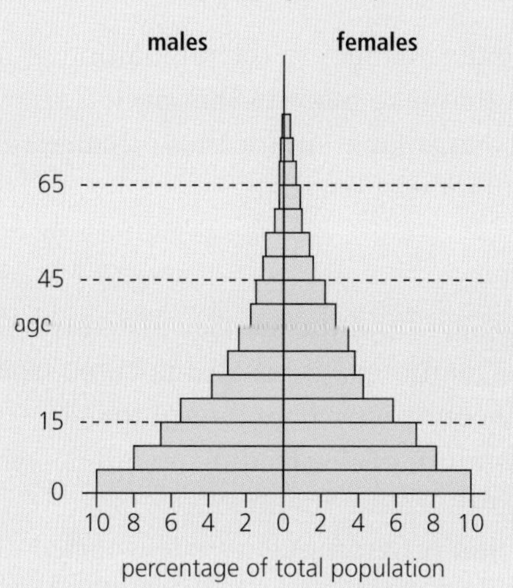

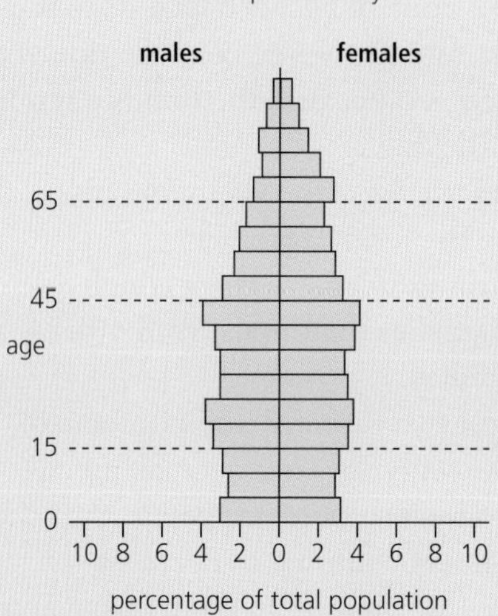

a Describe how the percentage of people in the population varies with age in
   i a developing country
   ii a developed country.

b These countries have a similar population size. Compare the two pyramids. State **one** difference between the populations
   i at under 15
   ii over 65.

c The pyramids can also be used to compare proportions of males and females in a population. State one way in which these pyramids are similar for people who live more than 65 years.

d With reference to **X** and **Y** chromosomes, explain the expected ratio of males to females at birth.

e This graph shows survival curves for developing and developed countries, based on samples of 10 000 people. The graph can be used to estimate the average life expectancy, defined as the age at which 50% of people in the sample are still alive.

6 This figure shows the causes of severe food shortages in the 1980s, 1990s and 2000s.

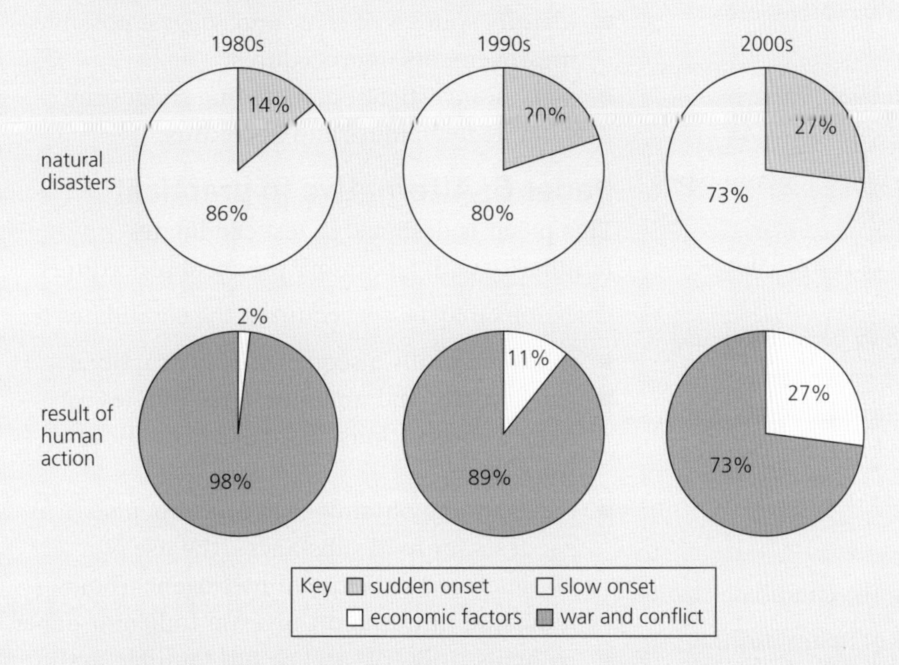

a **i** State two types of natural disaster that occur suddenly and may lead to severe food shortages.
   ii State **one** type of natural disaster that may take several years to develop.

b Explain how the increase in the human population may contribute to severe food shortages.

The quality and quantity of food available worldwide has been improved by artificial selection (selective breeding) and genetic engineering.

c Use a **named** example to outline how artificial selection is used to improve the quantity or quality of food.

d Define the term *genetic engineering*.

5.1 Practical assessment

The basis of scientific subjects is experimental: hypotheses are tested to establish theories and observations and measurements made to provide the factual background to science. Examiners believe that it is important that an assessment of a student's knowledge and understanding of biology should contain a practical component.

Schools' circumstances (e.g. the availability of resources) differ greatly, so two alternative ways of examining the relevant assessment are provided by University of Cambridge International Examinations (CIE). The two alternatives are:

- Paper 5 – practical test
- Paper 6 – alternative to practical (written paper).

The following points should be noted for both types of practical assessment:

- the same proportion of marks is available – 20% of the subject total
- the same practical skills are to be learned and developed
- the same benefits to theoretical understanding come from all practical work.

Paper 5: Practical test

The CIE specification states:

Exercises may be set requiring the candidates to:

- follow carefully a sequence of instructions
- use familiar, and unfamiliar, techniques to record observations and make deductions from them
- perform simple physiological experiments, e.g. tests for food substances and the use of hydrogencarbonate indicator, litmus and Universal Indicator paper
- use a scalpel or a razor blade, forceps, scissors and mounted needles skilfully
- use a hand lens of not less than ×6 magnification to recognise, observe and record familiar, and unfamiliar, biological specimens
- make a clear line drawing of a specimen provided, indicate the magnification of the drawing and label, as required
- perform simple arithmetical calculations.

Candidates may be required to do the following:

- record readings from apparatus
- describe, explain or comment on experimental arrangements and techniques
- complete tables of data
- draw conclusions from observations and/or from information given
- interpret and evaluate observations and experimental data
- plot graphs and/or interpret graphical information
- identify sources of error and suggest possible improvements in procedures
- plan an investigation, including suggesting suitable techniques and apparatus.

Paper 6: Alternative to practical

This paper is designed to test candidates' familiarity with laboratory practical procedures.

Questions may be set requiring the candidates to:

- follow carefully a sequence of instructions
- use familiar, and unfamiliar, techniques to record observations and make deductions from them
- recall simple physiological experiments, e.g. tests for food substances, the use of a potometer and the use of hydrogencarbonate indicator, litmus and Universal Indicator paper
- recognise, observe and record familiar, and unfamiliar, biological specimens
- make a clear line drawing from a photograph (or other visual representation) of a specimen, indicate the magnification of the drawing and label, as required
- perform simple arithmetical calculations
- record readings from apparatus
- describe, explain or comment on experimental arrangements and techniques
- complete tables of data
- draw conclusions from observations and/or from information given
- interpret and evaluate observations and experimental data
- plot graphs and/or interpret graphical information

- identify sources of error and suggest possible improvements in procedures
- plan an investigation, including suggesting suitable techniques and apparatus.

Throughout this book there are a number of examples of the type of experiment or exercise that might appear in either the practical test or the alternative to practical. These are listed in the table below, and should give you a great deal of help in dealing with whichever of these papers is used in your practical assessment.

List of practical exercises

Exercise	Spread reference	Comment
Measuring the size of objects	1.6	
Enzyme experiments and the scientific method	2.6	Includes drawing tables and graphs
Testing for biochemicals/foods	2.4	
Effectiveness of antibiotics	4.12	
Measurement of energy content of food	2.15	
Testing leaves for starch	2.7	Factors affecting photosynthesis
Measurement of the rate of photosynthesis	2.8	Includes analysis of experimental design
Photosynthesis and gas exchange	2.11	Use of hydrogencarbonate indicator
Plants and minerals	2.12	
Tracing xylem	2.22	Use of eosin
Bubble potometer	2.23	
Water loss from leaf surfaces	2.24	Use of cobalt chloride paper
Demonstration of respiration	2.36	Germination and respiration
The measurement of respiration	2.38	Respirometer / measurement of carbon dioxide released
Investigation of tropisms	2.53	Phototropism and gravitropism
Structure of flowers	3.2	
Experimental design: using models for pollination	3.3	
Conditions for germination	3.5	Gives references to enzymes, and the need for energy
Sampling populations	4.3	Use of quadrats

5.2 Laboratory equipment

You will find that the practical paper or the alternative to practical will be much more straightforward if you recognise certain standard pieces of equipment, and understand what they are used for.

Apparatus and materials

Safety equipment appropriate to the work being planned, but at least including eye protection such as safety spectacles or goggles.

3D image	Name and function	Diagram
	■ **Watch glass**: used for collection and evaporating liquids with no heat, and for immersing biological specimens in a liquid.	
	■ **Filter funnel**: used to separate solids from liquids, using a filter paper.	
	■ **Measuring cylinder**: used for measuring the volume of liquids.	
	■ **Thermometer**: used to measure temperature.	
The other very important measuring device in the laboratory is a balance (weighing machine).		
	■ **Spatula**: used to for handling solid chemicals; for example, when adding a solid to a liquid.	
	■ **Pipette**: used to measure and transfer small volumes of liquids.	
	■ **Stand, boss and clamp**: used to support the apparatus in place. This reduces the risk of dangerous spills. This is not generally drawn. If the clamp is merely to support a piece of appartus, it is usually represented by two crosses as shown.	
	■ **Bunsen burner**: used to heat the contents of other apparatus (e.g. a liquid in a test tube) or for **directly** heating solids.	HEAT
	■ **Tripod**: used to support apparatus above a Bunsen burner. **The Bunsen burner, tripod and gauze are the most common way of heating materials in school science laboratories.**	
	■ **Gauze**: used to spread out the heat from a Bunsen burner and to support the apparatus on a tripod.	
	■ **Test tube and boiling tube**: used for heating solids and liquids. They are also used to hold chemicals while other substances are added and mixed. They need to be put safely in a test tube rack.	
	■ **Evaporating dish**: used to collect and evaporate liquids with or without heating.	
	■ **Beaker**: used for mixing solutions and for heating liquids.	

Chemical reagents

In accordance with the COSHH (Control of Substances Hazardous to Health) regulations operative in the UK, a hazard appraisal of the list has been carried out. The following codes are used where relevant.

C = corrosive substance
F = highly flammable substance
H = harmful or irritating substance
O = oxidizing substance
T = toxic substance

Table of hazard symbols

Symbol	Description	Examples
	Oxidising These substances provide oxygen which allows other materials to burn more fiercely.	Bleach, sodium chlorate, potassium nitrate
	Highly flammable These substances easily catch fire.	Ethanol, petrol, acetone
	Toxic These substances can cause death. They may have their effects when swallowed or breathed in or absorbed through the skin.	Mercury, copper sulfate
	Harmful These substances are similar to toxic substances but less dangerous.	Dilute acids and alkalis
	Corrosive These substances attack and destroy living tissues, including eyes and skin.	Concentrated acids and akalis
	Irritant These substances are not corrosive but can cause reddening or blistering of the skin.	Ammonia, dilute acids and alkalis

Reagent	Use in biology	Spread reference
hydrogencarbonate indicator (bicarbonate indicator)	Detect changes in carbon dioxide concentration, for example in exhaled air following respiration	2.11, 2.38
iodine in potassium iodide solution (iodine solution)	Detection of starch	2.4, 2.7
Benedict's solution (or an alternative such as Fehling's)	Detection of a reducing sugar, such as glucose	2.4
Biuret reagent(s) (sodium or potassium hydroxide solution and copper sulfate solution)	Detection of protein	2.4
ethanol/methylated spirit	For dissolving lipids in testing for the presence of lipids. Also used to remove chlorophyll from leaves during starch testing.	2.4
cobalt chloride paper	Detects changes in water content	2.24
pH indicator paper or Universal Indicator solution or pH probes	Changes in pH during reactions such as digestion of fats to fatty acids and glycerol	
litmus paper	Qualitative detection of pH	
glucose	Change in water potential of solutions	
sodium chloride	Change in water potential of solutions	
aluminium foil or black paper	Foil can be used a heat reflector: black paper as a light absorber	2.12
a source of distilled or deionised water	Change in water potential of solutions. Also used in making up mineral nutrient solutions.	
eosin/red ink	To follow the pathway of water absorbed by plants	2.22
limewater	A liquid absorbent for carbon dioxide, for example in exhaled air	2.38

methylene blue	A stain for animal cells	1.6
potassium hydroxide	Removes carbon dioxide from the atmosphere, for example during experiments on conditions for photosynthesis	2.38, 2.7
sodium hydrogencarbonate (sodium bicarbonate)	Very mild alkali: use for adjusting pH of solutions for enzyme activity	
vaseline/petroleum jelly (or similar)	Blocks pores such as stomata, and so prevents water loss	2.24

Measurement of variables

In many investigations, biologists need to measure variable quantities such as volume, temperature, mass and time. It is very important to be able to read scales accurately and to choose the correct units for the quantities that have been measured. Some of the common measuring equipment used in biology laboratories is shown below.

Measuring temperature using a thermometer

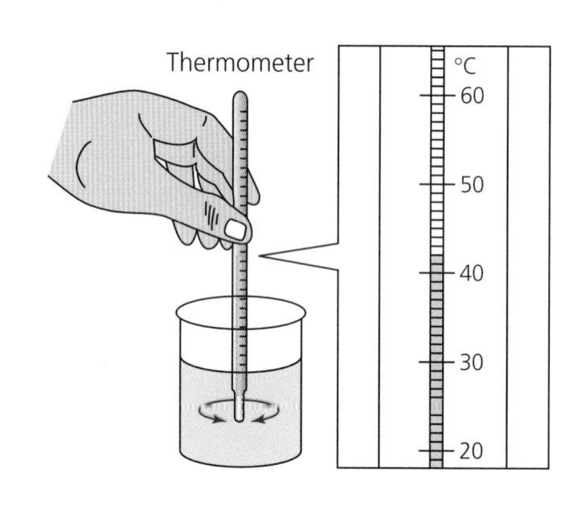

Normal temperature are measured on the **Celsius** scale sometimes called the **Centigrade** scale. The unit for temperature is the **degree Celsius** (°C).

The **scale** is worked out by checking how long the liquid column is firstly in melting ice and secondly in boiling water.

Column of coloured liquid: this gets **longer** as the liquid gets **hotter**, and **shorter** as the liquid gets **cooler**.

Bulb: this contains a coloured liquid.

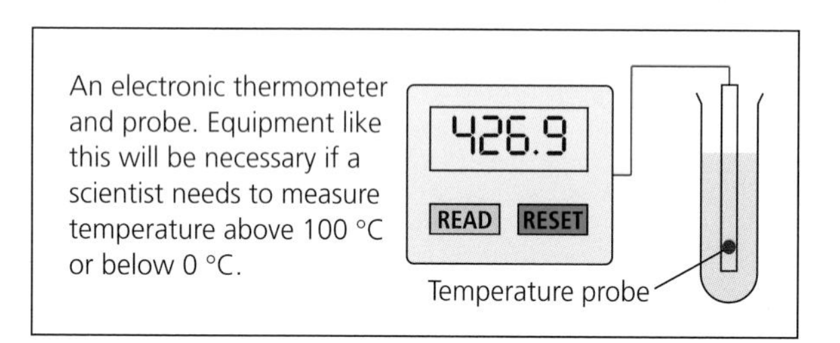

An electronic thermometer and probe. Equipment like this will be necessary if a scientist needs to measure temperature above 100 °C or below 0 °C.
426.9
READ
RESET
Temperature probe

Measuring mass using an electronic pan balance

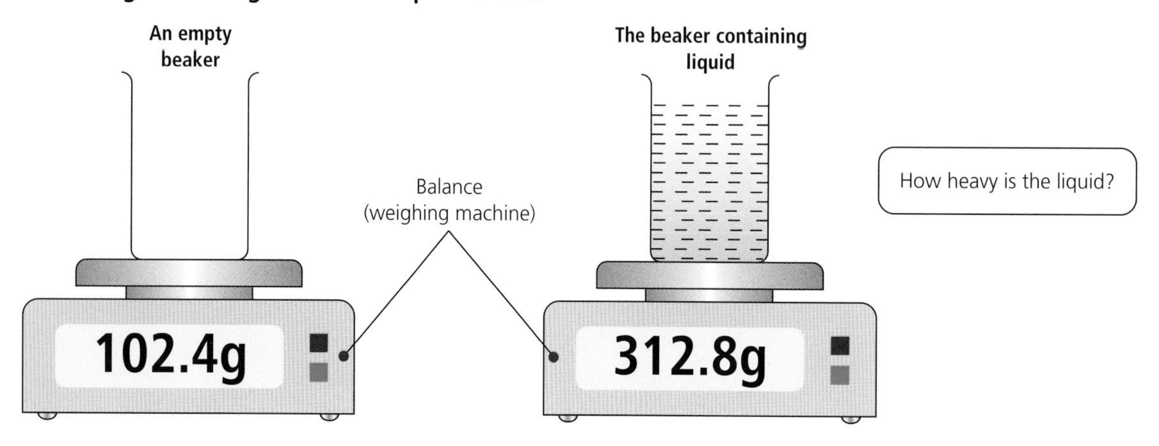

Measuring volume using a beaker or a measuring cylinder

BEAKER
It is not accurate to use a beaker because the scale is not fine enough.

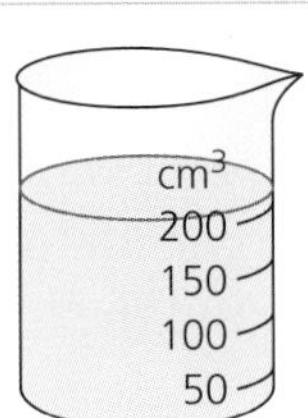

Is it cm^3 or ml? Some equipment is scaled in cm^3, and some is scaled in ml. It really doesn't matter – 1 cm^3 has exactly the same volume as 1 ml.

MEASURING CYLINDER

When using a measuring cylinder, stand the measuring cylinder on a level or bench, so that the liquid is level.

ml
100
90
80
70
60
50
40
30
20
10

ml
80
70
60
Liquid level

Make sure that you read the level carefully. You may notice that the surface of the fluid is curved; this ia called the **meniscus**.

Get your eye level with the liquid level. It may look as though there is a thick 'skin' on the water. This is because you are looking at the minute amount of liquid that has been drawn up the glass.

The volume of liquid is represented by the **bottom** of **the meniscus**.

Interestingly, plastic beakers and measuring cylinders often do not give rise to a meniscus!

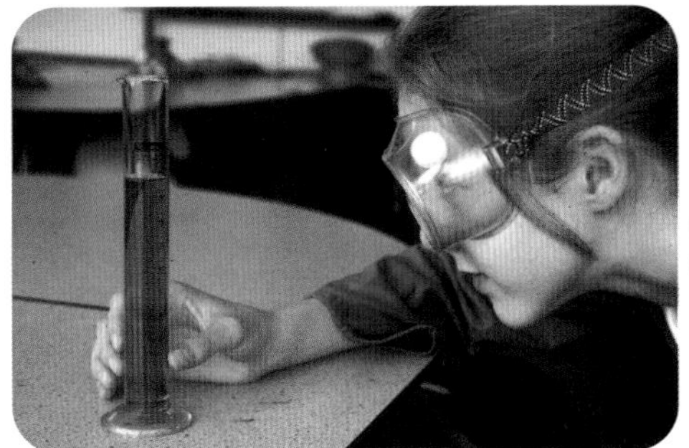

◀ Reading the measurement from a measuring cylinder by getting your eye level with the liquid level

This table gives a summary of the measuring equipment used in chemistry laboratories

Quantity	Units	Equipment
Volume (fluid)	Cubic centimetres (cm^3) Litres (dm^3) 1 litre = 1000 ml	Measuring cylinder or beaker
Temperature	Degrees Celsius (°C)	Thermometer
Time	Seconds (s) Minutes (min)	Stopclock (can be analogue or digital)
Mass	Grams (g) Kilograms (kg) 1 kg = 1000 g	Balance (usually top pan balances and electronic)

Answers to embedded questions

Unless otherwise indicated, the questions, example answers, marks awarded and/or comments that appear in this book were written by the author(s). In examination, the way marks would be awarded to answers like these may be different.

1.1 **1** About 2.2 billion years
2 Amino acids, fatty acids/lipids, sugars (nucleotides/nucleic acids)
3 REDGIRL, MRSGREN for example
4 To replace individuals that die/provide more individuals to colonise new environments/to combine genetic material from different individuals (provide variation)

1.2 **1** **a** Have well-developed carnassials (cutting) teeth
b Weasel and mink (same genus)
c The fox (the others are in the same family)
d They ingest their organic molecules in ready-made form
2 The wise man
3 Viruses do not show any of the characteristics of living organisms unless they are inside the cells of another organism.
4 **a** Specimen 1: has single bladders on its fronds;
Specimen 2: has serrated edges to its fronds;
Specimen 3: has paired bladders on its fronds;
Specimen 4: has a single large frond with wavy edges;
Specimen 5: has a visible root-like structure at the base;
b Q.1: has a root-like structure **specimen 5**
No visible root-like structure go to Q.2
Q.2: has bladders(air sacs) on frond go to Q.3
No bladders (air sacs) on frond go to Q.4
Q.3: single bladders **specimen 1**
Paired bladders **specimen 3**
Q.4: frond has serrated edges **specimen 2**
Frond has wavy edges **specimen 4**
There may be other acceptable answers offered.

1.3 **1** They have roots, for absorption of water and minerals, and a vascular system to transport these materials around the plant body.
2 Their gametes can be transferred without dependence on water.

1.4 **1** They ingest ready-made organic molecules; a backbone; arthropods
2 **a** They have segmented bodies/segments carry jointed limbs/they have a hard exoskeleton/
b Two pairs of wings/three pairs of legs/three major body sections/ metamorphosis in life cycle
insects (6, 3, 4, compound) spiders (8, 2, none, simple)
3 Life cycle has four distinct stages: egg, larva, pupa/chrysalis, adult. Adults can larvae can lead different lives, on different food sources – reproductive structures can develop during stage as pupa.

Characteristics and classification p.14

1

	1a	1b	2a	2b	3a	3b	4a	4b	5a	5b	name
A											
B	✓		✓		✓						Anopheles
C		✓						✓			Ornithodorus
D		✓					✓		✓		Pulex
E	✓			✓							Musca
F	✓		✓			✓					Periplaneta

2 **a** *Elephas* – genus; *maximus* – species;
b **i** jaguar/ leopard/lion/tiger OR Bobcat/European/Iberian Lynx
ii Acinonyx
3 **a** made of many cells – animal – vertebrate – skin covering – no scales on skin – mammal
c walk upright/communicate with speech

Cells and organisation pp.20–21

1

Cell	Tissue	Organ
Phagocyte, sperm, neurone	Epidermis, xylem, blood,	Liver, heart, leaf, ileum, ovary, brain, stem

2 Organelle – cell – tissue – organ – system – organism

3 Kilometre – metre – millimetre – micrometre;

$\frac{1000}{50} = 20$

4 a $(100 - \{23 + 13 + 0.5 + 24 + 8 + 11 + 0.1\}) = 20.4$

b check that 'other' is included

c $\frac{13}{100} \times 100\ 000\ 000\ 000\ 000$

$= 13\ 000\ 000\ 000\ 000$

5

structure	Liver cell	Palisade cell
Cell surface membrane	✓	✓
Chloroplasts	✕	✓
Cytoplasm	✓	✓
Cellulose cell wall	✕	✓
Nucleus	✓	✓
Starch granule	✕	✓
Glycogen granule	✓	✕
Large, permanent vacuole	✕	✓

6 Choose examples from Q.5. A good answer will include both plant and animal examples/both structural and biochemical examples.

7

Respiration	The release of energy from organic molecules
Sensitivity	The ability to detect changes in the environment
Growth	A change in size
Reproduction	The generation of offspring
Nutrition	The supply of food
Excretion	The removal of the waste products of metabolism
development	A change in shape or form

8

Cell membrane	g – controls the entry and exit of materials
Cell wall	c – contains cellulose and surrounds a plant cell
Chloroplasts	a – structures which contain chlorophyll
Chromosomes	e – coloured threads found inside the nucleus
Cytoplasm	d – jelly-like material which fills most of the cell
Mitochondria	f – the site of aerobic respiration
Nucleus	h – carries the genetic information and controls the activities of the cell
vacuole	b – cavity that is found only in plant cells

9 Red blood cell – b; white blood cell – g; cell lining bronchiole – f; motor nerve cell – e; palisade cell – i; root hair cell – a; phloem sieve tube – d; sperm cell – c; egg cell – h

10

Stomach	organ
Mitochondrion	organelle
Cat	organism
Onion bulb	organism
Carrot plant	organism
Chloroplast	organelle
Neurone	cell
Lung	organ
Brain	organ
Heart and circulation	system
Xylem	tissue
The lining of the lung	tissue

11 a – E; b – A; c – D; d – C; e – B

12 a Cells – tissues – phloem/xylem/blood – organ – systems – excretory system

b Specialized – red blood cell – division of labour – nervous – endocrine

c Palisade cell – chloroplasts – leaf – epidermis - xylem

2.2 1 A structure which allows some molecules/ particles to cross but prevents others from crossing. For example, the membrane lining the ileum allows glucose and amino acids to cross but prevents starch and large proteins from crossing.

2 A 'slope' between a high and a low concentration of a dissolved substance

3 Movement of gases (e.g. oxygen and carbon dioxide), food substances (e.g. glucose), wastes (e.g. urea), hormones (e.g. insulin)

4 Large surface area/thin membranes (short pathway for diffusion)/ability to set up and maintain diffusion gradients

5 cytoplasm; partially-permeable membrane; swell; lower; cell wall; glucose/amino acids; diffusion; concentration gradient; active transport; energy; against; carbon dioxide; diffusion; photosynthesis

6

	DIFFUSION	ACTIVE TRANSPORT
Is energy(ATP) used	no	yes
Direction of movement	With/down a concentration gradient	Can be against a concentration gradient
Are carrier molecules involved	No	Yes
Example	Oxygen movement from blood to air sacs	Uptake of mineral ions by root hair cells
Phagocytosis would be used to take up/ move large particles across a membrane		

c

diffusion and osmosis p.27

1 i turgid means 'fully inflated by water'

ii turgid plant cells push against nearby cells and together they support the plant

iii beaker A

iv the chip in beaker B

v the chip would slowly become firm and able to support the 10g weight. This is because water can enter the chip by osmosis (down a water potential gradient) so that the cells become turgid. The turgid cells can now support the chip.

2 a the graph should have concentration of sugar solution on the *x* axis and percentage change in length on the *y* axis. Check accuracy of points, and note where line crosses *x* axis.

b 0.24 mol per dm^3. At this concentration there is no net osmosis so the percentage change in length is zero.

c because by using percentage change any differences in initial length can be ignored.

2.3 1 Carbohydrates (source of energy)/lipids (energy store)/proteins (enzymes)/nucleic acids(genetic information)

2 Soluble molecules can be transported in the blood and across membranes/ insoluble molecules cannot be transported like this, and are excellent storage molecules.

3 Nucleic acids control the functions of cells as they carry the coded genetic information. Proteins regulate the activities of cells e.g. in the role as enzymes. Carbohydrates are the most commonly used source of energy, and keeping cells alive requires energy.

2.4 1 a Fat is solid at room temperature/oil is liquid at room temperature

b In the proportion of these elements – carbohydrates have a higher proportion of oxygen

Show backbone of glycerol with three fatty acids. Many different fats since there are many different fatty acids.

2 a A: washings from a laundry (contain starch); B: milk (contains lactose and protein),C: sweetened tea (gives positive test for sucrose); D: crushed potato (contains starch and small quantities of protein); E: urine from diabetic person (high concentration of glucose)

3 1: test that the reagents are working properly by trying the test on a 'known' substance. 2: to show that reagents are not contaminated, carry out the test on water – should have a negative result

2.5 **1** proteins; reactions/processes; catalysts; specific; denatured

2 Draw diagram in reverse(i.e. one molecule on active site becoming two products): any digestive enzyme would be a suitable example

2.6 **1** **b** 0.25/0.50/2.0/0.5/0.25/0.08

c Temperature on x axis, rate on y axis, maximum rate at 35 °C

Up to 35 °C rate increases as molecules gain more kinetic energy and so collide more often and with greater energy, but above this temperature the enzyme molecules begin to denature. Active site no longer 'recognises' substrate molecules and so rate decreases. 35 °C is the optimum temperature for this enzyme.

Enzymes and biological molecules p.36–37

1 **a** an enzyme is a biological catalyst – it can affect a biological reaction without being affected itself.

b check labels on axes/linear axes/accuracy of plot/title

c stomach (pH 2)

d rate would be zero/very low since boiling has denatured the enzyme

2 **a** excretion

b **i** pH

ii Enzyme concentration, substrate concentration, temperature

c $\frac{10}{17.4} = 0.57$

d check labels on axes (pH on x axis)/linear axes/accuracy of plot/title

e **i** there is an increase in rate (from 0.50 to 1.00) between pH 4 and pH 6, then a decrease to 0.57 from pH 6 to pH 8

ii rate of reaction falls as enzyme is denatured (loss of shape of active site) so fewer collisions between enzymes and substrate molecules

3 **a** **i** check labels on axes(pH on x axis)/linear axes/accuracy of plot/title

ii rate of reaction falls as enzyme is denatured (loss of shape of active site) so fewer collisions between enzymes and substrate molecules

b **i** different temperatures/inaccuracies in timing

ii reliable – repeat experiment and take means

valid - fix temperature

4 **a** check accuracy of plotting and drawing. Pie chart is the most visually informative – it is easier to see the proportions of the different elements.

b **i** water; **ii** protein; **iii** Water; **iv** Fats, **v** vitamins; **vi** Bones and teeth

5 **a** **i** is certainly correct, **ii** and **iii** are likely to be correct

b D is a control, to show that starch does not hydrolyse without the enzyme/saliva being present

c have a second control at 30 °C; run a tube with boiled cooled saliva at 30 °C;

d accept maltose or glucose (reducing sugar). Benedict's test is suitable – boiling with an equal volume of Benedict's Reagent. Change from pale blue to orange-red is a positive test for reducing sugar.

2.7 **1** Ethanol dissolves the chlorophyll and removes it from the leaf

2 **a** Light is necessary for photosynthesis

b Carbon dioxide is needed for photosynthesis

c Demonstrates that starch is only produced if chlorophyll is present

Carbon dioxide, light and chlorophyll

3 **a** Keep plant in dark for 48h – it will be 'destarched', and can be tested for the presence of starch

Could replace black paper with transparent paper

4 Could heat leaf to denature enzymes – leaf would then not be able to produce starch by photosynthesis

2.8 **1** They are the same size/have the same volume

2 **a** Show curve reaching a plateau at around 7 arbitrary units of light intensity (remember to label the axes)

b 3.5 arbitrary units

c Up to about 7 units – no benefit to grower in exceeding this

d Wavelength of light. Use coloured filters.

e Carbon dioxide concentration/ temperature/pH/type of leaf

f To improve validity of data

g To give results as volume produced *per g of plant tissue* i.e. eliminate another potential variable

h Input = temperature, outcome = volume of oxygen produced per minute/fixed = Carbon dioxide concentration/light intensity/ wavelength//pH/type of leaf

3 **a** That 'radioactive' CO_2 is treated in the same way as 'normal' CO_2 by the plant

b Otherwise, results obtained in this experiment could not be applied to other plants under 'normal' conditions

Respiration

b

2.10 **1** Light intensity/temperature/carbon dioxide concentration

2 Can produce acid rain, which is damaging to the mesophyll in leaves. However, higher temperatures and CO_2 concentration can increase the rate of photosynthesis

3 Magnesium; this ion is a part of the chlorophyll molecule

4 The factor furthest from its optimum i.e. which is actually reducing the rate to the greatest extent.

a Light intensity

b Temperature

Carbon dioxide concentration

c

2.11 **1** Hydrogencarbonate (bicarbonate) indicator: change from red to purple. Purple colour indicates that solution is less acidic – this is the situation that would apply as CO_2 is removed for photosynthesis

2 The point at which photosynthesis and respiration are balanced e.g. the amount of oxygen being used in respiration is exactly the same as the amount of oxygen being produced by photosynthesis. This is important since it shows when the plant is beginning to produce more carbohydrate than it is using up i.e. is beginning to make food for the crop.

3 Yes it is true, but it's not the whole story. Plants can use the carbohydrate made during photosynthesis to produce all organic molecules (such as aminao acids and lipids).

4 The plant might take up CO_2 to produce glucose. It would use some of this glucose for respiration, which would mean some CO_2 would be released to the atmosphere. This same CO_2 could then be taken up again as the plant continues to photosynthesise.

Photosynthesis and plant nutrition pp.50–51

1 **a** **i** that most of the water loss is from the lower surface

ii upper surface – waxy cuticle and absence of stomata limits water loss; lower surface – no waxy cuticle and presence of stomata allows water loss to occur more readily

b leaf might be damp/paper might be damp

c use more leaves/repeat experiment/ place paper on different parts of the leaf – in each case take mean results

d use a control piece of cobalt chloride paper, not attached to the leaf surface, and note time taken to turn pink

2 **a** osmosis – support – solvent/transport medium – photosynthesis

b roots – root hairs – surface area – ions/ minerals – magnesium – nitrate – active transport

c xylem – phloem – vascular

3 **a** **i** carbon dioxide and water

ii Oxygen

b **i** iodine solution

ii A – straw-brown, B – straw-brown, C – blue black, D – straw-brown

iii B – no chlorophyll present so no photosynthesis possible as light energy cannot be absorbed; D – although chlorophyll is present the black cover prevents light reaching the leaf so that no photosynthesis can take place

4 a i

Tube	Colour of indicator at start	Colour of indicator after 6 hours
A	Pinky red	yellow
B	Pinky red	yellow
C	Pinky red	yellow
D	Pinky red	Purple

ii tube A

1 respiration occurs;

2 carbon dioxide produced / added to water;

3 becomes acidic / more acidic / pH falls;

tube D

4 photosynthesis occurs;

5 carbon dioxide removed from water;

6 becomes alkaline / less acidic / pH rises;

b tube E

1 colour stays pinky red / does not change;

2 respiration and photosynthesis balance out;

3 carbon dioxide amount in water / pH does not change;

OR

1 colour goes purple;

2 photosynthesis more than respiration;

3 carbon dioxide amount in water drops / pH rises;

OR

1 colour goes yellow;

2 respiration more than photosynthesis;

3 carbon dioxide amount in water rises / pH falls;

2.13 1 Source of energy/raw materials for growth and repair/vitamins and minerals to enable other food substances to be used efficiently

2 c Autotrophic nutrition involves building organic molecules from simpler inorganic materials e.g. making starch from carbon dioxide and water. In heterotrophic nutrition, the organism obtains the organic molecules ready-made, as food.

3 A balanced diet contains all of the nutrients required for life in the correct proportions

4 Carbohydrates and fats

5 Enzymes (e.g. amylase), hormones (e.g. insulin), structures (e.g. keratin) and transporters (e.g. haemoglobin)

2.14 1 a 8400 kJ

b 1000 (10×100)

c $18900 - 10500 = 8400$ kJ (= 80%)

d i food 1 as it has the highest energy content per serving

ii $\frac{18900}{3800} = 4.97$ 'servings' i.e. 497 g

756

1 (p.57)

a A

b D

c C

d D

A

2 a Colostrum has more protein (400%), less fat (63%), less sugar (44%)

b 1 litre contains 10×100 cm^3, so contains 20g of protein

Any citrus fruit e.g. orange

2.15 1

56	20	1120	2489
52	20	1040	2000
54	20	1080	2298
Extra column			10,454 8,400 9,652

c 9502 kJ per g. This is more valid since it reduces the effect of any single result.

d show some means of reducing heat losses to the air, and ensuring that more of heat released by burning raises the temperature of the water. Perhaps burn in oxygen so that peanut burns completely.

e students repeated experiment using the same set of apparatus e.g. same volume of water/same thermometer. Students converted result to 'per g' to take account of different-sized peanuts.

Students could have taken readings over a period of time and plotted gradient of results to more accurately reflect heat transfer from burning peanut.

2.18 1 It grows, and it certainly is sensitive! The tooth also requires nourishment.

2 e Sailors had a very limited diet. It was hard to preserve fruit and vegetables, so the sailors rarely had any foods that contained vitamin C. Without vitamin C the sailors could not make the tough fibres that hold their teeth into their sockets – so their teeth fell out! This is a symptom of scurvy, and was prevented by eating limes or other citrus fruits.

3 a Caries begins when bacteria are able to penetrate the enamel, and infect the dentine of the tooth.

e By avoiding sweet foods (provide nutrients for bacteria, which convert them to acid) and acidic fizzy drinks. Limit caries by regular brushing of teeth to remove food and bacteria, or by adding fluoride to drinking water (strengthens enamel)

c

2.19 1 a B

b i Temperature

ii Beef placed in water

iii To make sure that the beef was sterile – there were no microbes to break down the beef

beef in stomach would have been at the right pH, the optimum temperature and so digestion would have been more complete.

2 a bile

emulsifies fats – increases their surface area so that the enzyme lipase can hydrolyse the fats to fatty acids and glycerol more efficiently

3 Mouth: amylase begins to hydrolyse starch to maltose

Stomach: amylase activity stops. Protease begins to hydrolyse proteins to peptides under acidic conditions

Duodenum/ileum: amylase hydrolyses more starch to maltose, and maltase hydrolyses maltose to glucose; proteases complete hydrolysis of peptides to amino acids

4 Bile (see Q.2); hydrochloric acid in stomach – optimum pH for protease.

2.20 1 There are two inputs – hepatic artery (oxygenated) and hepatic portal vein (digested foods) and one output -hepatic vein(assimilated foods to cells via vena cava)

2 b Glycogen and iron are stored. Amino acids are converted to proteins, and excess amino acids are converted to urea

3 Large surface area/lined by thin membrane/good blood supply(capillaries) to keep up concentration gradients by removing absorbed nutrients

4 c Goblet cells manufacture and secrete mucus, to protect the lining of the gut from digestive juices. Also found in the lining of the breathing passages, especially the trachea and bronchi.

5 b Excess water would be lost in the faeces – diarrhoea.

Animal nutrition and health pp.72–73

1 5090 – 6 – May 2009 – Q.1

a i detect starch with iodine solution, which gives a blue-black coloration;

ii The product is a reducing sugar – detect reducing sugar by boiling with Benedict's Solution, which gives an orange-red coloration

b i Check for accuracy of plot/'dip' corresponding to maximum rate of reaction at pH 4/axes labelled with correct units/key to distinguish between plots;

ii Optimum pH at 4.0/presence of salt increases rate of reaction over the whole of the pH range.

c Results will be more **reliable** if the experiments are repeated, and means are taken of collected results. Results will be **valid** if only one variable is changed in any experiment i.e. maintain temperature/concentration of enzyme/concentration of substrate.

2 own

a Hannah – she has only four fillings but Caitlin has eight

b There may be genetic variation between them/Caitlin may have eaten

more sweet, sticky foods, or drunk more fizzy, acidic soft drinks

c This removes gender and age from the list of input variables

d The cheek teeth crush food and so the sticky food spends longer there, and can fit into crevices between the teeth

e They may not have had so many cavities, and there would have been fewer teeth in the milk dentition

3 061-2-May 2008 – Q.3

pH	5.0	5.5	6.0	6.5	7.0	7.5	8.0	9.0
Time taken	20	15	8	4	1.25	1.25	3	8
Reaction rate	0.05	0.07	0.13	0.25	0.8	0.8	0.33	0.13

b check axes (pH on x axis)/axes linear/ labels on axes to include units/ accuracy of plot

c 7.0 – 7.5

d mouth/ileum – presence of sodium bicarbonate

e iodine test for starch/Benedicts test for reducing sugar

f temperature/concentration of enzyme/ concentration of starch solution

4 0610-2-May 2008- Q.3

a

Food	Enzyme	End products
starch	amylase	Simple sugars
Protein	protease	Amino acids
Fat	lipase	Fatty acids and glycerol

b excess glucose is stored as insoluble glycogen; excess amino acids are deaminated to urea and carbon dioxide

5 own

Many possible answers. Results appear invalid since no controls in evidence. Suitable controls (use other animals/ more subjects).

6 Own

Bounty	818
Maltesers	2016
Mars	1816
Milky way	1960
Minstrels	1775
Snickers	2016
Treets	2476
Twix	2016

b bar chart – bars should not be touching

c **i** Minstrels

ii Treets

d 1179

e 196.5 minutes; used to maintain temperature/keep nervous system working/digest food for example

f 39 minutes

g more – he could use some stored energy e.g. glycogen from the liver

2.21 **1** Low water potential – the dissolved particles tend to 'hang on' to the water molecules, so the water doesn't have much potential to leave.

2 Osmosis is the movement of water across a partially permeable membrane down a water potential gradient

3 As cytoplasm swells it presses against the cellulose cell wall. The cell wall stretches slightly, but pushes against nearby cells and so helps to keep the plant body supported (the cells are turgid)

4 **a** uptake involves enzymes or carriers – the higher temperature gives the moving molecules more kinetic energy

b energy (from respiration)is needed for this uptake

c this is an example of active transport the sugars were used up in respiration to provide the energy needed for the active transport

2.22 **1** Phloem and xylem. Cambium is the dividing tissue.

2 A part of the plant at which a substance is absorbed or manufactured. Leaves are sources of glucose, and roots are sources of water and mineral ions

3 Sometimes, e.g. towards autumn, the sugar is being stored – may be moving down to roots – or at other times, e.g the spring, it may be move up towards the growing points or flowers

4 To supply energy from respiration – growing points require energy for cell division, roots require energy for active uptake of ions

5 The coloured dye could be placed in a vase/pot and, if it moves up the xylem, it could change the colour of the flower petals.

2.24 1 (p.80)

d 2.4; 4.3; 35.9; 39.0

2 Lower surface is where the most stomata are located – water is lost as water vapour through these stomata, so leaf loses mass

3 So that they were not genetically different, with special adaptations to water conservation

4 Experiment to be repeated and mean values taken. Same amount of Vaseline used each time. Leaves of same size/surface area used

1 (p.81)

a transverse section is cut ACROSS the leaf

b i Photosynthesis

ii Carbon dioxide

iii Diffusion – the movement of molecules down a concentration gradient

c hairs prevent movement of water vapour/stomata are in pits to trap water vapour/thick waxy cuticle prevents water loss from outer surface

d for photosynthesis/support by turgidity/transport of ions up the xylem

sand dunes do not have very much water so it would be hard for the plant to replace any water losses

2 transpiration stream; carbon dioxide; stomata; water vapour; thicker waxy cuticle/leaf rolling/hairs on leaf surface

2.25 1 Because their surface area to volume ratio is too small for the uptake of gases/loss of wastes. As the organisms get larger they must develop a transport system to move substances to and from the surfaces used for exchange, such as the lungs and kidneys.

2 a red blood cells and plasma

b glucose/heat with Benedict's Reagent/positive result would be an orange-red precipitate

white blood cells

3 Red blood cells : function is transport of oxygen – adapted by absence of nucleus (so very flexible and with large surface area for uptake/loss of oxygen)and presence of haemoglobin (for binding to oxygen)

White blood cells : function in defence against disease – adapted by having irregularly-shaped nucleus (helps cells to squeeze through capillary walls) and cytoplasm which contains digestive enzymes (to break down microbes that have been engulfed)

4 a Jill: many red blood cells as an adaptation to the low oxygen concentrations at high altitude

b Jill: she has a low concentration of the white blood cells needed for defence against disease

c Jackie: low number of platelets, the fragments of cells needed for efficient blood clotting

d Jackie: low number of red blood cells – iron is needed for the production of haemoglobin in red blood cells

This eliminates any differences in blood cell count due to gender or age – these factors have become fixed variables

5 cells; tissues; epidermis; organ; specialised; red blood cell; division of labour

2.26 1 Arteries have elastic walls with layers of muscles (to cope with pulsing blood pressures, and to continue to push blood towards the tissues): veins have thinner walls (no pulse pressures to deal with) and valves (to prevent backflow of blood)

2 It passes through twice. Renal vein – vena cava – right atrium – right ventricle – pulmonary artery – (lungs) – pulmonary vein – left atrium – left ventricle – aorta – renal artery

2.27 1 a A: blood, B: tissue fluid,C: lymph

b Oxygen/glucose/amino acids. Carbon dioxide/urea

c Returns along lymph vessel and rejoins blood at thoracic duct close to right atrium (where blood pressure is very low)

c If water potential of blood is high e.g. if concentration of proteins or glucose in blood plasma is very low. Tissues would swell. It can be prevented by making sure that the diet provides sufficient protein.

2.28 **1** **a** Left atrium

b Blood is forced past the semi0lunar valves and out into the arteries

e Atrium only has to push blood down to ventricles (not very far) but ventricles must push blood out to the tissues

2 **a** By the closing of the valves during the cardiac cycle

Each beat includes 'lup-dup', so there is one 'lup' per cycle. Each cycle takes $\frac{60}{72}$ seconds i.e. there would be 0.8 seconds between successive 'lup' sounds.

2.29 **1** **a** Coronary arteries

Difficult to move blood past blockage – cells beyond the blockage could be deprived of oxygen/glucose, and so could die

2 **a** 23 and 46 years of age

b Blood cholesterol is higher in men, and death rate increases as blood cholesterol level increases

c Less than 4 arbitrary units

d There are other factors, as the risk never falls to zero

Male smoker, who eats a lot of fatty foods and takes little exercise

Circulation pp.94–95

1 **a** glucose – from gut to muscle/liver; amino acids – from gut to muscle/ liver; urea – from liver to kidney; oxygen – from lungs to tissues; carbon dioxide – from respiring tissues to lungs

b 8000 per mm^3 – may rise during infection

c $\frac{5\ 000\ 000}{8000} = 625{:}1$

d vehicle exhaust /burning domestic fuel/cigarette smoke

e falls, as carbon monoxide binds to red blood cells so these are not available for oxygen transport

f 120 days –replacement time for red blood cells

g worn out red blood cells are removed, and the iron in their haemoglobin is stored in the liver

2 **a** Aorta

b Pulmonary vein

c Pulmonary artery

d Hepatic vein

e Vena cava (from iliac vein)

f Hepatic portal vein

g Pulmonary artery

3 **a** it falls (from about 17 kPa to about 2 kPa)

b blood must be forced along the arteries to the respiring tissues

c valves

d they are surrounded by sphincters (rings of muscle) which can constrict or relax to regulate blood flow to different tissues

e one cell thick, so short diffusion distance/have pores which allow rapid transfer of solutes to tissues

4 **a** hepatic portal vein – hepatic vein – vena cava – right atrium – right ventricle – pulmonary artery – pulmonary vein – left atrium – left ventricle – aorta – muscle in leg

b pulmonary vein – left atrium – left ventricle – aorta – brain - vena cava – right atrium – right ventricle – pulmonary artery

5

rate	55	70	80	90	120	140	150	170
Output per beat	0.07	0.068	0.064	0.068	0.05	0.042	0.038	0.027

b accurate plot as directed

c **i** rising to peak then falling,

ii Output per beat falls as rate increases

d **i** there is no benefit in terms of total cardiac output,

ii Output per beat may be higher (heart muscle is more elastic in trained athletes)

e it is less likely that heart muscle will become fatigued/low rate means that there is further to go before the peak rate is reached

f to supply more oxygen to be transported by the blood to the respiring tissues

6 **a** **i** 5

i fewer than 2

b this has eliminated two other variables (gender and smoking0 which might influence the likelihood of developing CHD

7 a first set of data as line graph, second set as bar chart (check on bar chart that bars are not touching)

b i higher than normal blood cholesterol

ii 9.9 times more likely (3.2 × 3.1)

8 a low exercise/smoking

b high fat/cholesterol/salt

c being male!

2.30 1 They may be *harmful* by causing disease (i.e. as pathogens), by spoiling food, by damaging property (fungi destroy wooden houses)
They may be *helpful* by carrying out important processes such as fermentation (brewing and baking), by decomposition of animal and plant remains (carbon and nitrogen cycles) and by acting as recipients in genetic engineering.

2 Degenerative diseases e.g. hardening of the arteries;
Deficiency diseases e.g. rickets
Inherited e.g. sickle cell anaemia

3 In infected water (e.g. cholera);
By direct contact (contagious diseases) e.g. athlete's foot;
By animal vectors e.g. malaria;
In droplets in the air e.g. influenza (virus);
In contaminated food e.g. Salmonella food poisoning;
Via body fluids e.g. AIDS;

4 Signs are changes shown by the person who is ill – they can be observed by a doctor e.g. raised temperature; symptoms are experienced by the patient e.g. 'feeling ill'.
Typical disease symptoms are caused by multiplication of a pathogen or production of toxins, and the body's attempts to eliminate the pathogen.

5 Britain has a well-developed vaccination programme to minimise the effects of measles, and good care facilities for anyone who does contract the disease. Less economically developed countries often do not have these advantages.

6 Changes in sexual behaviour have increased the likelihood of contracting a sexually transmitted infection e.g. syphilis.

2.33 needed

1 Improve personal hygiene (less chance of athlete's foot);
Consume a balanced diet (less chance of a deficiency disease such as rickets;
Not smoking (reduce risk of lung cancer/ CHD);

2 Provision of safe drinking water, removal of refuse and provision of medical care for the unwell.

3 Viral (AIDS); bacterial (gonorrhoea)

4 AIDS i.e HIV infection – individual can moderate sexual behaviour, community can provide appropriate medication, scientists can work to develop new mediciens.

2.34 1 The skin is a barrier: it has waxy, bacteriocidal ('bacteria-killing')secretions to help prevent water-borne pathogens from crossing into the blood. Pathogens could feed on the body fluids and other tissues, multiplying and causing disease by destruction of cells or release of toxins.

2 a To prevent blood loss/entry of pathogens

b If the clot is internal it might block an important blood vessel and cut off oxygen and nutrient supply to tissues and organs

i It means that a small initial signal can quickly produce a very large response

ii 10 000 000 000 i.e. 10^{10}

3 During phagocytosis, pathogens or harmful particles are engulfed by white blood cells (phagocytes) – the cells form a vacuole around the pathogens (a diagram would help your description!). Phagocytes are very flexible, and contain digestive enzymes to break down the engulfed pathogens. Pathogens can escape phagocytes by 'hiding' inside body cells, or staying in areas where there are no phagocytes (such as in the gut).

2.35 1 By recognising molecules on the surface of the pathogen – these molecules are not found on the cells of the 'host'

2 Lymphocyte has a spherical nucleus, but phagocyte has an irregularly-shaped nucleus.

3 Pathogen is recognised, and infected animal makes B-lymphocytes to produce antibodies which help to destroy the pathogen. Some B-lymphocytes which can make these specific antibodies are kept in the plasma – they are 'memory cells' – so they are always ready if another attack by the same pathogen takes place.

2.37 **1** B

2 **c** Change of energy from one form to another. The conversion of light energy from the sun into chemical energy in foods is the key step in photosynthesis at the start of food chains. All is eventually released as heat.

3 **a** Time on x axis, lactic acid concentration on y axis: peak at 25 minutes

b 18 arbitrary units

c 15 minutes

d It is still being made in the muscles and then 'washed' into the blood

e Muscles

55 minutes (back to 18 arbitrary units)

2.38 **1** **a** So that they could calculate a mean value

Seeds: 0.40 mm per s; maggots: 0.52 mm per s. Maggots are more active than seeds, so require some energy for movement.

2 As a control: to show that no 'unknown' factor was responsible for the movement of the coloured liquid

3 **a** Temperature on x axis, oxygen consumption on y axis: peak at 35 °C. This corresponds to the optimum temperature of the enzymes involved in respiration

b Temperature is the independent/input variable, relative oxygen consumption is dependent/outcome variable

Mass of organism/surface area of filter paper/type of coloured liquid

4 To remove carbon dioxide from the air before it reaches the living organisms

5 **a** That no carbon dioxide remains in the air

6 Organisms in C have released CO_2. This is an acid gas, and so the indicator changes colour.

7 **f** No organisms in C – this would demonstrate that the indicator doesn't change colour due to some unknown factor

8 Walls of flask may become 'misty': organisms release water from respiration

2.39 **1** Thin, large surface area, close to good blood supply, moist, well-ventilated

2 **c** Along breathing passages (mouth/trachea/bronchus/bronchiole) into air sac. Cross membrane of air sac into capillary and into red blood cell. From red blood cell across wall of capillary and then across cell membrane of e.g. liver cell

3 Oxygen can be carried in greater quantities because haemoglobin is concentrated in the red blood cells. Red blood cells : function is transport of oxygen – adapted by absence of nucleus (so very flexible and with large surface area for uptake/loss of oxygen) and presence of haemoglobin (for binding to oxygen)

4 Respiration is the release of energy from food molecules. Gas exchange is the transfer of oxygen and carbon dioxide into and out of organisms to allow respiration to go on.

Gas exchange pp.118–119

1 **a** graph plotted appropriately – time on x axis

b Alan – he can get by with fewer breaths during periods of exercise

c 8:1 (4:0.5)

d 30 breaths per minute (each one takes 2 seconds)

e Increases both rate and depth of breathing

f Heart rate would increase to a maximum as exercise levels increased

g More oxygen (and glucose) are delivered to the respiring muscles

2 **a** The diaphragm

b They will collapse/deflate

c Exhalation/expiration

d No intercostal muscles shown

e 1 – rib; 2 – sternum; 3 – (external) intercostal muscle; 4 – backbone/vertebral column

f Breathing in involves contraction of the intercostal muscles and the diaphragm, so is 'active'; breathing out occurs when intercostals and diaphragm relax, so is 'passive'

3 a i marathon running uses all energy reserves (glucose, glycogen and fat) at about 4x the rate that walking does

ii blood glucose and muscle glycogen

iii fat carbohydrate

b i for muscle development and repair

ii to build up reserves of liver and muscle glycogen

c i $C_6H_{12}O_6 \longrightarrow 2C_3H_6O_3 + 2ATP$

ii sprinter can use anaerobic respiration since the production of lactate (a muscle toxin) is of less significance for a race which takes only a short time

iii liver glycogen is hydrolysed to glucose (influenced by the hormones adrenaline and glucagon), then is transported out of the liver into the blood. Blood glucose concentration remains constant since extra glucose is absorbed by the respiring muscles.

4 a any four from nutrition/reproduction/excretion/growth/development/sensitivity/locomotion

b **respiration** is the release of energy by the oxidation of foods, usually glucose; **breathing** is the process that removes carbon dioxide from the body and replaces it with oxygen. Breathing is a necessary process for aerobic respiration to continue.

5 a i smoking

ii smoking causes a further reduction of 3.4% - driving (and exposure to traffic exhaust fumes) causes a maximum reduction of 2.3%

iii day time drivers $= \frac{(5.712.3)}{2} = 4.0\%$,

night time drivers $= \frac{(4.411.0)}{2} = 2.7\%$.

Thus there is a greater reduction (by 1.3%) for day time drivers – more traffic so more traffic exhaust to provide carbon monoxide

b i nicotine – addictive/increases heart rate; tar – increases risk of lung cancer

ii fetus can be underweight as it is deprived of oxygen. Baby may also be nicotine dependent at birth

2.40 1 As these muscles contract they increase the volume of the chest. Lungs follow this change so air is drawn in.

2 Rate of breathing. Volume of air entering lungs at rest and during exercise. Volume of air that can be drawn in in excess of basic breathing. Maximum volume of lungs.

3 Aerobic respiration removes oxygen from inhaled air, and adds carbon dioxide to exhaled air.

2.41 1 Particles trigger the action of phagocytes, and mucus-secreting cells so more mucus is produced. Smoke dries out the lining of the lungs. Chemicals in smoke may trigger extra cell division and start growth of tumours in the lungs or airways.

2 Physical; body cannot function without the chemical. Psychological: user believes that chemical reduces stress or increases pleasure. By offering nicotine patches (to overcome physical addiction) and counselling (to overcome psychological addiction)

3 Increased blood pressure/damage to heart/bladder cancer

4 Cilia are lost and so cannot remove bacteria trapped in mucus

5 a 1000

b 6 cm^2

c 6000 cm^2

d 600 cm^2

Emphysema sufferers have fewer alveoli, so larger spaces/smaller surface areas for gas exchange.

2.42 1 Epidemiology is the study of the patterns of disease and the factors that affect them.

2 Working in a smoky or dusty environment. Working in an environment with airborne particles (e.g. in mining).

3 Approximately 24 times

4 People living close to power lines, and people living a great distance from power lines.
All of the same age – suggest 30 years of age.
Same sex
Same occupation
Similar diet as far as this can be controlled.

5 It would be unethical to deliberately make an individual smoke in the expectation of causing lung cancer.

2.43 **1** Excretion: the removal of the waste products of metabolism and substances in excess of requirements.
Osmoregulation: control of water balance in the body.

2 Carbon dioxide: respiration in cells – exhaled from lungs

e Urea: deamination in liver cells – removed by the kidneys

3 kidney; renal arteries; nephrons; bowman's; glucose; selective reabsorption; ureters; bladder; urethra

2.44 **1** **a** To keep the cell cytoplasm hydrated

b In food and drink
In urine, sweat and on the exhaled breath

2 In negative feedback a change in a factor causes a correction that cancels out this change

3 Transplant is more convenient for patient, and action is more natural for the body. Kidney may not be available, and may be rejected by the immune system.

2.45 **1** tissue; receptors; optimum; brain; sensory; negative feedback

2

Organ	Factor controlled
Liver	Blood glucose concentration
Lungs	Concentrations of oxygen and carbon dioxide
Kidney	Urea concentration
Skin	Temperature
Intestines	Availability of digested foods, such as glucose

3 Blood glucose level is detected by the pancreas, insulin is secreted if concentration is too high, liver is stimulated to remove glucose and store it as glycogen, blood glucose level returns to optimum.
Temperature in cabin is measured by thermal sensors, information is sent to on-board computer, electrical message is sent to cabin heaters, heaters are switched on or off to return temperature to optimum.

Excretion and homeostasis pp.133–135

a This is a condition in which the body temperature falls significantly below its normal level

b Old people have less efficient circulation, and also their temperature control systems work less well

c Children have a high surface area to volume ratio, so they lose heat more quickly than they can produce it

a any change in body temperature is detected by a sensor, and then corrective actions (shivering/sweating; constriction of skin blood vessels/dilation of skin blood vessels) return the temperature to normal body temperature

b shivering is rapid muscle contraction. This generates heat from respiration (and some friction)

c liver is the main metabolic centre of the body, so generates a great deal of heat (especially from respiration)

d in negative feedback a change from the optimum/norm causes actions which correct (cancel out) the original change

a Show artery connected to aorta, vein to vena cava and tube carrying urine to the entry to the bladder

b Rejection occurs because the body detects that the organ does not belong to the recipient, and the immune system produces more cells which attack the transplant

c The kidney tubule can select which materials will be lost and which ones will be returned from the urine to the blood. The useful materials are returned to the blood by active transport.

a It passes by diffusion i.e. down a concentration gradient into the dialysis fluid (the dialysis fluid has a very low urea concentration)

b i The transplanted kidney will carry out all of the functions of a normal kidney, and will not run the risk of infection that there is with an external connection to a dialysis machine;

ii The kidney might be rejected. This occurs because the body detects that the organ does not belong to the recipient, and the immune system produces more cells which attack the transplant

a i time on x axis/check axes are linear/check that y axis scale is appropriate to the 3 plots/accuracy of plots

ii A cools more rapidly (to 45 °C) than B (to 60 °C); B more slowly than C, and C a little more quickly (to 40 °C) than A

iii B is insulated with dry cotton wool, so loses heat slowly. A is 'naked' and C has a damp covering so some heat is used to evaporate water from this covering

b flasks had same surface area/ contained same volume of water/ had water at the same temperature at the beginning of the experiment/ environmental temperature was the same in each case

ii **accuracy** could be improved by taking readings more frequently; **reliability** could be improved by repeating the experiment a number of times and taking mean values for the readings.

a X – vena cava; Y – ureter; Z – urethra

b oxygen concentration falls because the kidney uses oxygen to release energy by aerobic respiration; urea concentration falls as urea is filtered into the urine at the Bowman's Capsules in the kidney; sodium concentration falls as excess sodium is removed by filtration in the kidney capsules

2.47 1 Similarities: both involved in coordination, both required for survival. Differences: nervous system is much quicker, relies on electrical (not chemical) messages. Nervous system is generally quicker and more localised, which means the effects are useful for short-term survival but not so useful for control of growth.

2 Long (messages carried over great distances), insulated (so no electrical interference), Nodes of Ranvier (allow rapid 'jumping' conduction), many connections with other neurones or muscle cells.

3 a Motor – information out of CNS, sensory – information towards CNS

Central – brain and spinal cord, peripheral – nerves leading to and from the CNS

4 a As a wave of electrical action potentials

By neurotransmitter molecules which can diffuse across the synapse

2.48 1 a Stretch receptor (in patellar tendon of the knee). As it stretches it produces an electrical potential

b The sensory neurone

c E (muscle which straightens the knee)

d Total distance covered – there and back – is about 60 cm. 100 m = 10 000 cm, so this distance would be covered in $\frac{60}{10\,000}$ s or 6 ms (milliseconds)

Impulse is slowed as it crosses the synapses

2.49 1 a 310 ms. The mean is valuable since it reduces the effect of any one result, which might not be a typical result.

b 110–150: 6 students; 160–200: 8 students; 210–250: 10 students; 260–300: 4 students; 310–350: 2 students (did you remember to include student 1?).

c i Shortest reaction time is 120 ms, so motorcycle would travel approximately 2 m

ii Longest reaction time is 330 ms, so motorcycle would travel approximately 5 m

In a conditioned reflex a second reflex replaces the natural one – here the 'bang' has replaced the need for the subject to observe a light.

Receptors and sensors pp.146–147

1 A **transducer** changes one form of energy (the stimulus) into another form of energy (usually an action potential).
  a Taste buds convert chemical messages into action potentials
  b Rod cells convert light energy into action potentials

2 cornea lens retina
  iris(diaphragm) pupil rods
  black and white cones
  colour high retina inverted
  smaller optic integration

3 a Ciliary muscles/suspensory ligaments
  b Independent is distance from eye, dependent is thickness of lens
  c Amount of light in room (could affect pupil size which might have an effect on lens shape)/size and colour of pencil (must make sure that the distance is the only independent variable
  d Repeat the measurements at each of the distances, and take a mean of the measurements

4 a A – retina; B – pupil; C – tear duct; D – choroid; E – conjunctiva; F – sclera
  b i The retina will not be damaged/ bleached by the high light intensity;
    ii Light – retina – brain – iris muscle
  c i 20 minutes;
    ii The 'two step' shape of the curve suggests that more than one factor is involved

5 a 55 years
  b Will be less able to accommodate to objects from different distances, so some images on retina will be blurred
  c Once over 60, there is little change in the ability to alter the shape of the lens
  d Cornea – accommodation will not be so efficient
  e Use model lenses, or lenses removed from animals slaughtered for food. Amount of smoke reaching lens is the independent variable, the measured hardness of the lens is the dependent variable. A suitable control would be a lens not subjected to cigarette smoke (to show that it doesn't harden without the smoke).

6 a i D;
    ii having an opposing action to (e.g. contracts when the antagonist relaxes)
  b automatic/rapid/not at conscious level/positive survival value; ii. across the spinal cord at a level lower than the chest
  c i adrenaline;
    ii increased rate of heartbeat/dilation of pupils/release of glucose into blood/increased breathing rate/ redistribution of blood to respiring muscles/skin turns pale

7 a 1 receptor / sensory;
    2 stimuli;
    3 tongue;
    4 nose;
  b i suspensory ligaments;
    ii becomes flatter / thinner / less curved / convex / rounded;
  c i 5;
    ii 2;
    iii 4;

Hormones p.151

1 adrenaline – glycogen – glucose – oxygen – deeper – faster – gut – muscles – pales – dilate – stands up – fight - flight

2 Male secretes more testosterone at puberty. Testosterone leads to aggressive/ territorial behaviour. As testosterone concentration in blood rises, it inhibits production of further testosterone. As a result, testosterone concentration slowly falls (as it is removed from the blood) and so the male does not become too aggressive.

2.52 1 a Stimulant : nicotine (which raises blood pressure); depressant: alcohol (increases reaction times – time to process information); narcotic: heroin

(mimics natural painkillers so gives feeling of euphoria)

b Physical; body cannot function without the chemical. Psychological: user believes that chemical reduces stress or increases pleasure.

c Tolerance : ability of body to function normally even at higher levels of drug. Means that even higher levels will be needed to have an effect on the user.

b Input = caffeine intake in coffee (e.g. number of cups drunk) Outcome: ability to perform a task involving co-ordination and concentration e.g. recognising different coloured shapes quickly. Fixed: age/gender/body mass of subjects. Concentration of caffeine in coffee. Time period for drinking coffee.

2 a Time on *x* axis, BAL on *y* axis. Note peak at one hour.

b i 95 mg alcohol per 100 cm^3 of blood

ii From about 0.6 to 2.9 hours i.e. 2.3 hours

iii Reaction times would be increased e.g. for braking, judgement of distance would be affected

c skin would appear red

d carbon dioxide on the breath, water in the urine (and probably as sweat). Long term use causes cirrhosis (hardening of/ loss of function of liver cells)

e weak bones and teeth, and a tendency to scurvy (bleeding of gums/loss of teeth/poor healing of cuts)

f shaking of the hands, as muscles contract uncontrollably. Alcohol would stop this, as it is an inhibitor of nervous stimulation of the muscles.

3 a i The likelihood of addiction

ii Increase = (49.6 – 4.2) = 45.4. Percentage increase

$$= \frac{45.4}{4.2} \times 100 = 10\,800\% \text{ increase}$$

iii Increased seizures of cocaine/ cannabis resin/LSD. Reduced seizures of heroin. Quantities of LSD are very much lower than of other drugs.

e i Suicide/blood infections/ malnutrition

ii Methadone gives the sense of euphoria without the physical dependence. Vitamin supplements counteract some of the dangers of malnutrition.

2.53 1 tip; auxin; swell; water; auxin; swell; phototropism; photosynthesis

2 d roots; copies/clones; fruiting; pests/ weeds; light; water; mineral nutrients; selective; seedless; fertilisation

3.2 1 f – 5; (b) – 6; (c) – 7; (d) – 1; (e) – 9; (f) – 4; (g) – 8; (h) – 2; (i) – 3; (j) – 10

2 b germinates; flower; reproduction; pollination; fertilisation; fruit; seed; dispersal

3.4 1 **Pollination** is the transfer of gametes (sex cells) from the anther to the stigma; **fertilisation** is the fusion of these gametes in the ovary.

2 a Label sepals and petals (anther and filament wither away)

The ovary

3 A seed is the part of a fruit that can develop, after germination, into a young plant

4 a Tomato – fruit; cucumber – fruit; Brussels sprout – neither fruit nor seed; baked bean – seed; runner bean – fruit; celery – neither seed nor fruit; pea – seed; grape - fruit

5 a Cover buds with muslin bags so that insects cannot reach flowers. Once flowers are formed, transfer pollen to stigma of half of the available flowers and do not transfer pollen to the other half (if plants are hermaphrodite, cut off anthers to prevent self-pollination). Re-cover flowers with muslin bags. Record which plants produce fruits.

b For self-pollination transfer pollen to stigma of same flower, for cross-pollination transfer to stigma of different flower.

Always make counts of fruits after the same period of time. Keep all plants under same environmental conditions, particularly with respect to soil minerals.

3.5 **1** water; micropyle; swell; testa/seed coat; oxygen; respiration; embryo; radicle; positive geotropism; anchorage; plumule; true; photosynthesis

2 Demonstrate that temperature and changes in pH have an effect on the rate of germination

Plant reproduction pp.170–171

1 small dull dry light
stamens style

2 **a** Anther – produces pollen grains; sepal – protects the flower when in bud; style – allows the passage of the pollen tube to the ovary; stigma – the surface on which the pollen lands during pollination

b wind pollinated is long (hangs outside petals) and feathery – makes it more likely to catch wind-blown pollen

c self pollination increases the chances of pollination (and so of eventual seed formation) but reduces the possibilities of genetic variation

3 **a** check for relative width/depth, labels to include epidermis/seed/jelly-like seed covering/flesh – ovary wall. Is scale included.

b one is soft/moist/seeds enclosed in jelly/no pedicel

c make sample of material (crush in water in mortar and pestle). Add Benedict's Reagent to equal volume of each sample (care – do not taste/ wear eye protection). Heat in boiling water bath for same time period (two minutes). Compare colour – deeper the brick red colour, the more concentrated the reducing sugar in the sample.

4 **a** **i** 1 pollination is the transfer of pollen to the stigma;
2 fertilisation is the fusion / joining of male and female / two gametes;
3 pollination needs a transfer agent, fertilisation does not / only pollination needs transfer agent;
4 pollination occurs before fertilisation / fertilisation cannot happen without pollination;
5 pollination is external (to the plant) and fertilisation is internal;

ii stigma;
iii ovule;

b (seed from) ovule;
(fruit from) ovary;

c (wind can) carry pollen / assists in pollination;
(wind can) disperse seeds / fruits;
(wind can) disperse scent (to attract pollinators);

5

Tube	Would seeds germinate?
A	NO
B	YES
C	NO
D	YES

ii germination is controlled by enzymes; enzymes work more efficiently at 30 °C – more collisions of substrate molecules with active sites

b **i** graph should have time on *x* axis dry mass on *y* axis/axes should be labelled, with units/ accuracy of plotting
ii 18 days
iii 30 days
iv stores of organic molecules (especially carbohydrate and/or fat) are hydrolysed and used to supply energy and raw materials for early stages of development.

3.6 **1** **a** allows joining of genetic material from two different parents

b the developmental stage which ensures that the individual has sex organs capable of producing gametes

this provides haploid cells ready for fertilisation

2 **a** from tubules in testis – along sperm ducts – through prostate gland – along urethra

from ovary into oviduct

3 Usually, in the first third of the oviduct

4 To replace individuals that have died, and to make sure that there is variation within a sexually-reproducing population

5 **a** **R**: oviduct/fallopian tube; **S**: vagina/ birth canal;

b i line **F** should be pointing to first third of oviduct;
ii line **I** should be pointing to the wall of the uterus;
c i **oestrogen** is produced by the **ovary**;
ii development of breasts/mammary glands; broadening of hips

6 **a** **A** – urethra; **B** – testis/testicle;
b **X** should be on top of testis;
c **i** only difference is that vas deferens/sperm duct has been cut
ii sperm cannot reach the urethra so cannot be released at ejaculation. Thus there can be no fertilisation.
d seminal fluid is still released from prostate gland, and ejaculated into the female. HIV may also be transmitted non-sexually in blood (and in saliva, although rarely).

7 **a** **i** ovulation
ii haploid means having only a single set of chromosomes in the nucleus
b

feature	egg cells	sperm cells
site of production	ovary	testis
relative size	large	small
numbers produced	few (hundreds)	many (millions)
mobility	not mobile	mobile (swims)

3.7 **1** **c** Menstruation is the release of the broken down lining of the uterus. Ovulation is the release of an ovum from the ovary. There must be a period of time after ovulation to offer the possibility of fertilisation, before menstruation can take place. Menstruation would include unfertilised ova.

2 **b** FSH – stimulates development of the follicle in the ovary; LH – stimulates release of ovum at ovulation, and development of corpus luteum from remains of follicle; oestrogen – repairs the lining of the uterus; progesterone – keeps the lining of the uterus ready for implantation if fertilisation occurs

3 Progesterone rises in concentration if fertilisation occurs – it then prevents ovulation by 'switching' off the release of FSH, so no more follicles develop. No follicles means no more ova, so no chance of pregnancy.

4 **a** menstruation – repair – receptive – pre-menstrual
b 28 days
At about 14 days

3.8 **1** Conception – implantation of the ball of cells formed as the zygote divides; copulation – another name for sexual intercourse, the hard penis is inserted into the vagina; fertilisation – the fusion of the male and female gametes

2 IVF – how many fertilised eggs should be implanted? Should a fertilised egg be implanted in the womb of a surrogate mother? Should an egg donor know who is the sperm donor?
AID - Should a female know who is the sperm donor? Should any attempt be made to match the sperm donor to the female (e.g. for physical characteristics/sporting abilities/intelligence)? When should a child born from AID be made aware of his/her father?

3.9 **1**

Female sterilisation	S
Vasectomy	S
Diaphragm	P
IUD	P
Pill	C
Spermicide	C
Female condom	P
Condom	P

2 Condom or diaphragm – these are temporary methods of contraception

3 Sterilisation is not reversible, and a change of sexual partner might involve a change in the desire for contraception

4 **c** There is so much variation in Human physiology, such as body temperature changes, that it is difficult to be sure when a woman has actually ovulated.

3.10 **1** The gestation period
2 It takes 42 – 43 divisions

3 Specialisation of cells, which is part of development. Specialisation involves a change in structure and function of cells, and development is a change in the function of an organ.

3.12 **1** uterus; oxytocin; progesterone; cervix; amniotic sac; vagina; oxygen; umbilical; placenta; afterbirth

2 Oxytocin- stimulates contractions of uterus; progesterone – concentration falls as labour begins; oestrogen – concentration rises as birth approaches (makes uterus more sensitive to oxytocin). Oxytocin also stimulates lactation, and oestrogen is responsible for the development of the breasts. Human Growth Hormone is one of the controls for growth, especially controlling production of protein in muscles.

3 **a** True
b False
c True
d False
e True
True

3.13 **1** Avoid unprotected sex (for AIDS, gonorrhoea and syphilis), avoid sharing syringes (AIDS), know sexual history of partner (for AIDS, gonorrhoea and syphilis)

2 Should offer testing for STDs, and be aware of treatment available for each type of STD. Should try to trace source of any outbreak, and work out special programmes for those particularly at risk (such as drug users).

3 AIDS (viral), gonorrhoea or syphilis (bacterial)

4 AIDS
Individuals: should know history of sexual partners, avoid unprotected sex
Communities: Should offer testing for AIDS, and be aware of treatment available for AIDS and any associated problems. Should try to trace source of any outbreak, and work out special programmes for those particularly at risk (such as drug users).
Scientists: should devise education programmes to try to limit infection, should try to develop vaccines to prevent infection, should try to develop drugs to manage the disease in people who have become infected

Human reproduction and growth pp.190–191

1 **a** time on x axis/ body temperature on y axis/menstruation and ovulation indicated

b **i** 36.1 – 36.8;
ii This is to make sure that there are no unexpected independent (input) variables

c **i** Menstruation – fall in temperature, ovulation – increase in temperature;
ii This can indicate the time when conception is most likely (a few days around ovulation) – this time should be avoided for contraception, or represents the best time for intercourse to increase the chance of conception

d the pill/diaphragm both are more reliable as there is much variation in temperature changes during the menstrual cycle

2 **a** penis becomes erect as spaces fill with blood/erect penis enters vagina/ friction between penis and vagina stimulates penis/sperm in seminal fluid is ejaculated

b **i** barrier method
ii condom collects seminal fluid/ sperm so no sperm are ejaculated into vagina, so no fertilisation can take place

c **i** seminal fluid, like other body fluids, may contain HIV. Without a condom the HIV can be transmitted from male to female. Infected female may transmit HIV to male who isn't wearing condom, so infection can be spread.
ii via an infected blood transfusion/ via sharing a needle for drug use with an infected person
iii helper cells in the immune system are infected/inactivated by HIV, so the production of white blood cells becomes uncoordinated

d **i** pain or burning when passing urine/creamy discharge from penis or vagina/inflammation of the testes
ii as **i**.
iii course of antibiotics

3 **a** it is not losing heat to the environment/it has a high rate of respiration, which generates heat
b **i** Fat is an excellent insulator against heat loss;
ii This prevents blood reaching the surface of the skin, where heat can be lost to the surroundings
c change in temperature is detected by sensor – thermostat controls action of heater - more/less heat is released to return temperature to normal;
ii Prevents heat loss to the environment/prevents baby drying out/allows nurse/mother to view the newborn baby
d **i** carbohydrate – supplies energy/ protein – supplies raw materials for growth;
ii Antibodies – baby may have less immunity to childhood diseases

4 **a** **i** maternal – V; fetus – Y
ii T is the umbilical cord. It shrivels and the remains fall off after birth.
iii fetus *receives* oxygen, soluble foods and antibodies, fetus excretes carbon dioxide and urea
b oestrogen – maintains female secondary sexual characteristics, including development of mammary glands/at birth makes uterus more sensitive to oxytocin (which leads to contraction of uterine muscle); progesterone – maintains the lining of the uterus and the blood supply to the developing fetus/level falls in preparation for birth.

3.14 **1** Inherited – eye colour; acquired – body mass;
2 Many possible examples
3 **a** a chromosome is a thread-like structure, carrying genetic information in the form of genes. Chromosomes are found in the nucleus, and only the nucleus of the sperm is passed on at fertilisation.
b chromosomes can be observed when cells divide. This is because the chromosomes shorten and condense (they can take up stains when they do) as part of cell division.

3.16 **1** Nucleotides
2 Adenine, guanine, cytosine and thymine.
3 Adenine – thymine and guanine – cytosine. Base pairing explains how DNA can be replicated (copied) accurately.
4 Replication is the copying of DNA i.e. it is DNA to DNA.
5 Haemoglobin – ability to transport oxygen in red blood;
Amylase – ability to digest starch in cells of salivary glands;
Keratin – structural strength in hair and nails;
Antibodies – immune response in lymphocytes;
6 Mitosis (or meiosis)
7 **a** X is thymine, Y is cytosine;
b replication
c the sequence in the DNA determines the amino acid sequence (and therefore the structure and function) in proteins.

3.17 **1** **a** They are physically damaged as they are forced along blood vessels, and they have no nucleus to control repair processes
b The liver
c Iron (actually Fe^{II}), and it is part of haemoglobin
i $5 \times 1\,000\,000 \times 5\,000\,000 = 25\,000\,000\,000\,000$ i.e. 25 million million
ii About 0.21 million million are replaced every day, so about 2.41 thousand million every second

3.18 **1** **Homologous pair** : chromosomes with the same genes in the same positions – one of the pair comes from the father and one from the mother. **Heterozygous** : a nucleus (or organism) carrying both alternative alleles of a gene e.g. Bb. **Homozygous** : a nucleus (or organism) carrying only one of the alternative alleles of a gene, although it does have two copies of that allele e.g. bb or BB.

2 A gene is a section of DNA that controls the production of a particular protein, and an allele is an alternative form of a gene

3 A dominant allele is expressed (i.e. is 'in control') in a heterozygous individual. In abbreviated form, the dominant allele is usually shown with a capital letter e.g. B is the dominant allele in the pair Bb

3.19 **1** Brown eyed parents are both Bb (Brown – B – is dominant to blue – b); gametes from each parent are B or b; at fertilisation one possible combination in the zygote is bb, a blue-eyed child.

2 Because the gametes are not produced in exactly equal numbers, and fertilisation is random (i.e. it is not certain that every possible combination will be produced in the expected numbers). The larger the size of the population studied, the closer actual results approach the expectation.

3 **d** This cross enables an experimenter (e.g. an animal breeder) to find out whether an individual showing the dominant characteristic is homozygous or heterozygous. Homozygous individuals might be more valuable, as they will be pure breeding for that characteristic.

3.20 **1** **a** a heterozygote, who has the recessive allele but does not show the recessive characteristic because the dominant allele is also present

show parents as Cc, for example. There is a $\frac{1}{4}$ probability that one of their children will have cystic fibrosis.

2 **a** she was Hh – she showed the condition, but must have had the h allele as she had children who were hh (unaffected)

b B was hh, C was Hh. Show gametes and possible zygotes.

c $\frac{1}{2}$ - remember that each 'event' (i.e. each child) is unaffected by the genotype of its siblings

Because affected individuals might die before they reproduce, so 'losing' the dominant allele from the population

3 **a** Show parents' genotype, possible gametes, possible combinations in zygotes

b $\frac{1}{2}$

c $\frac{1}{4}$: i.e. $\frac{1}{2}$ (female) $\times$ $\frac{1}{2}$ (blood group A)

codominance

3.21 **1** Parents XX (male) × XY (female)

Gametes X or X X or Y

Offspring genotypes XX XX XY XY

Offspring phenotypes male male female female

2 The mutant allele for colour blindness is carried on the X chromosome. The male has only one X chromosome, so whichever allele he has will be expressed (shown).

A colour blind female must be homozygous for the colour blind allele i.e. X^cX^c.

3 **a** the mother is a carrier because she 'carries' the allele for haemophilia but, since this is recessive, she does not show the condition in her phenotype.

b she has a 50% $\left(\frac{1}{2}\right)$ probability of having a son, and a 50% $\left(\frac{1}{2}\right)$ probability that the son will be haemophiliac.

Inheritance pp.208–209

1 **a** **i** recessive

ii parents 6 × 7 can be heterozygous and still have an affected homozygous child

b $\frac{1}{4}$

i 25%. Show results of cross between two heterozygotes.

2 Gene Meiosis Diploid Recessive Heterozygous

3 **a** haemoglobin cannot form probably within red blood cells, and cells become sickle shaped under slightly acidic conditions. Sickle cells are less efficient at oxygen transport.

b **i** P I^NI^S × I^NI^S

gametes I^N I^S I^N I^S

F_1 I^NI^N I^NI^S I^NI^S I^SI^S

Probability that child will be heterozygous is $\frac{1}{2}$ or 50%

c this advantage is noted in areas where the malarial parasite is common. People who are heterozygous only

have mild anaemia, and are protected from malaria since the malarial parasite cannot multiply in sickle-shaped red blood cells.

4 A Dominant Allele Heterozygous Genotype

b i white

ii P Rr × Rr

gametes R r R r

F_1 RR Rr Rr rr i.e $\frac{1}{4}$ are white

iii two possibilities

1 P Rr × rr

gametes R r r r

F_1 Rr Rr rr rr i.e. 1 red: 1 white

OR

2 P RR × rr

gametes R R r r

F_1 Rr Rr Rr Rr i.e. all red

5 a i TT must be 105;

ii Alleles of a gene;

iii To simulate the random nature of allele transfer

b i Red is dominant, Cross W:

ii Flowers from cross V are heterozygous i.e. Tt, so each could provide T or t at gamete formation

c i Tt × tt, gametes would be T or t and t or t, offspring would be Tt or tt;

ii Cube 1 T T T t t t and cube 2 t t t t t t

6 a parental genotypes $I^AI^O \times I^BI^O$

gametes I^A I^O I^B I^O

offspring genotype I^OI^O

b heterozygous genotype I^BI^O or I^AI^O

codominant alleles I^B and I^A

phenotype blood group O (blood group A and blood group B in parents would be acceptable answers)

3.22 1 discontinuous; continuous; genes; environmental; genotype; phenotype; phenotype; genotype; environment

2 Blood group –

a determined by genes alone;

b all others are affected by environment, such as nutrition

3 a category on x axis, number on y axis

b continuous – there are many categories

c gender/ability to roll tongue

3.23 1 A gene mutation is an alteration in the DNA on a single chromosome

a the cystic fibrosis allele

the sickle cell anaemia allele in regions where malaria is common

2 A cancer-causing factor. Some forms of radiation(u-v) can cause skin cancer, chemicals in tobacco smoke can cause lung cancer

3 Crossing-over during meiosis produces new combinations of alleles in gametes, independent assortment produces new combinations of chromosomes in gametes, and fertilisation has random combinations of gametes from mother and father

3.24 1 Adaptation is a change in structure, biochemistry or behaviour which makes an organism more suited to its environment

2 b Many possible examples – be careful that you explain how the adaptation gives the organism an advantage in its environment

c

3.25 1 a 32

b The banded snails are more obvious against the plain background

c i Banded (the background is plain)

ii Unbanded (the background has stripes/bands)

A: snails have many offspring, and there would not be enough food for all of them to live as adults; B: the snails have slightly-different appearances to one another – some have banded shells and some have plain shells; C: the pattern of the banded shells can provide camouflage in a grassy environment; D: the ones with plain shells are seen more easily by predators such as thrushes; E: the snails with banded shells survive longer, breed more often and so pass on their genes to the next generation

b

3.26 1 a Cut off anthers from 'receiving' flower; collect pollen from 'donor' flower on paintbrush; transfer pollen to stigma

of 'receiving' flower. Cover pollinated flowers to prevent pollination by natural means.

Use vegetative propagation

Variation pp.220–221

1 a take sample of DNA/compare/different patterns suggest different species (interbreed and cannot produce fertile offspring – not useful with large mammals)

b i discontinuous **ii** *Equus grevyi*

c i phenotype **ii** change in the type or amount of DNA

d i segmented limbs **ii** three body parts/three pairs of legs

e i they would appear horizontal so would reduce attacks by Tsetse flies

ii individuals with more stripes survive better than those with fewer – interbreed - off spring have even more stripes – even less likely to be attacked by Tsetse flies

2 a in bones and teeth, where calcium is found in a mammal's body

b increase risk of mutation/may change rate of cell division

3 a continuous

b i check that bars do not touch/blood group type on X axis

ii genes

c i change in the type or amount of DNA

ii ionising radiation (examples?)/some chemicals e.g tar in tobacco smoke

4 a i (1.2) 9; (1.3) 10

ii check that bars do not touch/mass of berry on X axis

b continuous – values are numerical/bar chart could be redrawn as a smooth curve/classes are arbitrary (choice of the person doing the measuring)

4.1 1 **Population**: all of the members of one species in a particular area; **community**: all of the populations of living organisms in one area; **ecosystem**: all of the living organisms and the non-living factors interacting together in one part of the environment

2 Light intensity/temperature/oxygen concentration/carbon dioxide concentration/humidity/availability of water

3 One may be food for the other (most likely plant being food for animal); insect may be responsible for pollination of flower; animal may help to disperse seeds/fruits of plant

4 d Food, shelter and a suitable breeding site

5
- Flying insects are 'cold-blooded' and so cannot remain active at the low temperatures found in Britain during the winter
- During UK winters temperatures are higher in Africa, so there is more active insect life available as food for the swallows

Hobbies depend on small birds (and large insects for food). Many populations of small birds migrate to Africa in the UK winter as more food is available to them there. The hobbies follow their food.

6 b Ecology is the study of living organisms in relation to their environment

7 a i June

ii April

iii The trees have many leaves, and these absorb the light before it can reach the ground layer

i There is plenty of light for them to photosynthesise before the trees have produced most of their leaves ii. By June the trees are in full leaf and so little light reaches the bluebells. Bluebells cannot photosynthesise without light, and so leaves die back

4.2 1 **Producer** : can trap light energy and produce organic compounds; **consumer**: must obtain its organic food molecules in a ready-made form, from another living organism; **decomposer**: obtain energy and raw materials from the wastes or remains of other living organisms. **Consumers** can be omitted : the cycling of raw materials can go on without consumers but not without producers or decomposers.

2 Whichever example you choose, make sure that there is a producer at the base,

and that you show the direction of energy flow with arrows.

3 Energy is lost from each transfer between stages e.g. when a lion eats an antelope, much of the antelope's energy content does not become part of the lion's body (a great deal is lost as heat during the lion's respiration). Because the energy transfer is quite low – only about 5-10% - there cannot be many links to the chain or the number of producers at the base would be enormous.

1 (p.226)

	Producers	Herbivores	Carnivores	Top carnivores
freshwater	Algae	Water fleas	Guppy, bass	Green heron
Seashore	Seaweed	Limpet	Crab	Gull
Coral reef	Phytoplankton	Zooplankton, coral, jellyfish	Anemone, parrotfish, sea turtle, butterfly fish, crab, octopus	shark

2 • Many possibilities e.g. number of parrotfish would increase, so coral might be eaten, so less food for butterfly fish, so zooplankton would increase. Main point to make is that food webs are very stable, and removal of any member makes them less stable.

3 energy; producer; light; chemical; herbivore; consumer; carnivore

4.4 1 bacteria; fungi; simple; amino acids; fats/lipids; enzymes; temperature; ph; sewage; food/buildings

2 a they can use it as fertiliser for growing plants

b to provide the moist conditions which supply water for decomposition (hydrolyses in digestion)

to make sure that oxygen is available, since aerobic respiration by microbes requires oxygen

4.5 1 a carbon dioxide

b glucose/starch/cellulose

c respiration

d photosynthesis

e diffusion

there was no oxygen to complete decomposition/it was too acid for decomposition to take place

4.6 1 a the stubble can be decomposed and release nitrates to the soil

b drainage increases the air content of the soil, providing oxygen for respiration by microbes. It also raises the temperature of the soil, helping the enzymes of decomposition

c peas and beans are legumes – they have a symbiotic relationship with bacteria that can 'fix' nitrogen from the atmosphere into ammonium salts and nitrates, which are excellent fertilisers

d increases the concentration of nitrate, phosphate and potassium – all mineral ions needed for plant growth

the compost can be decomposed to release nitrates and other minerals (and the decomposition also raises the soil temperature and moisture content)

2 drainage increases the air content of the soil, providing oxygen for respiration by microbes. It also raises the temperature of the soil, helping the enzymes of decomposition

4.7 1 a herbivore dies – is decomposed – nitrate is released – absorbed by a plant – converted to protein by plant – eaten by herbivore – digested to amino acids – assimilated into herbivore muscles as protein

c decomposition releases nitrates/ammonium salts for nitrogen cycle, and also releases carbon dioxide for the carbon cycle

2 energy; producer; light; chemical; herbivore; consumer; carnivore; photosynthesis; respiration; decomposers

3 a plants

b jackals and lions

c grass >sheep> jackal

d several jackals can bring down a large sheep; several jackals can search for food; female jackals can remain with pups while males hunt

e the jackals used other species as their prey e.g. dik-diks

f jugular vein/trachea/spine
g the collars would not be biodegradable i.e. they could not be decomposed by bacteria or fungi.

Ecosystems, decay and cycles pp.238–239

1 a (from top of web)
hawk spider
small bird flies
butterflies aphids mice
b i food web
ii hawk spider
c i pyramid shown with bean plant widest, then mice, then hawk
ii pyramid shown with single plant, many aphids(so wide bar), then single/few birds
2 a i many grass plants – fewer/single buffalo – many ticks – fewer oxpecker birds
ii broad base of grass plants (producer) – narrower bar for buffalo (herbivore/primary consumer)- even narrower for ticks (carnivore/secondary consumer) – narrowest for oxpeckers (top carnivore)
b consumers obtain their nutrients in ready-made organic form, whereas producers manufacture their nutrients from inorganic raw materials (most often by photosynthesis)
3 a Q is the absorption of nitrate, released by decomposers and nitrifying bacteria from animal and plant remains, R is the absorption of carbon dioxide to be used in the production of sugars by photosynthesis
b Sulphur dioxide and nitrogen oxides – both may produce acid rain which can lead to damage to leaf tissue/damage to calcium uptake by molluscs and crustaceans/leaching of minerals from soil
4 a pea plants are legumes, so can fix atmospheric nitrogen gas to nitrates in their root nodules. These nitrates are then available to the corn plants, which grow better as nitrate is usually a limiting factor for their growth.
b manure can be decomposed by bacteria, making nitrates and phosphates available to the crop plants. This is called 'organic' culture since no artificial inorganic fertilisers are used.
5 a i A – combustion B – respiration C – photosynthesis D – nutrition/ feeding
ii bacteria/ fungi
b increased combustion of fossil fuels increases carbon dioxide concentration/removal of forests takes away the trees which might otherwise reabsorb carbon dioxide
6 a i heat;
ii condensation / cooling of water vapour;
b i transpiration / evapo-transpiration;
ii 1 humidity;
2 temperature;
3 wind / air movement;
4 light / sunlight;
c i 1 reduced transpiration (in forest area);
2 leading to less water vapour (moving inland) / less clouds form;
3 thus less / no rainfall / less humid (inland);
ii 1 more surface runoff of rain water / flooding;
2 increased surface wind speed;
3 can result in greater erosion of soil / silting up of streams / rivers / landslides;
4 desertification;
5 destruction of habitats / disrupt food chains;
6 possible extinction of animal / plant species;
7 more carbon dioxide / less oxygen in atmosphere;

4.8 1 e Environmental resistance is any factor which limits the growth of a population, the availability of food, for example, or the presence of a predator
2 **Biotic** = 'living': an example of a biotic factor is the presence of a predator; **abiotic** = non-living: an example of an abiotic factor is temperature.
3 To improve yield of crops (e.g. by supplying fertiliser or removing weeds)or domestic animals (e.g by controlling disease).

4.12 1 a antiseptic is used to kill bacteria *outside* the body e.g. on kitchen surfaces, antibiotic is used to kill bacteria *inside* the body.

b antibiotic is a drug used to treat bacterial infection, antibody is produced by body's own cells (lymphocytes) to mark out invading cells.

c resistance is a condition in which an organism becomes unaffected by a drug or medicine, immunity is the ability to produce antibodies to recognise (and destroy) pathogens.

2 It must be non-toxic, must not react with the penicillin and must be dissolved by stomach juices (probably acids).

Bacteria p.251

1 $C_6H_{12}O_6 \longrightarrow 2C_2H_5OH + 2CO_2$

2 Vitamin A (retinene) is needed to produce visual pigment, so aids sight generally and night vision particularly;
Vitamin D and calcium are required for production of bones and teeth

a to prevent fungal/bacterial decay of bread so that it has a longer 'shelf life

b to make bread more attractive to consumers.

3 a i mycoprotein has less fat/more roughage/more carbohydrate

ii Less fat (less risk of obesity/ cardiovascular disease), more roughage (less risk of constipation/ cancer of the colon), less carbohydrate (less risk of obesity/ diabetes/tooth decay)

b i 100 – (49.0 + 9.2 + 19.5 + 20.6) = 1.7g

ii iron/calcium/vitamin B

c i glucose/amino acids;

ii To act as a filter to remove solid material/fungal mycelium;

iii Carbon dioxide

d i 28 – 30 °C;

ii By-product of respiration;

iii So that enzymes are not denatured/ are close to their optimum temperature;

iv The water jacket removes excess heat

4.14 1 g 2 – 3 – (1) – 6 – 4 – 7 – 5

4.14 1 New organisms can now fix nitrogen, and so (a) reduce the need for nitrate fertilisers and (b) allow new organism to manufacture more protein

2 Cystic fibrosis: the gene for the 'correct' protein can be carried in a virus into the lungs of a person with the 'faulty' gene. The virus infects the lung lining cells and replaces the faulty gene with the correct one. It is likely to be successful because the virus can be engineered to carry only the correct gene, and none which might be harmful.

4.16 1 This has been done to provide more land for agriculture, and to make mechanisation of agriculture more efficient.

2 Deforestation is the rapid destruction of woodland (for human commercial benefit). It may reduce soil fertility, cause flooding and landslides, affect the water and carbon cycles, cause climatic changes and increase the rate of species extinctions as habitats are lost.

4.18 1 a year on x axis, population on y axis

b 1400 million

c About 6500 million

Agricultural, industrial and medical revolutions have combined to increase birth rate and reduce death rate

1 (p.268)

a Algae > shrimps > fish > sea otter

The pesticides become more concentrated as they move up the food chain – for example, the otter eats many fish, and so accumulates the pesticide from all of those fish bodies. The pesticide might reach a lethal concentration in the bodies of the otters

2 a Distance downstream on x axis, other variables on y axis. Remember to use a key to distinguish the plots from one another.

b The bacteria had entered in the sewage

c The population falls as the water becomes cloudy with bacteria (no light for photosynthesis) then increases as population of bacteria becomes dispersed and as more nitrates are made available by decomposition of sewage

- **d** The 'rise' in fish numbers is misleading – the data is just showing that the fish are there (they have been there all the time, and haven't appeared from 'zero' fish)
- **e** Describe fall and then increase, giving actual values e.g. fall from 95% to 30% between 0m and 100m from sewage entry

Fall due to bacterial respiration, then rise as algae photosynthesise (releasing oxygen) and oxygen dissolves in the water from the atmosphere

4.20 1 Red squirrels are declining in number because of competition with grey squirrels, the fact that the red squirrels are susceptible to a viral disease, and the loss of suitable habitat. A suitable conservation strategy would make sure sufficient habitat was protected, and kept free of invasion by grey squirrels

2 Management to help species survival. Conservation is necessary because Humans are having an impact on other species populations, often by taking habitat from these species.

3 Many suitable examples, but the main point is that there is limited space in the Earth's habitats and Humans require some of that space for themselves. The only way that other species can survive is if Humans manage the environment for them, although this may benefit some species at the expense of others.

4.21 1
- **a** species on x axis and percentage on y axis. Check key for protein and lipid.
- **b** water
- **c** fish oils are important in the development of nervous tissue, and contain fat-soluble vitamins such as vitamin D
- **d** iron is part of haemoglobin in red blood cells, essential for oxygen transport.

2
- **a** they will have had time to breed to replace stocks
- **b** immature fish can escape from the net
- **c** may tag and release fish, and check if they are recaptured
- **d** reduce fishing with quotas/increase mesh size in nets/offer advice to fisherman about alternative species to catch

4.24 1 f A biomass fuel is one made from materials produced in photosynthesis

2 They remove CO_2 from the atmosphere, they may limit soil erosion, they may provide a habitat for some animal species, they may reduce the use of other woods for fuel

3
- **a** molasses can be added to cattle feed, dried yeast provides vitamins, and carbon dioxide can be used in fizzy drinks, waste sugar cane can be burned to fuel the distillation process
- **b** cellulose breaks down plant cell walls and so makes the cell contents accessible. Amylase hydrolyses starch to maltose/glucose, providing the substrate for alcoholic fermentation
- **c** genetic engineering has provided strains of yeast which can tolerate the high temperatures and high alcohol concentrations which occur during fermentation

This suggests that running one car is taking the landspace which could support 25 Humans on a subsistence diet. Is this the best use of crop-growing land?

4 Provide cheap source of fuel, and allow communities to get rid of much organic waste

5 The temperature is more stable there, they are easier to fill (gravity), and explosions are less dangerous!

6 The disinfectant would kill some of the active microbes and so slow down the production of biogas.

Human impact on ecosystems pp.281–283

1
- **a** increase in numbers from point of entry of raw sewage – reaches peak, then declines further downstream until it reaches the original population size
- **b** availability of nutrients supplied in the raw sewage – initial increase in population as more nutrients available, then decline as nutrient levels fall again and bacteria die

2 a i an organism which obtains its nutrients by eating other animals
ii the presence of fur or hair
b i fewer antelopes as land given over to farming/hunting or trapping by farmers
ii number of antelopes could increase, and feed on crop plants
c grass – antelope – wild dog – lion
d i the management of resources so that they are available for future generations
ii count their numbers – introduce protection measures (less hunting and remove some lions)/increase availability of food – count numbers to check success of scheme
e amino acids/proteins – ammonium compounds – nitrates (processes carried out by decomposers and nitrifying bacteria)

3 a shell enclosing body/combined body-foot
b species has lower case letter, genus has upper case(capital) letter
c sexual – gametes are released and fertilisation occurs
d i availability of oxygen in flowing water
ii change in shape or form
e many examples could be offered – check that conservation measures are appropriate to this species, and realistic

4 a check accuracy
b i sugar beet
ii wheat
c additional fertiliser – remove hedges – use machinery – weedkillers – pesticides
d fresh mass includes water, and this quantity is very variable
e each step in a food chain involves a loss of (approximately 90%) energy, so fewer steps means more available energy to human
f i $6O_2$
ii flat, thin and with palisade cells containing many chloroplasts
iii root hair cells have lower water potential than the surrounding soil solution, so water enters by osmosis from the soil solution
iv by diffusion down a concentration gradient from the air, through the stomata into the air spaces, then into solution on water covering of palisade cells

5 a i **developing country** – most people in younger age groups, with rapid fall of in numbers as ageing proceeds;
ii **developed country** – fairly constant population size up until age of 75 or thereabouts
b i under 15 the developing country has a much larger percentage of the population in this age group than the developed country
ii over 65 the developing country has a much smaller percentage of the population in this age group than the developed country
c in each case the percentage of females is greater than the percentage of males
d male can produce equal numbers of X and Y carrying sperm, with an equal probability of fertilising an X-carrying egg. Thus there should be 50% male (XY) and female (XX) babies.
e i **developing** – about 53 years, **developed** – about 75 years
ii developed countries enjoy better nutrition/better medical care (including immunisation programmes)

6 a i flooding/earthquakes/forest fires
ii drought/salination of soil
b increase in population means more people to feed and, perhaps, less land on which to grow food
c e.g. wheat: identify plants with more/larger seeds; self-pollinate these plants; collect seeds; grow plants; then repeat process of selection/growth
d the alteration of an organism's genotype/DNA for the benefit of humans.

Index